Lecture Notes in Artificial Intelligence 5327

Edited by R. Goebel, J. Siekmann, and W. Wahlster

Subseries of Lecture Notes in Computer Science

Paulo Cesar G. da Costa   Claudia d'Amato
Nicola Fanizzi   Kathryn B. Laskey
Kenneth J. Laskey   Thomas Lukasiewicz
Matthias Nickles   Michael Pool (Eds.)

Uncertainty Reasoning for the Semantic Web I

ISWC International Workshops
URSW 2005-2007
Revised Selected and Invited Papers

 Springer

Series Editors

Randy Goebel, University of Alberta, Edmonton, Canada
Jörg Siekmann, University of Saarland, Saarbrücken, Germany
Wolfgang Wahlster, DFKI and University of Saarland, Saarbrücken, Germany

Volume Editors

Paulo Cesar G. da Costa
Kathryn B. Laskey
George Mason University, Fairfax, VA 22030, USA; {pcosta,klaskey}@gmu.edu

Claudia d'Amato
Nicola Fanizzi
Università degli Studi di Bari, 70125 Bari, Italy; {claudia.damato,fanizzi}@di.uniba.it

Kenneth J. Laskey
MIRTE Corporation, McLean, VA 22102-7508, USA; klaskey@mitre.org

Thomas Lukasiewicz
Oxford University, Oxford OX1 3QD, UK; thomas.lukasiewicz@comlab.ox.ac.uk

Matthias Nickles
University of Bath, Bath, BA2 7AY, UK; nickles@gmx.net

Michael Pool
Convera, Inc., Vienna, VA 22182, USA; mpool@convera.com

Library of Congress Control Number: Applied for

CR Subject Classification (1998): I.2, H.3.5, H.5.3, B.2, F.3, F.4

LNCS Sublibrary: SL 7 – Artificial Intelligence

ISSN   0302-9743
ISBN-10   3-540-89764-X Springer Berlin Heidelberg New York
ISBN-13   978-3-540-89764-4 Springer Berlin Heidelberg New York

Springer is a part of Springer Science+Business Media

springer.com

© Springer-Verlag Berlin Heidelberg 2008
Printed in Germany

Typesetting: Camera-ready by author, data conversion by Scientific Publishing Services, Chennai, India
Printed on acid-free paper   SPIN: 12579013   06/3180   5 4 3 2 1 0

Preface

This volume contains the proceedings of the first three workshops on Uncertainty Reasoning for the Semantic Web (URSW), held at the International Semantic Web Conferences (ISWC) in 2005, 2006, and 2007. In addition to revised and strongly extended versions of selected workshop papers, we have included invited contributions from leading experts in the field and closely related areas.

With this, the present volume represents the first comprehensive compilation of state-of-the-art research approaches to uncertainty reasoning in the context of the Semantic Web, capturing different models of uncertainty and approaches to deductive as well as inductive reasoning with uncertain formal knowledge.

The World Wide Web community envisions effortless interaction between humans and computers, seamless interoperability and information exchange among Web applications, and rapid and accurate identification and invocation of appropriate Web services. As work with semantics and services grows more ambitious, there is increasing appreciation of the need for principled approaches to the formal representation of and reasoning under uncertainty. The term *uncertainty* is intended here to encompass a variety of forms of incomplete knowledge, including incompleteness, inconclusiveness, vagueness, ambiguity, and others. The term *uncertainty reasoning* is meant to denote the full range of methods designed for representing and reasoning with knowledge when Boolean truth values are unknown, unknowable, or inapplicable. Commonly applied approaches to uncertainty reasoning include probability theory, Dempster-Shafer theory, fuzzy logic and possibility theory, and numerous other methodologies.

A few Web-relevant challenges which are addressed by reasoning under uncertainty include:

Uncertainty of available information: Much information on the World Wide Web is uncertain. Examples include weather forecasts or gambling odds. Canonical methods for representing and integrating such information are necessary for communicating it in a seamless fashion.

Information incompleteness: Information extracted from large information networks such as the World Wide Web is typically incomplete. The ability to exploit partial information is very useful for identifying sources of service or information. For example, that an online service deals with greeting cards may be evidence that it also sells stationery. It is clear that search effectiveness could be improved by appropriate use of technologies for handling uncertainty.

Information incorrectness: Web information is also often incorrect or only partially correct, raising issues related to trust or credibility. Uncertainty representation and reasoning helps to resolve tension amongst information sources having different confidence and trust levels, and can facilitate the merging of controversial information obtained from multiple sources.

Uncertain ontology mappings: The Semantic Web vision implies that numerous distinct but conceptually overlapping ontologies will co-exist and interoperate. It is likely that in such scenarios ontology mapping will benefit from the ability to represent degrees of membership and/or likelihoods of membership in categories of a target ontology, given information about class membership in the source ontologies.

Indefinite information about Web services: Dynamic composability of Web services will require runtime identification of processing and data resources and resolution of policy objectives. Uncertainty reasoning techniques may be necessary to resolve situations in which existing information is not definitive.

Uncertainty is thus an intrinsic feature of many important tasks on the Web and the Semantic Web, and a full realization of the World Wide Web as a source of processable data and services demands formalisms capable of representing and reasoning under uncertainty. Unfortunately, none of these needs can be addressed in a principled way by current Web standards. Although it is to some degree possible to use semantic markup languages such as OWL or RDF(S) to represent qualitative and quantitative information about uncertainty, there is no established foundation for doing so, and feasible approaches are severely limited. Furthermore, there are ancillary issues such as how to balance representational power vs. simplicity of uncertainty representations, which uncertainty representation techniques address uses such as the examples listed above, how to ensure the consistency of representational formalisms and ontologies, etc.

In response to these pressing demands, in recent years several promising approaches to uncertainty reasoning on the Semantic Web have been proposed. The present volume covers a representative cross section of these approaches, from extensions to existing Web-related logics for the representation of uncertainty to approaches to inductive reasoning under uncertainty on the Web.

In order to reflect the diversity of the presented approaches and to relate them to their underlying models of uncertainty, the contributions to this volume are grouped as follows:

Probabilistic and Dempster-Shafer Models

Probability theory provides a mathematically sound representation language and formal calculus for rational degrees of belief, which gives different agents the freedom to have different beliefs about a given hypothesis. As this provides a compelling framework for representing uncertain, imperfect knowledge that can come from diverse agents, there are many distinct approaches using probability in the context of the Semantic Web. Classes of probabilistic models covered with the present volume are Bayesian Networks, probabilistic extensions to Description and First-Order Logics, and models based on the Dempster-Shafer theory (a generalization of the classical Bayesian approach).

Fuzzy and Possibilistic Models

Fuzzy formalisms allow for representing and processing degrees of truth about vague (or imprecise) pieces of information. In fuzzy description logics and ontology languages, concept assertions, role assertions, concept inclusions, and role inclusions have a degree of truth rather than a binary truth value. The present volume presents various approaches which exploit fuzzy logic and possibility theory in the context of the Semantic Web.

Inductive Reasoning and Machine Learning

Machine learning is supposed to play an increasingly important role in the context of the Semantic Web by providing various tasks such as the learning of ontologies from incomplete data or the (semi-)automatic annotation of data on the Web. Results obtained by machine learning approaches are typically uncertain. As a logic-based approach to machine learning, inductive reasoning provides means for inducing general propositions from observations (example facts). Papers in this volume exploit the power of inductive reasoning for the purpose of ontology learning, and project future directions for the use of machine learning on the Semantic Web.

Hybrid Approaches

This volume segment contains papers which either combine approaches from two or more of the previous segments, or which do not rely on any specific classical approach to uncertainty reasoning.

Acknowledgements. We would like to express our gratitude to the authors of this volume for their contributions and to the workshop participants for inspiring discussions, as well as to the members of the workshop Program Committees and the additional reviewers for their reviews and for their overall support.

September 2008

Paulo Cesar G. da Costa
Claudia d'Amato
Nicola Fanizzi
Kathryn B. Laskey
Kenneth J. Laskey
Thomas Lukasiewicz
Matthias Nickles
Michael Pool

Organization

First Workshop on Uncertainty Reasoning for the Semantic Web (URSW 2005)

Bill McDaniel   Adobe Systems, USA
Leo Obrst   MITRE Corporation, USA
Yung Peng   University of Maryland, Baltimore County, USA
Michael Pool   Information Extraction & Transport, Inc., USA
Masami Takikawa   Information Extraction & Transport, Inc., USA

Second Workshop on Uncertainty Reasoning for the Semantic Web (URSW 2006)

Organizing Committee

Paulo Cesar G. da Costa   George Mason University, USA
Francis Fung   Information Extraction & Transport, Inc., USA
Kathryn B. Laskey   George Mason University, USA
Kenneth J. Laskey   MITRE Corporation, USA
Michael Pool   Convera Technologies, Inc., USA

Program Committee

Ameen Abu-Hanna   Universiteit van Amsterdam, The Netherlands
Bill Andersen   Ontology Works, Inc., USA
Paulo Cesar G. da Costa   George Mason University, USA
Fabio G. Cozman   Universidade de Sao Paulo, Brazil
Bruce D'Ambrosio   Cleverset, Inc., USA
Francis Fung   Information Extraction & Transport, Inc., USA
Linda van der Gaag   Universiteit Utrecht, The Netherlands
Ivan Herman   C.W.I., The Netherlands
Markus Holi   Helsinki University of Technology, Finland
Kathryn B. Laskey   George Mason University, USA
Kenneth J. Laskey   MITRE Corporation, USA
Thomas Lukasiewicz   Università di Roma "La Sapienza," Italy
Anders Madsen   Hugin Expert A/S, Denmark
Scott Marshall   Universiteit van Amsterdam, The Netherlands
Trevor Martin   University of Bristol, UK
Kirk Martinez   University of Southampton, UK
Bill McDaniel   DERI, Ireland
Leo Obrst   MITRE Corporation, USA
Yung Peng   University of Maryland, Baltimore County, USA
Michael Pool   Convera Technologies, Inc., USA
Dave Robertson   University of Edinburgh, UK
Oreste Signore   ISTI-CNR, Italy
Masami Takikawa   Information Extraction & Transport, Inc., USA
Frans Voorbraak   Universiteit van Amsterdam, The Netherlands

Third Workshop on Uncertainty Reasoning for the Semantic Web (URSW 2007)

Organizing Committee

Paulo Cesar G. da Costa	George Mason University, USA
Fernando Bobillo	University of Granada, Spain
Claudia d'Amato	University of Bari, Italy
Nicola Fanizzi	University of Bari, Italy
Francis Fung	Information Extraction & Transport, Inc., USA
Thomas Lukasiewicz	Università di Roma "La Sapienza," Italy
Trevor Martin	University of Bristol, UK
Matthias Nickles	Technical University of Munich, Germany
Yun Peng	University of Maryland, Baltimore County, USA
Kathryn B. Laskey	George Mason University, USA
Kenneth J. Laskey	MITRE Corporation, USA
Michael Pool	Convera Technologies, Inc., USA
Pavel Smrž	Brno University of Technology, Czech Republic
Peter Vojtáš	Charles University, Czech Republic

Program Committee

Ameen Abu-Hanna	Universiteit van Amsterdam, The Netherlands
Fernando Bobillo	University of Granada, Spain
Paulo Cesar G. da Costa	George Mason University, USA
Fabio G. Cozman	Universidade de Sao Paulo, Brazil
Claudia d'Amato	University of Bari, Italy
Ernesto Damiani	University of Milan, Italy
Nicola Fanizzi	University of Bari, Italy
Francis Fung	Information Extraction & Transport, Inc., USA
Linda van der Gaag	Universiteit Utrecht, The Netherlands
Ivan Herman	C.W.I., The Netherlands
Kathryn B. Laskey	George Mason University, USA
Kenneth J. Laskey	MITRE Corporation, USA
Thomas Lukasiewicz	Università di Roma "La Sapienza," Italy
Anders L. Madsen	Hugin Expert A/S, Denmark
M. Scott Marshall	Universiteit van Amsterdam, The Netherlands
Trevor Martin	University of Bristol, UK
Bill McDaniel	DERI, Ireland
Matthias Nickles	Technical University of Munich, Germany
Leo Obrst	MITRE Corporation, USA
Yung Peng	University of Maryland, Baltimore County, USA
Michael Pool	Convera Technologies, Inc., USA
Livia Predoiu	University of Mannheim, Germany
Dave Robertson	University of Edinburgh, UK

Table of Contents

Inductive Reasoning and Machine Learning

Hybrid Approaches

Just Add Weights:
Markov Logic for the Semantic Web

Pedro Domingos[1], Daniel Lowd[1], Stanley Kok[1], Hoifung Poon[1],
Matthew Richardson[2], and Parag Singla[1]

[1] Department of Computer Science and Engineering
University of Washington
Seattle, WA 98195-2350, U.S.A.
{pedrod,lowd,koks,hoifung,parag}@cs.washington.edu
[2] Microsoft Research
Redmond, WA 98052
mattri@microsoft.com

Abstract. In recent years, it has become increasingly clear that the vision of the Semantic Web requires uncertain reasoning over rich, first-order representations. Markov logic brings the power of probabilistic modeling to first-order logic by attaching weights to logical formulas and viewing them as templates for features of Markov networks. This gives natural probabilistic semantics to uncertain or even inconsistent knowledge bases with minimal engineering effort. Inference algorithms for Markov logic draw on ideas from satisfiability, Markov chain Monte Carlo and knowledge-based model construction. Learning algorithms are based on the conjugate gradient algorithm, pseudo-likelihood and inductive logic programming. Markov logic has been successfully applied to problems in entity resolution, link prediction, information extraction and others, and is the basis of the open-source Alchemy system.

1 Introduction

The vision of the Semantic Web is that of a web of information that computers can understand and reason about, organically built with no central organization except for a common set of standards [1]. This promises the ability to answer more complex queries and build more intelligent and effective agents than ever before. The standard languages that have been introduced so far are generally special cases of first-order logic, allowing users to define ontologies, express a rich set of relationships among objects of different types, logical dependencies between them, etc.

Fulfilling this promise, however, requires more than purely logical representations and inference algorithms. Most things in the world have some degree of uncertainty or noise – future events, such as weather and traffic, are unpredictable; information is unreliable, either from error or deceit; even simple concepts such as "fruit" and "vegetable" are imprecisely and inconsistently applied. Any system that hopes to represent varied information about the world

P.C.G. da Costa et al. (Eds.): URSW 2005-2007, LNAI 5327, pp. 1–25, 2008.

must therefore acknowledge the uncertain, inconsist, and untrustworthy nature of that knowledge.

The Semantic Web project faces additional or exacerbated sources of uncertainty in a number of areas. Matching entities, ontologies and schemas is essential for linking data from different sources, but is also inherently uncertain. Moreover, data may contain false or contradictory information. To simply exclude noisy or untrusted sources is an inadequate solution since even trusted sources may have some errors and even noisy sources may have useful information to contribute. A final problem is incomplete information; when information is missing we may be able to conclude very little with certainty, but it would be a mistake to ignore the partial evidence entirely.

Markov logic is a simple yet powerful solution to the problem of integrating logic and uncertainty. Given an existing knowledge base in first-order logic, Markov logic attaches a weight to each formula. Semantically, weighted formulas are viewed as templates for constructing Markov networks. This yields a well-defined probability distribution in which worlds are more likely when they satisfy a higher-weight set of ground formulas. Intuitively, the magnitude of the weight corresponds to the relative strength of its formula; in the infinite-weight limit, Markov logic reduces to first-order logic. Since Markov logic is a direct extension of first-order logic, it does not invalidate or conflict with the existing Semantic Web infrastructure. With Markov logic, Semantic Web languages can be made probabilistic simply by adding weights to statements, and Semantic Web inference engines can be extended to perform probabilistic reasoning simply by passing the proof DAG (directed acylic graph), with weights attached, to a probabilistic inference system. Weights may be set by hand, inferred from various sources (e.g., trust networks), or learned automatically from data. We have also developed algorithms for learning or correcting formulas from data. Markov logic has already been used to efficiently develop state-of-the-art models for entity resolution, ontology induction, information extraction, social networks, collective classification, and many other problems important to the Semantic Web. All of our algorithms, as well as sample datasets and applications, are available in the open-source Alchemy system [16] (alchemy.cs.washington.edu).

In this chapter, we describe the Markov logic representation and give an overview of current inference and learning algorithms for it. We begin with some background on first-order logic and Markov networks.

2   First-Order Logic

A *first-order knowledge base (KB)* is a set of sentences or formulas in first-order logic [10]. Formulas are constructed using four types of symbols: constants, variables, functions, and predicates. Constant symbols represent objects in the domain of interest (e.g., people: Anna, Bob, Chris, etc.). Variable symbols range over the objects in the domain. Function symbols (e.g., MotherOf) represent mappings from tuples of objects to objects. Predicate symbols represent relations among objects in the domain (e.g., Friends) or attributes of objects (e.g.,

Smokes). An *interpretation* specifies which objects, functions and relations in the domain are represented by which symbols. Variables and constants may be *typed*, in which case variables range only over objects of the corresponding type, and constants can only represent objects of the corresponding type. For example, the variable x might range over people (e.g., Anna, Bob, etc.), and the constant C might represent a city (e.g, Seattle, Tokyo, etc.).

A *term* is any expression representing an object in the domain. It can be a constant, a variable, or a function applied to a tuple of terms. For example, Anna, x, and GreatestCommonDivisor(x, y) are terms. An *atomic formula* or *atom* is a predicate symbol applied to a tuple of terms (e.g., Friends$(x, MotherOf(Anna))$). Formulas are recursively constructed from atomic formulas using logical connectives and quantifiers. If F_1 and F_2 are formulas, the following are also formulas: $\neg F_1$ (negation), which is true iff F_1 is false; $F_1 \wedge F_2$ (conjunction), which is true iff both F_1 and F_2 are true; $F_1 \vee F_2$ (disjunction), which is true iff F_1 or F_2 is true; $F_1 \Rightarrow F_2$ (implication), which is true iff F_1 is false or F_2 is true; $F_1 \Leftrightarrow F_2$ (equivalence), which is true iff F_1 and F_2 have the same truth value; $\forall x \, F_1$ (universal quantification), which is true iff F_1 is true for every object x in the domain; and $\exists x \, F_1$ (existential quantification), which is true iff F_1 is true for at least one object x in the domain. Parentheses may be used to enforce precedence. A *positive literal* is an atomic formula; a *negative literal* is a negated atomic formula. The formulas in a KB are implicitly conjoined, and thus a KB can be viewed as a single large formula. A *ground term* is a term containing no variables. A *ground atom* or *ground predicate* is an atomic formula all of whose arguments are ground terms. A *possible world* (along with an interpretation) assigns a truth value to each possible ground atom.

A formula is *satisfiable* iff there exists at least one world in which it is true. The basic inference problem in first-order logic is to determine whether a knowledge base KB *entails* a formula F, i.e., if F is true in all worlds where KB is true (denoted by $KB \models F$). This is often done by *refutation*: KB entails F iff $KB \cup \neg F$ is unsatisfiable. (Thus, if a KB contains a contradiction, all formulas trivially follow from it, which makes painstaking knowledge engineering a necessity.) For automated inference, it is often convenient to convert formulas to a more regular form, typically *clausal form* (also known as *conjunctive normal form (CNF)*). A KB in clausal form is a conjunction of *clauses*, a clause being a disjunction of literals. Every KB in first-order logic can be converted to clausal form using a mechanical sequence of steps.[1] Clausal form is used in resolution, a sound and refutation-complete inference procedure for first-order logic [38].

Inference in first-order logic is only semidecidable. Because of this, knowledge bases are often constructed using a restricted subset of first-order logic with more desirable properties. The two subsets most commonly applied to the Semantic Web are *Horn clauses* and *description logics*. Horn clauses are clauses containing at most one positive literal. The Prolog programming language is based on Horn

---

[1] This conversion includes the removal of existential quantifiers by Skolemization, which is not sound in general. However, in finite domains an existentially quantified formula can simply be replaced by a disjunction of its groundings.

Table 1. Example of a first-order knowledge base and MLN. `Fr()` is short for `Friends()`, `Sm()` for `Smokes()`, and `Ca()` for `Cancer()`.

First-Order Logic	Clausal Form	Weight
"Friends of friends are friends."		
$\forall x \forall y \forall z\, Fr(x, y) \wedge Fr(y, z) \Rightarrow Fr(x, z)$	$\neg Fr(x, y) \vee \neg Fr(y, z) \vee Fr(x, z)$	0.7
"Friendless people smoke."		
$\forall x\, (\neg(\exists y\, Fr(x, y)) \Rightarrow Sm(x))$	$Fr(x, g(x)) \vee Sm(x)$	2.3
"Smoking causes cancer."		
$\forall x\, Sm(x) \Rightarrow Ca(x)$	$\neg Sm(x) \vee Ca(x)$	1.5
"If two people are friends, then either		
both smoke or neither does."	$\neg Fr(x, y) \vee Sm(x) \vee \neg Sm(y),$	1.1
$\forall x \forall y\, Fr(x, y) \Rightarrow (Sm(x) \Leftrightarrow Sm(y))$	$\neg Fr(x, y) \vee \neg Sm(x) \vee Sm(y)$	1.1

clause logic [21]. Prolog programs can be learned from databases by searching for Horn clauses that (approximately) hold in the data; this is studied in the field of inductive logic programming (ILP) [18]. Description logics are a decidable subset of first-order logic that is the basis of the Web Ontology Language (OWL) [7].

Table 1 shows a simple KB and its conversion to clausal form. Notice that, while these formulas may be *typically* true in the real world, they are not *always* true. In most domains it is very difficult to come up with non-trivial formulas that are always true, and such formulas capture only a fraction of the relevant knowledge. Thus, despite its expressiveness, pure first-order logic has limited applicability to practical AI problems. Many *ad hoc* extensions to address this have been proposed. In the more limited case of propositional logic, the problem is well solved by probabilistic graphical models such as Markov networks, described in the next section. We will later show how to generalize these models to the first-order case.

3   Markov Networks

A *Markov network* (also known as *Markov random field*) is a model for the joint distribution of a set of variables $X = (X_1, X_2, \ldots, X_n) \in \mathcal{X}$ [30]. It is composed of an undirected graph G and a set of potential functions ϕ_k. The graph has a node for each variable, and the model has a potential function for each clique in the graph. A potential function is a non-negative real-valued function of the state of the corresponding clique. The joint distribution represented by a Markov network is given by

$$P(X = x) = \frac{1}{Z} \prod_k \phi_k(x_{\{k\}}) \tag{1}$$

where $x_{\{k\}}$ is the state of the kth clique (i.e., the state of the variables that appear in that clique). Z, known as the *partition function*, is given by $Z = \sum_{x \in \mathcal{X}} \prod_k \phi_k(x_{\{k\}})$. Markov networks are often conveniently represented as *log-linear models*, with each clique potential replaced by an exponentiated weighted sum of features of the state, leading to

$$P(X = x) = \frac{1}{Z} \exp\left(\sum_j w_j f_j(x)\right) \tag{2}$$

A feature may be any real-valued function of the state. This chapter will focus on binary features, $f_j(x) \in \{0, 1\}$. In the most direct translation from the potential-function form (Equation 1), there is one feature corresponding to each possible state $x_{\{k\}}$ of each clique, with its weight being $\log \phi_k(x_{\{k\}})$. This representation is exponential in the size of the cliques. However, we are free to specify a much smaller number of features (e.g., logical functions of the state of the clique), allowing for a more compact representation than the potential-function form, particularly when large cliques are present. Markov logic will take advantage of this.

Inference in Markov networks is #P-complete [39]. The most widely used method for approximate inference in Markov networks is Markov chain Monte Carlo (MCMC) [11], and in particular Gibbs sampling, which proceeds by sampling each variable in turn given its Markov blanket. (The Markov blanket of a node is the minimal set of nodes that renders it independent of the remaining network; in a Markov network, this is simply the node's neighbors in the graph.) Marginal probabilities are computed by counting over these samples; conditional probabilities are computed by running the Gibbs sampler with the conditioning variables clamped to their given values. Another popular method for inference in Markov networks is belief propagation [52].

Maximum-likelihood or MAP estimates of Markov network weights cannot be computed in closed form but, because the log-likelihood is a concave function of the weights, they can be found efficiently (modulo inference) using standard gradient-based or quasi-Newton optimization methods [28]. Another alternative is iterative scaling [8]. Features can also be learned from data, for example by greedily constructing conjunctions of atomic features [8].

4 Markov Logic

A first-order KB can be seen as a set of hard constraints on the set of possible worlds: if a world violates even one formula, it has zero probability. The basic idea in Markov logic is to soften these constraints: when a world violates one formula in the KB it is less probable, but not impossible. The fewer formulas a world violates, the more probable it is. Each formula has an associated weight (e.g., see Table 1) that reflects how strong a constraint it is: the higher the weight, the greater the difference in log probability between a world that satisfies the formula and one that does not, other things being equal.

Definition 1. [36] *A Markov logic network (MLN) L is a set of pairs (F_i, w_i), where F_i is a formula in first-order logic and w_i is a real number. Together with a finite set of constants $C = \{c_1, c_2, \ldots, c_{|C|}\}$, it defines a Markov network $M_{L,C}$ (Equations 1 and 2) as follows:*

1. $M_{L,C}$ *contains one binary node for each possible grounding of each atom appearing in L. The value of the node is 1 if the ground atom is true, and 0 otherwise.*
2. $M_{L,C}$ *contains one feature for each possible grounding of each formula F_i in L. The value of this feature is 1 if the ground formula is true, and 0 otherwise. The weight of the feature is the w_i associated with F_i in L.*

Thus there is an edge between two nodes of $M_{L,C}$ iff the corresponding ground atoms appear together in at least one grounding of one formula in L. For example, an MLN containing the formulas $\forall x\, \text{Smokes}(x) \Rightarrow \text{Cancer}(x)$ (smoking causes cancer) and $\forall x \forall y\, \text{Friends}(x, y) \Rightarrow (\text{Smokes}(x) \Leftrightarrow \text{Smokes}(y))$ (friends have similar smoking habits) applied to the constants Anna and Bob (or A and B for short) yields the ground Markov network in Figure 1. Its features include $\text{Smokes}(\text{Anna}) \Rightarrow \text{Cancer}(\text{Anna})$, etc. Notice that, although the two formulas above are false as universally quantified logical statements, as weighted features of an MLN they capture valid statistical regularities, and in fact represent a standard social network model [47].

An MLN can be viewed as a *template* for constructing Markov networks. From Definition 1 and Equations 1 and 2, the probability distribution over possible worlds x specified by the ground Markov network $M_{L,C}$ is given by

$$P(X = x) = \frac{1}{Z} \exp\left(\sum_{i=1}^{F} w_i n_i(x)\right) \tag{3}$$

where F is the number of formulas in the MLN and $n_i(x)$ is the number of true groundings of F_i in x. As formula weights increase, an MLN increasingly resembles a purely logical KB, becoming equivalent to one in the limit of all infinite weights. When the weights are positive and finite, and all formulas are simultaneously satisfiable, the satisfying solutions are the modes of the distribution represented by the ground Markov network.

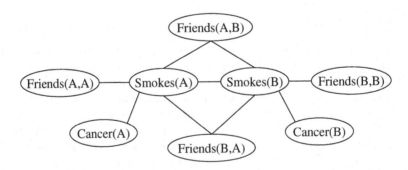

Fig. 1. Ground Markov network obtained by applying an MLN containing the formulas $\forall x\, \text{Smokes}(x) \Rightarrow \text{Cancer}(x)$ and $\forall x \forall y\, \text{Friends}(x, y) \Rightarrow (\text{Smokes}(x) \Leftrightarrow \text{Smokes}(y))$ to the constants Anna(A) and Bob(B)

Most importantly, Markov logic allows contradictions between formulas, which it resolves simply by weighing the evidence on both sides. This makes it well suited for merging multiple KBs. Markov logic also provides a natural and powerful approach to the problem of merging knowledge and data in different representations that do not align perfectly, as will be illustrated in the application section. Both of these tasks are also key to the success of the Semantic Web.

It is interesting to see a simple example of how Markov logic generalizes first-order logic. Consider an MLN containing the single formula $\forall x\, R(x) \Rightarrow S(x)$ with weight w, and $C = \{A\}$. This leads to four possible worlds: $\{\neg R(A), \neg S(A)\}$, $\{\neg R(A), S(A)\}$, $\{R(A), \neg S(A)\}$, and $\{R(A), S(A)\}$. From Equation 3 we obtain that $P(\{R(A), \neg S(A)\}) = 1/(3e^w + 1)$ and the probability of each of the other three worlds is $e^w/(3e^w + 1)$. (The denominator is the partition function Z; see Section 3.) Thus, if $w > 0$, the effect of the MLN is to make the world that is inconsistent with $\forall x\, R(x) \Rightarrow S(x)$ less likely than the other three. From the probabilities above we obtain that $P(S(A)|R(A)) = 1/(1 + e^{-w})$. When $w \to \infty$, $P(S(A)|R(A)) \to 1$, recovering the logical entailment.

It is easily seen that all discrete probabilistic models expressible as products of potentials, including Markov networks and Bayesian networks, are expressible in Markov logic. In particular, many of the models frequently used in AI can be stated quite concisely as MLNs, and combined and extended simply by adding the corresponding formulas. Most significantly, Markov logic facilitates the construction of non-i.i.d. models (i.e., models where objects are not independent and identically distributed).

When working with Markov logic, we typically make three assumptions about the logical representation: different constants refer to different objects (unique names), the only objects in the domain are those representable using the constant and function symbols (domain closure), and the value of each function for each tuple of arguments is always a known constant (known functions). These assumptions ensure that the number of possible worlds is finite and that the Markov logic network will give a well-defined probability distribution. These assumptions are quite reasonable in most practical applications, and greatly simplify the use of MLNs. After describing how each one can be relaxed, we will make these assumptions for the remainder of the chapter. See Richardson and Domingos [36] for further details on the Markov logic representation.

The unique names assumption can be removed by introducing the equality predicate ($\texttt{Equals}(x, y)$, or $x = y$ for short) and adding the necessary axioms to the MLN: equality is reflexive, symmetric and transitive; for each unary predicate P, $\forall x \forall y\, x = y \Rightarrow (P(x) \Leftrightarrow P(y))$; and similarly for higher-order predicates and functions [10]. This allows us to deal with instance and reference uncertainty, as illustrated in Section 7.1.

We can relax the domain closure assumption by introducing new constants to represent unknown objects. This works for any domain whose size is bounded. Markov logic can also be applied to a number of interesting infinite domains, such as when each node in the resulting infinite Markov network has a finite number of neighbors. See Singla and Domingos [43] for details on Markov logic in infinite domains.

Infinite domains can also be approximated as finite ones. Consider the transitive, anti-symmetric relation `AncestorOf`(x, y), meaning "x is an ancestor of y." In a logical KB, the rule "Everyone has an ancestor" is only valid in infinite or empty domains. In Markov logic, the rule can easily be applied to finite domains, so that worlds are more likely when more objects have an ancestor within the domain. Therefore, although Markov logic semantics are well-defined for many infinite domains, a finite approach suffices for most practical applications.

Let $H_{L,C}$ be the set of all ground terms constructible from the function symbols in L and the constants in L and C (the "Herbrand universe" of (L, C)). We can remove the known function assumption by treating each element of $H_{L,C}$ as an additional constant and applying the same procedure used to remove the unique names assumption. For example, with a function `G`(x) and constants `A` and `B`, the MLN will now contain nodes for `G(A)` = `A`, `G(A)` = `B`, etc. This leads to an infinite number of new constants, requiring the corresponding extension of MLNs. However, if we restrict the level of nesting to some maximum, the resulting MLN is still finite.

5   Inference

Recall that an MLN acts as a template for a Markov network. Therefore, we can always answer probabilistic queries using standard Markov network inference methods on the instantiated network. We have extended and adapted several of these standard methods to take particular advantage of the logical structure in a Markov logic network, yielding tremendous savings in memory and time. We describe these algorithms in this section.

For many queries, only a small subset of the instantiated Markov network is relevant. In such cases, we need not instantiate or even consider the entire MLN. The proof DAG from a logical inference engine can be used to generate the set of ground formulas and atoms relevant to a particular query. Together with the MLN weights, this can be used to generate a sub-network to answers the probabilistic query. In this way, Markov logic can easily be paired with traditional logical inference methods. This method, traditionally known as knowledge-based model construction (KBMC) [27], allows us to potentially reason efficiently over a very large knowledge base (like the Semantic Web), as long as only a small fraction of it is relevant to the query. In our descriptions, we will assume that inference is done over the entire MLN, but our methods apply to the sub-network case as well.

5.1   MAP/MPE Inference

In the remainder of this chapter, we assume that the MLN is in function-free clausal form for convenience, but these methods can be applied to other MLNs as well. A basic inference task is finding the most probable state of the world given some evidence. (This is known as MAP inference in the Markov network literature, and MPE inference in the Bayesian network literature.) Because of the form of Equation 3, in Markov logic this reduces to finding the truth assignment that maximizes the sum of weights of satisfied clauses. This can be done using any weighted satisfiability solver, and (remarkably) need not be more expensive than standard logical inference by model checking. (In fact, it can be faster, if some hard constraints are softened.) We have successfully used MaxWalkSAT, a weighted variant of the WalkSAT local-search satisfiability solver, which can solve hard problems with hundreds of thousands of variables in minutes [13]. MaxWalkSAT performs this stochastic search by picking an unsatisfied clause at random and flipping the truth value of one of the atoms in it. With a certain probability, the atom is chosen randomly; otherwise, the atom is chosen to maximize the sum of satisfied clause weights when flipped. This combination of random and greedy steps allows MaxWalkSAT to avoid getting stuck in local optima while searching. Pseudocode for MaxWalkSAT is shown in Algorithm 1. DeltaCost(v) computes the change in the sum of weights of unsatisfied clauses that results from flipping variable v in the current solution. Uniform(0,1) returns a uniform deviate from the interval $[0, 1]$.

One problem with this approach is that it requires propositionalizing the domain (i.e., grounding all atoms and clauses in all possible ways), which consumes memory exponential in the arity of the clauses. We have overcome this by developing LazySAT, a lazy version of MaxWalkSAT which grounds atoms and clauses only as needed [42]. This takes advantage of the sparseness of relational domains, where most atoms are false and most clauses are trivially satisfied. For example, in the domain of scientific research, most groundings of the atom Author(person, paper) are false, and most groundings of the clause Author(person1, paper) ∧ Author(person2, paper) ⇒ Coauthor(person1, person2) are satisfied. In LazySAT, the memory cost does not scale with the number of possible clause groundings, but only with the number of groundings that are potentially unsatisfied at some point in the search.

Algorithm 2 gives pseudo-code for LazySAT, highlighting the places where it differs from MaxWalkSAT. LazySAT maintains a set of *active atoms* and a set of *active clauses*. A clause is active if it can be made unsatisfied by flipping zero or more of its active atoms. (Thus, by definition, an unsatisfied clause is always active.) An atom is active if it is in the initial set of active atoms, or if it was flipped at some point in the search. The initial active atoms are all those appearing in clauses that are unsatisfied if only the atoms in the database are true, and all others are false. The unsatisfied clauses are obtained by simply going through each possible grounding of all the first-order clauses and materializing the groundings that are unsatisfied; search is pruned as soon as the partial grounding of a clause is satisfied. Given the initial active atoms, the definition of active clause requires that some clauses become active, and these are found using a similar process (with the difference that, instead of checking whether a ground clause is unsatisfied, we check whether it should be active). Each run of LazySAT is initialized by assigning random truth values to the active atoms. This

---

Algorithm 1. MaxWalkSAT(*weighted_clauses, max_flips, max_tries, target, p*)

$vars \leftarrow$ variables in *weighted_clauses*
for $i \leftarrow 1$ to *max_tries* **do**
   $soln \leftarrow$ a random truth assignment to *vars*
   $cost \leftarrow$ sum of weights of unsatisfied clauses in *soln*
   for $i \leftarrow 1$ to *max_flips* **do**
     if $cost \leq target$ **then**
       return "Success, solution is", *soln*
     end if
     $c \leftarrow$ a randomly chosen unsatisfied clause
     if Uniform(0,1) $< p$ **then**
       $v_f \leftarrow$ a randomly chosen variable from c
     else
       for each variable v in c **do**
         compute DeltaCost(v)
       end for
       $v_f \leftarrow v$ with lowest DeltaCost(v)
     end if
     $soln \leftarrow soln$ with v_f flipped
     $cost \leftarrow cost +$ DeltaCost(v_f)
   end for
end for
return "Failure, best assignment is", best *soln* found

---

differs from MaxWalkSAT, which assigns random values to all atoms. However, the LazySAT initialization is a valid MaxWalkSAT initialization, and we have verified experimentally that the two give very similar results. Given the same initialization, the two algorithms will produce exactly the same results.

At each step in the search, the variable that is flipped is activated, as are any clauses that by definition should become active as a result. When evaluating the effect on cost of flipping a variable v, if v is active then all of the relevant clauses are already active, and DeltaCost(v) can be computed as in MaxWalkSAT. If v is inactive, DeltaCost(v) needs to be computed using the knowledge base. This is done by retrieving from the KB all first-order clauses containing the atom that v is a grounding of, and grounding each such clause with the constants in v and all possible groundings of the remaining variables. As before, we prune search as soon as a partial grounding is satisfied, and add the appropriate multiple of the clause weight to DeltaCost(v). (A similar process is used to activate clauses.) While this process is costlier than using pre-grounded clauses, it is amortized over many tests of active variables. In typical satisfiability problems, a small core of "problem" clauses is repeatedly tested, and when this is the case LazySAT will be quite efficient.

At each step, LazySAT flips the same variable that MaxWalkSAT would, and hence the result of the search is the same. The memory cost of LazySAT is on the order of the maximum number of clauses active at the end of a run of flips. (The memory required to store the active atoms is dominated by the memory

---

Algorithm 2. LazySAT(*weighted_KB, DB, max_flips, max_tries, target, p* **)**

---

for $i \leftarrow 1$ to *max_tries* **do**

    active_atoms $\leftarrow$ atoms in clauses not satisfied by *DB*

    active_clauses $\leftarrow$ clauses activated by *active_atoms*

    soln $\leftarrow$ a random truth assignment to *active_atoms*

    cost $\leftarrow$ sum of weights of unsatisfied clauses in *soln*

    for $i \leftarrow 1$ to *max_flips* **do**

        if $cost \leq target$ **then**

            return "Success, solution is", *soln*

        end if

        $c \leftarrow$ a randomly chosen unsatisfied clause

        if $\text{Uniform}(0,1) < p$ **then**

            $v_f \leftarrow$ a randomly chosen variable from c

        else

            for each variable v in c **do**

                compute DeltaCost(v), using *weighted_KB* if $v \notin active\_atoms$

            end for

            $v_f \leftarrow v$ with lowest DeltaCost(v)

        end if

        if $v_f \notin active\_atoms$ **then**

            add v_f to *active_atoms*

            add clauses activated by v_f to *active_clauses*

        end if

        soln $\leftarrow$ *soln* with v_f flipped

        cost $\leftarrow$ *cost* + DeltaCost(v_f)

    end for

end for

return "Failure, best assignment is", best *soln* found

---

required to store the active clauses, since each active atom appears in at least one active clause.)

Experiments on entity resolution and planning problems show that this can yield very large memory reductions, and these reductions increase with domain size [42]. For domains whose full instantiations fit in memory, running time is comparable; as problems become larger, full instantiation for MaxWalkSAT becomes impossible.

5.2 Marginal and Conditional Probabilities

Another key inference task is computing the probability that a formula holds, given an MLN and set of constants, and possibly other formulas as evidence. By definition, the probability of a formula is the sum of the probabilities of the worlds where it holds, and computing it by brute force requires time exponential in the number of possible ground atoms. An approximate but more efficient alternative is to use Markov chain Monte Carlo (MCMC) inference [11], which samples a sequence of states according to their probabilities, and counting the fraction of sampled states where the formula holds. This can be extended to conditioning on other formulas by rejecting any state that violates one of them.

For the remainder of the chapter, we focus on the typical case where the evidence is a conjunction of ground atoms. In this scenario, further efficiency can be gained by applying a generalization of knowledge-based model construction [49]. This constructs only the minimal subset of the ground network required to answer the query, and runs MCMC (or any other probabilistic inference method) on it. The network is constructed by checking if the atoms that the query formula directly depends on are in the evidence. If they are, the construction is complete. Those that are not are added to the network, and we in turn check the atoms they depend on. This process is repeated until all relevant atoms have been retrieved. While in the worst case it yields no savings, in practice it can vastly reduce the time and memory required for inference. See Richardson and Domingos [36] for details.

One problem with applying MCMC to MLNs is that it breaks down in the presence of deterministic or near-deterministic dependencies (as do other probabilistic inference methods, e.g., belief propagation [52]). Deterministic dependencies break up the space of possible worlds into regions that are not reachable from each other, violating a basic requirement of MCMC. Near-deterministic dependencies greatly slow down inference, by creating regions of low probability that are very difficult to traverse. Running multiple chains with random starting points does not solve this problem, because it does not guarantee that different regions will be sampled with frequency proportional to their probability, and there may be a very large number of regions.

We have successfully addressed this problem by combining MCMC with satisfiability testing in the MC-SAT algorithm [32]. MC-SAT is a *slice sampling* MCMC algorithm. It uses a combination of satisfiability testing and simulated annealing to sample from the slice. The advantage of using a satisfiability solver (WalkSAT) is that it efficiently finds isolated modes in the distribution, and as a result the Markov chain mixes very rapidly. The slice sampling scheme ensures that detailed balance is (approximately) preserved.

MC-SAT is orders of magnitude faster than standard MCMC methods such as Gibbs sampling and simulated tempering, and is applicable to any model that can be expressed in Markov logic, including many standard models in statistical physics, vision, natural language processing, social network analysis, spatial statistics, etc.

Slice sampling [5] is an instance of a widely used approach in MCMC inference that introduces *auxiliary variables* to capture the dependencies between observed variables. For example, to sample from $P(X = x) = (1/Z) \prod_k \phi_k(x_{\{k\}})$, we can define $P(X = x, U = u) = (1/Z) \prod_k I_{[0,\phi_k(x_{\{k\}})]}(u_k)$, where ϕ_k is the kth potential function, u_k is the kth auxiliary variable, $I_{[a,b]}(u_k) = 1$ if $a \leq u_k \leq b$, and $I_{[a,b]}(u_k) = 0$ otherwise. The marginal distribution of X under this joint is $P(X = x)$, so to sample from the original distribution it suffices to sample from $P(x, u)$ and ignore the u values. $P(u_k | x)$ is uniform in $[0, \phi_k(x_{\{k\}})]$, and thus easy to sample from. The main challenge is to sample x given u, which is uniform

---

Algorithm 3. MC-SAT(*clauses, weights, num_samples* **)**

$x^{(0)} \leftarrow$ Satisfy(hard *clauses*)
for $i \leftarrow 1$ to *num_samples* **do**
  $M \leftarrow \emptyset$
  for all $c_k \in$ *clauses* satisfied by $x^{(i-1)}$ **do**
    With probability $1 - e^{-w_k}$ add c_k to M
  end for
  Sample $x^{(i)} \sim \mathcal{U}_{SAT(M)}$
end for

---

among all $\mathcal{X}$ that satisfies $\phi_k(x_{\{k\}}) \geq u_k$ for all k. MC-SAT uses SampleSAT [48] to do this. In each sampling step, MC-SAT takes the set of all ground clauses satisfied by the current state of the world and constructs a subset, M, that must be satisfied by the next sampled state of the world. (For the moment we will assume that all clauses have positive weight.) Specifically, a satisfied ground clause is included in M with probability $1 - e^{-w}$, where w is the clause's weight. We then take as the next state a uniform sample from the set of states $SAT(M)$ that satisfy M. (Notice that $SAT(M)$ is never empty, because it always contains at least the current state.) Algorithm 3 gives pseudo-code for MC-SAT. $\mathcal{U}_S$ is the uniform distribution over set S. At each step, all hard clauses are selected with probability 1, and thus all sampled states satisfy them. Negative weights are handled by noting that a clause with weight $w < 0$ is equivalent to its negation with weight $-w$, and a clause's negation is the conjunction of the negations of all of its literals. Thus, instead of checking whether the clause is satisfied, we check whether its negation is satisfied; if it is, with probability $1 - e^w$ we select all of its negated literals, and with probability e^w we select none.

It can be shown that MC-SAT satisfies the MCMC criteria of detailed balance and ergodicity [32], assuming a perfect uniform sampler. In general, uniform sampling is #P-hard and SampleSAT [48] only yields approximately uniform samples. However, experiments show that MC-SAT is still able to produce very accurate probability estimates, and its performance is not very sensitive to the parameter setting of SampleSAT.

We have applied the ideas of LazySAT to implement a lazy version of MC-SAT that avoids grounding unnecessary atoms and clauses. A working version of this algorithm is present in the open-source Alchemy system [16].

It is also possible to carry out lifted first-order probabilistic inference (akin to resolution) in Markov logic [3]. These methods speed up inference by reasoning at the first-order level about groups of indistinguishable objects rather than propositionalizing the entire domain. This is particularly applicable when the population size is given but little is known about most individual members.

6   Learning

In this section, we discuss methods for automatically learning weights, refining formulas, and constructing new formulas from data. Of course, learning is but one method for generating an MLN. In a distributed knowledge base such as the Semantic Web, formulas could come from many different sources and their weights could be set by the sources themselves or using credibility or trust propagation (e.g., [35]). When data is available, learning methods allow us to automatically adjust weights and refine or add formulas to an MLN.

6.1 Generative Weight Learning

MLN weights can be learned generatively by maximizing the likelihood of a relational database (Equation 3). This relational database consists of one or more "possible worlds" that form our training examples. Note that we can learn to generalize from even a single example because the clause weights are shared across their many respective groundings. This is essential when the training data is a single network, such as in the Semantic Web. The gradient of the log-likelihood with respect to the weights is

$$\frac{\partial}{\partial w_i} \log P_w(X\!=\!x) = n_i(x) - \sum_{x'} P_w(X\!=\!x')\, n_i(x') \qquad (4)$$

where the sum is over all possible databases x', and $P_w(X = x')$ is $P(X = x')$ computed using the current weight vector $w = (w_1, \ldots, w_i, \ldots)$. In other words, the ith component of the gradient is simply the difference between the number of true groundings of the ith formula in the data and its expectation according to the current model. In the generative case, even approximating these expectations tends to be prohibitively expensive or inaccurate due to the large state space. Instead, we maximize the pseudo-likelihood of the data, a widely-used alternative [2]. If x is a possible world (relational database) and x_l is the lth ground atom's truth value, the pseudo-log-likelihood of x given weights w is

$$\log P_w^*(X\!=\!x) = \sum_{l=1}^n \log P_w(X_l\!=\!x_l | MB_x(X_l)) \qquad (5)$$

where $MB_x(X_l)$ is the state of X_l's Markov blanket in the data (i.e., the truth values of the ground atoms it appears in some ground formula with). Computing the pseudo-likelihood and its gradient does not require inference, and is therefore much faster. Combined with the L-BFGS optimizer [20], pseudo-likelihood yields efficient learning of MLN weights even in domains with millions of ground atoms [36]. However, the pseudo-likelihood parameters may lead to poor results when long chains of inference are required.

In order to reduce overfitting, we penalize each weight with a Gaussian prior. We apply this strategy not only to generative learning, but to all of our weight learning methods, even those embedded within structure learning.

6.2 Discriminative Weight Learning

Discriminative learning is an attractive alternative to pseudo-likelihood. In many applications, we know *a priori* which atoms will be evidence and which ones will be queried, and the goal is to correctly predict the latter given the former. If we partition the ground atoms in the domain into a set of evidence atoms X and a set of query atoms Y, the *conditional likelihood (CLL)* of Y given X is $P(y|x) = (1/Z_x) \exp\left(\sum_{i \in F_Y} w_i n_i(x, y)\right) = (1/Z_x) \exp\left(\sum_{j \in G_Y} w_j g_j(x, y)\right)$, where F_Y is the set of all MLN clauses with at least one grounding involving a query atom, $n_i(x, y)$ is the number of true groundings of the ith clause involving query atoms, G_Y is the set of ground clauses in $M_{L,C}$ involving query atoms, and $g_j(x, y) = 1$ if the jth ground clause is true in the data and 0 otherwise. The gradient of the CLL is

$$\frac{\partial}{\partial w_i} \log P_w(y|x) = n_i(x, y) - \sum_{y'} P_w(y'|x) n_i(x, y')$$

$$= n_i(x, y) - E_w[n_i(x, y)] \qquad (6)$$

In the conditional case, we can approximate the expected counts $E_w[n_i(x, y)]$ using either the MAP state (i.e., the most probable state of y given x) or by averaging over several MC-SAT samples. The MAP approximation is inspired by the voted perceptron algorithm proposed by Collins [4] for discriminatively laerning hidden Markov models. We can apply a similar algorithm to MLNs using MaxWalkSAT to find the approximate MAP state, following the approximate gradient for a fixed number of iterations, and averaging the weights across all iterations to combat overfitting [40]. We get the best results, however, by applying a version of the scaled conjugate gradient algorithm [26]. We use a small number of MC-SAT samples to approximate the gradient and Hessian matrix, and use the inverse diagonal hessian as a preconditioner. See Lowd and Domingos [22] for more details and results.

6.3   Structure Learning

The structure of a Markov logic network is the set of formulas or clauses to which we attach weights. While this knowledge base is often specified by one or more experts, such knowledge is not always accurate or complete. In addition to learning weights for the provided clauses, we can revise or extend the MLN structure with new clauses learned from data. The inductive logic programming (ILP) community has developed many methods for learning logical rules from data. However, since an MLN represents a probability distribution, much better results are obtained by using an evaluation function based on pseudo-likelihood, rather than typical ILP ones like accuracy and coverage [14]. Log-likelihood or conditional log-likelihood are potentially better evaluation functions, but are vastly more expensive to compute. In experiments on two real-world datasets, our MLN structure learning algorithm found better MLN rules than the standard ILP algorithms CLAUDIEN [6], FOIL [34], and Aleph [45], and even a hand-written knowledge base.

MLN structure learning can start from an empty network or from an existing KB. Either way, we have found it useful to start by adding all unit clauses (single atoms) to the MLN. The weights of these capture (roughly speaking) the marginal distributions of the atoms, allowing the longer clauses to focus on modeling atom dependencies. To extend this initial model, we either repeatedly find the best clause using beam search and add it to the MLN, or add all "good" clauses of length l before trying clauses of length $l + 1$. Candidate clauses are formed by adding each predicate (negated or otherwise) to each current clause, with all possible combinations of variables, subject to the constraint that at least one variable in the new predicate must appear in the current clause. Hand-coded clauses are also modified by removing predicates.

Recently, Mihalkova and Mooney [25] introduced BUSL, an alternative, bottom-up structure learning algorithm for Markov logic. Instead of blindly constructing candidate clauses one literal at a time, they let the training data guide and constrain clause construction. First, they use a propositional Markov network structure learner to generate a graph of relationships among atoms. Then they generate clauses from paths in this graph. In this way, BUSL focuses on clauses that have support in the training data. In experiments on three datasets, BUSL evaluated many fewer candidate clauses than our top-down algorithm, ran more quickly, and learned more accurate models.

We are currently investigating further approaches to learning MLNs, including automatically inventing new predicates (or, in statistical terms, discovering hidden variables) [15].

7   Applications

We have already applied Markov logic to a variety of problems relevant to the Semantic Web, including link prediction and collective classification, for filling in missing attributes and relationships; entity resolution, for matching equivalent entities that have different names; information extraction, for adding structure to raw or semi-structured text; and other problems [36,40,14,41,32,33]. Even our simple Friends and Smokers example touches on link prediction, collective classification, and social network analysis. In this section, we will show in detail how Markov logic can be used to build state-of-the-art models for entity resolution and information extraction, and present experimental results on real-world citation data.

Others have also applied Markov logic in a variety of areas. A system based on it recently won a competition on information extraction for biology [37]. Cycorp has used it to make parts of the Cyc knowledge base probabilistic [24]. The CALO project is using it to integrate probabilistic learning and inference across many components [9]. Of particular relevance to the Semantic Web is the recent work of Wu and Weld [51] on automatically refining the Wikipedia infobox ontology.

7.1   Entity Resolution

The application to entity resolution illustrates well the power of Markov logic [41]. Entity resolution is the problem of determining which observations (e.g., database records, noun phrases, video regions, etc.) correspond to the same real-world objects. This is an important and difficult task even on small, well-defined, and well-maintained databases. In the Semantic Web, automatically determining which objects, fields, and types are equivalent becomes much harder since the data may come from many different sources with varied quality. Manual annotation does not scale, so automatically determining these relationships is essential for maintaining connectedness in the Semantic Web.

Entity resolution is typically done by forming a vector of properties for each pair of observations, using a learned classifier (such as logistic regression) to predict whether they match, and applying transitive closure. Markov logic yields an improved solution simply by applying the standard logical approach of removing the unique names assumption and introducing the equality predicate and its axioms: equality is reflexive, symmetric and transitive; groundings of a predicate with equal constants have the same truth values; and constants appearing in a ground predicate with equal constants are equal. This last axiom is not valid in logic, but captures a useful statistical tendency. For example, if two papers are the same, their authors are the same; and if two authors are the same, papers by them are more likely to be the same. Weights for different instances of these axioms can be learned from data. Inference over the resulting MLN, with entity properties and relations as the evidence and equality atoms as the query, naturally combines logistic regression and transitive closure. Most importantly, it performs *collective* entity resolution, where resolving one pair of entities helps to resolve pairs of related entities.

As a concrete example, consider the task of deduplicating a citation database in which each citation has author, title, and venue fields. We can represent the domain structure with eight relations: Author(bib, author), Title(bib, title), and Venue(bib, venue) relate citations to their fields; HasWord(author/title/venue, word) indicates which words are present in each field; SameAuthor (author, author), SameTitle(title, title), and SameVenue(venue, venue) represent field equivalence; and SameBib(bib, bib) represents citation equivalence. The truth values of all relations except for the equivalence relations are provided as background theory. The objective is to predict the SameBib relation.

We begin with a logistic regression model to predict citation equivalence based on the words in the fields. This is easily expressed in Markov logic by rules such as the following:

$$\text{Title}(b1, t1) \land \text{Title}(b2, t2) \land \text{HasWord}(t1, +\text{word})$$
$$\land \text{HasWord}(t2, +\text{word}) \Rightarrow \text{SameBib}(b1, b2)$$

The '+' operator here generates a separate rule (and with it, a separate learnable weight) for each constant of the appropriate type. When given a positive weight, each of these rules increases the probability that two citations with a particular title word in common are equivalent. We can construct similar rules for other fields. Note that we may learn negative weights for some of these rules, just as logistic regression may learn negative feature weights. Transitive closure consists of a single rule:

$$\text{SameBib}(b1, b2) \wedge \text{SameBib}(b2, b3) \Rightarrow \text{SameBib}(b1, b3)$$

This model is similar to the standard solution, but has the advantage that the classifier is learned in the context of the transitive closure operation.

We can construct similar rules to predict the equivalence of two fields as well. The usefulness of Markov logic is shown further when we link field equivalence to citation equivalence:

$$\text{Author}(b1, a1) \wedge \text{Author}(b2, a2) \wedge \text{SameBib}(b1, b2) \Rightarrow \text{SameAuthor}(a1, a2)$$
$$\text{Author}(b1, a1) \wedge \text{Author}(b2, a2) \wedge \text{SameAuthor}(a1, a2) \Rightarrow \text{SameBib}(b1, b2)$$

The above rules state that if two citations are the same, their authors should be the same, and that citations with the same author are more likely to be the same. The last rule is not valid in logic, but captures a useful statistical tendency.

Most importantly, the resulting model can now perform *collective* entity resolution, where resolving one pair of entities helps to resolve pairs of related entities. For example, inferring that a pair of citations are equivalent can provide evidence that the names *AAAI-06* and *21st Natl. Conf. on AI* refer to the same venue, even though they are superficially very different. This equivalence can then aid in resolving other entities.

Experiments on citation databases like Cora and BibServ.org show that these methods can greatly improve accuracy, particularly for entity types that are difficult to resolve in isolation as in the above example [41]. Due to the large number of words and the high arity of the transitive closure formula, these models have thousands of weights and ground millions of clauses during learning, even after using canopies to limit the number of comparisons considered. Learning at this scale is still reasonably efficient: preconditioned scaled conjugate gradient with MC-SAT for inference converges within a few hours [22].

7.2   Information Extraction

In this citation example, it was assumed that the fields were manually segmented in advance. The goal of information extraction is to extract database records starting from raw text or semi-structured data sources. This has many applications for the Semantic Web, including using the vast amount of unstructured information on the Web to bootstrap the Semantic Web. Information extraction could also be used to segment labeled fields, such as "name," into more specific fields, such as "first name," "last name," and "title."

Traditionally, information extraction proceeds by first segmenting each candidate record separately, and then merging records that refer to the same entities. Such a pipeline architecture is adopted by many AI systems in natural language processing, speech recognition, vision, robotics, etc. Markov logic allows us to perform the two tasks jointly [33]. This enables us to use the segmentation of one candidate record to help segment similar ones. For example, resolving a well-segmented field with a less-clear one can disambiguate the latter's boundaries. We will continue with the example of citations, but similar ideas could be applied to other data sources, such as Web pages or emails.

The main evidence predicate in the information extraction MLN is $\texttt{Token}(\texttt{t}, \texttt{i},$ $\texttt{c})$, which is true iff token $\texttt{t}$ appears in the $\texttt{i}$th position of the $\texttt{c}$th citation. A token can be a word, date, number, etc. Punctuation marks are not treated as separate tokens; rather, the predicate $\texttt{HasPunc}(\texttt{c}, \texttt{i})$ is true iff a punctuation mark appears immediately after the $\texttt{i}$th position in the $\texttt{c}$th citation. The query predicates are $\texttt{InField}(\texttt{i}, \texttt{f}, \texttt{c})$ and $\texttt{SameCitation}(\texttt{c}, \texttt{c}')$. $\texttt{InField}(\texttt{i}, \texttt{f}, \texttt{c})$ is true iff the $\texttt{i}$th position of the $\texttt{c}$th citation is part of field $\texttt{f}$, where $\texttt{f} \in \{\texttt{Title}, \texttt{Author}, \texttt{Venue}\}$, and inferring it performs segmentation. $\texttt{SameCitation}(\texttt{c}, \texttt{c}')$ is true iff citations $\texttt{c}$ and $\texttt{c}'$ represent the same publication, and inferring it performs entity resolution.

Our segmentation model is essentially a hidden Markov model (HMM) with enhanced ability to detect field boundaries. The observation matrix of the HMM correlates tokens with fields, and is represented by the simple rule

$$\texttt{Token}(+\texttt{t}, \texttt{i}, \texttt{c}) \Rightarrow \texttt{InField}(\texttt{i}, +\texttt{f}, \texttt{c})$$

If this rule was learned in isolation, the weight of the (t, f)th instance would be $\log(p_{tf}/(1 - p_{tf}))$, where p_{tf} is the corresponding entry in the HMM observation matrix. In general, the transition matrix of the HMM is represented by a rule of the form

$$\texttt{InField}(\texttt{i}, +\texttt{f}, \texttt{c}) \Rightarrow \texttt{InField}(\texttt{i} + 1, +\texttt{f}', \texttt{c})$$

However, we (and others, e.g., [12]) have found that for segmentation it suffices to capture the basic regularity that consecutive positions tend to be part of the same field. Thus we replace $\texttt{f}'$ by $\texttt{f}$ in the formula above. We also impose the condition that a position in a citation string can be part of at most one field; it may be part of none.

The main shortcoming of this model is that it has difficulty pinpointing field boundaries. Detecting these is key for information extraction, and a number of approaches use rules designed specifically for this purpose (e.g., [17]). In citation matching, boundaries are usually marked by punctuation symbols. This can be incorporated into the MLN by modifying the rule above to

$$\texttt{InField}(\texttt{i}, +\texttt{f}, \texttt{c}) \wedge \neg \texttt{HasPunc}(\texttt{c}, \texttt{i}) \Rightarrow \texttt{InField}(\texttt{i} + 1, +\texttt{f}, \texttt{c})$$

The $\neg \texttt{HasPunc}(\texttt{c}, \texttt{i})$ precondition prevents propagation of fields across punctuation marks. Because propagation can occur differentially to the left and right, the MLN also contains the reverse form of the rule. In addition, to account for commas being weaker separators than other punctuation, the MLN includes versions of these rules with $\texttt{HasComma}()$ instead of $\texttt{HasPunc}()$.

Finally, the MLN contains rules capturing a variety of knowledge about citations: the first two positions of a citation are usually in the author field, and the middle one in the title; initials (e.g., "J.") tend to appear in either the author or the venue field; positions preceding the last non-venue initial are usually not part of the title or venue; and positions after the first venue keyword (e.g., "Proceedings", "Journal") are usually not part of the author or title.

By combining this segmentation model with our entity resolution model from before, we can exploit relational information as part of the segmentation process. In practice, something a little more sophisticated is necessary to get good

Table 2. CiteSeer entity resolution: cluster recall on each section

Approach	Constr.	Face	Reason.	Reinfor.
Fellegi-Sunter	84.3	81.4	71.3	50.6
Lawrence et al. (1999)	89	94	86	79
Pasula et al. (2002)	93	97	96	94
Wellner et al. (2004)	95.1	96.9	93.7	94.7
Joint MLN	96.0	97.1	95.1	96.7

results on real data. In Poon and Domingos [33], we define predicates and rules specifically for passing information between the stages, as opposed to just using the existing InField() outputs. This leads to a "higher bandwidth" of communication between segmentation and entity resolution, without letting excessive segmentation noise through. We also define an additional predicate and modify rules to better exploit information from similar citations during the segmentation process. See [33] for further details.

We evaluated this model on the CiteSeer and Cora datasets. For entity resolution in CiteSeer, we measured *cluster recall* for comparison with previously published results. Cluster recall is the fraction of clusters that are correctly output by the system after taking transitive closure from pairwise decisions. For entity resolution in Cora, we measured both cluster recall and pairwise recall/precision. In both datasets we also compared with a "standard" Fellegi-Sunter model (see [41]), learned using logistic regression, and with oracle segmentation as the input.

In both datasets, joint inference improved accuracy and our approach outperformed previous ones. Table 2 shows that our approach outperforms previous ones on CiteSeer entity resolution. (Results for Lawrence et al. (1999) [19], Pasula et al. (2002) [29] and Wellner et al. (2004) [50] are taken from the corresponding papers.) This is particularly notable given that the models of [29] and [50] involved considerably more knowledge engineering than ours, contained more learnable parameters, and used additional training data.

Table 3 shows that our entity resolution approach easily outperforms Fellegi-Sunter on Cora, and has very high pairwise recall/precision.

Table 3. Cora entity resolution: pairwise recall/precision and cluster recall

Approach	Pairwise Rec./Prec.	Cluster Recall
Fellegi-Sunter	78.0 / 97.7	62.7
Joint MLN	94.3 / 97.0	78.1

8   The Alchemy System

The inference and learning algorithms described in the previous sections are publicly available in the open-source Alchemy system [16]. Alchemy makes it possible to define sophisticated probabilistic models with a few formulas, and to add probability to a first-order knowledge base by learning weights from a

Table 4. A comparison of Alchemy, Prolog and BUGS

Aspect	Alchemy	Prolog	BUGS
Representation	First-order logic + Markov nets	Horn clauses	Bayes nets
Inference	Model checking, MCMC	Theorem proving	MCMC
Learning	Parameters and structure	No	Parameters
Uncertainty	Yes	No	Yes
Relational	Yes	Yes	No

relevant database. It can also be used for purely logical or purely statistical applications, and for teaching AI. From the user's point of view, Alchemy provides a full spectrum of AI tools in an easy-to-use, coherent form. From the researcher's point of view, Alchemy makes it possible to easily integrate a new inference or learning algorithm, logical or statistical, with a full complement of other algorithms that support it or make use of it.

Alchemy can be viewed as a declarative programming language akin to Prolog, but with a number of key differences: the underlying inference mechanism is model checking instead of theorem proving; the full syntax of first-order logic is allowed, rather than just Horn clauses; and, most importantly, the ability to handle uncertainty and learn from data is already built in. Table 4 compares Alchemy with Prolog and BUGS [23], one of the most popular toolkits for Bayesian modeling and inference.

9   Current and Future Research Directions

We are actively researching better learning and inference methods for Markov logic, as well as extensions of the representation that increase its generality and power.

Exact methods for learning and inference are usually intractable in Markov logic, but we would like to see better, more efficient approximations along with the automatic application of exact methods when feasible.

One method of particular interest is lifted inference. In short, we would like to reason with clusters of nodes for which we have exactly the same amount of information. The inspiration is from lifted resolution in first order logic, but must be extended to handle uncertainty. Prior work on lifted inference such as [31] and [3] mainly focused on exact inference which can be quite slow. We have recently extended loopy belief propagation, an approximate inference method for probabilistic graphical models, to perform lifted inference in Markov logic networks [44]. When the amount of evidence is limited, this can speed up inference by many orders of magnitude.

We are also working to develop a general framework for decision-making in relational domains. This can be accomplished in Markov logic by adding utility weights to formulas and finding the settings of all action predicates that jointly maximize expected utility. Decision-making is key to the original Semantic Web vision, which called for intelligent agents to act on the information they gathered.

Numerical attributes must be discretized to be used in Markov logic, but we have recently introduced methods to incorporate continuous random variables and features [46]. Continuous values could be useful in a variety of Semantic Web problems, such as incorporating numeric features into similarities for entity resolution, ontology alignment, or schema matching.

Current work also includes semi-supervised learning, and learning with incomplete data in general. The large amount of unlabeled data on the Web is an excellent resource that, properly exploited, could help bootstrap or enrich the Semantic Web.

10   Conclusion

The Semantic Web must deal with uncertainty from many sources, including inconsistent knowledge bases, incorrect or untrustworthy information, missing data, different ontologies and schemas, and more. Markov logic is a simple yet powerful approach for adding probability to logical representations such as those already used by the Semantic Web: Given a set of formulas, just add weights. We have developed a series of learning and inference algorithms for it, and successfully applied them in a number of domains. These algorithms are included in the open-source Alchemy system (available at alchemy.cs.washington.edu). We hope that Markov logic and its implementation in Alchemy will be of use to Semantic Web researchers and practitioners who wish to have the full spectrum of logical and statistical inference and learning techniques at their disposal, without having to develop every piece themselves.

Acknowledgements

This research was partly supported by DARPA grant FA8750-05-2-0283 (managed by AFRL), DARPA contract NBCH-D030010, NSF grant IIS-0534881, ONR grants N00014-02-1-0408 and N00014-05-1-0313, a Sloan Fellowship and NSF CAREER Award to the first author, and a Microsoft Research fellowship awarded to the second author. The views and conclusions contained in this document are those of the authors and should not be interpreted as necessarily representing the official policies, either expressed or implied, of DARPA, NSF, ONR, or the United States Government.

References

1. Berners-Lee, T., Hendler, J., Lassila, O.: The Semantic Web. Scientific American 284(5), 34–43 (2001)
2. Besag, J.: Statistical analysis of non-lattice data. The Statistician 24, 179–195 (1975)
3. Braz, R., Amir, E., Roth, D.: Lifted first-order probabilistic inference. In: Proceedings of the Nineteenth International Joint Conference on Artificial Intelligence, Edinburgh, UK, pp. 1319–1325. Morgan Kaufmann, San Francisco (2005)

4. Collins, M.: Discriminative training methods for hidden Markov models: Theory and experiments with perceptron algorithms. In: Proceedings of the 2002 Conference on Empirical Methods in Natural Language Processing, Philadelphia, PA, pp. 1–8. ACL (2002)

5. Damien, P., Wakefield, J., Walker, S.: Gibbs sampling for Bayesian non-conjugate and hierarchical models by auxiliary variables. Journal of the Royal Statistical Society, Series B 61 (1999)

6. De Raedt, L., Dehaspe, L.: Clausal discovery. Machine Learning 26, 99–146 (1997)

7. Dean, M., Schreiber, G., Bechhofer, S., van Harmelen, F., Hendler, J., Horrocks, I., McGuinness, D.L., Patel-Schneider, P.F., Stein, L.A.: OWL web ontology language reference (2004), http://www.w3.org/TR/owl-ref/

8. Della Pietra, S., Della Pietra, V., Lafferty, J.: Inducing features of random fields. IEEE Transactions on Pattern Analysis and Machine Intelligence 19, 380–392 (1997)

9. Dietterich, T., Bao, X.: Integrating multiple learning components through Markov logic. In: Proceedings of the Twenty-Third National Conference on Artificial Intelligence, Chicago, IL. AAAI Press, Menlo Park (2008)

10. Genesereth, M.R., Nilsson, N.J.: Logical Foundations of Artificial Intelligence. Morgan Kaufmann, San Mateo (1987)

11. Gilks, W.R., Richardson, S., Spiegelhalter, D.J. (eds.): Markov Chain Monte Carlo in Practice. Chapman and Hall, London (1996)

12. Grenager, T., Klein, D., Manning, C.D.: Unsupervised learning of field segmentation models for information extraction. In: Proceedings of the Forty-Third Annual Meeting on Association for Computational Linguistics, Ann Arbor, Michigan, pp. 371–378. Association for Computational Linguistics (2005)

13. Kautz, H., Selman, B., Jiang, Y.: A general stochastic approach to solving problems with hard and soft constraints. In: Gu, D., Du, J., Pardalos, P. (eds.) The Satisfiability Problem: Theory and Applications, pp. 573–586. American Mathematical Society, New York (1997)

14. Kok, S., Domingos, P.: Learning the structure of Markov logic networks. In: Proceedings of the Twenty-Second International Conference on Machine Learning, Bonn, Germany, pp. 441–448. ACM Press, New York (2005)

15. Kok, S., Domingos, P.: Statistical predicate invention. In: Proceedings of the Twenty-Fourth International Conference on Machine Learning, Corvallis, OR, pp. 433–440. ACM Press, New York (2007)

16. Kok, S., Sumner, M., Richardson, M., Singla, P., Poon, H., Lowd, D., Domingos, P.: The Alchemy system for statistical relational AI. Technical report, Department of Computer Science and Engineering, University of Washington, Seattle, WA (2007), http://alchemy.cs.washington.edu

17. Kushmerick, N.: Wrapper induction: Efficiency and expressiveness. Artificial Intelligence 118(1-2), 15–68 (2000)

18. Lavrač, N., Džeroski, S.: Inductive Logic Programming: Techniques and Applications. Ellis Horwood, Chichester (1994)

19. Lawrence, S., Bollacker, K., Giles, C.L.: Autonomous citation matching. In: Proceedings of the Third International Conference on Autonomous Agents. ACM Press, New York (1999)

20. Liu, D.C., Nocedal, J.: On the limited memory BFGS method for large scale optimization. Mathematical Programming 45(3), 503–528 (1989)

21. Lloyd, J.W.: Foundations of Logic Programming. Springer, Berlin (1987)

22. Lowd, D., Domingos, P.: Efficient weight learning for Markov logic networks. In: Kok, J.N., Koronacki, J., Lopez de Mantaras, R., Matwin, S., Mladenič, D., Skowron, A. (eds.) PKDD 2007. LNCS, vol. 4702, pp. 200–211. Springer, Heidelberg (2007)

23. Lunn, D.J., Thomas, A., Best, N., Spiegelhalter, D.: WinBUGS – a Bayesian modeling framework: concepts, structure, and extensibility. Statistics and Computing 10, 325–337 (2000)

24. Matuszek, C., Witbrock, M.: Personal communication (2006)

25. Mihalkova, L., Mooney, R.: Bottom-up learning of Markov logic network structure. In: Proceedings of the Twenty-Fourth International Conference on Machine Learning, Corvallis, OR, pp. 625–632. ACM Press, New York (2007)

26. Møller, M.: A scaled conjugate gradient algorithm for fast supervised learning. Neural Networks 6, 525–533 (1993)

27. Ngo, L., Haddawy, P.: Answering queries from context-sensitive probabilistic knowledge bases. Theoretical Computer Science 171, 147–177 (1997)

28. Nocedal, J., Wright, S.: Numerical Optimization. Springer, New York (2006)

29. Pasula, H., Marthi, B., Milch, B., Russell, S., Shpitser, I.: Identity uncertainty and citation matching. In: Advances in Neural Information Processing Systems 14. MIT Press, Cambridge (2002)

30. Pearl, J.: Probabilistic Reasoning in Intelligent Systems: Networks of Plausible Inference. Morgan Kaufmann, San Francisco (1988)

31. Poole, D.: First-order probabilistic inference. In: Proceedings of the Eighteenth International Joint Conference on Artificial Intelligence, Acapulco, Mexico, pp. 985–991. Morgan Kaufmann, San Francisco (2003)

32. Poon, H., Domingos, P.: Sound and efficient inference with probabilistic and deterministic dependencies. In: Proceedings of the Twenty-First National Conference on Artificial Intelligence, Boston, MA, pp. 458–463. AAAI Press, Menlo Park (2006)

33. Poon, H., Domingos, P.: Joint inference in information extraction. In: Proceedings of the Twenty-Second National Conference on Artificial Intelligence, Vancouver, Canada, pp. 913–918. AAAI Press, Menlo Park (2007)

34. Quinlan, J.R.: Learning logical definitions from relations. Machine Learning 5, 239–266 (1990)

35. Richardson, M., Agrawal, R., Domingos, P.: Trust management for the Semantic Web. In: Fensel, D., Sycara, K.P., Mylopoulos, J. (eds.) ISWC 2003. LNCS, vol. 2870, pp. 351–368. Springer, Heidelberg (2003)

36. Richardson, M., Domingos, P.: Markov logic networks. Machine Learning 62, 107–136 (2006)

37. Riedel, S., Klein, E.: Genic interaction extraction with semantic and syntactic chains. In: Proceedings of the Fourth Workshop on Learning Language in Logic, Bonn, Germany, pp. 69–74. IMLS (2005)

38. Robinson, J.A.: A machine-oriented logic based on the resolution principle. Journal of the ACM 12, 23–41 (1965)

39. Roth, D.: On the hardness of approximate reasoning. Artificial Intelligence 82, 273–302 (1996)

40. Singla, P., Domingos, P.: Discriminative training of Markov logic networks. In: Proceedings of the Twentieth National Conference on Artificial Intelligence, Pittsburgh, PA, pp. 868–873. AAAI Press, Menlo Park (2005)

41. Singla, P., Domingos, P.: Entity resolution with Markov logic. In: Proceedings of the Sixth IEEE International Conference on Data Mining, Hong Kong, pp. 572–582. IEEE Computer Society Press, Los Alamitos (2006)

42. Singla, P., Domingos, P.: Memory-efficient inference in relational domains. In: Proceedings of the Twenty-First National Conference on Artificial Intelligence, Boston, MA. AAAI Press, Menlo Park (2006)
43. Singla, P., Domingos, P.: Markov logic in infinite domains. In: Proceedings of the Twenty-Third Conference on Uncertainty in Artificial Intelligence, Vancouver, Canada, pp. 368–375. AUAI Press (2007)
44. Singla, P., Domingos, P.: Lifted first-order belief propagation. In: Proceedings of the Twenty-Third National Conference on Artificial Intelligence, Chicago, IL. AAAI Press, Menlo Park (2008)
45. Srinivasan, A.: The Aleph manual. Technical report, Computing Laboratory, Oxford University (2000)
46. Wang, J., Domingos, P.: Hybrid Markov logic networks. In: Proceedings of the Twenty-Third National Conference on Artificial Intelligence, Chicago, IL. AAAI Press, Menlo Park (2008)
47. Wasserman, S., Faust, K.: Social Network Analysis: Methods and Applications. Cambridge University Press, Cambridge (1994)
48. Wei, W., Erenrich, J., Selman, B.: Towards efficient sampling: Exploiting random walk strategies. In: Proceedings of the Nineteenth National Conference on Artificial Intelligence, San Jose, CA. AAAI Press, Menlo Park (2004)
49. Wellman, M., Breese, J.S., Goldman, R.P.: From knowledge bases to decision models. Knowledge Engineering Review 7 (1992)
50. Wellner, B., McCallum, A., Peng, F., Hay, M.: An integrated, conditional model of information extraction and coreference with application to citation matching. In: Proceedings of the Twentieth Conference on Uncertainty in Artificial Intelligence, Banff, Canada, pp. 593–601. AUAI Press (2004)
51. Wu, F., Weld, D.: Automatically refining the Wikipedia infobox ontology. In: 17th International World Wide Web Conference (WWW 2008), Beijing,China (2008)
52. Yedidia, J.S., Freeman, W.T., Weiss, Y.: Generalized belief propagation. In: Leen, T., Dietterich, T., Tresp, V. (eds.) Advances in Neural Information Processing Systems 13, pp. 689–695. MIT Press, Cambridge (2001)

Semantic Science: Ontologies, Data and Probabilistic Theories

David Poole[1], Clinton Smyth[2], and Rita Sharma[2]

[1] Department of Computer Science,
University of British Columbia
http://www.cs.ubc.ca/spider/poole/
[2] Georeference Online Ltd.,
http://www.georeferenceonline.com/

Abstract. This chapter overviews work on *semantic science*. The idea is that, using rich ontologies, both observational data and theories that make (probabilistic) predictions on data are published for the purposes of improving or comparing the theories, and for making predictions in new cases. This paper concentrates on issues and progress in having machine accessible scientific theories that can be used in this way. This paper presents the grand vision, issues that have arisen in building such systems for the geological domain (minerals exploration and geohazards), and sketches the formal foundations that underlie this vision. The aim is to get to the stage where: any new scientific theory can be tested on all available data; any new data can be used to evaluate all existing theories that make predictions on that data; and when someone has a new case they can use the best theories that make predictions on that case.

1 Introduction

The aim of the semantic web (Berners-Lee and Fischetti, 1999; Berners-Lee et al., 2001) is that the world's information is available in a machine-understandable form. This chapter overviews what we call *semantic science*, the application of semantic technology and reasoning under uncertainty to the practice of science. Semantic science requires machine-understandable information of three sorts: ontologies to define vocabulary, data about observations of the world, and theories that make predictions on such data.

Our idea of semantic science is that scientists can publish data and theories that can inter-operate by virtue of using common ontologies. The theories can be judged by how well they predict unseen data and can be used for new cases.

An ontology (Smith, 2003b) is a formal specification of the meaning of the vocabulary used in an information system. Ontologies are needed so that information sources can inter-operate at a semantic level.

There has been recent success in publishing scientific data that adheres to ontologies (McGuinness et al., 2007). Publishing data with respect to well-defined ontologies can allow for semantic inter-operation of the data sets. Meaningful queries can be made against multiple data sets that were collected separately.

P.C.G. da Costa et al. (Eds.): URSW 2005-2007, LNAI 5327, pp. 26–40, 2008.

Data repositories include the Community Data Portal (http://cdp.ucar.edu/) and the Virtual Solar-Terrestrial Observatory (http://vsto.hao.ucar.edu/index.php).

Science operates by making refutable theories (Popper, 1959). These theories[1] are judged by their predictions, by their usefulness, and by their elegance or plausibility. Theories make (probabilistic) predictions about new cases. Theories may require arbitrary computations to make predictions; indeed many real theories need enormous computational resources. Semantic science aims to provide an infrastructure to test theories on data, and to make theories available for new cases.

Theories need to refer to ontologies as they need to inter-operate with data. Theories specify what data they can make predictions about, and make predictions that can be checked against the relevant data and applied to new cases. It is the ontologies that allow the inter-operation of the data and the theories. Theories can be tested against all of the relevant data sets, and data can be used to discriminate theories.

Given access to the theories, and information about how they perform on the available data sets, practitioners can use the best theories to make predictions on new cases. This thus promises to form a new basis for expert systems.

We have been working on two instances of the semantic science framework in two domains in earth sciences (Smyth et al., 2007), namely minerals exploration in the MINEMATCH® system (http://www.georeferenceonline.com/minematch/) and landslides in the HAZARDMATCH™ system. MineMatch contains about 25,000 descriptions of mineral occurrences (called *instances*) that are described at various levels of abstraction and detail using multiple taxonomies, including the British Geological Survey rock classification scheme (http://www.bgs.ac.uk/bgsrcs/) and the Micronex taxonomy of minerals (http://micronex.golinfo.com). We are currently moving to OWL representations of the ontologies. We also work with more than 100 deposit models (these form the theories about where to find particular minerals), including those described by the US Geological Survey (http://minerals.cr.usgs.gov/team/depmod.html) and the British Columbia Geological Survey (http://www.em.gov.bc.ca/Mining/Geolsurv/MetallicMinerals/MineralDepositProfiles/). Similarly, HazardMatch uses tens of thousands of spatial instances (polygons) described using standard taxonomies of environmental modeling such as rock type, geomorphology and geological age. There are currently about 10 models of different landslide types that are derived from published models. We can compare the prediction of the models to known cases and new cases.

Semantic science allows for a diversity of theories. Each theory will specify what data it is prepared to make predictions about. Some theories may be competing and some may be complementary. For example, there may be multiple theories that predict whether a patient has cancer. If they make different predictions in some cases, they can be compared by how well they predict the available data. There may be other theories that make predictions about the type(s) of

---

[1] Theories are often called hypotheses, laws or models depending on how well established they are. This distinction is redundant in the semantic science realm where we can test how well these actually perform on data.

cancer for patients with cancer. These theories are not applicable for patients who don't have cancer. When making predictions, a doctor may use an ensemble of multiple complementary theories: e.g., one to predict whether the patient has cancer and another to predict the type of cancer if cancer is present.

Theories can make predictions in different forms. A theory could make, e.g., a definitive prediction, a probabilistic prediction, a range prediction, or a qualitative prediction. Users can use whatever criteria they like to judge the theories, and use whichever theory or mix of theories they like. For different evaluation criteria, there will be ways to judge the theories on the criteria. We anticipate that probabilistic predictions will be the most useful, as it is probabilities that one gets from data, and probabilities are what is needed (with utilities) to make decisions. However, there are many cases where users will be reluctant to use probabilistic theories (see below). Scientists who wish to judge a theory by elegance or simplicity, as well as fit to data, are free to do so; they can use published data to determine its accuracy and whatever criteria they like to evaluate elegance or simplicity.

We mean science in the broadest sense. We can imagine having theories about what apartment someone would like, or theories about what companies will make the best investments, or theories about diseases and symptoms. Search engines such as Google are being used for diagnosis (Tang and Ng, 2006). It is arguably better to be able to specify symptoms unambiguously using an ontology. Measures such as pagerank (Page et al., 1999) measure popularity. Fortunately, searches for diagnostic tend to return authoritative sites. Scientists, however, should be suspicious of popularity and authority as a basis for prediction. We should base our predictions on the empirical evidence. Building an infrastructure for this is the aim of semantic science.

Figure 1 shows the relationship between ontologies, data and theories. The data depends on the world and the ontology. The theories depend on the ontology, indirectly on the world (if a human is designing the theory), and directly

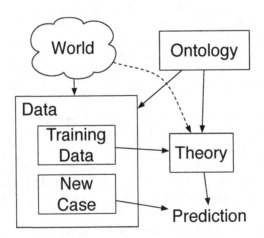

Fig. 1. Ontologies, Data and Theories in Semantic Science

on some of the data (as we would expect that the best theories would be based on as much data as possible). Given a new case, a theory can be used to make a prediction. The real world is more complicated, as there are many theories, many ontologies, and lots of data, and they all evolve in time.

This work is complementary to providing services and other tools to scientists, e.g., using the Semantic Grid (De Roure et al., 2005). We expect that the semantic grid will be important for implementing the ideas in this paper.

This chapter is based on Poole et al. (2008).

2   Background

2.1   Ontologies

In philosophy, *ontology* is the study of what exists. In AI, *ontology* (Smith, 2003b) has come to mean a specification of the meaning of the symbols (or of the data) in an information system. In particular, an ontology makes a commitment to what entities and relationships are being modelled, specifies what vocabulary will be used for the entities and relationships, and gives axioms that restrict the use of the vocabulary. The axioms have two purposes: to rule out uses of the terms that are inconsistent with the intended interpretation, and to allow for inference to derive conclusions that are implicit in the use of the vocabulary.

An ontology can be any specification, formal or informal, of the meaning of the symbols. This can be in the head of the person who created the data, or can be stated in some language. Without an ontology, we do not have information, but just a sequence of bits. The simplest form of an ontology is a database schema with informal natural language descriptions of the attributes and the constants. Formal ontologies allow machine understandable specifications.

An ontology written in a language such as OWL (McGuinness and van Harmelen, 2004) specifies individuals, classes and relationships and the vocabulary used to express them. Sometimes classes and relationships are defined in terms of more primitive classes and relationships, but ultimately they are grounded out into primitive classes and relationships that are not actually defined. For example, an ontology could specify that the term "building" will represent buildings. The ontology will not *define* a building, but will give some properties that restrict the use of the term.

Ontologies date back to Aristotle (350 B.C.), who defined terms using what has been called an Aristotelian definition (Berg, 1982; Smith, 2003a). An Aristotelian definition of A is of the form "An A is a B such that C", where B is the immediate super-class of A and C is a condition that defines how A is special. Aristotle called the B the *genus* and C the *differentia* (Sowa, 2000, p. 4).

To build Aristotelian definitions, we will use what we call the *multi-dimensional* design pattern (Alexander et al., 1977), where the differentia in the Aristotelian definition are built from multiple properties. To define the conditions for a class, we need to think about what properties distinguish this class from the other subclasses of the super-class. Each of these properties defines a (local) dimension. The domain of each property is the most general class for which it makes sense. In the multi-dimensional design pattern, classes are only defined in terms of values of properties. The subclass relation can be derived from this.

There is not a fixed number of dimensions that distinguish all individuals. Rather, dimensions come into existence at different levels of abstraction. For example, the dimensions size and weight may appear for physical individuals, but are not applicable for abstract concepts. "Number of units" may be a dimension for apartment buildings but may not be applicable for other buildings such as sewage plants, where other dimensions may be applicable.

This idea is due to Aristotle:

> "If genera are different and co-ordinate, their differentiae are themselves different in kind. Take as an instance the genus 'animal' and the genus 'knowledge'. 'With feet', 'two-footed', 'winged', 'aquatic', are differentiae of 'animal'; the species of knowledge are not distinguished by the same differentiae. One species of knowledge does not differ from another in being 'two-footed'." (Aristotle, 350 B.C.)

Example 1. Geologists define rocks along three major dimensions: genesis (sedimentary, igneous or metamorphic), composition and texture (Gillespie and Styles, 1999). Particular rocks, such as granite and limestone, are defined by particular values in each dimension (or some subset of the dimensions). Rock taxonomies built using this approach that commit to splitting rock sub-type based on these dimensions in a certain order (usually genesis first, then composition, then texture) do not conveniently represent the sub-types that occur in real data (Struik et al., 2002). For example, if the aforementioned order of splitting the taxonomy is used, there is no convenient single place in the taxonomy for the class of rocks with a particular texture, independent of its members' genesis or composition. The multi-dimensional ontologies seem to be the natural specification, and they also integrate well with probabilities (see Section 4.2).

2.2   Data and Ontologies

Scientists produce lots of data, and science cannot be carried out without data. By data, we mean information about a domain that is produced from sensing.

In linguistics the Sapir-Whorf Hypothesis (Sapir, 1929; Whorf, 1940), says essentially that people's perception and thought are determined by what can be described in their language. The Sapir-Whorf Hypothesis is controversial in linguistics, but a stronger version of this hypothesis should be uncontroversial in information systems:

> What is stored and communicated by an information system is constrained by the representation and the ontology used by the information system.

The reason that this should be less controversial is that the representation and the ontology represent the language of thought or mentalese (Fodor, 1975; Pinker, 1994), not just the language of communication.

As an example, suppose the world produces a deterministic sequence of coin tosses: head, tail, head, tail, head, tail, etc. If the representation and the ontology does not specify the time of each observation or which is the next coin toss in the sequence, that information will have been lost in translating the observation into the internal representation. The best prediction would be to predict heads with probability of 0.5. As another example, if some data adheres to an ontology that specifies that a house is a residential building, then, by definition, all of the observed houses are residential buildings, and so the data cannot refute the fact that houses are residential buildings.

This hypothesis has a number of implications:

- An ontology mediates how perceptions of the world are stored and communicated.
- If there is no distinction in the ontology, there will be no distinction in the data. For example, if an ontology does not have any sub-types of "granite", and does not record the information needed to distinguish between types of granite, the data will not record any sub-types of granite and none can be discovered.
- Ontologies must come before data. This may be confusing as much work is done on building ontologies for existing data sets. This activity should be seen as reconstructing the ontology that was used to create the data set.

  Note that this does not imply that finding regularities in data cannot be used to evolve ontologies; we are claiming that the ontology for each data set comes logically before that data set. This frequently occurs in research when a data set may record the output of a sensor where it is unknown what the senor actually measures (i.e., the meaning of the sensor report is unknown). The initial ontology will then specify the meaning is just a real number, perhaps with some range and precision. Later ontologies may give the output a name.

Some people have argued that uncertainty should be explicitly represented in an ontology because of the inherent uncertainty in data (Pool et al., 2005; da Costa et al., 2005; Laskey et al., 2007). While we believe that it is essential to model the uncertainty in data, we don't believe actual probability values should be in the ontology[2]. The main reason is the ontology is logically prior to the data, but the models of uncertainty in the data are logically posterior to the data: it is only by seeing (some of) the data, that we can estimate the uncertainty (i.e., we want the uncertainty to reflect the posterior distribution after we have seen some data). Because the probabilities are posterior to the data, they should change as data comes in, and so should not be part of the stable foundation of the data that an ontology needs to be. Another way to think about it is that the ontologies define the vocabulary; they do not make empirical claims. Saying that

---

[2] An ontology will contain the vocabulary to express probability distributions. We need the vocabulary to express continuous and discrete conditional probability distributions, e.g., using PR-OWL (da Costa et al., 2005). The ontologies need to be rich enough to express what scientists want to state in theories.

a granite is an igneous, felsic, course rock is not an empirical claim, it just defines what a granite is. Theories make empirical (testable) claims. A specification of a probability is an empirical claim, even if the probability is theory-based (e.g., based on symmetries) and not data-summaries. Thus the probability should not be in the ontology. Note that our claim that probabilities do not belong in definitions is an empirical claim, and is not part of the definition of semantic science.

2.3   Theories

We would argue that theories are best described in terms of probabilities (Polya, 1954) for two main reasons:

- Probabilities summarize the empirical content of data. In particular, we want predictions that can be evaluated against the empirical evidence, and so can be optimized with respect to the evidence. Probability distributions optimize most of the common evaluation criteria, and other predictions (such as the mean or the mode) can be derived from the probability distribution.
- Probabilities, together with utilities, are what is needed to make decisions.

Like data, theories need to adhere to ontologies. There are a number of reasons:

- Theories make predictions on data that adhere to an ontology. To allow semantic interoperability between the data and the theories, they should adhere to a common ontology.
- People should be allowed to disagree about how the world works without disagreeing about the meaning of the terms. If two people have different theories, they should first agree on the terminology (for otherwise they would not know they have a disagreement)—this forms the ontology—and then they should give their theories. Their theories can then be compared to determine what their disagreement is. It is by creating these disagreements, and testing them on data, that science progresses.

Theories can expand the ontology by hypothesizing unobserved objects or properties (hidden variables) that help explain the observations. By expanding the ontology, other theories can refer to the theoretical constructs, and they could appear in data. For example, a theory could postulate that the data is better explained by having a new form of cancer; other theories could refer to this type of cancer and this new type of cancer could even be recorded in data. In this way the theories and the vocabulary can evolve as science advances.

Semantic interoperability can only be achieved by adhering to common ontologies. A community needs to agree on an ontology to make sure they use the same terminology for the same things. However, a community need not, and we argue should not, agree on the probabilities, as people may have different prior knowledge and have access to different data, and the probabilities should change as more data comes in.

To make a prediction, we usually use many theories. Theories that individuals produce are typically very narrow, only making predictions in very narrow cases.

The theories that are put together to make a predictions form a theory *ensemble*. We judge individual theories by how well they fit into ensembles. An example of a theory ensemble is "when the speed of the objects involved is less than 70% of the speed of light, use Newtonian mechanics, otherwise use Einstein's theory of relativity". This ensemble is another theory that may work better than either of the composite theories in practice. Producing such ensembles is of a different sort than producing the base theories, and so should be separated. Rather than dismissing these theories as trivial, they form the basis of prediction for new cases. Virtually all predictions in complex cases will rely on theory ensembles.

The structure of probabilistic theories does not necessarily follow the structure of the ontology. For example, an ontology of lung cancer should specify what lung cancer is, but whether someone will have lung cancer depends on many factors of the particular case and not just on other parts of ontologies (e.g., whether they have other cancers and their work history that includes when they worked in bars that allowed smoking). As another example, the probability that a room will be used as a living room depends not just on properties of that room, but on the properties of other rooms in an apartment.

There are major challenges in building probabilistic theories using ontologies based on languages such as OWL. The main challenge is that OWL sees the world in terms of individuals, classes and properties, while probability theory is in terms of random variables. Section 4.2 discusses how to construct random variables from ontologies.

3 Pragmatic Considerations

The MineMatch and HazardMatch systems we have been developing have multiple instances that describe entities and their properties at particular locations on Earth, and models (theories) that make predictions about these locations. The systems are used in various modes:

– In instance-to-models matching, one instance is compared to multiple models. Finding the most likely models for the instance can be used to determine what is the most likely mineral to occur at a location or what types of landslides are predicted to occur at a particular place. In both of these cases, the instance is a place whose description is compared to the models.
– In model-to-instances matching, one model is compared to multiple instances. This can be used to find the location(s) that are most likely to have landslides or contain particular minerals.
– In instance-to-instances matching, one instance is compared to multiple instances to find which other instances are most like this instance.
– In model-to-models matching, one model is compared to multiple models to find which other models are most like this model.

These applications have a number of features that we believe will be shared by many scientific disciplines:

- The instances are heterogeneous, described at various levels of abstraction (using more general or less general terms) and detail (described in terms of parts and sub-parts or not). Similarly, the models use various levels of abstraction and detail. Sometimes the distinctions that are in the instance descriptions are not required by the models, and sometimes the instance descriptions do not make distinctions that are needed by the models.
- The experts often do not publish probabilities in their models, and are reluctant to have probabilities in the system. There are a number of reasons for this. First, they may have very few data points for any model, so that the probabilities will not be based on anything meaningful. Second, the people who want to make decisions (those who want to decide whether to try to mine an area profitably, or insurance companies that decide on insurance premiums) will want to use their own prior probabilities, and may take into account more information than is used in the system.
- The problem domains are afflicted by combinatorial complexity; there many possible model combinations, and very large data collections for assessment. It is difficult to find those few areas that are most likely to contain ore-grade minerals or be susceptible to landslides, and to provide explanations that can be used for further analysis.
- The models are "positive"; there are models of where to find a particular mineral, but people do not publish models of where the mineral is absent. Similarly for landslides; there are models of where particular types of landslides are likely to occur, but not models of where landslides are unlikely to occur.
- The models are neither covering, disjoint nor independent. Often the models are variants of each other. Starting from one model, people produce variants of that model to suit their own purpose. A model does not include all of the cases where the phenomenon it is modelling may occur; it only about a specific context.

4   Foundations of Probabilistic Theories

In this section, we describe the logical and probabilistic foundations for building theories, and relate them to pragmatic choices that we have used in our fielded systems.

4.1   Role of Models in Decision Making

The Bayesian view of using models for decision making is that we would like to make a probabilistic prediction of x for a new case based on a description d of that case. Thus we want $P(x|d)$. The role of the models is to provide a framework for this prediction.

In terms of probabilities, we can use models as intermediaries:

$$P(x|d) = \sum_{m \in Models} P(x|m \wedge d)P(m|d)$$

where $Models$ is a set of mutually exclusive and covering hypotheses. Thus, for each model, we need to decide what it predicts, and how likely it is based on the description, d, of the current case. Typically models are rich enough to convey the information about the rest of the description, and so we assume $P(x|m \wedge d) = P(x|m)$.

In Bayesian modelling, we try to determine what features best predict (in unseen data) the phenomenon of interest, and then build probabilistic models in terms of these features.

Typically, we do not have $P(m|d)$ which specifies how likely the model is given the description, but instead have predictions of the model, i.e., $P(d|m)$. These two quantities are related by Bayes' theorem:

$$P(m|d) = \frac{P(d|m)P(m)}{P(d)}$$

That is, we often have causal or consequential knowledge and want to do evidential reasoning. For example, we model the symptoms of chicken pox with $P(fever|ch\_pox)$ but want $P(ch\_pox|fever)$. These are related by Bayes' theorem:

$$P(ch\_pox|fever) = \frac{P(fever|ch\_pox) \times P(ch\_pox)}{P(fever)}$$

The reason that we want to store causal or consequential knowledge is that it is more stable to changing contexts. You would expect the symptoms of chicken pox to be stable; they would be the same whether the patient was at home, in a school or in a hospital. However, the probability that someone with a fever has chicken pox would be different in these three contexts, as the prevalence of fever and chicken pox is different in these three contexts.

This has an impact on how diagnostic a feature is. Suppose $fever$ and $spots$ are common given chicken pox, e.g., $P(fever|ch\_pox) = 0.9$, $P(spots|ch\_pox) = 0.9$. Suppose $fever$ has many causes and $spots$ has few. Then spots is more diagnostic of $chicken\ pox$, i.e., $P(ch\_pox|spots) > P(ch\_pox|fever)$, as $P(fever) > P(spots)$.

Note also that the probabilities needed for the prediction, namely $P(x|m)$ are of the same form as $P(d|m)$—they all specify what the model predicts. Rather than making a model to be for a particular feature, a model makes predictions about all of its features.

4.2 Probabilities, Ontologies and Existence

There seems to be a fundamental mismatch between the random variable formalization of probability theory and the formalization of modern ontologies in terms of individuals, classes and properties. Probabilistic models typically assume we know what random variables exist at modelling time, but what individuals exists is often unknown at modelling time. Interestingly, a large body of research on Bayesian modelling (e.g., Bayesian networks) and modern research into ontologies both have their roots in the expert systems of the 1970's and 1980's

(Henrion et al., 1991). Both fields have advanced our understanding of reasoning, and part of our research is to bring these together.

We can reconcile these views by having properties of individuals correspond to random variables. This complicates the probabilistic modelling as the individuals typically only become known at run-time, and so the random variables are unknown at modelling time. This has spurred a body of research in first-order probabilistic models or relational probabilistic models (e.g., Poole (1993), Getoor and Taskar (2007), Kersting and De Raedt (2007), Laskey (2008), Lukasiewicz (2008)). It is even possible to be unsure about the existence of an individual, and so unsure about the existence of a random variable (Poole, 2007).

When dealing with probabilities and individuals we need to deal with three types of uncertainty:

- the probability of existence (Poole, 2007) — the probability that an individual that fits a description actually exists.
- the probability distribution over the types of an individual. This is complicated when there are complex interrelations between classes that can be the types of the individuals.
- the probability of property values. Functional properties give a random variable for each individual with a non-zero probability of being in the class that is the domain of the property. Non-functional properties have a Boolean random variable for each value in the range and each individual with a non-zero probability of being in the domain of the property.

Aristotelian definitions, where a class is defined in terms of its immediate superclass and differentia, provide a way to reduce the second case to the third case. The differentia are described in terms of property values with appropriate domains. By having a probability distribution over the values of the properties (perhaps conditioned on other variable assignments), we can induce a probability distribution over the classes. Note that Aristotelian definitions are general: any class hierarchy can be represented by Aristotelian definitions by introducing new properties.

For example, a granite can be defined as a rock with the property *genesis* having value *igneous*, property *composition* having value *felsic*, and *texture* is *coarse*. By having a probability distribution over the values of *genesis*, a probability distribution over the value of *composition*, and a probability distribution over the values of *texture*, we can determine the prior probability that a rock is a granite.

Note that the probabilistic formulation is complicated by existence prerequisites: only individuals that exist have properties, and only individuals in the class that is domain of a property can have values for that property.

4.3   Bayesian Modelling Meets Pragmatism

Bayesian modelling of scientific reasoning seems like the appropriate formulation of the role of theories or models in science. However, the pragmatic considerations discussed above lead us to not adopt it directly, although it remains the gold standard. The theories (or models) in our fielded systems are based on qualitative probabilistic matching (Smyth and Poole, 2004; Poole and Smyth, 2005; Lukasiewicz and Schellhase, 2007), with the following properties:

- Rather than using probabilities that experts do not want to give, and cannot judge the output from, we use qualitative probabilities, using a 5-point scale (*always, usually, sometimes, rarely, never*) that is derived from the terminology used in published papers. These qualitative probabilities act like log-probabilities, where the values add rather than multiply (Pearl, 1989; Darwiche and Goldszmidt, 1994).
- The models need to be fleshed out for each instance. Models refer to multiple individuals, but they do not refer to the named individuals in the instances. Models specify roles that can be filled by the instance individuals. The predictions of the model for an instance can only be determined given a role assignment that specifies which instance individuals fill the roles in the model.
- Rather than averaging over all possibilities and role assignments, we choose the most likely ones.
- We allow for diverse data about instances and models at multiple levels of abstraction and detail. We also require prior probabilities of the descriptions; we do not assume that we can get the probability of a description from the set of models (as we could if the models were exclusive and covering).
- The explanations for the answers are as important as the answers themselves.

5   Conclusions

This paper has presented the big picture of what we see as semantic science as well as the pragmatic considerations that have gone into our fielded systems that are a first try at realizing our vision. This view of semantic science is meant to complement other views that provide ontologically-based views of data (McGuinness et al., 2007) and ontology-based services (De Roure et al., 2005).

There are many challenges in building the semantic science vision, including how to construct theories, how to determine what theories are useful in making predictions in a particular case, and in finding the data about which a theory makes predictions. The growing interest in scientific ontologies, the desire for scientists (and their funders) to make their data and theories as widely used as possible, and the desire for users to have the best predictions, indicates that this semantic science vision should succeed.

Acknowledgements

Thanks to Jacek Kisyński, Mark Crowley, and the anonymous reviewers for valuable comments. This research was funded by NSERC and Georeference Online.

Bibliography

Alexander, C., Ishikawa, S., Silverstein, M., Jacobson, M., Fiksdahl-King, I., Angel, S.: A Pattern Language. Oxford University Press, New York (1977)

Aristotle (350 B.C.). Categories. Translated by E. M. Edghill, http://www. classicallibrary.org/Aristotle/categories/

Berg, J.: Aristotle's theory of definition. In: ATTI del Convegno Internazionale di Storia della Logica, San Gimignano, pp. 19–30 (1982), http://ontology.buffalo.edu/bio/berg. pdf

Berners-Lee, T., Fischetti, M.: Weaving the Web: The original design and ultimate destiny of the World Wide Web, by its inventor, Harper Collins, San Francisco, CA (1999)

Berners-Lee, T., Hendler, J., Lassila, O.: The semantic web: A new form of web content that is meaningful to computers will unleash a revolution of new possibilities. Scientific American, 28–37 (2001)

da Costa, P.C.G., Laskey, K.B., Laskey, K.J.: PR-OWL: A Bayesian ontology language for the semantic web. In: Proceedings of the ISWC Workshop on Uncertainty Reasoning for the Semantic Web, Galway, Ireland (2005), http://sunsite.informatik. rwth-aachen.de/Publications/CEUR-WS//Vol-173/

Darwiche, A., Goldszmidt, M.: On the relation between kappa calculus and probabilistic reasoning. In: UAI 1994, pp. 145–153 (1994)

De Roure, D., Jennings, N.R., Shadbolt, N.R.: The semantic grid: Past, present and future. Procedings of the IEEE 93(3), 669–681 (2005), http://www.semanticgrid.org/ documents/semgrid2004/semgrid2004.pdf

Fodor, J.A.: The Language of Thought. Harvard University Press, Cambridge (1975)

Getoor, L., Taskar, B. (eds.): Introduction to Statistical Relational Learning. MIT Press, Cambridge (2007)

Gillespie, M.R., Styles, M.T.: BGS rock classification scheme, 2nd edn., RR 99-06, British Geological Survey. Classification of igneous rocks. Research Report, vol. 1 (1999), http://www.bgs.ac.uk/bgsrcs/

Henrion, M., Breese, J., Horvitz, E.: Decision analysis and expert systems. AI Magazine 12(4), 61–94 (1991)

Kersting, K., De Raedt, L.: Bayesian logic programming: Theory and tool. In: Getoor, L., Taskar, B. (eds.) An Introduction to Statistical Relational Learning. MIT Press, Cambridge (2007)

Laskey, K.B., Wright, E.J., da Costa, P.C.G.: Envisioning uncertainty in geospatial information. In: UAI Applications Workshop 2007 The 5th Bayesian Modeling Applications Workshop (2007), http://ite.gmu.edu/~klaskey/uai07workshop/ AppWorkshopProceedings/UAIAppWorkshop/paper3.pdf

Laskey, K.B.: MEBN: A language for first-order Bayesian knowledge bases. Artificial Intelligence 172(2-3) (2008), http://www.sciencedirect. com/science/article/B6TYF4PTMXXP-1/2/ce6bcf1c5a5fecfd805501056e9b62a1, doi:10.1016/j.artint.2007.09.006

Lukasiewicz, T.: Expressive probabilistic description logics. Artificial Intelligence 172(6-7), 852–883 (2008)

Lukasiewicz, T., Schellhase, J.: Variable-strength conditional preferences for ranking objects in ontologies. Journal Web Semantics 5(3), 180–194 (2007)

McGuinness, D., Fox, P., Cinquini, L., West, P., Garcia, J., Benedict, J.L., Middleton, D.: The virtual solar-terrestrial observatory: A deployed semantic web application case study for scientific research. In: Proceedings of the Nineteenth Conference on Innovative Applications of Artificial Intelligence (IAAI 2007), Vancouver, BC, Canada (2007), http://www.ksl.stanford.edu/KSL_Abstracts/KSL-07-01.html

McGuinness, D.L., van Harmelen, F.: OWL web ontology language overview. W3C Recommendation, W3C (2004) (February 10, 2004), http://www.w3.org/TR/owl-features/

Page, L., Brin, S., Motwani, R., Winograd, T.: The pagerank citation ranking: Bringing order to the web. Technical Report SIDL-WP-1999-0120, Stanford InfoLab (1999), http://dbpubs.stanford.edu/pub/1999-66

Pearl, J.: Probabilistic semantics for nonmonotonic reasoning: A survey. In: Brachman, R.J., Levesque, H.J., Reiter, R. (eds.) KR 1989, Toronto, pp. 505–516 (1989)

Pinker, S.: The Language Instinct, Harper Collins, New York (1994)

Polya, G.: Mathematics and Plausible Reasoning. Patterns of Plausible Inference, vol. II. Princeton University Press, Princeton (1954)

Pool, M., Fung, F., Cannon, S., Aikin, J.: Is it worth a hoot? Qualms about OWL for uncertainty reasoning. In: Proceedings of the ISWC Workshop on Uncertainty Reasoning for the Semantic Web (2005), http://sunsite.informatik.rwth-aachen.de/Publications/CEUR-WS//Vol-173/

Poole, D.: Probabilistic Horn abduction and Bayesian networks. Artificial Intelligence 64(1), 81–129 (1993)

Poole, D.: Logical generative models for probabilistic reasoning about existence, roles and identity. In: 22nd AAAI Conference on AI, AAAI 2007 (2007), http://www.cs.ubc.ca/spider/poole/papers/AAAI07-Poole.pdf

Poole, D., Smyth, C.: Type uncertainty in ontologically-grounded qualitative probabilistic matching. In: Godo, L. (ed.) ECSQARU 2005. LNCS (LNAI), vol. 3571, pp. 763–774. Springer, Heidelberg (2005), http://www.cs.ubc.ca/spider/poole/papers/Poole-Smyth-ecsqaru2005.pdf

Poole, D., Smyth, C., Sharma, R.: Semantic science and machine-accessible scientific theories. In: AAAI Spring Symposium on Semantic Science Knowledge Integration, Stanford, CA (2008)

Popper, K.: The Logic of Scientific Discovery. Basic Books, New York (1959)

Sapir, E.: The status of linguistics as a science. Language 5(209) (1929)

Smith, B.: The logic of biological classification and the foundations of biomedical ontology. In: Westerståhl, D. (ed.) 10th International Conference in Logic Methodology and Philosophy of Science. Elsevier-North-Holland, Oviedo (2003a), http://ontology.buffalo.edu/bio/logic_of_classes.pdf

Smith, B.: Ontology. In: Floridi, L. (ed.) Blackwell Guide to the Philosophy of Computing and Information, pp. 155–166. Blackwell, Oxford (2003b), http://ontology.buffalo.edu/smith/articles/ontology_pic.pdf

Smyth, C., Poole, D.: Qualitative probabilistic matching with hierarchical descriptions. In: KR 2004, Whistler, BC, Canada (2004), http://www.cs.ubc.ca/spider/poole/papers/KR04SmythC.pdf

Smyth, C., Poole, D., Sharma, R.: Semantic e-science and geology. In: AAAI 2007 Semantic e-Science workshop (2007), http://www.cs.ubc.ca/spider/poole/papers/SmythPooleSharmaSemSci2007.pdf

Sowa, J.F.: Knowledge Representation: Logical, Philosophical, and Computational Foundations. Brooks Cole Publishing Co., Pacific Grove (2000)

Struik, L., Quat, M., Davenport, P., Okulitch, A.: A preliminary scheme for multihierarchical rock classification for use with thematic computer-based query systems. Current Research 2002-D10, Geological Survey of Canada (2002), http://daks.ucdavis.edu/~ludaesch/289F-SQ06/handouts/GSC_D10_2002.pdf

Tang, H., Ng, J.H.K.: Googling for a diagnosis–use of google as a diagnostic aid: internet based study. BMJ (2006), doi:10.1136/bmj.39003.640567.AE

Whorf, B.L.: Science and linguistics. Technology Review 42(6), 229–231, 247–248 (1940)

Probabilistic Dialogue Models for Dynamic Ontology Mapping

Paolo Besana and Dave Robertson

Centre for Intelligent Systems and Applications
University of Edinburgh

Abstract. Agents need to communicate in order to accomplish tasks that they are unable to perform alone. Communication requires agents to share a common ontology, a strong assumption in open environments where agents from different backgrounds meet briefly, making it impossible to map all the ontologies in advance. An agent, when it receives a message, needs to compare the foreign terms in the message with all the terms in its own local ontology, searching for the most similar one. However, the content of a message may be described using an interaction model: the entities to which the terms refer are correlated with other entities in the interaction, and they may also have prior probabilities determined by earlier, similar interactions. Within the context of an interaction it is possible to predict the set of possible entities a received message may contain, and it is possible to sacrifice recall for efficiency by comparing the foreign terms only with the most probable local ones. This allows a novel form of dynamic ontology matching.

1 Introduction

Agents collaborate and communicate to perform tasks that they cannot accomplish alone. To communicate means to exchange messages, that convey meanings encoded into signs for transmission. To understand a message, a receiver should be able to map the signs in the message to meanings aligned with those intended by the transmitter.

Therefore agents should agree on the terminology used to describe the domain of the interaction: for example, if an agent wants to buy a particular product from a seller, it must be able to specify the properties of the products unambiguously. Ontologies specify the terminology used to describe a domain [4].

However, a shared ontology can be a strong assumption in an open environment, such as a Peer-to-Peer system: agents may come from different backgrounds, and have different ontologies, designed for their specific needs [13].

In this sort of environment, communication implies translation. The standard approach is to find mappings between the ontologies, creating a sort of bilingual dictionary. Many different techniques have been developed for ontology mapping, but in an open environment it is impossible to know which agents will take part in the interactions; therefore it is impossible to anticipate which ontologies should be mapped.

P.C.G. da Costa et al. (Eds.): URSW 2005-2007, LNAI 5327, pp. 41–51, 2008.

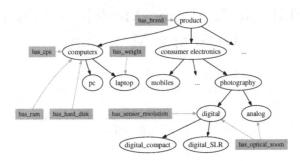

Fig. 1. Fragment of buyer a_b ontology

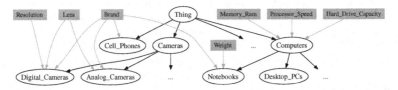

Fig. 2. Fragment of seller a_s ontology

Agents have to map ontologies dynamically when needed. Mapping full ontologies is a time-consuming task: in the standard process, each term in one ontology is compared with all the terms in the other ontology, and the most similar term is the mapping.

However, agents may meet infrequently, for a single interaction on a specific topic. A full ontology mapping would be a waste of resources: mapping only "foreign" terms that have appeared in the conversation can be more convenient.

Comparing a foreign term in a message with all the terms in the ontology can still be costly. Yet, the entities referred by the signs in the message are not randomly chosen: the dialogue has a meaning because entities are related. For example, if the conversation is about the purchase of a laptop, entities related to cars are unlikely to appear. It is reasonable to compare the signs in the message with entities about laptops, rather than compare with all the entities indiscriminately.

This paper shows how to extract, represent, and use knowledge about the relations and properties of the entities in an interaction to support dynamic ontology mapping.

2   Example Scenario

The example scenario is a purchase interaction between the buyer and seller agents a_b and a_s. In the dialogue, agent a_b asks a_s about a laptop he needs. The seller a_s inquires about properties of the product in order to make an offer.

The two agents do not share the same ontology: the buyer uses the one in Figure 1 and the seller the one in Figure 2. In the figures the ovals are classes and the grey boxes are properties. The classes are structured in taxonomies, and the domains of the properties are shown by grey arrows.

```
a((buyer(S),B) ::=
ask(Prd) ⇒ a(vendor,S) ← want(Prd)
then
a(neg_buy(Prd,S),B).

a(neg_buy(Prd,S),B) ::=
 / ask(Attr) ⇐ a(neg_vend,S)                                        \
 | then                                                             |
 |  / inform(Attr,Val) ⇒ a(neg_vend,S) ← required(Prd,Attr,Val) \  |
 |  | or                                                         |  |
 |  \ dontcare(Attr) ⇒ a(neg_vend,S)                             /  |
 | then                                                             |
 \ a(neg_buy(Prd,S),B)                                              /
or
 / propose(Prd,Price,Const) ⇐ a(neg_vend,S)          \
 | then                                               |
 |  / accept ⇒ a(neg_vend,S) ← afford(Prd,Price) \   |
 |  | then                                        |   |
 |  \ ack ⇐ a(neg_vend,S)                         /   |
 | or                                                 |
 \ reject ⇒ a(neg_vend,S)                             /
or

sorry ⇐ a(neg_vend,S)
```

In the interaction, the agent a_b initially takes the role of buyer: it first sends a request to agent a_s for the product it wants to buy (found satisfying want(Prd)) and then becomes a negotiating buyer, waiting for a reply.

The agent a_s receives the request: if it has the product, it selects the attributes the buyer needs to specify and becomes a negotiating seller; otherwise it says sorry. As a negotiating seller, a_s recursively extracts the attributes from the list and asks about them to a_b, creating a filter with the received information. The buyer agent receives the request, and if it cares and knows about the value of each attribute (if it can satisfy required(Prd, Attr, Val)), replies with it, otherwise it sends a dontcare message. When the list of attributes is empty, a_s sends an offer using the created filter. The agent a_b accepts the offer if it can afford the price (afford(Prd, Price) must be satisfied) or rejects it.

Fig. 3. LCC dialogue fragment used by the buyer agent

3 Communication

An approach to communication, for which Electronic Institution [11] is an example, focuses on the interaction itself, using norms, laws and conventions to define the expected behaviours of the agents, without specifying their mental state.

As described in [12], norms and conventions form the skeleton for many human coordinated activities, and they work similarly in agents' societies: they provide a template for actions, and simplify the decision-making process, dictating the course of action to be followed in certain situations.

3.1 Lightweight Coordination Calculus

In this paper, interactions are modelled using the *Lightweight Coordination Calculus* [7,8], that borrows notions from Electronic Institutions.

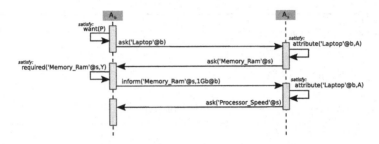

Fig. 4. A sequence diagram representing a protocol run between a_b and a_s

The *Lightweight Coordination Calculus* (LCC) is an executable specification language adapted to peer-to-peer workflow and has been used in applications such as business process enactment [5] and e-science service integration [1]. It is also used to represent interactions between peers in the OpenKnowledge[1] project [10].

LCC is based on process calculus: protocols are declarative scripts written in Prolog and circulated with messages. Agents execute the protocols they receive by applying *rewrite rules* to expand the state and find the next move.

It uses roles for agents and constraints on message sending to enforce the social norms. The basic behaviours are to send ($\Rightarrow$) or to receive ($\Leftarrow$) a message. More complex behaviours are expressed using connectives: **then** creates sequences, **or** creates choices. Common knowledge can be stored in the protocol.

Figure 3 shows and explains the LCC protocol used by the buyer for the interaction in the example scenario. Figure 4 represents the sequence diagram of the exchanged messages and of the constraints satisfied during a run of the protocol for the purchase of a laptop.

3.2 Communication and Contexts

The agents execute the protocols inside a separate "box": in theory, it is possible to write a protocol that can be run without requiring any specific knowledge from the agent. It requires that the constraints are satisfied with the information available in the common knowledge.

The "box" in which a protocol is run can be compared to the idea of *context* described by Giunchiglia: in [3] he defines a context c_i as "*partial*" and "*approximate*" theory of the world, represented by the triplet $\langle L_i, A_i, \Delta_i \rangle$. In the tuple, L_i is the language local to the context, A_i is the set of axioms of the context, and Δ_i is the inference engine local to the context. Moreover, a reasoner can connect a deduction in one context with a deduction in another using *bridge rules*.

For the protocol run context $c_r = \langle L_r, A_r, \Delta_r \rangle$, the language L_r is composed by all the terms that can be introduced by the agents involved in the interaction (terms are tagged with their origin: a_b introduces '*Laptop*' satisfying **want(Prd)**, and therefore **Prd** is replaced throughout the protocol with the tagged value '**Laptop'**@a_b); the axioms A_r are the role clauses together with the axioms in the common knowledge and Δ_r is the protocol expansion engine.

---

[1] http://www.openk.org

Even though protocols can be autonomous from the agent, they become useful only if they can exploit the agents' knowledge, that is if it is possible to bridge the reasoning between the interaction context c_r and in the agent's local context c_a, for example accessing the peer's database of products in order to query the availability or the price of a product. This is accomplished using a bridge rule that connects the constraints in the protocol with the predicates in the agent's local knowledge:

$$\frac{c_r : \kappa_p(W_1, ..., W_n)}{c_a : \kappa_a(Y_1, ..., Y_m)} \tag{1}$$

where κ_p is a formula of a protocol constraint and κ_a is a formula in the agent's local knowledge, that can be satisfied only by using its own language L_a.

4   Ontology Mapping

In traditional ontology mapping, the bridges should be valid for any value from L_r and L_a in two contexts c_r and c_a:

$$\forall W_1...W_n \in L_r, \ \exists Y_1...Y_n \in L_a. \ c_i : \kappa_p(W_1, ..., W_n) \rightarrow c_j : \kappa_q(Y_1, ..., Y_m) \tag{2}$$

That is, for any value of $W_1, ..., W_n$ in κ_p, it is possible to find the values for $Y_1, ..., Y_n$ so that κ_a is equivalent to κ_p. In the example scenario, the mappings should cover the possible requests from the buyer agent a_b for buying any element in its ontology (see figure 1), such as mobile phones, analog cameras and so on - even if these interactions never take place.

This is a strong requirement: it implies that it is possible to find a corresponding term in L_a for every term in L_r, and this may not always be the case. Static *ontology mapping* tries to achieve this. An ontology mapping function receives two ontologies and returns the relations between their entities:

$map : O_1 \times O_2 \rightarrow \Omega$

where Ω contains all the binary relations r (*equivalence, similarity, generalisation, specialisation, etc*) between entities in O_1 and O_2.

The existence of inconsistencies in the ontologies undermines the possibility of satisfying the definition in Expression 2. Mapping systems use various methods to verify the relations between terms: detailed reviews of these approaches can be found in [9,6].

5   Dynamic Ontology Mapping: Motivation

As said in the introduction, it is possible to limit the mappings to those needed to handle the occurring interactions, and there is no need to guarantee complete equivalence between the languages. Therefore an agent needs to map at the minimum the terms that appear in κ_p in order to satisfy κ_a :

$$\exists W_1...W \in L_r, Y_1...Y_n \in L_a. \ c_r : \kappa_p(W_1, ..., W_n) \wedge c_a : \kappa_a(Y_1, ..., Y_m) \tag{3}$$

This is a much weaker requirement: we need to find the values for $Y_1, ..., Y_n$ so that κ_a is valid for the given instances of $W_1, ..., W_n$. In the example, it means that only the mappings required for buying the laptop are needed.

Not every grounding of the variables is meaningful: some will make κ_a more similar to κ_p than others. The mapping function:

$$singlemap : t_{L_r} \times L_a \rightarrow t_{L_a}$$

is the "oracle" used to search the best possible mapping to make the bridge in Expression 1 meaningful. It does this by comparing t_{L_r} with all the terms in L_a.

The values for the variables $W_1, ..., W_n$ in κ_p are introduced by received messages (for example, the first `ask(Attr)` in Figure 4 introduces `'Memory_Ram'@s`), by satisfying constraints (for example, `want(P)` introduces `'Laptop'@b`) or when a role is invoked with parameters. Only terms introduced by received messages can be defined in other ontologies and require mapping.

Suppose an agent receives a message $m_k (..., w_i, ...)$, where $w_i \notin L_a$ is the foreign term. The task of the oracle is to find what entity or concept, represented in the agent's ontology by the term t_m, was encoded in w_i by the transmitter. Not all the comparisons between w_i and terms $t_j \in L_a$ are useful: the aim of this work is to specify a method for choosing the smallest set $\Gamma \subseteq L_a$ of terms to compare with w_i, given a probability of finding the matching term $t_m \in L_a$. We assume that t_m exists and that there is a single best match.

Let $p(t_j)$ be the probability that the entity represented by $t_j \in L_a$ was used in W_i inside m_k. The oracle will find t_m if $t_m \in \Gamma$ with probability:

$$p(t_m \in \Gamma) = \sum_{t_j \in \Gamma} p(t_j)$$

If all terms are equiprobable, then $p(t_m \in \Gamma)$ will be proportional to $|\Gamma|$. For example, if $|L_a| = 1000$, then $p(t_j) = 0.001$. Setting $|\Gamma| = 800$ yields $p(t_m \in \Gamma) = 0.8$, and there is no strategy for choosing the elements to add to Γ.

Instead, if the probability is distributed unevenly, as described in section 6, and we keep the most likely terms, discarding the others, we can obtain a higher probability for a smaller Γ. For example, suppose that $p(t_j)$ is distributed approximately according to Zipf's law (an empirical law mainly used in language processing that states that the frequency of a word in corpora is inversely proportional to its rank):

$$p(k; s; N) = \frac{1/k^s}{\sum_{n=1}^{N} 1/n^s}$$

where k is the rank of the term, s is a parameter (which we set to 1 to simplify the example), and N is the number of terms in the vocabulary. The probability of finding t_m becomes:

$$p(t_m \in \Gamma) = \frac{\sum_{k=1}^{|\Gamma|} 1/k}{\sum_{n=1}^{|L_a|} 1/n}$$

For $|L_a| = 1000$, then $p(t_m \in \Gamma) = 0.70$ for $|\Gamma| = 110$, and maybe more remarkably $p(t_m \in \Gamma) = 0.5$ for $|\Gamma| = 25$, as shown in Figure 5.

Therefore, given a probability distribution for the terms, it is possible to trade off a decrement in the probability of finding the matching term t_m in Γ with an important reduction of comparisons made by the oracle.

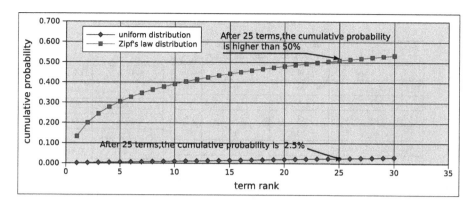

Fig. 5. uniform distribution vs Zipf's distribution of terms in a message

The core issue dealt with by this paper is how to create and assign probabilities to the entities that can be used in a message $m_k (\ldots, w_i, \ldots)$. Intuitively, the type of interaction, the specific topic and the messages already exchanged bind w_i to a set of possible expected entities.

In particular, this paper shows how an interaction model as LCC forms a framework that enforces relations between the entities: the roles provide a first filter for them. For example, messages in a buyer role will likely refer to entities like products, prices, and attributes of the products. Different runs of the same protocol tend to follow the same path, adding predictability to the interaction.

6   Modelling the Interactions

6.1   Asserting the Possible Values

The solution proposed is a model that stores and updates properties of the entities used to instantiate each variable W_i in different runs of the same protocol.

As seen in Section 3.1, the variables are replaced by values during protocol execution, and therefore it is not possible to refer to them directly. A variable W_i is a *slot* A (an argument position) in an LCC *node* N (that can be a message, a constraint or a role header) inside a role R, and it is represented as $\langle N, A \rangle_R$. For example, the variable Prd appears in $\langle \texttt{want}, 1 \rangle_b$, where b means buyer.

In general, the possible values for the slot $\langle N_i, A \rangle_R$ are modelled by M assertions, each assigning a probability to the hypothesis that the matching entity for the slot belongs to a set Ψ:

$$A_j^{\langle N_i, A \rangle_R} \doteq Pr\left(\langle N_i, A \rangle_R \in \Psi | c \right) \tag{4}$$

The probability can be made dependent on the value of another slot. Therefore the assertion is in the form of a posterior probability: the element c can become a constraint on the value of another slot. The probability can also be independent from any other slot: in this case the element c becomes the **true** constant and can be omitted.

Table 1. Mappings for $\langle \text{ask}, 1 \rangle_{\text{nb}}$ and $\langle \text{ask}, 1 \rangle_{\text{s}}$

$\langle \text{ask}, 1 \rangle_{\text{nb}}$	$\langle \text{want}, 1 \rangle_{\text{b}} =$ Laptop	$\langle \text{want}, 1 \rangle_{\text{b}} =$ digital_SLR	Total
has_brand	4	5	9
has_cpu	6	0	6
has_ram	6	0	6
has_hard_disk	4	0	4
has_weight	3	1	4
has_optical_zoom	0	5	5
has_sensor_resolution	0	6	6
Total	**23**	**17**	**40**

$\langle \text{ask}, 1 \rangle_{\text{s}}$	Total
Digital_Cameras	40
Cell_Phones	30
Laptops	20
PC_Desktops	10
Total	**100**

How Assertions are Obtained. Assertions are created and updated every time a protocol is executed. Let's suppose that the agents a_s and a_b have already used the protocol in different interactions with other agents. The agent a_b has used it 12 times with different vendors: 6 times searching for a laptop, and 6 times seeking a digital camera. In total, a_b has received the message ask(Attr) that inquired about properties of the requested product 40 times. The content of the slot in the received messages has been mapped to the entities from its own ontology (see figure 1) with the frequencies in table 1. The seller agent a_s has used the protocol 100 times with different buyers, receiving the message ask(Prd) every time. The content of the slot has been mapped to entities in its own ontology (see figure 2) with the frequencies in table 1. The frequencies of the mappings are used to compute the probabilities in the assertions dynamically.

Assertions About Entities. Assertions can simply be about the prior probability of entities in a slot, disregarding the values of other slots in the protocol run:

$$A_j^{\langle \text{N}_\text{i}, \text{a} \rangle_\text{R}} \doteq Pr(\langle \text{N}_\text{i}, \text{a} \rangle_\text{R} \in \{e_q\}) = p_j$$

In the scenario, assertions about $\langle \text{ask}, 1 \rangle_{\text{nb}}$ are:

$$A_1^{\langle \text{ask}, 1 \rangle_{\text{nb}}} \doteq Pr(\langle \text{ask}, 1 \rangle_{\text{nb}} \in \{\text{``}has\_brand\text{''}\}) = \tfrac{9}{40} = 0.225$$
$$A_7^{\langle \text{ask}, 1 \rangle_{\text{nb}}} \doteq Pr(\langle \text{ask}, 1 \rangle_{\text{nb}} \in \{\text{``}has\_sensor\_resolution\text{''}\}) = \tfrac{6}{40} = 0.15$$

More precise assertions can be about the posterior probability of the entity N_i given the values of previous slots N_{i-d}:

$$A_j^{\langle \text{N}_\text{i}, \text{a} \rangle_\text{R}} \doteq Pr(\langle \text{N}_\text{i}, \text{a} \rangle_\text{R} \in \{e_q\} \mid \langle \text{N}_{\text{i}-\text{d}}, \text{a} \rangle_\text{R} = e_k) = p_j$$

In the example scenario, we have:

$$A_{10}^{\langle ask, 1 \rangle_{\text{nb}}} \doteq Pr(\langle \text{ask}, 1 \rangle_{\text{nb}} \in \{\text{``}has\_brand\text{''}\} \mid \langle \text{want}, 1 \rangle_{\text{b}} = \text{``}Laptop\text{''}) = \tfrac{4}{23} = 0.174$$
$$A_{11}^{\langle ask, 1 \rangle_{\text{nb}}} \doteq Pr(\langle \text{ask}, 1 \rangle_{\text{nb}} \in \{\text{``}has\_cpu\text{''}\} \mid \langle \text{want}, 1 \rangle_{\text{b}} = \text{``}Laptop\text{''}) = \tfrac{6}{23} = 0.260$$

Assertions About Properties and Relations. Assertions can also be about ontological relations between the entities in the slot and other entities. The possible relations depend on the expressivity of the ontology: if it is a simple list of allowed terms, it will not be possible to verify any relation; if it is a taxonomy, subsumption can be found; for a richer ontology, more complex relations such as domain or range can be found. The assertions about the probabilities of ontological relations are obtained by generating hypotheses about different relations and counting the frequencies of the proved ones.

The hypotheses can be about an ontological relation between the entity in the slot and an entity e_k in the agent's ontology:

$$A_j^{\langle \mathbb{N}_i, \mathtt{a}\rangle_R} \doteq Pr\left(\langle \mathbb{N}_i, \mathtt{a}\rangle_R \in \{X | rel\left(X, e_k\right)\}\right) = p_j$$

In the example scenario, the seller can prove some relations between the entities in $\langle \mathtt{ask}, 1\rangle_s$ and other entities in its ontology (see figure 2):

$$A_1^{\langle \mathtt{ask}, 1\rangle_s} \doteq Pr\left(\langle \mathtt{ask}, 1\rangle_s \in \{X | subClass(X, \text{``Computers''})\}\right) = \tfrac{30}{100} = 0.3$$

The assertions can also regard the relation with another slot in the protocol:

$$A_j^{\langle \mathbb{N}_i, \mathtt{a}\rangle_R} \doteq Pr\left(\langle \mathbb{N}_i, \mathtt{a}\rangle_R \in \{X | rel\left(X, \langle \mathbb{N}_{i-d}, \mathtt{a}_k\rangle_R\right)\}\right) = p_j$$

In the example scenario the buyer can prove the relation between $\langle ask, 1\rangle_{nb}$ and $\langle \mathtt{want}, 1\rangle_b$ in its ontology (see figure 1):

$$A_{20}^{\langle \mathtt{ask}, 1\rangle_{nb}} \doteq Pr\left(\langle \mathtt{ask}, 1\rangle_{nb} \in \{X | hasDomain\left(X, \langle \mathtt{want}, 1\rangle_b\right)\}\right) = 1.0$$

which means that the domain of the entity in the $\langle \mathtt{ask}, 1\rangle_{nb}$ in the negotiator clause is always the content of the first slot in the node $\mathtt{want}$ in the $\mathtt{buyer}$ role.

Assertion Reliability. Assertions that assign probabilities to entities work correctly in well known and stationary situations. But interactions can have different content, such as the purchase of a different product, and the probabilities of entities can change over time (for example, a type of product may go out of fashion). Assertions about ontological relations can work on new content, but sometimes they can overfit the actual relations in interactions.

6.2   Using Assertions

When a known protocol about a role R is used and the message $m_k\left(\ldots, w_i, \ldots\right)$ arrives, the system computes the probability distribution for the terms in $\langle m_k, i\rangle_R$: all the assertions relative to the slot are selected and instantiated if needed.

In the example in Figure 4, a_b receives the message $\mathtt{ask('Memory\_Ram'@s)}$, and $\langle \mathtt{want}, 1\rangle_b$ contains '$Laptop'@b$. Thus, the assertions about $\langle \mathtt{ask}, 1\rangle_{nb}$ are:

$$A_1^{\langle \mathtt{ask}, 1\rangle_{nb}} \doteq Pr\left(\langle \mathtt{ask}, 1\rangle_{nb} \in \{\text{``}has\_brand\text{''}\}\right) = 0.225$$

$$\overset{..}{A}_6^{\langle \mathtt{ask}, 1\rangle_{nb}} \doteq Pr\left(\langle \mathtt{ask}, 1\rangle_{nb} \in \{\text{``}has\_optical\_zoom\text{''}\}\right) = 0.125$$

$$\overset{...}{A}_{10}^{\langle ask, 1\rangle_{nb}} \doteq Pr\left(\langle \mathtt{ask}, 1\rangle_{nb} \in \{\text{``}has\_brand\text{''}\} | true\right) = 0.174$$

$$\overset{...}{A}_{20}^{\langle \mathtt{ask}, 1\rangle_{nb}} \doteq Pr\left(\langle \mathtt{ask}, 1\rangle_{nb} \in \{\text{``}has\_brand\text{''}, \text{``}has\_cpu\text{''}, \text{``}has\_ram\text{''} \ \text{``}has\_hard\_disk\text{''}, \text{``}has\_weight\text{''}\}\right) = 1.0$$

The assertions can be generated using different strategies, and they assign probabilities to overlapping sets that can have one or more elements. The motivation of the work is to select the most likely entities for a slot in order to reach a given probability of finding the mapping, and therefore we need to assign to the terms the probabilities computed with the assertions.

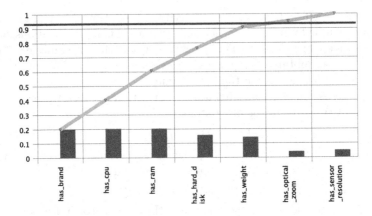

Fig. 6. Probability distribution for the terms in slot $\langle ask, 1 \rangle_{nb}$

This requires two steps. First, probabilities given to sets are uniformly distributed among the members: according to the *principle of indifference*, the probability of mutually exclusive elements in a set should be evenly distributed. Then, the probability of an entity t_i is computed by summing all its probabilities and dividing it by the sum of all the probabilities about the slot:

$$p(t_i) = \frac{\sum A_j^{\langle N,A \rangle R}(\langle N,A \rangle_R \in \{t_i\})}{\sum A_k^{\langle N,A \rangle R}} \tag{5}$$

In the example above, the entities will have the probabilities:

$P(has\_brand) = \frac{A_1 + A_{10} + A_{20}/5}{A_1 + \ldots + A_{20}} = \frac{0.225 + 0.174 + 0.2}{3} = 0.2$

...

$P(has\_sensor\_resolution) = \frac{A_7}{A_1 + \ldots + A_{20}} = \frac{0.15}{3} = 0.05$

The probabilities of terms related to the interactions have higher probabilities than those of unrelated terms. As shown in Figure 6, using the first four terms for the set Γ of terms to compare with the term 'Memory_Ram'@s in the received message yields a probability of finding the mapping in Γ greater than 0.8.

7   Conclusion

In this paper we showed an approach for dynamic ontology mapping that exploits knowledge about interactions to reduce the waste of resources normally employed to verify unlikely similarities between unrelated terms in different ontologies.

The traditional approaches aim at finding all the possible mappings between the ontologies, so any possible interaction can occur. As shown in Section 5, our goal is pragmatic: only the mappings required for the interactions that take place need to be found. For an agent, this means that only the terms in received messages and defined in external ontologies will be mapped.

In the standard approach, an ontology mapper oracle compares these "foreign" terms with all the terms in the agent's ontology, although most of the compared terms are not related. However, the terms that appear in messages are not all equally probable: given the context of the interaction, some will be more likely than others. The use of protocols allows us to collect consistent information about the mappings used during an interaction: in Section 6 we show first how to create and update a probabilistic model of the content of the messages and then how to use the model to select what are the most likely entities contained in a message, so that the mapper oracle can focus on them, improving the efficiency. While in [2] we gave a first evaluation of the framework, we are currently focussing on the effect that predictions have on the precision and recall of an ontology matcher.

References

1. Barker, A., Mann, B.: Agent-based scientific workflow composition. In: Astronomical Data Analysis Software and Systems XV, vol. 351, pp. 485–488 (2006)
2. Besana, P., Robertson, D.: How service choreography statistics reduce the ontology mapping problem. In: ISWC 2007 (2007)
3. Giunchiglia, F.: Contextual reasoning. Technical report, IRST, Istituto per la Ricerca Scientifica e Tecnologica (1992)
4. Gruber, T.R.: A translation approach to portable ontology specifications. Knowledge Acquisition 5(2), 199–220 (1993)
5. Guo, L., Robertson, D., Chen-Burger, Y.: A novel approach for enacting the distributed business workflows using bpel4ws on the multi-agent platform. In: IEEE Conference on E-Business Engineering, pp. 657–664 (2005)
6. Kalfoglou, Y., Schorlemmer, M.: Ontology mapping: the state of the art. The Knowledge Engineering Review 18(1), 1–31 (2003)
7. Robertson, D.: A lightweight coordination calculus for agent systems. In: Declarative Agent Languages and Technologies, pp. 183–197 (2004)
8. Robertson, D., Walton, C., Barker, A., Besana, P., Chen-Burger, Y., Hassan, F., Lambert, D., Li, G., McGinnis, J., Osman, N., Bundy, A., McNeill, F., van Harmelen, F., Sierra, C., Giunchiglia, F.: Models of interaction as a grounding for peer to peer knowledge sharing. Advances in Web Semantics 1 (in press)
9. Shvaiko, P., Euzenat, J.: A survey of schema-based matching approaches. Journal on Data Semantics 4, 146–171 (2005)
10. Siebes, R., Dupplaw, D., Kotoulas, S., de Pinninck, A.P., van Harmelen, F., Robertson, D.: The openknowledge system: an interaction-centered approach to knowledge sharing. In: Meersman, R., Tari, Z. (eds.) OTM 2007, Part I. LNCS, vol. 4803, pp. 381–390. Springer, Heidelberg (2007)
11. Sierra, C., Rodriguez Aguilar, J., Noriega, P., Arcos, J., Esteva, M.: Engineering multi-agent systems as electronic institutions. European Journal for the Informatics Professional 4 (2004)
12. Wooldridge, M.: An Introduction to Multiagent Systems. John Wiley and Sons, Chichester (2002)
13. Zaihrayeu, I.: Towards Peer-to-Peer Information Management Systems. PhD thesis, University of Trento, Italy (2006)

An Approach to Probabilistic Data Integration for the Semantic Web*

Andrea Calì[1,2] and Thomas Lukasiewicz[2,**]

[1] Oxford-Man Institute of Quantitative Finance, University of Oxford
Blue Boar Court, 9 Alfred Street, Oxford OX1 4EH, UK
andrea.cali@oxford-man.ox.ac.uk
[2] Computing Laboratory, University of Oxford
Wolfson Building, Parks Road, Oxford OX1 3QD, UK
{andrea.cali,thomas.lukasiewicz}@comlab.ox.ac.uk

Abstract. Probabilistic description logic programs are a powerful tool for knowledge representation in the Semantic Web, which combine description logics, normal programs under the answer set or well-founded semantics, and probabilistic uncertainty. The task of data integration amounts to providing the user with access to a set of heterogeneous data sources in the same fashion as when querying a single database, that is, through a global schema, which is a common representation of all the underlying data sources. In this paper, we make use of probabilistic description logic programs to model expressive data integration systems for the Semantic Web, where constraints are expressed both over the data sources and the global schema. We describe different types of probabilistic data integration, which aim especially at applications in the Semantic Web.

Keyword: Probabilistic data integration, Semantic Web, probabilistic description logic programs, description logics, normal programs, answer set semantics, well-founded semantics, probabilistic uncertainty.

1 Introduction

Recent research on knowledge representation has focused especially on the *Semantic Web*, which is an extension of the current Web by standards and technologies that help machines to understand the information on the Web so that they can support richer discovery, data integration, navigation, and automation of tasks [1]. The nature of the vast amount of data that are present on the Web is such that most information that we can retrieve is conflicting, overlapping with other information, or incomplete.

The Semantic Web as it is now consists of hierarchical layers, at different levels of abstraction; the *Ontology layer* is the one that had the fastest development lately, and it adopts the *Web Ontology Language (OWL)* [2] in different dialects of different expressiveness, namely, *OWL Lite*, *OWL DL*, and *OWL Full*. Integrating data [3] and

---

* This paper is a significantly extended and revised version of a position paper that appeared in: *Proceedings URSW-2006. CEUR Workshop Proceedings* 218, CEUR-WS.org, 2006.
** Alternative address: Institut für Informationssysteme, Technische Universität Wien, Favoritenstraße 9-11, 1040 Wien, Austria; email: lukasiewicz@kr.tuwien.ac.at.

mapping ontologies onto one another [4] are the major issues in this field. Besides the ontology layer, further layers such as the *Rules*, *Logic*, and *Proof layers* aim at providing support for sophisticated reasoning capabilities [5]. In particular, there is a large body of work on integrating rules and ontologies, which is a key requirement of the layered architecture of the Semantic Web. Rules can be build on top of ontologies, that is, rules may use the vocabulary of an ontology; another possible approach is to build ontologies on top of rules. Both types of integration are realized in recent hybrid integrations of rules and ontologies, called *description logic programs* (or *dl-programs*), which have the form $KB = (L, P)$, where L is a description logic knowledge base and P is a finite set of rules involving either queries to L in a loose coupling [6,7] or concepts and roles from L as unary and binary predicates, respectively, in a tight coupling [8].

In Web search engines, which have to deal with pieces of data that are intrinsically inconsistent, and which provide uncertain information, statistical methods are commonly applied. Differently, research on probabilistic approaches to the Semantic Web is fairly recent. An important recent forum for approaches to uncertainty reasoning in the Semantic Web is the annual *Workshop on Uncertainty Reasoning for the Semantic Web (URSW)*; there also exists a W3C Incubator Group on *Uncertainty Reasoning for the World Wide Web*. There are especially probabilistic extensions of description logics [9,10], of Web ontology languages [11,12] (see also [13]), and of dl-programs [14,15] (to encode ambiguous information, such as "John is a student with the probability 0.7 and a teacher with the probability 0.3", which is very different from vague / fuzzy / imprecise information, such as "John is tall with the degree of truth 0.7"). In particular, probabilistic extensions of the loosely (resp., tightly) coupled dl-programs in [6,7] (resp., [8]) have been proposed in [14] (resp., [15]). Important related works combine standard answer set programming with probabilities [16] and positive logic programs with Bayesian networks [17]. The approach of probabilistic dl-programs [14] is especially promising, since it is extremely expressive and flexible, and it is able to generalize answer set programming, Bayesian networks, and the independent choice logic [18]. In these probabilistic dl-programs, logic is nicely blended with probability; we indeed show that this allows us to naturally employ dl-programs in the declarative specification of data integration systems in the Semantic Web. The techniques found in [19] provide tractable algorithms for reasoning tasks in probabilistic dl-programs; in particular, they provide polynomial data complexity algorithms for processing queries under a novel semantics, called the *total well-founded semantics*.

Data integration is a general problem that is among the main goals of the Semantic Web. In a data integration system for the Semantic Web, heterogeneous data sources in the Semantic Web are presented to the user as a single database, which can be queried through a common representation of all the stored information. Such a common representation is called the *global schema*, and is usually virtual (that is, rather than being stored at the global level, the pieces of data reside at the sources). The crucial issues in data integration are *(i)* the *declarative* representation of the relationship between the sources and the global schema, called *mapping*, and *(ii)* the algorithms to answer queries that are posed over the global schema.

In this paper, we start from the results of [19] about tight query processing in (loosely coupled) probabilistic dl-programs, and we provide a natural way of employing such dl-programs in the integration of data over the Semantic Web. We show different natural ways of specifying a data integration framework with probabilistic dl-programs, allowing the use of the efficient reasoning service that are available for such dl-programs. In particular, we show three different ways of specifying a probabilistic mapping with probabilistic dl-programs:

- modeling *trust probabilities*: here, we use probabilities to represent different levels of trust that we have with respect to different data sources;
- modeling *error probability*: here, probabilities are used to represent the information that is stored in data sources that have a certain probability of error; such uncertainty is modeled with rules of a probabilistic dl-program, which encode the proper inference in case of error or in case of non-error; and
- *purely probabilistic mappings*: this is the case when different pieces of overlapping information are collected from the data sources, each with an assigned probability.

Furthermore, we show that the flexibility of probabilistic dl-programs allows us to represent situations where it is necessary to represent a sort of fine-grained probability defined on single tuples (ground facts) at the data sources. This is a common scenario in the literature about probabilistic data, with many applications in practice. Our modeling is possible because, roughly speaking, probabilistic dl-programs define probabilities on each single ground atom of the Herbrand base of the program. In such scenarios, probabilistic dl-programs "filter" the probabilities through the mapping, thus giving a global representation of the source data that takes into account their uncertainty.

Finally, we remark that probabilistic dl-programs are able to deal with cases of data integration where constraints (expressible with probabilistic dl-rules) are enforced on the sources and on the global schema. This is particularly useful when integrating sources that are represented intensionally by a local ontology.

The rest of this paper is organized as follows. In Section 2, we recall the tractable description logic *DL-Lite*. Section 3 describes (loosely coupled) dl-programs under the answer set and the well-founded semantics. In Section 4, we describe (loosely coupled) probabilistic dl-programs under the answer set and the total well-founded semantics. Section 5 presents our approach to probabilistic data integration for the Semantic Web on top of these probabilistic dl-programs. In Section 6, we summarize our main results and give an outlook on future research.

2   Description Logics

In this section, we recall the syntax and the semantics of *DL-Lite*, a tractable description logic especially suited for representing large amounts of data. Intuitively, description logics model a domain of interest in terms of concepts and roles, which represent classes of individuals resp. binary relations between classes of individuals. While we restrict ourselves to *DL-Lite* here, the approach continues to be valid for the variants of *DL-Lite* in [20], since the reasoning algorithms can be easily extended to such variants.

2.1   Syntax

We first define concepts and axioms and then knowledge bases and conjunctive queries in *DL-Lite*. We assume pairwise disjoint sets **A**, **R**, and **I** of *atomic concepts, abstract* *roles*, and *individuals*, respectively. We use $\mathbf{R}^-$ to denote the set of all inverses R^- of roles $R \in \mathbf{R}$. A *basic concept* B is either an atomic concept $A \in \mathbf{A}$ or an exists restriction $\exists R$, where $R \in \mathbf{R} \cup \mathbf{R}^-$. An *axiom* is either (1) a concept inclusion axiom $B \sqsubseteq \phi$, where B is a basic concept, and ϕ is either a basic concept B or its negation $\neg B$, or (2) a *functionality axiom* (funct R), where $R \in \mathbf{R} \cup \mathbf{R}^-$, or (3) a concept membership axiom $B(a)$, where B is a basic concept and $a \in \mathbf{I}$, or (4) a role membership axiom $R(a,c)$, where $R \in \mathbf{R}$ and $a, c \in \mathbf{I}$. A *(description logic) knowledge base* L is a finite set of axioms. A *conjunctive query* over L is of the form $Q(\mathbf{x}) = \exists \mathbf{y} \, (conj(\mathbf{x}, \mathbf{y}))$, where $\mathbf{x}$ and $\mathbf{y}$ are tuples of distinct variables, and $conj(\mathbf{x}, \mathbf{y})$ is a conjunction of assertions $B(z)$ and $R(z_1, z_2)$, where B and R are basic concepts and roles from $\mathbf{R}$, respectively, and z, z_1, and z_2 are individuals from $\mathbf{I}$ or variables in $\mathbf{x}$ or $\mathbf{y}$.

Example 1. A university database may use a description logic knowledge base L to characterize students and exams. For example, suppose that (1) every bachelor student is a student, (2) every master student is a student, (3) professors are not students, (4) only students give exams and only exams are given, and (5) *john* is a student, *mary* is a master student, *java* is an exam, and *john* has given it. These relationships are encoded by the following axioms in L:

(1) *bachelor_student* $\sqsubseteq$ *student*; (2) *master_student* $\sqsubseteq$ *student*;
(3) *professor* $\sqsubseteq \neg$*student*; (4) $\exists$*given* $\sqsubseteq$ *student*; $\exists$*given*$^{-1}$ $\sqsubseteq$ *exam*;
(5) *student*(*john*); *master_student*(*mary*); *exam*(*java*); *given*(*john*, *java*) .

2.2 Semantics

The semantics of *DL-Lite* is defined as usual in first-order logics. An *interpretation* $\mathcal{I} = (\Delta^{\mathcal{I}}, \cdot^{\mathcal{I}})$ consists of a nonempty *domain* $\Delta^{\mathcal{I}}$ and a mapping $\cdot^{\mathcal{I}}$ that assigns to each $A \in \mathbf{A}$ a subset of $\Delta^{\mathcal{I}}$, to each $o \in \mathbf{I}$ an element of $\Delta^{\mathcal{I}}$ (such that $o_1 \neq o_2$ implies $o_1^{\mathcal{I}} \neq o_2^{\mathcal{I}}$; that is, we make the *unique name assumption*), and to each $R \in \mathbf{R}$ a subset of $\Delta^{\mathcal{I}} \times \Delta^{\mathcal{I}}$. We extend $\cdot^{\mathcal{I}}$ to all concepts and roles, and we define the *satisfaction* of an axiom F in $\mathcal{I}$, denoted $\mathcal{I} \models F$, as usual. A tuple $\mathbf{c}$ of individuals from $\mathbf{I}$ is an *answer* for a conjunctive query $Q(\mathbf{x}) = \exists \mathbf{y} \, (conj(\mathbf{x}, \mathbf{y}))$ to a description logic knowledge base L iff for every $\mathcal{I} = (\Delta^{\mathcal{I}}, \cdot^{\mathcal{I}})$ that satisfies all $F \in L$, there exists a tuple $\mathbf{o}$ of elements from $\Delta^{\mathcal{I}}$ such that all assertions in $conj(\mathbf{c}, \mathbf{o})$ are satisfied in $\mathcal{I}$. In *DL-Lite*, computing all such answers is possible in polynomial time in the data complexity.

3 Description Logic Programs

We adopt the (loosely coupled) description logic programs (or dl-programs) of [6,7], which consist of a description logic knowledge base L and a generalized normal program P, which may contain queries to L (called dl-queries) in rule bodies. We remark that these dl-programs can also be extended by queries to other formalisms, such as RDF theories. We now first define the syntax of dl-programs and then their answer set and their well-founded semantics. Note that in contrast to [6,7], we assume here that dl-queries may be conjunctive queries to L.

3.1 Syntax

We assume a function-free first-order vocabulary Φ with finite nonempty sets of constant and predicate symbols Φ_c and Φ_p, respectively, and a set of variables $\mathcal{X}$. We assume that *(i)* Φ_c is a subset of $\mathbf{I}$ (since the constants in Φ_c may occur in concept and role assertions of dl-queries) and that *(ii)* Φ and $\mathbf{A}$ (resp., $\mathbf{R}$) have no unary (resp., binary) predicate symbols in common (and thus dl-queries are the only interface between L and P). A *term* is a constant symbol from Φ or a variable from $\mathcal{X}$. If p is a predicate symbol of arity $k \geqslant 0$ from Φ, and $t_1, \ldots, t_k$ are terms, then $p(t_1, \ldots, t_k)$ is an *atom*. A *literal* is an atom a or a default-negated atom $not\, a$. A *(normal) rule r* is of the form

$$a \leftarrow b_1, \ldots, b_k, not\, b_{k+1}, \ldots, not\, b_m, \tag{1}$$

where $a, b_1, \ldots, b_m$ are atoms and $m \geqslant k \geqslant 0$. We call a the *head* of r, denoted $H(r)$, while the conjunction $b_1, \ldots, b_k, not\, b_{k+1}, \ldots, not\, b_m$ is the *body* of r; its *positive* (resp., *negative*) part is $b_1, \ldots, b_k$ (resp., $not\, b_{k+1}, \ldots, not\, b_m$). We define $B(r)$ as the union of $B^+(r) = \{b_1, \ldots, b_k\}$ and $B^-(r) = \{b_{k+1}, \ldots, b_m\}$. A *(normal) program P* is a finite set of normal rules. We say P is *positive* iff it is "*not*"-free.

A *dl-query* $Q(\mathbf{t})$ is a conjunctive query. A *dl-atom* has the form $DL[S_1 \uplus p_1, \ldots, S_m \uplus p_m; Q(\mathbf{t})]$, where each S_i is a concept or role, p_i is a unary resp. binary predicate symbol, $Q(\mathbf{t})$ is a dl-query, and $m \geqslant 0$. We call $p_1, \ldots, p_m$ its *input predicate symbols*. Intuitively, $\uplus$ increases S_i by the extension of p_i. A *(normal) dl-rule r* is of the form (1), where any $b \in B(r)$ may be a dl-atom. A *(normal) dl-program* $KB = (L, P)$ consists of a description logic knowledge base L and a finite set of dl-rules P. We say $KB = (L, P)$ is *positive* iff P is positive. *Ground terms, atoms, literals*, etc., are defined as usual. We denote by $ground(P)$ the set of all ground instances of dl-rules in P relative to Φ_c.

Example 2. A dl-program $KB = (L, P)$ is given by L as in Example 1 and P consisting of the following dl-rules, which express that (1) the relation of propaedeutics enjoys the transitive property, (2) if a student has given an exam, then he/she has given all exams that are propaedeutic to it, (3) if two students have a given exam in common, then they have given the same exam, and (4) *unix* is propaedeutic for *java*, and *java* is propaedeutic for *programming_languages*:

(1) $propaedeutic(X, Z) \leftarrow propaedeutic(X, Y), propaedeutic(Y, Z)$;

(2) $given\_prop(X, Z) \leftarrow DL[given(X, Y)], propaedeutic(Z, Y)$;

(3) $given\_same\_exam(X, Y) \leftarrow DL[given \uplus given\_prop; \exists Z(given(X, Z) \wedge given(Y, Z))]$;

(4) $propaedeutic(unix, java); propaedeutic(java, programming\_languages)$.

3.2 Answer Set Semantics

The *Herbrand base* HB_Φ is the set of all ground atoms constructed from constant and predicate symbols in Φ. An *interpretation* I is any $I \subseteq HB_\Phi$. We say I is a *model* of $a \in HB_\Phi$ under a description logic knowledge base L, denoted $I \models_L a$, iff $a \in I$. We say I is a *model* of a ground dl-atom $a = DL[S_1 \uplus p_1, \ldots, S_m \uplus p_m; Q(\mathbf{c})]$ under L, denoted $I \models_L a$, iff $L \cup \bigcup_{i=1}^m A_i(I) \models Q(\mathbf{c})$, where $A_i(I) = \{S_i(\mathbf{e}) \mid p_i(\mathbf{e}) \in I\}$. We say I is a *model* of a ground dl-rule r iff $I \models_L H(r)$ whenever $I \models_L B(r)$, that is, $I \models_L a$

for all $a \in B^+(r)$ and $I \not\models_L a$ for all $a \in B^-(r)$. We say I is a *model* of a dl-program $KB = (L, P)$, denoted $I \models KB$, iff $I \models_L r$ for all $r \in ground(P)$.

Like ordinary positive programs, each positive dl-program KB has a unique least model, denoted M_{KB}, which naturally characterizes its semantics. The *answer set semantics* of general dl-programs is then defined by a reduction to the least model semantics of positive ones, using a reduct that generalizes the ordinary Gelfond-Lifschitz reduct [21] and removes all default-negated atoms in dl-rules: For dl-programs $KB = (L, P)$, the *dl-reduct* of P relative to L and an interpretation $I \subseteq HB_\Phi$, denoted P_L^I, is the set of all dl-rules obtained from $ground(P)$ by *(i)* deleting each dl-rule r such that $I \models_L a$ for some $a \in B^-(r)$, and *(ii)* deleting from each remaining dl-rule r the negative body. An *answer set* of KB is an interpretation $I \subseteq HB_\Phi$ such that I is the unique least model of (L, P_L^I). A dl-program is *consistent* iff it has an answer set.

The answer set semantics of dl-programs has several nice features. In particular, for dl-programs $KB = (L, P)$ without dl-atoms, it coincides with the ordinary answer set semantics of P. Answer sets of a general dl-program KB are also minimal models of KB. Furthermore, positive and locally stratified dl-programs have exactly one answer set, which coincides with their canonical minimal model.

3.3 Well-Founded Semantics

Rather than associating with every dl-program a (possibly empty) set of two-valued interpretations, the well-founded semantics associates with every dl-program a unique three-valued interpretation.

A *classical literal* is either an atom a or its negation $\neg a$. For sets $S \subseteq HB_\Phi$, we define $\neg S = \{\neg a \mid a \in S\}$. We define $Lit_\Phi = HB_\Phi \cup \neg HB_\Phi$. A set of ground classical literals $S \subseteq Lit_\Phi$ is *consistent* iff $\{a, \neg a\} \not\subseteq S$ for all $a \in HB_\Phi$. A *three-valued interpretation* is any consistent $I \subseteq Lit_\Phi$. We define the well-founded semantics of dl-programs $KB = (L, P)$ via a generalization of the operator γ^2 for ordinary normal programs. We define the operator γ_{KB} as follows. For every $I \subseteq HB_\Phi$, we define $\gamma_{KB}(I)$ as the least model of the positive dl-program $KB^I = (L, P_L^I)$. The operator γ_{KB} is anti-monotonic, and thus the operator γ_{KB}^2 (defined by $\gamma_{KB}^2(I) = \gamma_{KB}(\gamma_{KB}(I))$, for every $I \subseteq HB_\Phi$) is monotonic and has a least and a greatest fixpoint, denoted $lfp(\gamma_{KB}^2)$ and $gfp(\gamma_{KB}^2)$, respectively. Then, the *well-founded semantics* of the dl-program KB, denoted $WFS(KB)$, is defined as $lfp(\gamma_{KB}^2) \cup \neg(HB_\Phi - gfp(\gamma_{KB}^2))$.

As an important property, the well-founded semantics for dl-programs approximates their answer set semantics. That is, for all consistent dl-programs KB and $\ell \in Lit_\Phi$, it holds that $\ell \in WFS(KB)$ iff ℓ is true in every answer set of KB.

4 Probabilistic Description Logic Programs

We now recall the (loosely coupled) probabilistic dl-programs from [14]. We first define the syntax of probabilistic dl-programs and then their answer set semantics. Informally, they consist of a dl-program (L, P) and a probability distribution μ over a set of total choices B. Every total choice B along with the dl-program (L, P) then defines a set of Herbrand interpretations of which the probabilities sum up to $\mu(B)$.

4.1 Syntax

We now define the syntax of probabilistic dl-programs and queries addressed to them. We first define choice spaces and probabilities on choice spaces.

A *choice space* C is a set of pairwise disjoint and nonempty sets $A \subseteq HB_\Phi$. Any $A \in C$ is called an *alternative* of C, and any element $a \in A$ is called an *atomic choice* of C. Intuitively, every alternative $A \in C$ represents a random variable and every atomic choice $a \in A$ one of its possible values. A *total choice* of C is a set $B \subseteq HB_\Phi$ such that $|B \cap A| = 1$ for all $A \in C$ (and thus $|B| = |C|$). Intuitively, every total choice B of C represents an assignment of values to all the random variables. A *probability* μ on a choice space C is a probability function on the set of all total choices of C. Intuitively, every probability μ is a probability distribution over the set of all variable assignments. Since C and all its alternatives are finite, μ can be defined by *(i)* a mapping $\mu \colon \bigcup C \rightarrow [0, 1]$ such that $\sum_{a \in A} \mu(a) = 1$ for all $A \in C$, and *(ii)* $\mu(B) = \Pi_{b \in B} \mu(b)$ for all total choices B of C. Intuitively, *(i)* defines a probability over the values of each random variable of C, and *(ii)* assumes independence between the random variables.

A *probabilistic dl-program* $KB = (L, P, C, \mu)$ consists of a dl-program (L, P), a choice space C such that *(i)* $\bigcup C \subseteq HB_\Phi$ and *(ii)* no atomic choice in C coincides with the head of any $r \in ground(P)$, and a probability μ on C. Intuitively, since the total choices of C select subsets of P, and μ is a probability distribution on the total choices of C, every probabilistic dl-program is the compact encoding of a probability distribution on a finite set of normal dl-programs. Observe here that P is fully general and not necessarily stratified or acyclic. An *event* α is any Boolean combination of atoms (that is, constructed from atoms via the Boolean operators "$\wedge$" and "$\neg$"). A *conditional event* is of the form $\beta | \alpha$, where α and β are events. A *probabilistic query* has the form $\exists (\beta | \alpha)[r, s]$, where $\beta | \alpha$ is a conditional event, and r and s are variables.

Example 3. Consider $KB = (L, P, C, \mu)$, where L and P are as in Examples 1 and 2, respectively, except that the following two (probabilistic) rules are added to P:

$$friends(X, Y) \leftarrow given\_same\_exam(X, Y), DL[master\_student(X)],$$
$$DL[master\_student(Y)], choice_m ;$$
$$friends(X, Y) \leftarrow given\_same\_exam(X, Y), DL[bachelor\_student(X)],$$
$$DL[bachelor\_student(Y)], choice_b .$$

Let $C = \{\{choice_m, not\_choice_m\}, \{choice_b, not\_choice_b\}\}$, and let the probability μ on C be given by $\mu \colon choice_m, not\_choice_m, choice_b, not\_choice_b \mapsto 0.9, 0.1, 0.7, 0.3$. Here, the new rules express that if two master (resp., bachelor) students have given the same exam, then there is a probability of 0.9 (resp., 0.7) that they are friends. Note that probabilistic facts can be encoded by rules with only atomic choices in their body. Our wondering about the entailed tight interval for the probability that *john* and *bill* are friends can then be expressed by the probabilistic query $\exists (friends(john, bill))[R, S]$.

4.2 Answer Set Semantics

We now define a probabilistic answer set semantics of probabilistic dl-programs, and the notions of consistency and tight answers.

Given a probabilistic dl-program $KB = (L, P, C, \mu)$, a *probabilistic interpretation* Pr is a probability function on the set of all $I \subseteq HB_\Phi$. We say Pr is an *answer set* of KB iff *(i)* every interpretation $I \subseteq HB_\Phi$ with $Pr(I) > 0$ is an answer set of $(L, P \cup \{p \leftarrow \mid p \in B\})$ for some total choice B of C, and *(ii)* $Pr(\bigwedge_{p \in B} p) = \mu(B)$ for every total choice B of C. Informally, Pr is an answer set of $KB = (L, P, C, \mu)$ iff *(i)* every interpretation $I \subseteq HB_\Phi$ of positive probability under Pr is an answer set of the dl-program (L, P) under some total choice B of C, and *(ii)* Pr coincides with μ on the total choices B of C. We say KB is *consistent* iff it has an answer set Pr.

Given a ground event α, the *probability* of α in a probabilistic interpretation Pr, denoted $Pr(\alpha)$, is the sum of all $Pr(I)$ such that $I \subseteq HB_\Phi$ and $I \models \alpha$. We say $(\beta|\alpha)[l, u]$, where α and β are ground events, and $l, u \in [0, 1]$, is a *tight consequence* of a consistent probabilistic dl-program KB under the answer set semantics iff l (resp., u) is the infimum (resp., supremum) of $Pr(\alpha \wedge \beta) / Pr(\alpha)$ subject to all answer sets Pr of KB with $Pr(\alpha) > 0$. Note that this infimum (resp., supremum) is naturally defined as 1 (resp., 0) iff no such Pr exists. The *tight answer* for a probabilistic query $Q = \exists(\beta|\alpha)[r, s]$ to KB under the answer set semantics is the set of all ground substitutions θ (for the variables in Q) such that $(\beta|\alpha)[r, s]\theta$ is a tight consequence of KB under the answer set semantics. For ease of presentation, since the tight answers for probabilistic queries $Q = \exists(\beta|\alpha)[r, s]$ with non-ground $\beta|\alpha$ can be reduced to the tight answers for probabilistic queries $Q = \exists(\beta|\alpha)[r, s]$ with ground $\beta|\alpha$, we consider only the latter type of probabilistic queries in the following.

4.3   Total Well-Founded Semantics

We now recall the total well-founded semantics for probabilistic dl-programs, which is defined for all probabilistic dl-programs (as opposed to the answer set semantics, which is only defined for consistent probabilistic dl-programs) and for all probabilistic queries to probabilistic dl-programs (as opposed to the previous well-founded semantics of [14], which is only defined for a very limited class of probabilistic queries).

More precisely, given a probabilistic dl-program $KB = (L, P, C, \mu)$ and a probabilistic query $Q = \exists(\beta|\alpha)[r, s]$ with ground $\beta|\alpha$, the tight answer θ for Q to KB under the previous well-founded semantics of [14] *exists* iff both ground events $\alpha \wedge \beta$ and α are defined in every $S = WFS(L, P \cup \{p \leftarrow \mid p \in B\})$ such that B is a total choice of C. Here, a ground event ϕ is *defined* in S iff either $I \models \phi$ for every interpretation $I \supseteq S \cap HB_\Phi$, or $I \not\models \phi$ for every interpretation $I \supseteq S \cap HB_\Phi$. If α is false in every $WFS(L, P \cup \{p \leftarrow \mid p \in B\})$ such that B is a total choice of C, then the tight answer is defined as $\theta = \{r/1, s/0\}$; otherwise, the tight answer (if it exists) is defined as $\theta = \{r/\frac{u}{v}, s/\frac{u}{v}\}$, where u (resp., v) is the sum of all $\mu(B)$ such that *(i)* B is a total choice of C and *(ii)* $\alpha \wedge \beta$ (resp., α) is true in $WFS(L, P \cup \{p \leftarrow \mid p \in B\})$.

We define the total well-founded semantics for probabilistic dl-programs as follows.

Definition 1 (Total Well-Founded Semantics). Let $KB = (L, P, C, \mu)$ be a probabilistic dl-program, and let $Q = \exists(\beta|\alpha)[r, s]$ be a probabilistic query with ground $\beta|\alpha$. Let a (resp., b^-) be the sum of all $\mu(B)$ such that *(i)* B is a total choice of C and *(ii)* $\alpha \wedge \beta$ is true (resp., false) in $WFS(L, P \cup \{p \leftarrow \mid p \in B\})$. Let c (resp., d^-) be the sum of all $\mu(B)$ such that *(i)* B is a total choice of C and *(ii)* $\alpha \wedge \neg\beta$ is true (resp., false) in

$WFS(L, P \cup \{p \leftarrow \mid p \in B\})$. Let $b = 1 - b^-$ and $d = 1 - d^-$. Then, *the tight answer θ for Q to KB under the total well-founded semantics ($TWFS(KB)$)* is defined by

$$\theta = \begin{cases} \{r/1,\ s/0\} & \text{if } b = 0 \text{ and } d = 0; \\ \{r/0,\ s/0\} & \text{if } b = 0 \text{ and } d \neq 0; \\ \{r/1,\ s/1\} & \text{if } b \neq 0 \text{ and } d = 0; \\ \{r/\frac{a}{a+d},\ s/\frac{b}{b+c}\} & \text{otherwise.} \end{cases} \tag{2}$$

We finally report some results from [19] on the complexity of tight query processing in probabilistic dl-programs under the total well-founded semantics.

Tight query processing in probabilistic dl-programs $KB = (L, P, C, \mu)$ in *DL-Lite* (where L is in *DL-Lite*) under $TWFS(KB)$ can be done in polynomial time in the data complexity. This result follows from the facts that (a) computing the well-founded semantics of a normal dl-program and (b) conjunctive query processing in *DL-Lite* can both be done in polynomial time in the data complexity. Here, $|C|$ is bounded by a constant, since C and μ define the probabilistic information of P, which is fixed as a part of the program in P, while the ordinary facts in P are the variable input. Computing tight answers is EXP-complete in the combined complexity.

5 Probabilistic Data Integration

Integrating data from different sources is a crucial issue in the Semantic Web. In this section, we show how probabilistic dl-programs can be employed as a formalism for data integration in the Semantic Web. We first give some general definitions.

A *data integration system* (in its most general form) [22] $I = (G, S, M)$ consists of *(i)* a *global* (or *mediated*) *schema G*, which represents the domain of interest of the system, *(ii)* a *source schema S*, which represents the data sources that take part in the system, and *(iii)* a *mapping M*, which establishes a relation between the source schema and the global schema. Here, G is purely virtual, while the data are stored in S. The mapping M can be specified in different ways, which is a crucial aspect in a data integration system. In particular, when every data structure in G is defined through a view over S, the mapping is said to be *GAV (global-as-view)*, while when very data structure in S is defined through a view over G the mapping is *LAV (local-as-view)*. A mixed approach, called *GLAV* [23,24], associates views over G to views over S.

5.1 Modeling Data Integration Systems

In our framework, we assume that the global schema G, the source schema S, and the mapping M are each encoded by a probabilistic dl-program. More formally, we partition the vocabulary Φ into the sets Φ_G, Φ_S, and Φ_c: *(i)* the symbols in Φ_G are of arity at least 1 and represent the global predicates, *(ii)* the symbols in Φ_S are of arity at least 1 and represent source predicates, and *(iii)* the symbols in Φ_c are constants. Let $\mathbf{A}_G$ and $\mathbf{R}_G$ be disjoint denumerable sets of atomic concepts and abstract roles, respectively, for the global schema, and let $\mathbf{A}_S$ and $\mathbf{R}_S$ (disjoint from $\mathbf{A}_G$ and $\mathbf{R}_G$) be similar sets for the source schema. We also assume a denumerable set of individuals $\mathbf{I}$ that is disjoint from the set of all concepts and roles and a superset of Φ_c. A *probabilistic data integration system* $PI = (KB_G, KB_S, KB_M)$ consists of a probabilistic dl-program $KB_G = (L_G, P_G, C_G, \mu_G)$ for the global schema, a probabilistic dl-program $KB_S = (L_S, P_S, C_S, \mu_S)$ for the source schema, and a probabilistic dl-program $KB_M = (\emptyset, P_M, C_M, \mu_M)$ for the mapping:

- KB_G (resp., KB_S) is defined over the predicates, constants, concepts, roles, and individuals of the global (resp., source) schema, and it encodes ontological, rule-based, and probabilistic relationships in the global (resp., source) schema.
- KB_M is defined over the predicates, constants, concepts, roles, and individuals of the global and the source schema, and it encodes a probabilistic mapping between the predicates, concepts, and roles of the source and those of the global schema.

Our probabilistic dl-rules permit a specification of the mapping that can freely use global and source predicates together in rules, thus having a formalism that generalizes LAV and GAV in some way. Moreover, with a simple technicality, we are able to partly model GLAV systems. In GLAV data integration systems, the mapping is specified by means of rules of the form $\psi \leftarrow \varphi$, where ψ is a conjunction of atoms of G and φ is a conjunction of atoms of S. We introduce an auxiliary atom α that contains all the variables of ψ; moreover, let $\psi = \beta_1 \wedge \ldots \wedge \beta_m$. We model the GLAV mapping rule with the following rules:

$$\beta_1 \leftarrow \alpha$$
$$\vdots$$
$$\beta_m \leftarrow \alpha$$
$$\alpha \leftarrow \varphi$$

What our framework does not allow is having rules that are *unsafe*, that is, having existentially-quantified variables in the head.

Note also that correct and tight answers to probabilistic queries on the global schema are formally defined relative to the probabilistic dl-program $KB = (L, P, C, \mu)$, where $L = L_G \cup L_S$, $P = P_G \cup P_S \cup P_M$, $C = C_G \cup C_S \cup C_M$, and $\mu = \mu_G \cdot \mu_S \cdot \mu_M$. Informally, KB is the result of merging KB_G, KB_S, and KB_M. In a similar way, the probabilistic dl-program KB_S of the source schema S can be defined by merging the probabilistic dl-programs $KB_{S_1}, \ldots, KB_{S_1}$ of $n \geqslant 1$ source schemas $S_1, \ldots, S_n$.

The fact that the mapping is probabilistic allows for a high flexibility in the treatment of the uncertainty that is present when pieces of data come from heterogeneous sources whose informative content may be inconsistent and/or redundant relative to the global schema G, which in general incorporates constraints. Some different types of probabilistic mappings that can be modeled in our framework are summarized below.

5.2 Types of Probabilistic Mappings

In addition to expressing probabilistic knowledge about the global schema and about the source schema, the probabilities in probabilistic dl-programs can especially be used for specifying the probabilistic mapping in the data integration process. We distinguish three different types of probabilistic mappings, depending on whether the probabilities are used as *trust*, *error*, or *mapping probabilities*.

The simplest way of probabilistically integrating several data sources is to weigh each data source with a *trust probability* (which all sum up to 1). This is especially useful when several redundant data sources are to be integrated. In such a case, pieces of data from different data sources may easily be inconsistent with each other.

Example 4. Suppose that we want to obtain a weather forecast for a certain place by integrating the potentially different weather forecasts of several weather forecast institutes. For ease of presentation, suppose that we only have three weather forecast institutes A, B, and C. In general, one trusts certain weather forecast institutes more than others. In our case, we suppose that our trust in the institutes A, B, and C is expressed by the trust probabilities 0.6, 0.3, and 0.1, respectively. That is, we trust most in A, medium in B, and less in C. In general, the different institutes do not use the same data structure to represent their weather forecast data. For example, institute A may use a single relation *forecast*(*place, date, weather, temperature, wind*) to store all the data, while B may have one relation *forecast_place*(*date, weather, temperature, wind*) for each place, and C may use several different relations *forecast_weather* (*place, date, weather*), *forecast_temperature*(*place, date, temperature*), and *forecast_wind* (*place, date, wind*). Suppose the global schema G has the relation *forecast_rome_global*(*date, weather, temperature, wind*), which may e.g. be posted on the web by the tourist information of Rome. The probabilistic mapping of the source schemas of A, B, and C to the global schema G can then be specified by the following $KB_M = (\emptyset, P_M, C_M, \mu_M)$:

$$P_M = \{forecast\_rome\_global(D, W, T, M) \leftarrow forecast(rome, D, W, T, M),\ inst_A;$$
$$forecast\_rome\_global(D, W, T, M) \leftarrow forecast\_rome(D, W, T, M),\ inst_B;$$
$$forecast\_rome\_global(D, W, T, M) \leftarrow forecast\_weather(rome, D, W),$$
$$forecast\_temperature(rome, D, T),\ forecast\_wind(rome, D, M),\ inst_C\};$$

$$C_M = \{\{inst_A, inst_B, inst_C\}\};$$

$$\mu_M : \ inst_A,\ inst_B,\ inst_C \ \mapsto \ 0.6,\ 0.3,\ 0.1.$$

The mapping assertions state that the first, second, and third rule above hold with the probabilities 0.6, 0.3, and 0.1, respectively. This is motivated by the fact that three institutes may generally provide conflicting weather forecasts, and our trust in the institutes A, B, and C are given by the trust probabilities 0.6, 0.3, and 0.1, respectively.

A more complex way of probabilistically integrating several data sources is to associate each data source (or each derivation) with an *error probability*.

Example 5. Suppose that we want to integrate the data provided by the different sensors in a sensor network. For example, suppose that we have a sensor network measuring the concentration of ozone in several different positions of a certain town, which may e.g. be the basis for the common hall to reduce or forbid individual traffic. Suppose that each sensor $i \in \{1, \ldots, n\}$ with $n \geqslant 1$ is associated with its position through *sensor*(*i, position*) and provides its measurement data in a single relation *reading$_i$* (*date, time, type, result*). Each such reading may be erroneous with the probability e_i. That is, any tuple returned (resp., not returned) by a sensor $i \in \{1, \ldots, n\}$ may not hold (resp., may hold) with probability e_i. Suppose that the global schema contains a single relation $reading(position, date, time, type, result)$. Then, the probabilistic mapping of the source schemas of the sensors $i \in \{1, \ldots, n\}$ to the global schema G can be specified by the following probabilistic dl-program $KB_M = (\emptyset, P_M, C_M, \mu_M)$:

$$
\begin{aligned}
P_M = {} & \{aux_i(P, D, T, K, R) \leftarrow reading_i(D, T, K, R),\, sensor(i, P) \mid i \in \{1, \ldots, n\}\} \cup \\
& \{reading(P, D, T, K, R) \leftarrow aux_i(P, D, T, K, R),\, not\_error_i \mid i \in \{1, \ldots, n\}\} \cup \\
& \{reading(P, D, T, K, R) \leftarrow not\, aux_i(P, D, T, K, R),\, error_i \mid i \in \{1, \ldots, n\}\} \,;
\end{aligned}
$$

$$
C_M = \{\{error_i, not\_error_i\} \mid i \in \{1, \ldots, n\}\} \,;
$$

$$
\mu_M : \quad error_1, not\_error_1, \ldots, error_n, not\_error_n \mapsto e_1,\, 1-e_1, \ldots, e_n,\, 1-e_n \,.
$$

Note that if there are two sensors j and k for the same position, and they both return the same tuple as a reading, then this reading is correct with the probability $1 - e_j e_k$ (since it may be erroneous with the probability $e_j e_k$). Note also that this modeling assumes that the errors of the sensors are independent from each other, which can be achieved by eventually unifying atomic choices. For example, if the sensor j depends on the sensor k, then j is erroneous when k is erroneous, and thus the atomic choices $\{error_j, not\_error_j\}$ and $\{error_k, not\_error_k\}$ are merged into the new atomic choice $\{error_j error_k,\ not\_error_j error_k,\ not\_error_j not\_error_k\}$.

When integrating several data sources, it may be the case that the relationships between the source schema and the global schema are *purely probabilistic*.

Example 6. Suppose that we want to integrate the schemas of two libraries, and that the global schema contains the predicate symbol *logic_programming*, while the source schemas contain only the concepts *rule-based_systems* resp. *deductive_databases* in their ontologies. These three concepts are overlapping to some extent, but they do not exactly coincide. For example, a randomly chosen book from *rule-based_systems* (resp., *deductive_databases*) may belong to the area *logic_programming* with the probability 0.7 (resp., 0.8). The probabilistic mapping from the source schemas to the global schema can then be expressed by the following $KB_M = (\emptyset, P_M, C_M, \mu_M)$:

$$
\begin{aligned}
P_M = {} & \{logic\_programming(X) \leftarrow DL[rule\text{-}based\_systems(X)],\ choice_1 \,; \\
& \quad logic\_programming(X) \leftarrow DL[deductive\_databases(X)],\ choice_2\} \,;
\end{aligned}
$$

$$
C_M = \{\{choice_1, not\_choice_1\}, \{choice_2, not\_choice_2\}\} \,;
$$

$$
\mu_M : \quad choice_1, not\_choice_1, choice_2, not\_choice_2 \mapsto 0.7,\, 0.3,\, 0.8,\, 0.2 \,.
$$

5.3 Deterministic Mappings on Probabilistic Data

Finally, we briefly describe an approach to use probabilistic dl-programs to model probabilistic data, such as those in [25].

Example 7. Suppose that the weather in Oxford can be sunny, cloudy, or rainy with probabilities 0.2, 0.45, and 0.35, respectively, and similar probabilities are assigned for other cities. This setting is analogous to the "classical" one of probabilistic data, where there is a probability distribution over ground facts. In such a case, the choice space is $C = \{\{weather(oxford, sunny),\ weather(oxford, cloudy),\ weather(oxford,$

rainy)}, . . .}, and the probability is μ: *weather*(*oxford*, *sunny*), *weather*(*oxford*, *cloudy*), *weather*(*oxford*, *rainy*) $\mapsto$ 0.2, 0.4, 0.3. A mapping rule such as

$$candidate\_destination(L) \leftarrow weather(L, sunny)$$

can now express the fact that a destination is a candidate for a day-trip if it has sunny weather. While the mapping is purely deterministic, the probability distributions on the sets of atomic choices of the choice space enforce, by virtue of the mapping, a probability distribution on the ground facts of the global schema. Our framework is able to capture this situation, providing a framework for query answering over uncertain data.

6   Conclusion

We have considered tractable probabilistic dl-programs for the Semantic Web, which combine tractable description logics, normal programs under the answer set and the well-founded semantics, and probabilities. Based on the results of [19], we have introduced a framework to model and represent data integration systems for the Semantic Web, which are capable of taking into account uncertainty in the mappings and the data sources. We have shown that probabilistic dl-programs are capable of modeling probabilistic mappings, and also data where for each tuple a probability is defined.

Our future research will focus on considering more expressive forms of rules, and in particular rules that have existentially quantified variables in the head, similarly to tuple-generating dependencies in database theory. Furthermore, we plan to develop top-k query techniques for the presented framework, which is especially important in the case of integrating large sets of data in the Semantic Web.

Acknowledgments. Andrea Calì was supported by the Engineering and Physical Sciences Research Council (EPSRC) project "Schema Mappings and Automated Services for Data Integration and Exchange" (EP/E010865/1). Thomas Lukasiewicz was supported by the German Research Foundation (DFG) under the Heisenberg Programme and by the Austrian Science Fund (FWF) under the project P18146-N04. We thank the reviewers of this paper and its URSW-2006 position version for their constructive and useful comments, which helped to improve this work.

References

1. Berners-Lee, T.: Weaving the Web, Harper, San Francisco, CA (1999)
2. W3C: OWL Web Ontology Language Overview, W3C Recommendation (February 10, 2004), http://www.w3.org/TR/2004/REC-owl-features-20040210/
3. van Keulen, M., de Keijzer, A., Alink, W.: A probabilistic XML approach to data integration. In: Proceedings ICDE 2005, pp. 459–470. IEEE Computer Society, Los Alamitos (2005)
4. Pan, R., Ding, Z., Yu, Y., Peng, Y.: A Bayesian network approach to ontology mapping. In: Gil, Y., Motta, E., Benjamins, V.R., Musen, M.A. (eds.) ISWC 2005. LNCS, vol. 3729, pp. 563–577. Springer, Heidelberg (2005)

5. Horrocks, I., Patel-Schneider, P.F.: Reducing OWL entailment to description logic satisfiability. In: Fensel, D., Sycara, K.P., Mylopoulos, J. (eds.) ISWC 2003. LNCS, vol. 2870, pp. 17–29. Springer, Heidelberg (2003)

6. Eiter, T., Ianni, G., Lukasiewicz, T., Schindlauer, R., Tompits, H.: Combining answer set programming with description logics for the Semantic Web. Artif. Intell. 172(12/13), 1495–1539 (2008)

7. Eiter, T., Lukasiewicz, T., Schindlauer, R., Tompits, H.: Well-founded semantics for description logic programs in the Semantic Web. In: Antoniou, G., Boley, H. (eds.) RuleML 2004. LNCS, vol. 3323, pp. 81–97. Springer, Heidelberg (2004)

8. Lukasiewicz, T.: A novel combination of answer set programming with description logics for the Semantic Web. In: Franconi, E., Kifer, M., May, W. (eds.) ESWC 2007. LNCS, vol. 4519, pp. 384–398. Springer, Heidelberg (2007)

9. Giugno, R., Lukasiewicz, T.: P-$\mathcal{SHOQ}$(D): A probabilistic extension of $\mathcal{SHOQ}$(D) for probabilistic ontologies in the Semantic Web. In: Flesca, S., Greco, S., Leone, N., Ianni, G. (eds.) JELIA 2002. LNCS, vol. 2424, pp. 86–97. Springer, Heidelberg (2002)

10. Lukasiewicz, T.: Expressive probabilistic description logics. Artif. Intell. 172(6/7), 852–883 (2008)

11. da Costa, P.C.G.: Bayesian Semantics for the Semantic Web. PhD thesis, George Mason University, Fairfax, VA, USA (2005)

12. da Costa, P.C.G., Laskey, K.B.: PR-OWL: A framework for probabilistic ontologies. In: Proceedings FOIS 2006, pp. 237–249. IOS Press, Amsterdam (2006)

13. Udrea, O., Deng, Y., Hung, E., Subrahmanian, V.S.: Probabilistic ontologies and relational databases. In: Meersman, R., Tari, Z. (eds.) OTM 2005. LNCS, vol. 3760, pp. 1–17. Springer, Heidelberg (2005)

14. Lukasiewicz, T.: Probabilistic description logic programs. Int. J. Approx. Reasoning 45(2), 288–307 (2007)

15. Calì, A., Lukasiewicz, T.: Tightly integrated probabilistic description logic programs for the Semantic Web. In: Dahl, V., Niemelä, I. (eds.) ICLP 2007. LNCS, vol. 4670, pp. 428–429. Springer, Heidelberg (2007)

16. Baral, C., Gelfond, M., Rushton, J.N.: Probabilistic reasoning with answer sets. In: Lifschitz, V., Niemelä, I. (eds.) LPNMR 2004. LNCS, vol. 2923, pp. 21–33. Springer, Heidelberg (2003)

17. Kersting, K., De Raedt, L.: Bayesian logic programs. CoRR cs.AI/0111058 (2001)

18. Poole, D.: The independent choice logic for modelling multiple agents under uncertainty. Artif. Intell. 94(1–2), 7–56 (1997)

19. Lukasiewicz, T.: Tractable probabilistic description logic programs. In: Prade, H., Subrahmanian, V.S. (eds.) SUM 2007. LNCS, vol. 4772, pp. 143–156. Springer, Heidelberg (2007)

20. Calvanese, D., De Giacomo, G., Lembo, D., Lenzerini, M., Rosati, R.: Data complexity of query answering in description logics. In: Proceedings KR 2006, pp. 260–270. AAAI Press, Menlo Park (2006)

21. Gelfond, M., Lifschitz, V.: Classical negation in logic programs and deductive databases. New Generation Computing 17, 365–387 (1991)

22. Lenzerini, M.: Data integration: A theoretical perspective. In: Proceedings PODS 2002, pp. 233–246. ACM Press, New York (2002)

23. Friedman, M., Levy, A.Y., Millstein, T.D.: Navigational plans for data integration. In: Proceedings AAAI 1999, pp. 67–73. AAAI Press/ MIT Press (1999)

24. Calì, A.: Reasoning in data integration systems: Why LAV and GAV are siblings. In: Zhong, N., Raś, Z.W., Tsumoto, S., Suzuki, E. (eds.) ISMIS 2003. LNCS, vol. 2871, pp. 562–571. Springer, Heidelberg (2003)

25. Dalvi, N.N., Suciu, D.: Management of probabilistic data: Foundations and challenges. In: Proceedings PODS 2007, pp. 1–12. ACM Press, New York (2007)

Rule-Based Approaches for Representing Probabilistic Ontology Mappings[*]

Andrea Calì[1,2], Thomas Lukasiewicz[2,**], Livia Predoiu[3],
and Heiner Stuckenschmidt[3]

[1] Oxford-Man Institute of Quantitative Finance, University of Oxford
Blue Boar Court, 9 Alfred Street, Oxford OX1 4EH, UK
andrea.cali@oxford-man.ox.ac.uk
[2] Computing Laboratory, University of Oxford
Wolfson Building, Parks Road, Oxford OX1 3QD, UK
{andrea.cali,thomas.lukasiewicz}@comlab.ox.ac.uk
[3] Institut für Informatik, Universität Mannheim
68159 Mannheim, Germany
{livia,heiner}@informatik.uni-mannheim.de

Abstract. Using mappings between ontologies is a common way of approaching the semantic heterogeneity problem on the Semantic Web. To fit into the landscape of Semantic Web languages, a suitable logic-based representation formalism for mappings is needed, which allows to reason with ontologies and mappings in an integrated manner, and to deal with uncertainty and inconsistencies in automatically created mappings. We analyze the requirements for such a formalism, and propose to use frameworks that integrate description logic ontologies with probabilistic rules. We compare two such frameworks and show the advantages of using the probabilistic extensions of their deterministic counterparts. The two frameworks that we compare are tightly coupled probabilistic dl-programs, which tightly combine the description logics behind OWL DL resp. OWL Lite, disjunctive logic programs under the answer set semantics, and Bayesian probabilities, on the one hand, and generalized Bayesian dl-programs, which tightly combine the DLP-fragment of OWL Lite with Datalog (without negation and equality) based on the semantics of Bayesian networks, on the other hand.

Keyword: Representing probabilistic ontology mappings, rule languages, Semantic Web, uncertainty, inconsistency, probabilistic description logic programs, description logics, disjunctive logic programs, answer set semantics, Bayesian probabilities, Bayesian description logic programs, Datalog, Bayesian networks.

1 Introduction

The problem of aligning heterogeneous ontologies via semantic mappings has been identified as one of the major challenges of Semantic Web technologies. In order to ad-

---

[*] This paper is a significantly extended and revised version of a paper that appeared in: *Proceedings URSW-2007. CEUR Workshop Proceedings* 327, CEUR-WS.org, 2008.

[**] Alternative address: Institut für Informationssysteme, Technische Universität Wien, Favoritenstraße 9-11, 1040 Wien, Austria; email: lukasiewicz@kr.tuwien.ac.at.

P.C.G. da Costa et al. (Eds.): URSW 2005-2007, LNAI 5327, pp. 66–87, 2008.

dress this problem, a number of languages for representing semantic relations between elements in different ontologies as a basis for reasoning and query answering across multiple ontologies have been proposed [1]. In the presence of real world ontologies, it is unrealistic to assume that mappings between ontologies are created manually by domain experts, since existing ontologies (e.g., in the area of medicine) contain thousands of concepts and hundreds of relations. Recently, a number of heuristic methods for matching elements from different ontologies have been proposed that support the creation of mappings between different languages by suggesting candidate mappings (e.g., [2]). These methods rely on linguistic and structural criteria. The resulting mapping either contains a fair amount of errors or only covers a small part of the ontologies involved [3,4]. To leverage the weaknesses of the individual methods, it is common practice to combine the results of a number of matching components or even the results of different matching systems to achieve a better coverage of the problem [2].

This means that automatically created mappings often contain uncertain hypotheses and errors that need to be dealt with, as briefly summarized as follows:

- mapping hypotheses are often oversimplifying, since most matchers only support very simple semantic relations (mostly equivalence between individual elements);
- there may be conflicts between different hypotheses for semantic relations from different matching components and often even from the same matcher;
- semantic relations are only given with a degree of confidence in their correctness.

If we want to use the resulting mapping, we have to find a way to deal with these uncertainties and errors in a suitable way. We argue that the most suitable way of dealing with uncertainties in mappings is to provide means to explicitly represent uncertainties in the target language that encodes the mappings. In this paper, we address the problem of designing a mapping representation language that is capable of representing the kinds of uncertainty mentioned above. We propose two approaches to such a language, which are based on an integration of ontologies and rules under probabilistic uncertainty, and compare them regarding the necessary representation requirements.

We choose rules for the representation of mappings, because they are an intuitive means for this task. Thinking in if-then statements (e.g., if an instance belongs to a certain concept in ontology O_1, then it belongs to a certain concept in ontology O_2) is very straight-forward and easily comprehensible also to people with few background in logics. Furthermore, reasoning with rules has the advantage that instance retrieval can generally be performed more efficiently than with description logics. Another advantage of rule languages is that they allow to formulate meta-modeling statements while with description logics this is generally not possible. In this way, the distinction between concepts and their instances is flattened, that is, instances can also be concepts at the same time and vice versa [5]. As we want to use a rule language for the representation of mappings, we need a language that provides a tight integration of a rule language and a description logic on the formal level.

There is a large body of work on rules for the Semantic Web and on integrating ontologies and rules for the Semantic Web; see especially [6] and [7], respectively, for an overview. Here, we consider two frameworks, namely, (i) tightly coupled dl-programs [8], which integrate the description logics behind OWL DL resp. OWL Lite and disjunctive logic programs under the answer set semantics, and (ii) generalized dl-programs, which are an integration of the DLP-fragment of description logics [9] and Datalog. For both formalisms, we provide a formal representation and show how mappings can be represented in these frameworks. Note that both formalisms are decidable and allow the arbitrary usage of description logic concepts and roles in the rules component.

Other works explore formalisms for *uncertainty reasoning in the Semantic Web* (an important recent forum for approaches to uncertainty in the Semantic Web is the annual *Workshop on Uncertainty Reasoning for the Semantic Web (URSW)*; there also exists a W3C Incubator Group on *Uncertainty Reasoning for the World Wide Web*). There are especially probabilistic extensions of description logics [10], Web ontology languages [11,12], and description logic programs [13] (to encode ambiguous information, such as "John is a student with the probability 0.7 and a teacher with the probability 0.3", which is very different from vague/fuzzy information, such as "John is tall with the degree of truth 0.7"). However, to our knowledge, none of these formalisms have been used for representing uncertain mappings. We provide probabilistic extensions of the above-mentioned tightly coupled dl-programs and generalized dl-programs, which are based on general Bayesian probabilities and Bayesian networks [14], respectively.

The main contributions of this paper can be briefly summarized as follows.

- We show how tightly coupled dl-programs can be used for representing and reasoning with ontologies and deterministic mappings between ontologies. We introduce tightly coupled probabilistic dl-programs, and show how they can be used for representing and reasoning with ontologies and uncertain mappings between them.
- We introduce generalized dl-programs and generalized Bayesian dl-programs, and show how they can be used for representing and reasoning with ontologies and deterministic and uncertain mappings between ontologies, respectively.
- We give a detailed comparison of the features of the two deterministic and the two probabilistic formalisms with respect to representing and reasoning with ontologies and deterministic resp. uncertain ontology mappings in the Semantic Web.

The rest of this paper is structured as follows. In Section 2, we define the requirements that a formal language has to fulfill for representing mappings between ontologies in the Semantic Web. In Section 3, we recall the description logics behind OWL DL and OWL Lite as well as the DLP-fragment [9]. We also provide an example scenario consisting of two ontologies. In Section 4, we present tightly coupled dl-programs and generalized dl-programs, and show how they can be used for representing deterministic mappings between ontologies. In Section 5, we present tightly coupled probabilistic dl-programs and generalized Bayesian dl-programs. We also provide an example scenario, and show how uncertain mappings can be represented in both formalisms and how reasoning can be performed. We finally conclude with Section 6, where we discuss the representation features of both formalisms and give an outlook on future research.

2   Representation Requirements

The problem of ontology matching can be defined as follows [2]. Ontologies are theories encoded in a certain language L. In this work, we assume that ontologies are encoded in OWL DL, OWL Lite, or the DLP-fragment of OWL Lite. For each ontology O

in language L, we denote by $Q(O)$ the matchable elements of the ontology O. Given two ontologies O and O', the task of matching is now to determine correspondences between the matchable elements in the two ontologies. In general, correspondences are 5-tuples (id, e, e', r, n) such that

- id is a unique identifier for referring to the correspondence;
- $e \in Q(O)$ and $e' \in Q(O')$ are matchable elements from the two ontologies;
- $r \in R$ is a semantic relation;
- n is a degree of confidence in the correctness of the correspondence.

In this paper, we only consider semantic relations r which can be interpreted as an implication. We consider two formal languages for representing and combining correspondences that are produced by different matching components or systems. From the above general description of automatically generated correspondences between ontologies, we can derive a number of requirements for such a formal language for representing the results of multiple matchers as well as the contained uncertainties:

- *Tight integration of mapping and ontology language:* The semantics of the language used to represent the correspondences between elements in different ontologies has to be tightly integrated with the semantics of the ontology language used (here OWL). This is important if we want to use the correspondences to reason across different ontologies in a semantically coherent way. This also means that the interpretation of the mapped elements depends on the definitions in the ontologies. Failing this requirement comes along with not that nice semantic properties.
- *Support for mappings refinement:* The language should be expressive enough to allow the user to refine oversimplifying correspondences from the matching system. This is important to be able to provide a precise account of the true semantic relation between elements in the mapped ontologies. In particular, this requires the ability to describe correspondences that include several elements from the two ontologies.
- *Support for repairing inconsistencies:* Inconsistent mappings are a major problem for the combined use of ontologies because they can cause inconsistencies in the mapped ontologies, which can make logical reasoning impossible, since everything can be derived from an inconsistent ontology. The mapping language should be able to represent and reason about inconsistent mappings in an approximate fashion.
- *Representation and combination of confidence:* The confidence values provided by matching systems is an important indicator for the uncertainty to be taken into account. The mapping language should be able to use these confidence values when reasoning with mappings. In particular, it should be able to represent the confidence in a mapping rule and to combine confidence values on a sound formal basis.
- *Decidability and efficiency of instance reasoning:* An important use of ontology mappings is the exchange of data across different ontologies. In particular, we normally want to be able to ask queries using the vocabulary of one ontology and receive answers that do not only consist of instances of this ontology but also of ontologies connected through ontology mappings. To support this, query answering in the combined formalism consisting of ontology language and mapping language has to be decidable. Furthermore, to be able to handle large amounts of data in the Semantic Web, there should be efficient algorithms for answering queries.

In the following, we consider two different logic formalisms and investigate their ability to fulfill the above representation requirements.

3 Representing Ontologies

In this section, we first recall the two expressive description logics $\mathcal{SHOIN}(\mathbf{D})$ and $\mathcal{SHIF}(\mathbf{D})$ as well as the DLP-fragment of $\mathcal{SHIF}(\mathbf{D})$. We then illustrate along an example scenario how they are used to represent ontologies.

3.1 The Description Logics $\mathcal{SHOIN}(\mathbf{D})$ and $\mathcal{SHIF}(\mathbf{D})$

The description logics $\mathcal{SHOIN}(\mathbf{D})$ and $\mathcal{SHIF}(\mathbf{D})$ are important in the Semantic Web, since they stand behind the Web ontology languages OWL DL and OWL Lite [15], respectively. Intuitively, description logics model a domain of interest in terms of concepts and roles, which represent classes of individuals and binary relations between classes of individuals, respectively. A description logic knowledge base encodes especially subset relationships between concepts and between roles, and the membership of individuals to concepts and of pairs of individuals to roles.

Syntax. We first describe the syntax of $\mathcal{SHOIN}(\mathbf{D})$. We assume a set of *elementary datatypes* and a set of *data values*. A *datatype* is either an elementary datatype or a set of data values (*datatype oneOf*). A *datatype theory* $\mathbf{D} = (\Delta^{\mathbf{D}}, \cdot^{\mathbf{D}})$ consists of a *datatype domain* $\Delta^{\mathbf{D}}$ and a mapping $\cdot^{\mathbf{D}}$ that assigns to each elementary datatype a subset of $\Delta^{\mathbf{D}}$ and to each data value an element of $\Delta^{\mathbf{D}}$. The mapping $\cdot^{\mathbf{D}}$ is extended to all datatypes by $\{v_1, \ldots\}^{\mathbf{D}} = \{v_1^{\mathbf{D}}, \ldots\}$. Let $\mathbf{A}$, $\mathbf{R}_A$, $\mathbf{R}_D$, and $\mathbf{I}$ be pairwise disjoint (denumerable) sets of *atomic concepts*, *abstract roles*, *datatype roles*, and *individuals*, respectively. We denote by $\mathbf{R}_A^-$ the set of *inverses* R^- of all $R \in \mathbf{R}_A$.

A *role* is any element of $\mathbf{R}_A \cup \mathbf{R}_A^- \cup \mathbf{R}_D$. *Concepts* are inductively defined as follows. Every $\phi \in \mathbf{A}$ is a concept, and if $o_1, \ldots, o_n \in \mathbf{I}$, then $\{o_1, \ldots, o_n\}$ is a concept (*oneOf*). If ϕ, ϕ_1, and ϕ_2 are concepts and if $R \in \mathbf{R}_A \cup \mathbf{R}_A^-$, then also $(\phi_1 \sqcap \phi_2)$, $(\phi_1 \sqcup \phi_2)$, and $\neg \phi$ are concepts (*conjunction, disjunction*, and *negation*, respectively), as well as $\exists R.\phi$, $\forall R.\phi$, $\geqslant nR$, and $\leqslant nR$ (*existential, value, atleast*, and *atmost restriction*, respectively) for an integer $n \geqslant 0$. If D is a datatype and $U \in \mathbf{R}_D$, then $\exists U.D$, $\forall U.D$, $\geqslant nU$, and $\leqslant nU$ are concepts (*datatype existential, value, atleast*, and *atmost restriction*, respectively) for an integer $n \geqslant 0$. We write $\top$ and $\bot$ to abbreviate the concepts $\phi \sqcup \neg\phi$ and $\phi \sqcap \neg\phi$, respectively, and we eliminate parentheses as usual.

An *axiom* has one of the following forms: (1) $\phi \sqsubseteq \psi$ (*concept inclusion axiom*), where ϕ and ψ are concepts; (2) $R \sqsubseteq S$ (*role inclusion axiom*), where either $R, S \in \mathbf{R}_A \cup \mathbf{R}_A^-$ or $R, S \in \mathbf{R}_D$; (3) $\mathrm{Trans}(R)$ (*transitivity axiom*), where $R \in \mathbf{R}_A$; (4) $\phi(a)$ (*concept membership axiom*), where ϕ is a concept and $a \in \mathbf{I}$; (5) $R(a, b)$ (resp., $U(a, v)$) (*role membership axiom*), where $R \in \mathbf{R}_A$ (resp., $U \in \mathbf{R}_D$) and $a, b \in \mathbf{I}$ (resp., $a \in \mathbf{I}$ and v is a data value); and (6) $a = b$ (resp., $a \neq b$) (*equality* (resp., *inequality*) *axiom*), where $a, b \in \mathbf{I}$. Axioms of the form (1)–(3) (resp., (4) and (5)) are also called *TBox* (resp., *ABox*) axioms. A *(description logic) knowledge base* L is a finite set of axioms. For decidability, number restrictions in L are restricted to simple abstract roles [16].

The syntax of $\mathcal{SHIF}(\mathbf{D})$ is as the above syntax of $\mathcal{SHOIN}(\mathbf{D})$, but without the oneOf constructor and with the atleast and atmost constructors limited to 0 and 1.

Semantics. An *interpretation* $\mathcal{I} = (\Delta^{\mathcal{I}}, \cdot^{\mathcal{I}})$ relative to a datatype theory $\mathbf{D} = (\Delta^{\mathbf{D}}, \cdot^{\mathbf{D}})$ consists of a nonempty (*abstract*) *domain* $\Delta^{\mathcal{I}}$ disjoint from $\Delta^{\mathbf{D}}$, and a mapping $\cdot^{\mathcal{I}}$ that assigns to each atomic concept $\phi \in \mathbf{A}$ a subset of $\Delta^{\mathcal{I}}$, to each individual $o \in \mathbf{I}$ an element of $\Delta^{\mathcal{I}}$, to each abstract role $R \in \mathbf{R}_A$ a subset of $\Delta^{\mathcal{I}} \times \Delta^{\mathcal{I}}$, and to each datatype role $U \in \mathbf{R}_D$ a subset of $\Delta^{\mathcal{I}} \times \Delta^{\mathbf{D}}$. We extend $\cdot^{\mathcal{I}}$ to all concepts and roles, and we define the *satisfaction* of an axiom F in an interpretation $\mathcal{I} = (\Delta^{\mathcal{I}}, \cdot^{\mathcal{I}})$, denoted $\mathcal{I} \models F$, as usual [15]. We say $\mathcal{I}$ *satisfies* the axiom F, or $\mathcal{I}$ is a *model* of F, iff $\mathcal{I} \models F$. We say $\mathcal{I}$ *satisfies* a knowledge base L, or $\mathcal{I}$ is a *model* of L, denoted $\mathcal{I} \models L$, iff $\mathcal{I} \models F$ for all $F \in L$. We say L is *satisfiable* iff L has a model. An axiom F is a *logical consequence* of L, denoted $L \models F$, iff every model of L satisfies F.

3.2   The DLP-Fragment of $\mathcal{SHIF}(\mathbf{D})$

The *description logic programming fragment* (or *DLP-fragment*) of $\mathcal{SHIF}(\mathbf{D})$ [9] lies in the expressive intersection of description logics and logic programs. Hence, it is possible to translate an ontology in the DLP-fragment into a logic program and vice versa without loss of declarative semantics. This process, that is, the bidirectional translation from the description logic syntax to the logic programming syntax and vice versa, has been called *DLP-fusion* [9]. It provides a basis for achieving interoperability.

We now describe the restrictions that are imposed on $\mathcal{SHIF}$ to obtain the DLP-fragment. First, negation, equality, inequality, the atleast constructor, and the atmost constructor are disallowed. Note that some combinations of the atleast and the atmost constructors in $\mathcal{SHIF}$ can be modeled by other constructors. Another restriction on $\mathcal{SHIF}$ is based on a distinction between the *body* ϕ and the *head* ψ of concept inclusion axioms $\phi \sqsubseteq \psi$: the head is constructed from atomic concepts via conjunctions and value restrictions, while the body is constructed from atomic concepts via conjunctions, disjunctions, and existential restrictions. In addition, one allows concept inclusion axioms $\top \sqsubseteq \forall R.\psi$, where $R \in \mathbf{R}_A \cup \mathbf{R}_A^-$, and ψ is a head concept. Furthermore, one allows only concept membership axioms $\psi(a)$, where ψ is a head concept.

Although the DLP-fragment has a restricted expressivity, it has several advantages. In particular, a large amount of existing ontologies lie within the DLP-fragment. Moreover, reasoning in the DLP-fragment is not only decidable, but also has a much lower complexity than reasoning in $\mathcal{SHIF}(\mathbf{D})$ and $\mathcal{SHOIN}(\mathbf{D})$ in theory and practice.

3.3   Example Scenario

We consider a retrieval scenario where two different peers provide information about publications based on two different bibliographic ontologies O_1 and O_2.

Ontology O_1: (1) technical reports have topics as keywords and persons as authors; (2) publications are not technical reports; (3) every book is a publication; (4) every article is a publication; (5) every collection is a publication; (6) a publication is either a book or an article or a collection; (7) books are not articles; (8) books are not collections; (9) articles are not collections; (10) publications have topics as keywords and authors as publications. These relationships are expressed by the following TBox axioms:

(1) $Technical\_Report \sqsubseteq \forall keyword.Topic \sqcap \forall author.Person;$
(2) $Publication \sqsubseteq \neg Technical\_Report;$
(3) $Book \sqsubseteq Publication;$ (4) $Article \sqsubseteq Publication;$ (5) $Collection \sqsubseteq Publication;$
(6) $Publication \sqsubseteq Book \sqcup Article \sqcup Collection;$ (7) $Book \sqsubseteq \neg Article;$
(8) $Book \sqsubseteq \neg Collection;$ (9) $Article \sqsubseteq \neg Collection;$
(10) $Publication \sqsubseteq \forall keyword.Topic \sqcap \forall author.Person.$

Note that the ontology O_1' obtained from O_1 by removing the axioms (2) and (6)–(9) lies in the DLP-fragment. The ABox shown below belongs to O_1 and O_1':

(11) $Book(b_1);$ (12) $Article(a_1);$ (13) $Collection(c_1);$ (14) $Technical\_Report(t_1);$
(15) $keyword(b_1, artificial\_intelligence);$ $keyword(a_1, artificial\_intelligence);$
(16) $keyword(c_1, software\_engineering);$ $keyword(t_1, artificial\_intelligence);$
(17) $author(b_1, John);$ $author(a_1, Paul);$ $author(b_1, Michael);$ $author(t_1, Peter).$

Ontology O_2: (1) every paper is a publication; (2) every proceedings is a publication; (3) papers are not proceedings; (4) the role *includes* relates proceedings with papers; (5) proceedings include at least 5 different entities (that is, papers); (6) publications are not publishers; (7) publications are not persons; (8) publications are not subjects; (9) persons are not subjects; (10) persons are not publishers; (11) the role *published_by* relates publications with publishers; (12) the role *about* relates publications with subjects; (13) the role *author* relates publications with persons. These relationships are expressed by the following TBox axioms:

(1) $Paper \sqsubseteq Publication;$ (2) $Proceedings \sqsubseteq Publication;$ (3) $Paper \sqsubseteq \neg Proceedings;$
(4) $\top \sqsubseteq \forall includes.Paper;$ $\top \sqsubseteq \forall includes^{-1}.Proceedings;$
(5) $Proceedings \sqsubseteq \geqslant 5\ includes;$
(6) $Publication \sqsubseteq \neg Publisher;$ (7) $Publication \sqsubseteq \neg Person;$
(8) $Publication \sqsubseteq \neg Subject;$ (9) $Person \sqsubseteq \neg Subject;$ (10) $Person \sqsubseteq \neg Publisher;$
(11) $\top \sqsubseteq \forall published\_by.Publisher;$ $\top \sqsubseteq \forall published\_by^{-1}.Publication;$
(12) $\top \sqsubseteq \forall about.Subject;$ $\top \sqsubseteq \forall about^{-1}.Publication;$
(13) $\top \sqsubseteq \forall author.Person;$ $\top \sqsubseteq \forall author^{-1}.Publication.$

Note again that the ontology O_2' obtained from O_2 by removing the axioms (3) and (5)–(10) lies in the DLP-fragment. The ABox shown below belongs to O_2 and O_2':

(14) $includes(proc_1, p_1);$
(15) $about(p_1, artificial\_intelligence),$ $about(p_2, artificial\_intelligence);$
(16) $author(p_1, Mary),$ $author(p_2, Elizabeth);$
(17) $published\_by(p_1, Springer);$ $published\_by(proc_1, Springer);$
(18) $paper(p_2).$

Since the above two ontologies describe overlapping domains, a user may want to query the information represented by both of them in an integrated manner. For example, a user may be looking for all technical reports in both ontologies, or a user may be interested in all publications about artificial intelligence in both ontologies. In the following, we investigate two formal information integration frameworks that allow dealing with ontologies and with mappings encoded in a probabilistic rule language at the same time. At first, however, in Section 4, we look at their deterministic variants.

4 Representing Deterministic Ontology Mappings

The integration task is stated as follows. Let $O = O_1 \cup O_2$ be the union of two ontologies O_1 and O_2 with overlapping domains. Let O_1, O_2, and O be encoded in the description logics L_1, L_2, and $L = L_1 \cup L_1$, respectively. Let the logic program P represent a set of deterministic mappings. How can we then reason with O and P such that queries reflect the integrated knowledge? In the following, we present two formal frameworks that aim at integrating description logic ontologies with mapping rules.

For this purpose, we assume a first-order vocabulary Φ with finite nonempty sets of constant and predicate symbols, but no function symbols. We use Φ_c to denote the set of all constant symbols in Φ. We also assume a set of data values $\mathbf{V}$ (relative to a datatype theory $\mathbf{D} = (\Delta^{\mathbf{D}}, \cdot^{\mathbf{D}})$) and pairwise disjoint (denumerable) sets $\mathbf{A}$, $\mathbf{R}_A$, $\mathbf{R}_D$, and $\mathbf{I}$ of atomic concepts, abstract roles, datatype roles, and individuals, respectively, as in Section 3. We assume that (i) Φ_c is a subset of $\mathbf{I} \cup \mathbf{V}$, and that (ii) Φ and $\mathbf{A}$ (resp., $\mathbf{R}_A \cup \mathbf{R}_D$) may have unary (resp., binary) predicate symbols in common. Let $\mathcal{X}$ be a set of variables. A *term* is either a variable from $\mathcal{X}$ or a constant symbol from Φ. An *atom* is of the form $p(t_1, \ldots, t_n)$, where p is a predicate symbol of arity $n \geqslant 0$ from Φ, and $t_1, \ldots, t_n$ are terms. Such an atom is *ground* iff $t_1, \ldots, t_n$ are constant symbols.

4.1 Tightly Coupled DL-Programs

We now recall the *tightly coupled* approach to *disjunctive description logic programs* (or *tightly coupled dl-programs*) $KB = (L, P)$ under the answer set semantics from [8], where KB consists of a description logic knowledge base L and a disjunctive logic program P. Their semantics is defined in a modular way as in [7], but it allows for a much tighter integration of L and P. Note that we do not assume any structural separation between the vocabularies of L and P. The main idea behind the semantics is to interpret P relative to Herbrand interpretations that are compatible with L, while L is interpreted relative to general first-order interpretations. Thus, we modularly combine the standard semantics of logic programs and of description logics, which allows for building on the standard techniques and results of both areas. As another advantage, the novel dl-programs are decidable, even when their components of logic programs and description logic knowledge bases are both very expressive. See especially [8] for further details on the novel approach to dl-programs and for a comparison to related works.

Syntax. A *literal* l is an atom p or a default-negated atom *not* p. A *disjunctive rule* (or simply *rule*) r is an expression of the form

$$\alpha_1 \vee \cdots \vee \alpha_k \leftarrow \beta_1, \ldots, \beta_n, not\ \beta_{n+1}, \ldots, not\ \beta_{n+m}, \qquad (1)$$

where $\alpha_1, \ldots, \alpha_k, \beta_1, \ldots, \beta_{n+m}$ are atoms and $k, m, n \geqslant 0$. We call $\alpha_1 \vee \cdots \vee \alpha_k$ the *head* of r, while the conjunction $\beta_1, \ldots, \beta_n, not\ \beta_{n+1}, \ldots, not\ \beta_{n+m}$ is its *body*. We define $H(r) = \{\alpha_1, \ldots, \alpha_k\}$ and $B(r) = B^+(r) \cup B^-(r)$, where $B^+(r) = \{\beta_1, \ldots, \beta_n\}$ and $B^-(r) = \{\beta_{n+1}, \ldots, \beta_{n+m}\}$. A *disjunctive program* P is a finite set of disjunctive rules of the form (1). We say P is *positive* iff $m = 0$ for all disjunctive rules (1) in P. We say P is a *normal program* iff $k \leqslant 1$ for all disjunctive rules (1) in P.

A *tightly coupled disjunctive description logic program* (or simply *tightly coupled dl-program*) $KB = (L, P)$ consists of a description logic knowledge base L and a disjunctive program P. It is *positive* (resp., *normal*) iff P is positive (resp., normal).

Semantics. We now define the answer set semantics of tightly coupled dl-programs as a generalization of the answer set semantics of ordinary disjunctive logic programs. In the sequel, let $KB = (L, P)$ be a tightly coupled dl-program.

A *ground instance* of a rule $r \in P$ is obtained from r by replacing every variable that occurs in r by a constant symbol from Φ_c. We denote by $ground(P)$ the set of all ground instances of rules in P. The *Herbrand base* relative to Φ, denoted HB_Φ, is the set of all ground atoms constructed with constant and predicate symbols from Φ. We use DL_Φ to denote the set of all ground atoms in HB_Φ that are constructed from atomic concepts in $\mathbf{A}$, abstract roles in $\mathbf{R}_A$, and datatype roles in $\mathbf{R}_D$.

An *interpretation* I is any subset of HB_Φ. Informally, every such I represents the Herbrand interpretation in which all $a \in I$ (resp., $a \in HB_\Phi - I$) are true (resp., false). We say an interpretation I is a *model* of a description logic knowledge base L, denoted $I \models L$, iff $L \cup I \cup \{\neg a \mid a \in HB_\Phi - I\}$ is satisfiable. We say I is a *model* of a ground atom $a \in HB_\Phi$, or I *satisfies* a, denoted $I \models a$, iff $a \in I$. We say I is a *model* of a ground rule r, denoted $I \models r$, iff $I \models \alpha$ for some $\alpha \in H(r)$ whenever $I \models B(r)$, that is, $I \models \beta$ for all $\beta \in B^+(r)$ and $I \not\models \beta$ for all $\beta \in B^-(r)$. We say I is a *model* of a set of rules P iff $I \models r$ for every $r \in ground(P)$. We say I is a *model* of a tightly coupled dl-program $KB = (L, P)$, denoted $I \models KB$, iff I is a model of both L and P.

We now define the answer set semantics of tightly coupled dl-programs by generalizing the ordinary answer set semantics of disjunctive logic programs. We generalize the definition via the FLP-reduct [17] (which coincides with the answer set semantics defined via the Gelfond-Lifschitz reduct [18]). Given a dl-program $KB = (L, P)$, the *FLP-reduct* of KB relative to an interpretation $I \subseteq HB_\Phi$, denoted KB^I, is the dl-program (L, P^I), where P^I is the set of all $r \in ground(P)$ such that $I \models B(r)$. An interpretation $I \subseteq HB_\Phi$ is an *answer set* of KB iff I is a minimal model of KB^I. A dl-program KB is *consistent* (resp., *inconsistent*) iff it has an (resp., no) answer set.

We finally define the notions of *cautious* (resp., *brave*) *reasoning* from tightly coupled dl-programs under the answer set semantics as follows. A ground atom $a \in HB_\Phi$ is a *cautious* (resp., *brave*) *consequence* of a tightly coupled dl-program KB under the answer set semantics iff every (resp., some) answer set of KB satisfies a.

4.2   Generalized DL-Programs

We next present generalized description logic programs (or generalized dl-programs), which informally generalize the DLP-fragment of $\mathcal{SHIF}(\mathbf{D})$ by Datalog rules.

Syntax. A *generalized description logic program* (or *generalized dl-program*) $KB = (L, P)$ consists of a knowledge base L in the DLP-fragment of $\mathcal{SHIF}(\mathbf{D})$ and a logic program P in Datalog (without negation), which consist of a set of *rules* r of the form

$$h \leftarrow \beta_1, \ldots, \beta_m, \tag{2}$$

where $h, \beta_1, \ldots, \beta_m$ are atoms and $m \geqslant 0$. We call the atom h (resp., the conjunction $\beta_1, \ldots, \beta_m$) the *head* (resp., *body*) of r. Note that all the variables in r are implicitly universally quantified. We say r is a *fact* iff $i = 0$ and h is ground.

Semantics. We interpret generalized dl-programs as Datalog programs, via the translation of L into its Datalog equivalent (as L lies in the intersection of $\mathcal{SHIF}(\mathbf{D})$ and Datalog). Due to the absence of negation, this also corresponds to the semantics of first-order logics. Note that when L is translated into Datalog, we only obtain 2-ary predicates and the variable graph of the body of each rule is connected and acyclic.

4.3 Example Scenario Cont'd

We now compare the two formalisms above regarding the representation requirements stated in Section 2. For this purpose, we first show how tightly coupled and generalized dl-programs $KB = (L, P)$ can be used for representing (possibly inconsistent) mappings (without confidence values) between two ontologies. Intuitively, L encodes the union of the two ontologies, while P encodes the mappings between the ontologies.

Tightly coupled and generalized dl-programs $KB = (L, P)$ naturally represent two heterogeneous ontologies O_1 and O_2, and mappings between O_1 and O_2 as follows. The description logic knowledge base L is the union of two independent description logic knowledge bases L_1 and L_2, which encode the ontologies O_1 and O_2, respectively. Here, we assume that L_1 and L_2 have signatures $\mathbf{A}_1, \mathbf{R}_{A,1}, \mathbf{R}_{D,1}, \mathbf{I}_1$ and $\mathbf{A}_2, \mathbf{R}_{A,2}, \mathbf{R}_{D,2}, \mathbf{I}_2$, respectively, such that $\mathbf{A}_1 \cap \mathbf{A}_2 = \emptyset$, $\mathbf{R}_{A,1} \cap \mathbf{R}_{A,2} = \emptyset$, $\mathbf{R}_{D,1} \cap \mathbf{R}_{D,2} = \emptyset$, and $\mathbf{I}_1 \cap \mathbf{I}_2 = \emptyset$. Note that this can easily be achieved for any pair of ontologies by a suitable renaming, e.g., as done below by using the prefix 'O_i:'. A mapping between elements e_1 and e_2 from L_1 and L_2, respectively, is then represented by a simple rule $e_2(\mathbf{x}) \leftarrow e_1(\mathbf{x})$ in P, where $e_1 \in \mathbf{A}_1 \cup \mathbf{R}_{A,1} \cup \mathbf{R}_{D,1}$, $e_2 \in \mathbf{A}_2 \cup \mathbf{R}_{A,2} \cup \mathbf{R}_{D,2}$, and $\mathbf{x}$ is a suitable variable vector. Informally, such a rule encodes that every instance of (the concept or role) e_1 in O_1 is also an instance of (the concept or role) e_2 in O_2. Note that demanding the signatures of L_1 and L_2 to be disjoint guarantees that the set of rules that represents ontology mappings is stratified as long as there are no cyclic mappings.

The simple mappings above are the kind of mappings usually found by common matching tools. Both tightly coupled and generalized dl-programs allow to represent such simple mappings. Examples of such mapping rules found by a specific typical matcher m between the ontologies in our example scenario are the following ones:

(1) $O_1 : Publication(x) \leftarrow O_2 : Publication(x);$
(2) $O_1 : Article(x) \leftarrow O_2 : Paper(x);$
(3) $O_1 : Technical\_Report(x) \leftarrow O_2 : Paper(x);$
(4) $O_1 : Person(x) \leftarrow O_2 : Person(x);$
(5) $O_1 : Book(x) \leftarrow O_2 : Proceedings(x);$
(6) $O_1 : Collection(x) \leftarrow O_2 : Proceedings(x);$
(7) $O_1 : keyword(x, y) \leftarrow O_2 : about(x, y);$
(8) $O_1 : author(y, x) \leftarrow O_2 : author(x, y).$

These mappings are very simple Horn rules without negation and with only one body atom. They can be expressed with both tightly coupled and generalized dl-programs.

These simple mappings can be refined, e.g., if we want to add in the mapping rule (7) that only those tuples in the relation "about" are allowed to be mapped to the relation "keyword" that have a subject which is also a topic in O_2. In this way, the integration does not consider relations where topics are involved that do not occur in O_1:

(9) $O_1 : keyword(x, y) \leftarrow O_2 : about(x, y) \land O_2 : Subject(y) \land O_1 : Topic(y)$.

Such refinements are possible in both tightly coupled and generalized dl-programs.

We next consider the mapping rules (2) and (3). In O_1, there is an axiom that declares *Publication* and *Technical_Report* as being disjoint and another axiom that declares *Article* as being a subclass of *Publication*. Thus, these two rules produce an inconsistency, because a concept of O_2 is mapped to two disjoint concepts of O_1 at the same time. In [19], a method for detecting such inconsistent mappings is presented which can be used for this purpose. There are different approaches for resolving this inconsistency. The most straightforward one is to drop mappings until no inconsistency is present anymore. Peng and Xu [20] have proposed a more suitable method for dealing with inconsistencies in terms of a relaxation of the mappings. In particular, they propose to replace a number of conflicting mappings by a single mapping that includes a disjunction of the conflicting concepts. In this example, we would replace the two mapping rules (2) and (3) by the following one:

(10) $O_1 : Article(x) \lor O_1 : Technical\_Report(x) \leftarrow O_2 : Paper(x)$.

This new mapping rule resolves the inconsistency and can be represented in tightly coupled dl-programs, but not in generalized dl-programs. More specifically, for a particular paper p in the ontology O_2, it imposes the existence of two partial answer sets

$$\{O_1 : Article(p),\ O_2 : Paper(p)\};$$
$$\{O_1 : Technical\_Report(p),\ O_2 : Paper(p)\}.$$

None of these answer sets is invalidated by the disjointness constraints imposed by the ontology O_1. However, we can only deduce $O_2 : Paper(p)$ cautiously, the other atoms can be deduced bravely. More generally, with such rules, instances that are only available in the ontology O_2 cannot be classified with certainty.

We can solve this issue by refining the rules again and make use of nonmonotonic negation which again can be used only in the framework of tightly coupled dl-programs and not in the framework of generalized dl-programs. In particular, we can extend the body of the original mappings with the following additional requirement:

$O_1 : Article(x) \leftarrow O_2 : Paper(x) \land O_2 : published\_by(x, y);$
$O_1 : Technical\_Report(x) \leftarrow O_2 : Paper(x) \land not\ O_2 : published\_by(x, y).$

This refinement of the mapping rules resolves the inconsistency and also provides a more correct mapping because background information has been added. A drawback of this approach is the fact that it requires manual post-processing of mappings because the additional background information is not obvious. In the next section, we present a probabilistic extension of tightly coupled dl-programs that allows us to directly use confidence estimations of matching engines to resolve inconsistencies and to combine the results of different matchers.

With both tightly coupled and generalized dl-programs, it is possible to refine mappings positively by adding constraints by means of additional conjuncts in the body. With tightly coupled dl-programs, it is additionally possible to refine mappings by additional nonmonotonic negated conjuncts in the body. Another refinement possibility supported by tightly coupled dl-programs, but not by generalized dl-programs is relaxing the antecedent of the rule by means of additional disjuncts in the head.

5   Representing Probabilistic Ontology Mappings

In this section, we present probabilistic extensions of tightly coupled and generalized dl-programs, called *tightly coupled probabilistic dl-programs* and *generalized Bayesian dl-programs*, respectively. Intuitively, they extend the rule component of tightly coupled and generalized dl-programs by Bayesian probabilities.

5.1   Tightly Coupled Probabilistic DL-Programs

We now present a *tightly coupled* approach to *probabilistic disjunctive description logic programs* (or *tightly coupled probabilistic dl-programs*) under the answer set semantics. Differently from [13] (in addition to being a tightly coupled approach), the probabilistic dl-programs here also allow for disjunctions in rule heads. Similarly to the probabilistic dl-programs in [13], they are defined as a combination of dl-programs with Poole's ICL [21], but using the tightly coupled disjunctive dl-programs of [8] (see Section 4.1), rather than the loosely coupled dl-programs of [7]. The ICL is based on ordinary acyclic logic programs P under different "choices", where every choice along with P produces a first-order model, and one then obtains a probability distribution over the set of all first-order models by placing a probability distribution over the different choices. We use the tightly coupled disjunctive dl-programs under the answer set semantics of [8], instead of ordinary acyclic logic programs under their canonical semantics (which coincides with their answer set semantics).

Syntax. We now define the syntax of tightly coupled probabilistic dl-programs and queries to them. We first introduce choice spaces and probabilities on choice spaces.

A *choice space* C is a set of pairwise disjoint and nonempty sets $A \subseteq HB_\Phi - DL_\Phi$. Any $A \in C$ is an *alternative* of C and any element $a \in A$ an *atomic choice* of C. Intuitively, every alternative $A \in C$ represents a random variable and every atomic choice $a \in A$ one of its possible values. A *total choice* of C is a set $B \subseteq HB_\Phi - DL_\Phi$ such that $|B \cap A| = 1$ for all $A \in C$ (and thus $|B| = |C|$). Intuitively, every total choice B of C represents an assignment of values to all the random variables. A *probability* μ on a choice space C is a probability function on the set of all total choices of C. Intuitively, every probability μ is a probability distribution over the set of all variable assignments. Since C and all its alternatives are finite, μ can be defined by (i) a mapping $\mu: \bigcup C \to [0, 1]$ such that $\sum_{a \in A} \mu(a) = 1$ for all $A \in C$, and (ii) $\mu(B) = \Pi_{b \in B} \mu(b)$ for all total choices B of C. Intuitively, (i) defines a probability over the values of each random variable of C, and (ii) assumes independence between the random variables.

A *tightly coupled probabilistic disjunctive description logic program* (or *tightly coupled probabilistic dl-program*) $KB = (L, P, C, \mu)$ consists of a tightly coupled dl-program (L, P), a choice space C such that no atomic choice in C coincides with the head of any rule in $ground(P)$, and a probability μ on C. Intuitively, since the total choices of C select subsets of P, and μ is a probability distribution on the total choices of C, every probabilistic dl-program is the compact representation of a probability distribution on a finite set of disjunctive dl-programs. Observe here that P is fully general and not necessarily stratified or acyclic. We say KB is *normal* iff P is normal. A *probabilistic query* to KB has the form $\exists (c_1(x) \vee \cdots \vee c_n(x))[r, s]$, where x, r, s is a tuple of variables, $n \geqslant 1$, and each $c_i(x)$ is a conjunction of atoms constructed from predicate and constant symbols in Φ and variables in x. Note that the above probabilistic queries can also be easily extended to conditional expressions as in [13].

Semantics. We now define an answer set semantics of probabilistic dl-programs, and we introduce the notions of consistency, consequence, tight consequence, and correct and tight answers for queries to tightly coupled probabilistic dl-programs.

Given a tightly coupled probabilistic dl-program $KB = (L, P, C, \mu)$, a *probabilistic interpretation Pr* is a probability function on the set of all $I \subseteq HB_\Phi$. We say Pr is an *answer set* of KB iff (i) every $I \subseteq HB_\Phi$ with $Pr(I) > 0$ is an answer set of $(L, P \cup \{p \leftarrow \mid p \in B\})$ for some total choice B of C, and (ii) $Pr(\bigwedge_{p \in B} p) = \sum_{I \subseteq HB_\Phi, B \subseteq I} Pr(I) = \mu(B)$ for every total choice B of C. Informally, Pr is an answer set of $KB = (L, P, C, \mu)$ iff (i) every $I \subseteq HB_\Phi$ of positive probability under Pr is an answer set of the dl-program (L, P) under some total choice B of C, and (ii) Pr coincides with μ on the total choices B of C. We say KB is *consistent* iff it has an answer set Pr.

We define the notions of consequence and tight consequence as follows. Given a probabilistic query $\exists (q(x))[r, s]$, the *probability* of $q(x)$ in a probabilistic interpretation Pr under a variable assignment σ, denoted $Pr_\sigma(q(x))$ is defined as the sum of all $Pr(I)$ such that $I \subseteq HB_\Phi$ and $I \models_\sigma q(x)$. We say $(q(x))[l, u]$ (where $l, u \in [0, 1]$) is a *consequence* of KB, denoted $KB \mid\hspace{-3pt}\sim (q(x))[l, u]$, iff $Pr_\sigma(q(x)) \in [l, u]$ for every answer set Pr of KB and every variable assignment σ. We say $(q(x))[l, u]$ (where $l, u \in [0, 1]$) is a *tight consequence* of KB, denoted $KB \mid\hspace{-3pt}\sim_{tight} (q(x))[l, u]$, iff l (resp., u) is the infimum (resp., supremum) of $Pr_\sigma(q(x))$ subject to all answer sets Pr of KB and all σ. A *correct* (resp., *tight*) *answer* to a probabilistic query $\exists (q(x))[r, s]$ is a ground substitution θ (for the variables x, r, s) such that $(q(x))[r, s]\theta$ is a consequence (resp., tight consequence) of KB.

Example Scenario Cont'd. We now show how tightly coupled probabilistic dl-programs $KB = (L, P, C, \mu)$ can be used for representing (possibly inconsistent) mappings with confidence values between two ontologies. Here, (i) L is the union of two description logic knowledge bases L_1 and L_2 encoding two ontologies O_1 and O_2, respectively, and (ii) P, C, and μ encode the mappings between the two ontologies O_1 and O_2, where confidence values are encoded as error probabilities to combine mappings produced by different matchers, and inconsistencies are resolved via trust probabilities (in addition to using disjunctions and nonmonotonic negations in P).

More concretely, we interpret the confidence value as an *error probability* and state that the probability that a mapping introduces an error is $1 - p$. Conversely, the probability that a mapping correctly describes the semantic relation between elements of the different ontologies is $1 - (1 - p) = p$. This means that we can use the confidence value p as a probability for the correctness of a mapping. The indirect formulation is chosen, because it allows us to combine the results of different matchers in a meaningful way. In particular, if we assume that the error probabilities of two matchers are independent, we can calculate the joint error probability of two matchers that have found the same mapping rule as $(1 - p_1) \cdot (1 - p_2)$. This means that we can get a new probability for the correctness of the rule found by two matchers which is $1 - (1 - p_1) \cdot (1 - p_2)$. This way of calculating the joint probability meets the intuition that a mapping is more likely to be correct if it has been discovered by more than one matcher because $1 - (1 - p_1) \cdot (1 - p_2) \geqslant p_1$ and $1 - (1 - p_1) \cdot (1 - p_2) \geqslant p_2$.

In addition, when merging inconsistent results of different matching systems, we weigh each matching system and its result with a (user-defined) *trust probability*, which describes our confidence in its quality. All these trust probabilities sum up to 1. For example, the trust probabilities of the matching systems m_1, m_2, and m_3 may be 0.6, 0.3, and 0.1, respectively. That is, we trust most in m_1, medium in m_2, and less in m_3.

We illustrate this approach along our example scenario. The following rules in P encode the mappings that have been produced by two matchers m_1 and m_2:

(1) $O_1 : Publication(x) \leftarrow O_2 : Publication(x), m_{1,1}$;
(2) $O_1 : Article(x) \leftarrow O_2 : Paper(x), m_{1,2}$;
(3) $O_1 : Technical\_Report(x) \leftarrow O_2 : Paper(x), m_{2,1}$;
(4) $O_1 : Person(x) \leftarrow O_2 : Person(x), m_{1,3}$;
(5) $O_1 : Person(x) \leftarrow O_2 : Person(x), m_{2,2}$;
(6) $O_1 : Book(x) \leftarrow O_2 : Proceedings(x), m_{2,3}$;
(7) $O_1 : keyword(x, y) \leftarrow O_2 : about(x, y), m_{1,4}$;
(8) $O_1 : author(y, x) \leftarrow O_2 : author(x, y), m_{2,4}$.

More concretely, every rule contains a conjunct $m_{i,j}$, identifying with i and j the matching system that has created it and the mapping, respectively. Thus, mappings (1), (2), (4), and (7) have been found by m_1, while mappings (3), (5), (6), and (8) have been found by m_2. These additional conjuncts $m_{i,j}$ are atomic choices of the choice space C_i and link probabilities (which are specified in the probability μ_i on the choice space C_i) to the rules. The two choice spaces C_1 and C_2 of the matchers are

$$C_1 = \{\{m_{1,i}, not\_m_{1,i}\} \mid i \in \{1, 2, 3, 4\}\}, \; C_2 = \{\{m_{2,j}, not\_m_{2,j}\} \mid j \in \{1, 2, 3, 4\}\}.$$

They come along with the probabilities μ_1 and μ_2 on C_1 and C_2, respectively, which assign the corresponding confidence value p to each atomic choice $m_{1,i}$ and the complement $1 - p$ to the atomic choice $not\_m_{1,i}$ (and the same holds for $m_{2,j}$ and $not\_m_{2,j}$). For example, we have $\mu_1(m_{1,1}) = 0.9$ and $\mu_1(not\_m_{1,1}) = 0.1$. Because the probability value of each atomic choice is determined by the probability value of the other atomic choice in an alternative, we restrict the presentation of the probability to only one element of each alternative: $\mu_1(m_{1,2}) = 0.62$, $\mu_1(m_{1,3}) = 0.73$, $\mu_1(m_{1,4}) = 0.84$. $\mu_2(m_{2,1}) = 0.94$, $\mu_2(m_{2,2}) = 0.96$, $\mu_2(m_{2,3}) = 0.72$, and $\mu_2(m_{2,4}) = 0.93$.

The benefits of this explicit treatment of uncertainty becomes clear when we now try to merge the mappings of m_1 with the mappings of m_2. Note that the mappings (2) and (3) produce an inconsistency, since the same concept of the source ontology O_2 (here, $O_2 : Paper$) is mapped to two disjoint concepts of the target ontology O_1. Note also that the mappings (4) and (5) are identical and found by each of the matchers.

Directly merging these two mappings as they are is not a good idea for two reasons. First, by adding mappings (2) and (3), we encounter an inconsistency problem as mentioned above. Therefore, rules (2) and (3) cannot contribute to a model of the knowledge base. Second, a simple merge does not account for the fact that the mappings (4) and (5) are identical and have been found by both matchers, and should thus be strengthened. Here, the mapping rule has the same status as any other rule in the mapping and each instance of $O_2 : Person$ has two probabilities at the same time.

Suppose we associate with m_1 and m_2 the trust probabilities 0.55 and 0.45, respectively. Based on the interpretation of confidence values as error probabilities, and on the use of trust probabilities when resolving inconsistencies between rules, we can now define a merged mapping set, which consists of the mappings (1), (4), (5), (6), (7), and (8) from above and the following two rules instead of (2) and (3):

$$(2') \ O_1 : Article(x) \leftarrow O_2 : Paper(x), m_{1,2}, sel\_m_{1,2};$$
$$(3') \ O_1 : Technical\_Report(x) \leftarrow O_2 : Paper(x), m_{2,1}, sel\_m_{2,1}.$$

The choice space C and the probability μ on C are obtained from $C_1 \cup C_2$ and $\mu_1 \cdot \mu_2$ (which is the product of μ_1 and μ_2, that is, $(\mu_1 \cdot \mu_2)(B_1 \cup B_2) = \mu_1(B_1) \cdot \mu_2(B_2)$ for all total choices B_1 of C_1 and B_2 of C_2), respectively, by adding the alternative $\{sel\_m_{1,2}, sel\_m_{2,1}\}$ and the probabilities $\mu(sel\_m_{1,2}) = 0.55$ and $\mu(sel\_m_{2,1}) = 0.45$ for resolving the inconsistency between the first two rules.

It is not difficult to verify that, due to the independent combination of alternatives, the last two rules encode that the rule $O_1 : Person(x) \leftarrow O_2 : Person(x)$ holds with the probability $1 - (1 - \mu(m_{1,3})) \cdot (1 - \mu(m_{2,2})) = 0.9892$, as desired. Informally, any randomly chosen instance of $Person$ of the ontology O_2 is also an instance of $Person$ of the ontology O_1 with the probability 0.9892. In contrast, if the mapping rule would have been discovered only by m_1 (resp., m_2), such an instance of $Person$ of O_2 would be an instance of $Person$ of O_1 with the probability 0.73 (resp., 0.96).

A probabilistic query Q asking for the probability that a specific publication pub in the ontology O_2 is an instance of the concept $Article$ of the ontology O_1 is given by $Q = \exists(Article(pub))[R, S]$. The tight answer θ to Q is given by $\theta = \{R/0, S/0\}$, if pub is not an instance of the concept $Paper$ in the ontology O_2 (since there is no mapping rule that maps another concept than $Paper$ to the concept $Article$). If pub is an instance of the concept $Paper$, however, then the tight answer to Q is given by $\theta = \{R/0.341, S/0.341\}$ (as $\mu(m_{1,2}) \cdot \mu(sel\_m_{1,2}) = 0.62 \cdot 0.55 = 0.341$). Informally, pub belongs to the concept $Article$ with the probabilities 0 resp. 0.341.

5.2 Generalized Bayesian DL-Programs

In [22], we have proposed *Bayesian dl-programs* for information integration and retrieval. We now define *generalized Bayesian dl-programs*, which extend Bayesian dl-programs to the generalized dl-programs presented in Section 4.2. More concretely, the logic programming component is extended to full Datalog (without equality and negation). That is, the predicates are allowed to be of a higher arity than 2 and the dependency graph of the rules does not need to be connected and fully acyclic anymore.

Syntax. A *generalized Bayesian dl-program* is a 4-tuple $KB = (L, P, \mu, Comb)$, where (i) (L, P) is a generalized dl-program, (ii) μ associates with each rule $r\colon h \leftarrow b_1, \ldots, b_n$ in $ground(P)$ and every truth valuation $v\colon \{b_1, \ldots, b_n\} \to \{\text{false}, \text{true}\}$ of the body atoms of r a probability function $\mu(r, v)$ over all truth valuations $w\colon \{h\} \to \{\text{false}, \text{true}\}$ of the head atom of r, and (iii) $Comb$ is a *combining rule*, which defines how rules $r \in ground(P)$ with the same head atom can be combined to obtain a single rule.

Semantics. Each generalized Bayesian dl-program $KB = (L, P, \mu, Comb)$ encodes the structure of a Bayesian network BN and provides a complete specification of its conditional probability distributions. We now describe the translation from KB to BN.

We first translate (L, P) into its Datalog equivalent DE. We say a ground atom a is *active* iff it belongs to the canonical model of DE. We say $r \in ground(DE)$ is *active* iff all its atoms are active. Every active ground atom then corresponds to a node in BN, and the dependencies between the active ground atoms that are encoded in the active rules $r \in ground(DE)$ correspond to the parent relationships in BN. For this reason, we also implicitly assume that the set of all active rules in $ground(DE)$ is acyclic.

The function μ is the conditional probability density of each of the random variables that are represented by the direct influence relationship between ground atoms encoded by the active rules in $ground(P)$. In terms of a Bayesian network, each of these functions is translated to links connecting the node representing the possible instantiations of the head with the nodes representing the instantiations of the different atoms in the body. Note that rules with empty bodies are facts for which the a-priori probability density is given in the same way. Note also that the function μ is implicitly extended to all active ground instances of rules $r\colon h \leftarrow b_1, \ldots, b_n$ in the Datalog equivalent of L, by assuming that $\mu(r, v)\colon h, \neg h \mapsto 1, 0$ iff $v(b_i) = \text{true}$ for all $i \in \{1, \ldots, n\}$. If an active ground atom h can be deduced by only one active rule $r \in ground(DE)$, then its conditional probabilities are given by the distributions attached via μ to this rule. If, however, we have at least two active rules $r \in ground(DE)$ with the same head h, then the conditional probabilities of h need to consider all these rules. For this purpose, the combining rule $Comb$ generates a joint conditional distribution from the individual ones of the involved rules. More concretely, $Comb$ maps a finite set of conditional probability densities $\{p(h|a_{i,1}, \ldots, a_{i,n_i}) \mid m \geqslant i \geqslant 1, n_i \geqslant 0\}$, $m \geqslant 1$, to the conditional probability density $p(h|b_1, \ldots, b_n)$ with $\{b_1, \ldots, b_n\} = \bigcup_{i=1}^{m} \{a_{i,1}, \ldots, a_{i,n_i}\}$. Different combining rules are allowed. The simplest combining rule is the maximum of the conditional probability densities, which we use in the following, as it fulfills our purposes. More sophisticated ways of combining rules are, e.g., variations of noisy-or.

Example Scenario Cont'd. We now show how we can use generalized Bayesian dl-programs $KB = (L, P, \mu, Comb)$ for reasoning with ontologies and uncertain mappings between them. Here, (i) L is the union of two description logic knowledge bases L_1 and L_2 encoding two ontologies O_1 and O_2, respectively, and (ii) P, μ, and $Comb$ represent the mappings between the two ontologies O_1 and O_2. Note that μ associates with the mapping rules in P conditional probability distributions, and $Comb$ is the combination rule, which in our example simply corresponds to the maximum, as mentioned above.

We illustrate this approach along our example scenario. The following rules in P (including the conditional probability distributions via μ) encode the mappings between O_1' and O_2' that have been found by a matcher m:

(1) $O'_1 : Publication(x) \overset{(0.9,0.2)}{\longleftarrow} O'_2 : Publication(x);$

(2) $O'_1 : Article(x) \overset{(0.7,0.2)}{\longleftarrow} O'_2 : Paper(x);$

(3) $O'_1 : Person(x) \overset{(0.9,0.2)}{\longleftarrow} O'_2 : Person(x);$

(4) $O'_1 : Collection(x) \overset{(0.7,0.2)}{\longleftarrow} O'_2 : Proceedings(x);$

(5) $O'_1 : keyword(x,y) \overset{(0.7,0.2)}{\longleftarrow} O'_2 : about(x,y);$

(6) $O'_1 : author(y,x) \overset{(0.7,0.2)}{\longleftarrow} O'_2 : author(x,y).$

Here, we use an intuitive graphical representation of P and μ. This is possible because the rules contain only one head atom and one body atom: we write only the probabilities p_1 and p_2 for the true head atom given the true resp. false body atom. The probabilities of the false head atom given the true resp. false body atom are then $1 - p_1$ resp. $1 - p_2$. For example, the mapping (1) says that (i) each publication in O'_2 is also a publication in O'_1 with the probability 0.9, and (ii) each non-publication in O'_2 (that is, each element of $\neg Publication$) in O'_2 is a publication in O'_1 with the probability 0.2.

In order to reason with the ontologies and the mappings, the ontologies need to be translated into their logic programming syntax. The translation is shown below:

Translation of O'_1:

(1a) $Topic(y) \leftarrow Technical\_Report(x) \wedge keyword(x,y);$

(1b) $Person(y) \leftarrow Technical\_Report(x) \wedge author(x,y);$

(3) $Publication(x) \leftarrow Book(x);$ (4) $Publication(x) \leftarrow Article(x);$

(5) $Publication(x) \leftarrow Collection(x);$

(10a) $Topic(y) \leftarrow Publication(x) \wedge keyword(x,y);$

(10b) $Person(y) \leftarrow Publication(x) \wedge author(x,y).$

Translation of O'_2:

(1) $Publication(x) \leftarrow Paper(x);$ (2) $Publication(x) \leftarrow Proceedings(x);$

(4a) $Paper(y) \leftarrow includes(x,y);$ (4b) $Proceedings(x) \leftarrow includes(x,y);$

(11a) $Publisher(y) \leftarrow published\_by(x,y);$ (11b) $Publication(x) \leftarrow published\_by(x,y);$

(12a) $Subject(y) \leftarrow about(x,y);$ (12b) $Publication(x) \leftarrow about(x,y);$

(13a) $Person(y) \leftarrow author(x,y);$ (13b) $Publication(x) \leftarrow author(x,y).$

As described above, for every ground instance of such a rule, the probability of the head atom being true is 1, if all the body atoms are also true, and 0, otherwise.

The processing of queries posed to a generalized Bayesian dl-program KB consists of two steps of a so-called *knowledge-based model construction*. If we consider KB without probability densities attached, then we obtain a logic program KB', called the *corresponding logic program* of KB. In the first step, the least Herbrand model M of KB' is deduced by means of logic programming. The Bayesian network that corresponds to KB is then created by means of the ground atoms in M.

The construction of the Bayesian network BN for KB is briefly described as follows: for each fact f of KB, a node in BN is created, and the probability density of f is attached to the node. For each ground instance r of a rule that has been used for properly deriving the (ground) head atom h, there are then two possibilities:

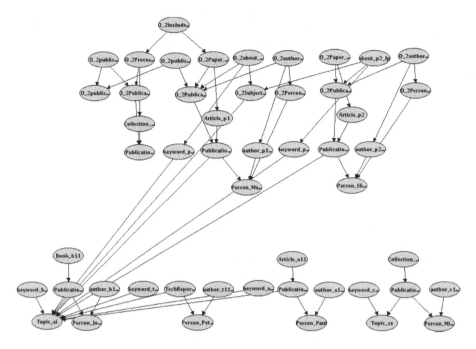

Fig. 1. The Bayesian network for the example scenario

- a node that corresponds to h does not exist in BN: Then, such a node is created and for each body atom of the rule r, an arc from the corresponding node in BN (which already exist in BN, since each time a ground rule r can derive a new head atom, the body atoms are already in M) to the newly created node is created. The probability density of this rule in KB is attached to this node.
- a node that corresponds to h exists in BN: Then, for each body atom in r, which has a corresponding node in BN, but no arc from this node to the node that corresponds to h, such an arc is created. Afterwards, the combining rule is applied to the probability density of r and the node corresponding to h. Thus, the node corresponding to h is equipped with a probability density which considers the probability densities of r and the rules that already have derived the same head atom h.

The resulting Bayesian network BN can be used for answering any *probabilistic query*, which is an expression of the form $? - Q_1, \dots, Q_n | E_1, \dots, E_m$ with atoms Q_i and E_j. Such an expression asks for the probability of the conditional event $Q_1, \dots,$ $Q_n | E_1, \dots, E_m$, that is, the probability that the Q_i's are true given that the E_j's are true. We distinguish between ground and non-ground queries. In the latter case, it is asked for the probability of each valid grounding of the query. Such queries are processed by first computing all valid groundings and then asking for the probability of each grounding in the corresponding Bayesian network. Non-ground queries can be used for information retrieval where the probabilities are used for ranking.

Fig. 1 shows the Bayesian network for the example scenario. The subgraph in the upper part of the figure encodes O_2' and the mappings to O_1', while the lower part of the figure encodes O_1'. Note that there is only one already existing node in O_1' that participates in a couple of mappings. This node is O_1' : $Topic(artificial\_intelligence)$.

For example, consider the probabilistic query $Q(x) = ? - O_1'$: $Publication(x)$, O_1' : $keyword(x, artificial\_intelligence)$. We then obtain the instances $Q(a_1)$, $Q(b_1)$, $Q(p_1)$, and $Q(p_2)$. Since a_1 and b_1 stem from O_1', they belong for sure to O_1' : $Publication$ and O_1' : $\exists keyword.\{artificial\_intelligence\}$. Hence, they are delivered with the probability 1, while p_1 and p_2 are delivered with the probabilities 0.4731 and 0.7470, respectively. We see here very nicely how we can rank answers to queries.

6   Conclusion

In this work, we have presented two rule-based approaches for representing uncertain mappings based on probabilistic concepts. The approaches discussed are tightly coupled probabilistic dl-programs, on the one hand, and generalized Bayesian dl-programs, on the other hand. Both approaches have been designed to represent probabilistic mappings between description logic ontologies, but differ with respect to the kinds of ontologies and mappings supported. We have discussed both approaches in terms of (i) the kinds of ontologies supported, (ii) how rule-based mappings and ontologies are linked, (iii) how rule-based mappings are extended with a probabilistic semantics, and (iv) the reasoning methods supported by the languages. We now turn back to the requirements for mapping languages that have been defined earlier and summarize the characteristics of the two approaches with respect to these requirements.

- *Tight integration of mapping and ontology language:* Both approaches support the tight integration of rules and ontologies in the sense that concepts and relations from ontologies can occur in the head as well as the body of rules, which can therefore be used to represent mappings between different ontologies. But this tight integration is achieved in different ways. As for generalized Bayesian dl-programs, the integration is achieved by restricting the expressiveness of the ontologies to the DLP-fragment of $\mathcal{SHIF}(\mathbf{D})$. This enables us to translate the overall model into a Datalog model with a corresponding semantics. Tightly coupled probabilistic dl-programs do not limit the expressive power of the ontologies. The integration is achieved in terms of a novel semantics, which consists of a general first-order semantics of the ontological part and a Herbrand semantics of the rules part. From a representational point of view, this is a big advantage, since it allows us to connect arbitrary OWL ontologies using this framework. This advantage comes at the price of a new semantics, which is not immediately supported by existing tools and systems (but can be easily implemented on top of existing tools and systems).
- *Support for mappings refinement and repairing inconsistencies:* This second requirement is concerned with the expressiveness of the rule language for representing mappings. As mentioned above, the expressiveness of generalized Bayesian dl-programs is restricted to plain Datalog. That is, mappings can only state a relation between conjunctions of concepts and relations in the source ontology to one concept or one relation in the target ontology. This approach provides limited possibilities for refining mappings. As discussed in the paper, however, refinement is primarily needed to deal with inconsistencies. As the language does not support

negation, inconsistencies cannot occur anyway, significantly reducing the need for refinement support. This is different for the case of tightly coupled probabilistic dl-programs. Here, we allow full negation in the ontologies and non-monotonic negation in rule bodies. This means that mappings can introduce inconsistency in the model which we have to deal with in order to be able to get a meaningful probability distribution for the overall model. As we have shown in Section 4.3, the rule language supports the resolution of inconsistencies using refinement of the body or generalization of the head of mapping rules. As we have shown in Section 5, inconsistencies can also be resolved probabilistically in tightly coupled probabilistic dl-programs. In any case, however, there is a need for mechanisms for detecting inconsistencies and determining their cause, which is a difficult problem in itself.

- *Representation and combination of confidence:* Both approaches explicitly address the problem of representing and combining confidence values in the framework of probability. The concrete model used, however, is rather different for the two approaches. Generalized Bayesian dl-programs are based on a complete definition of the probability distribution and require the confidence to be expressed in terms of a complete probability distribution over the terms in a mapping rule: we need the probability for the truth of the head given all combinations of truth values for the body. While for simple mapping rules, such as those normally produced by automatic matching systems, this is not too much of a problem, and the corresponding probabilities can be determined either manually or by appropriate statistical estimations. In the presence of complex rules with many terms in the rule body, the number of probabilities needed grows exponentially, making it much harder to acquire the corresponding knowledge. Tightly coupled probabilistic dl-programs address this problem by allowing an incomplete specification of the probability distribution. As a result, we only have to specify one probability for each mapping rule, directly showing its confidence. This also means, however, that in such a more general modeling, the probability of statements in the overall model may often only be determined up to an interval which contains the true probability. Using a suitable definition of the choice space, the approach also allows to express a general confidence in a source that provides mapping information. This is a clear advantage in practical settings where we might want to combine the results of different matching systems. In summary, generalized Bayesian dl-programs are suitable for situations where only simple mappings have to be represented, and there is no need to distinguish between different sources of information. In more complex situations, the model used by tightly coupled probabilistic dl-programs is the more suitable one.

- *Decidability and efficiency of instance reasoning:* The different choices that the compared approaches make with respect to the expressive power of the logical formalism, on the one hand, and the probabilistic model, on the other hand, has of course a strong influence on the efficiency of reasoning in these formalisms. For both approaches, the problem of probabilistic query answering, which is central in the context of probabilistic mapping representation is decidable and corresponding methods for computing answers have been described. When it comes to efficiency, however, there are differences. Generalized Bayesian dl-programs have been designed for efficient query processing, not only from a theoretical, but also from a technical point. They allow the use of the most efficient technologies currently available. In contrast to that, tightly coupled probabilistic dl-programs in their full generality have been designed to maximize expressiveness of the logical as well as the probabilistic model. The advantages of this choice have already been discussed in the preceding paragraphs. With respect to reasoning, this does not only mean that reasoning in general is very expensive in this framework, but also that we cannot directly rely on existing optimized systems. However, there are also data-tractable special cases of tightly coupled probabilistic dl-programs [23].

We conclude that the two approaches compared in this paper represent two extremes with respect to trading off representation and reasoning. Both approaches are useful in certain situations. In cases where we are concerned with rather weak ontologies, e.g., plain taxonomies, RDF Schemas, or thesauri, the simpler approach will often be sufficient for representing ontologies and mappings between them, and we can benefit from the better immediate computational properties. In other cases, where expressive ontologies have to be connected by complex mappings, the simple approach will not be sufficient anymore. Here, we have to revert to the more expressive approach.

An important topic for future research is the exploration of the space between these two extreme approaches, and to try to find more balanced trade-offs between expressiveness and efficiency that address the needs of real problems as directly as possible. Some first steps in this direction have been done by the identification of tractable subsets of tightly coupled probabilistic dl-programs [23]. Corresponding work will benefit from work on combining description logics with expressive rule languages that is currently being done in connection with the development of the OWL 2.0 standard.

Acknowledgments. Andrea Calì was supported by the Engineering and Physical Sciences Research Council (EPSRC) under the project "Schema Mappings and Automated Services for Data Integration and Exchange" (EP/E010865/1) and by the European Union under the STREP FET project TONES (FP6-7603). Thomas Lukasiewicz was supported by the German Research Foundation (DFG) under the Heisenberg Programme and by the Austrian Science Fund (FWF) under the project P18146-N04. Heiner Stuckenschmidt and Livia Predoiu were supported by an Emmy-Noether Grant of the German Research Foundation (DFG). We thank the reviewers of this paper and its URSW-2007 abstract for their useful comments, which helped to improve this work.

References

1. Serafini, L., Stuckenschmidt, H., Wache, H.: A formal investigation of mapping language for terminological knowledge. In: Proceedings IJCAI 2005, pp. 576–581. Professional Book Center (2005)
2. Euzenat, J., Shvaiko, P.: Ontology Matching. Springer, Heidelberg (2007)
3. Euzenat, J., Mochol, M., Shvaiko, P., Stuckenschmidt, H., Svab, O., Svatek, V., van Hage, W.R., Yatskevich, M.: First results of the ontology alignment evaluation initiative 2006. In: Proceedings ISWC 2006 Workshop on Ontology Matching (2006)
4. Euzenat, J., Stuckenschmidt, H., Yatskevich, M.: Introduction to the ontology alignment evaluation 2005. In: Proceedings K-CAP 2005 Workshop on Integrating Ontologies (2005)
5. Motik, B.: Reasoning in Description Logics using Resolution and Deductive Databases. PhD thesis, University of Karlsruhe, Karlsruhe, Germany (2006)

6. RIF Working Group, http://www.w3.org/2005/rules/
7. Eiter, T., Ianni, G., Lukasiewicz, T., Schindlauer, R., Tompits, H.: Combining answer set programming with description logics for the Semantic Web. Artif. Intell. 172(12/13), 1495–1539 (2008)
8. Lukasiewicz, T.: A novel combination of answer set programming with description logics for the Semantic Web. In: Franconi, E., Kifer, M., May, W. (eds.) ESWC 2007. LNCS, vol. 4519, pp. 384–398. Springer, Heidelberg (2007)
9. Grosof, B.N., Horrocks, I., Volz, R., Decker, S.: Description logic programs: Combining logic programs with description logics. In: Proceedings WWW 2003, pp. 48–57. ACM Press, New York (2003)
10. Giugno, R., Lukasiewicz, T.: P-$\mathcal{SHOQ}(\mathbf{D})$: A probabilistic extension of $\mathcal{SHOQ}(\mathbf{D})$ for probabilistic ontologies in the Semantic Web. In: Flesca, S., Greco, S., Leone, N., Ianni, G. (eds.) JELIA 2002. LNCS, vol. 2424, pp. 86–97. Springer, Heidelberg (2002)
11. da Costa, P.C.G.: Bayesian Semantics for the Semantic Web. PhD thesis, George Mason University, Fairfax, VA, USA (2005)
12. da Costa, P.C.G., Laskey, K.B.: PR-OWL: A framework for probabilistic ontologies. In: Proceedings FOIS 2006, pp. 237–249. IOS Press, Amsterdam (2006)
13. Lukasiewicz, T.: Probabilistic description logic programs. Int. J. Approx. Reasoning 45(2), 288–307 (2007)
14. Jensen, F.V.: Introduction to Bayesian Networks. Springer, Heidelberg (1996)
15. Horrocks, I., Patel-Schneider, P.F.: Reducing OWL entailment to description logic satisfiability. In: Fensel, D., Sycara, K.P., Mylopoulos, J. (eds.) ISWC 2003. LNCS, vol. 2870, pp. 17–29. Springer, Heidelberg (2003)
16. Horrocks, I., Sattler, U., Tobies, S.: Practical reasoning for expressive description logics. In: Ganzinger, H., McAllester, D., Voronkov, A. (eds.) LPAR 1999. LNCS, vol. 1705, pp. 161–180. Springer, Heidelberg (1999)
17. Faber, W., Leone, N., Pfeifer, G.: Recursive aggregates in disjunctive logic programs: Semantics and complexity. In: Alferes, J.J., Leite, J. (eds.) JELIA 2004. LNCS, vol. 3229, pp. 200–212. Springer, Heidelberg (2004)
18. Gelfond, M., Lifschitz, V.: Classical negation in logic programs and disjunctive databases. New Generation Comput. 9(3/4), 365–386 (1991)
19. Meilicke, C., Stuckenschmidt, H., Tamilin, A.: Repairing ontology mappings. In: Proceedings AAAI 2007, pp. 1408–1413. AAAI Press, Menlo Park (2007)
20. Wang, P., Xu, B.: Debugging ontology mapping: A static method. Computing and Informatics 27(1), 21–36 (2008)
21. Poole, D.: The independent choice logic for modelling multiple agents under uncertainty. Artif. Intell. 94(1/2), 7–56 (1997)
22. Predoiu, L., Stuckenschmidt, H.: A probabilistic framework for information integration and retrieval on the Semantic Web. In: Proceedings InterDB 2007 Workshop on Database Interoperability (2007)
23. Calì, A., Lukasiewicz, T., Predoiu, L., Stuckenschmidt, H.: Tightly integrated probabilistic description logic programs for representing ontology mappings. In: Hartmann, S., Kern-Isberner, G. (eds.) FoIKS 2008. LNCS, vol. 4932, pp. 178–198. Springer, Heidelberg (2008)

PR-OWL: A Bayesian Ontology Language
for the Semantic Web

Paulo Cesar G. da Costa[1], Kathryn B. Laskey[1], and Kenneth J. Laskey[2,*]

[1] School of Information Technology and Engineering,
George Mason University
4400 University Drive
Fairfax, VA 22030-4444 USA
{pcosta,klaskey}@gmu.edu
[2] MITRE Corporation, M/S H305
7515 Colshire Drive
McLean, VA 22102-7508 USA
klaskey@mitre.org

Abstract. This paper addresses a major weakness of current technologies for the Semantic Web, namely the lack of a principled means to represent and reason about uncertainty. This not only hinders the realization of the original vision for the Semantic Web, but also creates a barrier to the development of new, powerful features for general knowledge applications that require proper treatment of uncertain phenomena. We present PR-OWL, a probabilistic extension to the OWL web ontology language that allows legacy ontologies to interoperate with newly developed probabilistic ontologies. PR-OWL moves beyond the current limitations of deterministic classical logic to a full first-order probabilistic logic. By providing a principled means of modeling uncertainty in ontologies, PR-OWL can be seen as a supporting tool for many applications that can benefit from probabilistic inference within an ontology language, thus representing an important step toward the W3C's vision for the Semantic Web. In order to fully present the concepts behind PR-OWL, we also cover Multi-Entity Bayesian Networks (MEBN), the Bayesian first-order logic supporting the language, and UnBBayes-MEBN, an open source GUI and reasoner that implements PR-OWL concepts. Finally, a use case of PR-OWL probabilistic ontologies is illustrated here in order to provide a grasp of the potential of the framework.

1 A Deterministic View of a Probabilistic World

Uncertainty is ubiquitous. If the Semantic Web vision [1] is to be realized, a sound and principled means of representing and reasoning with uncertainty will be required. Our broad objective is to address this need by developing a Bayesian framework for probabilistic ontologies and plausible reasoning services. As an initial step toward our

---

* The author's affiliation with The MITRE Corporation is provided for identification purposes only, and is not intended to convey or imply MITRE's concurrence with, or support for, the positions, opinions or viewpoints expressed by the author.

P.C.G. da Costa et al. (Eds.): URSW 2005-2007, LNAI 5327, pp. 88–107, 2008.

objective, we introduce PR-OWL, a probabilistic extension to the Web ontology language OWL.

Current generation Semantic Web technology is based on classical logic, and lacks adequate support for plausible reasoning. For example, OWL, a W3C Recommendation [2], has no built-in support for probabilistic information and reasoning. This is understandable, given that OWL is rooted in web language predecessors (i.e. XML, RDF) and traditional knowledge representation formalisms (e.g.. Description Logics [3]). This historical background somewhat explains the lack of support for uncertainty in OWL. Nevertheless, it is a serious limitation for a language intended for environments where one cannot simply ignore incomplete information.

A similar historical progression occurred in Artificial Intelligence (AI). From its inception, AI has struggled with how to cope with incomplete information. Although probability theory was initially neglected due to tractability concerns, graphical probability languages changed things dramatically [4]. Probabilistic languages have evolved from propositional to full first-order expressivity (e.g., [5, 6]), and have become the technology of choice for reasoning under uncertainty in an open world [7]. Clearly, the Semantic Web will pose similar uncertainty-related issues as those faced by AI. Thus, just as AI has moved from a deterministic paradigm to embrace probability, a similar path appears promising for ontology engineering.

This path is not yet being followed. The lack of support for representing and reasoning with uncertain, incomplete information seriously limits the ability of current Semantic Web technologies to meet the requirements of the Semantic Web. Our work is an initial step toward changing this situation. We aim to establish a framework that enables full support for uncertainty in the field of ontology engineering and, as a consequence, for the Semantic Web. In this work, we focus on extending OWL so it can represent uncertainty in a principled way.

In Section 2 we present related work on the subject. Then, we start Section 3 with an example illustrating the limitations of BNs in terms of expressiveness and how those are addressed in MEBN logic. Section 4 conveys the definition of a probabilistic ontology used in this work. In Section 5, we present the PR-OWL probabilistic ontology language, its main concepts, and an overview of its structure. PR-OWL is implemented in UnBBayes-MEBN, a Java-Based, open source system that is briefly explained in Section 6. In order to provide a general idea of the potential use for the PR-OWL/MEBN framework, Section 7 discusses how the SOA model can benefit from the expressivity and flexibility of a probabilistic ontology system.

2   Related Research

One of the main reasons why Semantic Web research is still focused on deterministic approaches has been the limited expressivity of traditional probabilistic languages. There is a current line of research focused on extending OWL so it can represent probabilistic information contained in a Bayesian Network (e.g. [8, 9]). The approach involves augmenting OWL semantics to allow probabilistic information to be represented via additional markups. The result would be a probabilistic annotated ontology that could then be translated to a Bayesian network (BN). Such a translation would be based on a set of translation rules that would rely on the probabilistic information attached to individual concepts and properties within the annotated ontology. BNs provide an elegant mathematical structure for modeling complex relationships among hypotheses while keeping a relatively simple visualization of these relationships. Yet, the limited attribute-value representation of BNs makes them unsuitable for problems requiring greater expressive power.

Another popular option for representing uncertainty in OWL has been to focus on OWL-DL, a decidable subset of OWL that is based on Description Logics [3]. Description Logics are a family of knowledge representation formalisms that represent the knowledge of an application domain (the "world") by first defining the relevant concepts of the domain (its terminology), and then using these concepts to specify properties of objects and individuals occurring in the domain (the world description).

Description logics are highly effective and efficient for the classification and subsumption problems they were designed to address. However, their ability to represent and reason about other commonly occurring kinds of knowledge is limited. One restrictive aspect of DL languages is their limited ability to represent constraints on the instances that can participate in a relationship. As an example, suppose we want to express that for a carnivore to be a threat to another carnivore in a specific type of situation it is mandatory that the two individuals of class Carnivore involved in the situation are not the same. Making sure the two carnivores are different in a specific situation is only possible in DL if we actually create/specify the tangible individuals involved in that situation. Indeed, stating that two "fillers" (i.e. the actual individuals of class Carnivore that will "fill the spaces" of concept carnivore in our statement) are not equal without specifying their respective values would require constructs such as *negation* and *equality role-value-maps*, which cannot be expressed in description logic. While equality role-value-maps provide useful means to specify structural properties of concepts, their inclusion makes the logic undecidable [10].

Although the above approaches are promising where applicable, a definitive solution for the Semantic Web requires a general-purpose formalism that gives ontology designers a range of options to balance tractability against expressiveness.

Pool and Aiken [11] developed an OWL-based interface for the relational probabilistic toolset Quiddity*Suite, developed by IET, Inc. Their constructs provide a very expressive method for representing uncertainty in OWL ontologies. Their work is similar in spirit to ours, but is specialized to the Quiddity*Suite toolset. We focus on the more general problem of enabling probabilistic ontologies for the SW. We employ Multi-Entity Bayesian Networks (MEBN) as our underlying logical basis, thus providing full first-order expressiveness.

3 Multi-entity Bayesian Networks

The acknowledged standard for logically coherent reasoning under uncertainty is probability theory. Probability theory provides a principled representation of uncertainty, a logic for combining prior knowledge with observations, and a learning theory for refining the ontology as evidence accrues.

Bayesian networks provide an elegant framework for implementing probability theory, but as we explained in the previous section, the limited expressiveness of their attribute-value representation limit their applicability, excluding many real-world problems relevant to the SW. To understand this limitation, consider a relational database in which some entries are uncertain. A BN can represent only probabilities for a single table, and treats the rows of the table independently of each other. For example, in a logistics control system, the "Truck" table might include information such as type, capacity, current delivery schedule, geographical position, and whether it is suitable and available for a given delivery. Assuming such a system being used to track possible deliverers within a 100-kilometer range of the depot area, a BN might represent the probability of a given truck being an optimal delivery option as a function of its availability given the traffic along major local routes and its suitability for the upcoming weather (e.g. preparation for snow conditions). If a truck is currently within the predefined range, the BN of Figure 1 could estimate the probability of this truck of being an optimal deliverer given the abovementioned variables.

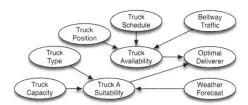

Fig. 1. One Delivery Truck Scenario

However, this BN cannot represent relational information such as the increase in the probability of being an optimal deliverer for all trucks that are within the 100-km range. To incorporate this kind of knowledge in a coherent manner, we need to combine *relational* knowledge (e.g., trucks that are within the same predefined area) with attribute-value knowledge (e.g., heavy traffic and bad weather conditions decrease the probability of being an optimal deliverer for trucks farther from the base and for those not prepared for inclement weather).

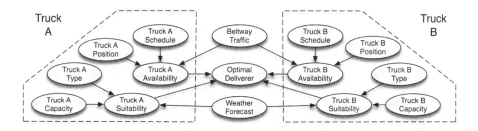

Fig. 2. One Delivery Truck Scenario

Figures 2 and 3 show that as the number of trucks within the 100-km range changes, one must build different BNs. Since this number is variable, multiple BNs must be constructed for each specific case.

MEBN logic [5, 6], which is the logical basis for PR-OWL, combines Bayesian probability theory with classical First Order Logic to overcome BNs' limitations in expressiveness. Probabilistic knowledge is expressed as a set of MEBN fragments (MFrags) organized into MEBN Theories. An MFrag is a knowledge structure that

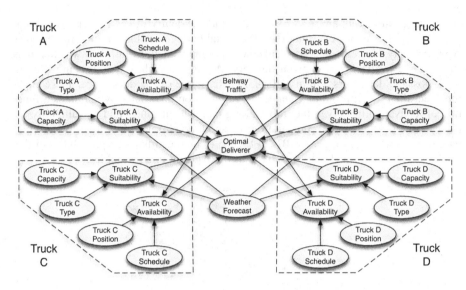

Fig. 3. Delivery Scenario with Four Trucks

represents probabilistic knowledge about a collection of related hypotheses. Hypotheses in an MFrag may be *context* (must be satisfied for the probability definitions to apply), *input* (probabilities are defined in other MFrags), or *resident* (probabilities defined in the MFrag itself). An MFrag can be instantiated to create as many instances of the hypotheses as needed (e.g., an instance of all the "Truck X nodes" created for each Truck within the predefined range). Instances of different MFrags may be combined to form complex probability models for specific situations. An MTheory is a collection of MFrags that satisfies consistency constraints ensuring the existence of a unique joint probability distribution over instances of the hypotheses in its MFrags.

MEBN inference begins when a query is posed to assess the degree of belief in a target random variable given a set of evidence random variables. We start with a generative MTheory, add a set of finding MFrags representing problem-specific information, and specify the target nodes for our query. The first step in MEBN inference is to construct a situation-specific Bayesian network (SSBN), which is a Bayesian network constructed by creating and combining instances of the MFrags in the generative MTheory. When each MFrag is instantiated, instances of its random variables are created to represent known background information, observed evidence, and queries of interest to the decision maker. If there are any random variables with undefined distributions, then the algorithm proceeds by instantiating their respective home MFrags. The process of retrieving and instantiating MFrags continues until there are no remaining random variables having either undefined distributions or unknown values. An SSBN may contain any number of instances of each MFrag, depending on the number of entities and their interrelationships. Next, a standard Bayesian network inference algorithm is applied. Finally, the answer to the query is obtained by inspecting the posterior probabilities of the target nodes. For the Delivery Truck example, the MTheory depicted in Figure 4 could be used for building each of the BNs in the previous figures as well as BNs for any number of Trucks within the predefined range.

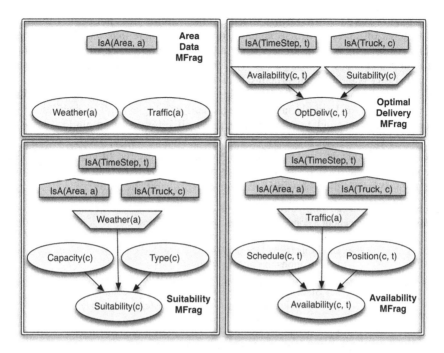

Fig. 4. The Delivery Truck MTheory

To draw generalizations about individuals related in various ways, we need first-order expressive power. Description logics are attractive because they provide limited first-order expressivity, yet certain categories of reasoning problem, such as classification and subsumption, are decidable. The ontology language P-SHOQ(D) [12] extends the description logic SHOQ(D) to represent probabilistic information.

We have chosen to base PR-OWL on MEBN logic because of its expressiveness: MEBN can express a probability distribution over models of any finitely axiomatizable first-order theory. As a consequence, there are no guarantees that exact reasoning with a PR-OWL ontology will be efficient or even decidable. On the other hand, a future objective is to identify restricted sub-languages of PR-OWL specialized to classes of problems for which efficient exact or approximate reasoning algorithms exist. There has been a great deal of research on classes of problems for which efficient probabilistic algorithms exist (e.g., Naïve Bayes classification, in which features are modeled as conditionally independent given an object's class). This research can inform the development of restrictions on MEBN theories that lead to efficient inference on particular kinds of problem. It is our view that a general-purpose language for the Semantic Web should be as expressive as possible, while providing a means for ontology engineers to stay within a tractable subset of the language when warranted by the application.

4 Probabilistic Ontologies

The usual workaround for representing probabilities in deterministic languages like OWL is to show probability information as annotations. This means that numerical information is stored as text strings. Because this solution does not convey the structural features of a probabilistic domain theory, it is no more than a palliative. This is no minor shortcoming. Researchers have stressed the importance of structural information in probabilistic models (see [13]). For instance, Shafer ([14], pages 5-9) stated that probability is more about structure than it is about numbers.

A major concept behind PR-OWL is that of probabilistic ontologies. Probabilistic ontologies go beyond simply annotating standard ontologies with probabilities, providing a logically sound formalism to express all relevant uncertainties about the entities and relationships that exist in a domain. This not only provides a consistent representation of uncertain knowledge that can be reused by different probabilistic systems, but also allows applications to perform plausible reasoning with that knowledge. PR-OWL uses the following definition of a probabilistic ontology [15]:

Definition 1. A *probabilistic ontology* is an explicit, formal knowledge representation that expresses knowledge about a domain of application. This includes:

       1a) Types of entities that exists in the domain;
       1b) Properties of those entities;
       1c) Relationships among entities;
       1d) Processes and events that happen with those entities;
       1e) Statistical regularities that characterize the domain;
       1f) Inconclusive, ambiguous, incomplete, unreliable, and dissonant knowledge;
       1g) Uncertainty about all the above forms of knowledge;
where the term entity refers to any concept (real of fictitious, concrete or abstract) that can be described and reasoned about within the domain of application.   ☐

Probabilistic Ontologies provide a principled, structured and sharable way to comprehensively describe knowledge about a domain and the uncertainty regarding that knowledge. They also expand the possibilities of standard ontologies by introducing the requirement of a proper representation of the statistical regularities and the uncertain evidence about entities in a domain of application. Ideally, the representation is in a format that can be read and processed by a computer.

5 PR-OWL

PR-OWL is an extension that enables OWL ontologies to represent complex Bayesian probabilistic models in a way that is flexible enough to be used by diverse Bayesian probabilistic tools based on different probabilistic technologies. That level of flexibility can only be achieved using the underlying semantics of first-order Bayesian logic, which is not a part of the standard OWL semantics and abstract syntax. Therefore, it seems clear that PR-OWL can only be realized via extending the semantics and abstract syntax of OWL. However, in order to make use of those extensions, it is necessary to develop new tools supporting the extended syntax and implied semantics of each extension. Such an effort would require commitment from diverse developers and workgroups, which falls outside our present scope.

Therefore, in this initial work our intention is to create an upper ontology to guide the development of probabilistic ontologies. Daconta *et al.* define an upper ontology as a set of integrated ontologies that characterizes a set of basic commonsense knowledge notions [16]. In its current form, PR-OWL is an upper ontology of basic notions related to representing uncertainty in a principled way using OWL syntax. If PR-OWL were to become a W3C Recommendation, this collection of notions would be formally incorporated into the OWL language as a set of constructs that can be employed to build probabilistic ontologies.

The PR-OWL upper ontology for probabilistic systems consists of a set of classes, subclasses and properties that collectively form a framework for building probabilistic ontologies. The first step toward building a probabilistic ontology in compliance with our definition is to import into any OWL editor an OWL file containing the PR-OWL classes, subclasses, and properties.

From our definition, it is clear that nothing prevents a probabilistic ontology from being "partially probabilistic". That is, a knowledge engineer can choose the concepts he/she wants to include in the "probabilistic part" of the ontology, while writing the other concepts in standard OWL. In this case, the "probabilistic part" refers to the concepts written using PR-OWL definitions and that collectively form a complete or partial MTheory. There is no need for all the concepts in a probabilistic ontology to be probabilistic, but at least some have to form a valid complete or partial MTheory. Of course, only the concepts that are part of the MTheory will be subject to the advantages of the probabilistic ontology over a deterministic one.

The subtlety here is that legacy OWL ontologies can be upgraded to probabilistic ontologies only with respect to concepts for which the modeler wants to have uncertainty represented in a principled manner, make plausible inferences from that uncertain evidence, or to learn its parameters from incoming data via Bayesian learning. While the first two are direct consequences of using a probabilistic knowledge representation, the latter is a specific advantage of the Bayesian paradigm, where learning falls into the same conceptual framework as knowledge representation.

The ability to perform probabilistic reasoning with incomplete or uncertain information conveyed through an ontology is a major advantage of PR-OWL. However, it should be noted that in some cases solving a probabilistic query might be intractable or even undecidable. In fact, providing the means to ensure decidability was the reason why the W3C defined three different version of the OWL language. While OWL Full is more expressive, it enables an ontology to represent knowledge that can lead to undecidable queries. OWL-DL imposes some restrictions to OWL in order to eliminate these cases. Similarly, restrictions of PR-OWL could be developed that limit expressivity to avoid undecidable queries or guarantee tractability. Possible restrictions to be considered for an eventual PR-OWL Lite include (*i*) constraining the language to classes of problems for which tractable exact or approximate algorithms exist; (*ii*) restrict the representation of the conditional probability tables (CPT) to express a tractable and expressive subset of first-order logic; and/or (*iii*) to employ a standard semantic web language syntax to represent the CPTs (e.g. RDF). As an initial step, we chose to focus on the most expressive version of PR-OWL, which does not have expressivity restrictions and provides the ability to represent CPTs in multiple formats.

An overview of the general concepts involved in the definition of an MTheory in PR-OWL is depicted in Figure 5. In this diagram, the ovals represent general classes; and arrows represent major relationships between classes. A probabilistic ontology

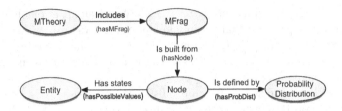

Fig. 5. Overview of a PR-OWL MTheory Concepts

must have at least one individual of class MTheory, which is a label linking a group of MFrags that collectively form a valid MTheory. In actual PR-OWL syntax, that link is expressed via the object property hasMFrag (which is the inverse of object property isMFragIn).

Individuals of class MFrag are comprised of nodes, which can be resident, input, or context nodes (not shown in the picture). Each individual of class Node is a random variable and thus has a mutually exclusive and collectively exhaustive set of possible states. In PR-OWL, the object property hasPossibleValues links each node with its possible states, which are individuals of class Entity. Finally, random variables (represented by the class Nodes in PR-OWL) have unconditional or conditional probability distributions, which are represented by class Probability Distribution and linked to its respective nodes via the object property hasProbDist.

The scheme in Figure 5 is intended to present just a general view and thus fails to show many of the intricacies of an actual PR-OWL representation of an MTheory. Figure 6 shows an expanded version conveying the main elements in Figure 5, their subclasses, the secondary elements that are needed for representing an MTheory and the reified relationships that were necessary for expressing the complex structure of a Bayesian probabilistic model using OWL syntax.

Reification of relationships in PR-OWL is necessary because of the fact that properties in OWL are binary relations (i.e. link two individuals or an individual and a value), while many of the relations in a probabilistic model include more than one individual (i.e. N-ary relations). The use of reification for representing N-ary relations on the Semantic Web is covered by a working draft from the W3C's Semantic Web Best Practices Working Group [17].

Although the scheme in Figure 6 shows all the elements needed to represent a complete MTheory, it is clear that any attempt at a complete description would render the diagram cluttered and incomprehensible. A complete account of the classes, properties and the code of PR-OWL that define an upper ontology for probabilistic systems is given in [15]. These definitions can be used to represent any MTheory.

In its current stage, PR-OWL contains only the basic elements needed to represent any MTheory. Such a representation could be used by a Bayesian tool (acting as a probabilistic ontology reasoner) to perform inferences to answer queries and/or to learn from newly incoming evidence via Bayesian learning.

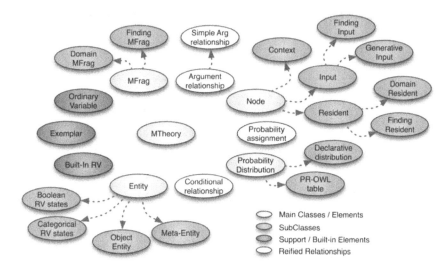

Fig. 6. Elements of a PR-OWL Probabilistic Ontology

6 UnBBayes-MEBN

Building MFrags and all its elements into a PO is a difficult, tedious and error prone process that demands deep knowledge of PR-OWL's syntax, semantics and data structure. Creating a PO without a PR-OWL software tool would be very difficult. An ordinary OWL ontology can be built using a graphical ontology editor such as Protégé. To add probability information, PR-OWL definitions can be imported into Protégé (from http://www.pr-owl.org/pr-owl.owl), making the task of building a PO a bit easier because it is not necessary to remember all information and OWL tags that should be inserted. However, the input of information is not intuitive and the user has to know many technical terms as hasPossibleValues, isNodeFrom, hasParents, etc (Figures 7 and 8).

UnBBayes-MEBN [18] is a PR-OWL GUI and reasoner that addresses all the above issues. It provides a GUI designed to allow building a PO in an intuitive way, enforcing the consistency of an MTheory without the requirement of deep knowledge of the PR-OWL specification.

As an example of the process of creating an MFrag, a click on the "R" icon and another click anywhere in the editing panel will create a resident node, as shown in Figure 9. After that, clicking on the "+" button allows the user to fill a name and to add the states of the node. All the remaining tasks required by PR-OWL syntax (e.g. filling the terms as isResidentNodeIn, etc.) are automatically performed by UnBBayes. Figure 10 shows how UnBBayes allows a more adequate and better visualization of the MTheory and MFrags being created, as well as their nodes. In short, it is not difficult to perceive the advantages of building POs with the GUI implemented in UnBBayes.

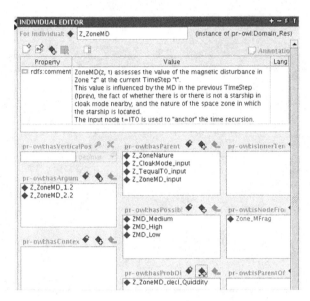

Fig. 7. Node ZoneMD specification with Protégé

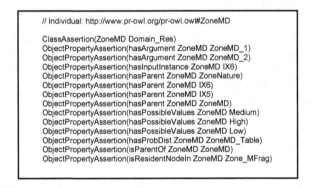

Fig. 8. Node ZoneMD specification in OWL syntax (Manchester)

Implementing a complex logic such as MEBN while focusing on the usability requirements of an (probabilistic) ontology editor requires making trade-offs between
performance, decidability, expressivity, and ease of use. In other words, the complexity of the logic and the fact that it is still in development imply that any implementation has to include alternative algorithms and optimizations to make a working,
feasible tool. UnBBayes-MEBN is no exception to this rule, and many of the design
decisions were based on the above-cited constraints.

Probabilistic ontologies in UnBBayes-MEBN are saved in the PR-OWL format,
which is an extension of OWL format. UnBBayes-MEBN uses the Java open source
Protégé application programming interface (API) for importing and saving OWL
files. UnBBayes-MEBN provides support for MEBN input/output operations using
the Protégé-OWL editor, which is based on the class JenaOWLModel. Protégé uses

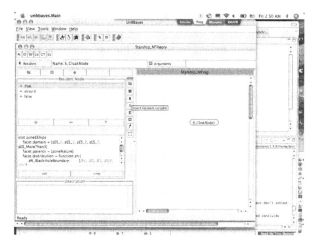

Fig. 9. Node specification with UnBBayes-MEBN

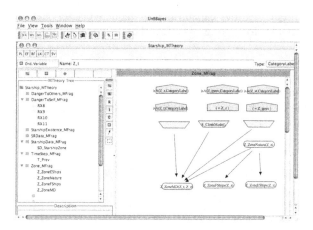

Fig. 10. General view of the MTheory and MFrag in UnBBayes

the Jena API for various tasks, in particular for parsing OWL/RDF files. Since the structure and algorithms behind UnBBayes-MEBN are outside of the scope of this Chapter, the interested reader should refer to [19, 20] for more information.

7 The Role of Probabilistic Ontologies in SOA

Service Oriented Architecture (SOA) has become the leading approach for accessing and using distributed resources developed by independent entities and working with independently developed vocabularies and associated semantics. The advent of SOA marks a transformation from a mostly data-driven Web, with little interaction between requesters and providers of information, into an environment in which information and other resources are accessed and used in a much more dynamic, interactive, and unpredictable fashion.

The supporting technology for the SOA model is composed of XML-based standards and protocols focused on providing a shared understanding of the available services. Currently, accepted standards for developing solutions based on Web Services (the most prevalent implementation of SOA) include SOAP, a message structure used for exchanging XML serializations of content and message handling instructions in a decentralized, distributed environment [21], and the Web Services Description Language (WSDL), which represents messages exchanged when invoking a Web Service [22]. However, these XML-based structures do not have the ability to explicitly formalize the underlying semantics of a given Web Service description, rendering them insufficient to ensure a common understanding of the described Web Service. As pointed out by Paolucci et al. [23], two identical XML descriptions may have different meanings depending on who uses them and when. Different providers and consumers will have perspectives aligned with their respective domains; thus, a common understanding of a given Web Service can be reached only at the semantic level, where the different perspectives and knowledge can be matched.

Not surprisingly, the need for semantic-aware resource descriptions is widely recognized, and is being addressed by work focused on enabling Web Service providers to describe the properties and capabilities of their Web Services in unambiguous, computer-interpretable form (e.g. OWL-S [24], WSMO [25], SWSL [26], and SAWSDL [27]).

This section argues that progress on both SW and SOA is hampered by the lack of support for uncertainty in common ontology formalisms. We postulate that probabilistic ontologies can fill a key gap in semantic matching technology, thus facilitating widespread usage of Web Services for efficient resource sharing in open and distributed environments.

7.1 Uncertainty Present in SOA

In order to envision the applicability of POs in SOAs, it is necessary to first understand what kind of uncertainties might be present in a service-oriented environment. As defined in the SOA reference model [28], SOA is a paradigm for bringing together needs and capabilities to address those needs. It requires establishing an execution context (EC), which is an alignment of all technical and policy-related aspects, including vocabularies, protocols, licensing, quality of service (QoS), etc. Much of this specific information is contained in or linked to the service description and/or the description of underlying capabilities. Considering the complexity involved, many forms of uncertainty can be present within a given execution context. For example, uncertainty may arise in the description content (e.g. information is annotated with its source, but there is no way to verify whether the identity of the source is correct), in the way information is captured as part of a description (e.g. information is annotated with its source, but there is no indication of whether it is raw data or what processing has been applied), or in the applicability of information to current need (e.g., information on recording equipment does not indicate whether the recorded data fall within a reasonable range for the recording conditions). An ontology that represents statistical information can enable a reasoner to draw inferences about the missing information. For example, consider a report that a device has recorded an ambient temperature of 5 degrees Celsius at Rio de Janeiro's Tom Jobin International Airport (GIG) on 23 January. This is a highly unlikely, but not impossible, temperature reading for January near Rio. Statistical information about climate, sensor reliability, and data recording error rates, if represented in the relevant domain ontologies, could be used to draw inferences about the about the likely temperature at GIG on 23 January and could appropriately account for the possibility of various kinds of error. If a Web service was used to access the temperature, a service description that included or referenced a representation of such uncertainties would enable decisions that could improve the effective and appropriate use of the available capabilities. Such an uncertainty feature is not incorporated in current Web Service specifications.

A typical Web Services scenario is often described in terms of the publish-find-bind triangle: (1) a service provider publishes a service description, (2) a consumer searches a service registry for a service satisfying his criteria, analyzes the returned information (or link to information) on the message structure to be exchanged and the address to exchange it, and (3) interacts with the service to retrieve the resources needed. In this triangle, there are implicit, unspoken challenges for which a principled representation of uncertainty is needed. For example:

- The service provider, typically playing the role of description publisher, has to choose a vocabulary with which to describe the service (or some other resource related to the service), thus setting the properties by which to describe that class of item. Service providers attempt to define the "right" set and structure of properties that make visible what they wish to highlight as discriminators for those looking for services. The consumer, on the other hand, has her own criteria to satisfy but must know and understand the semantics of the service provider vocabulary because these are the properties used to describe the generic service class and its instances. The consumer must understand and use this vocabulary or there must be a known and accessible mapping between the properties used for description and those more naturally used by the consumer as search categories. There are many opportunities for uncertainty about intended meanings of the service class properties, the use of those properties to describe service instances, and the relationship to consumer search criteria.

- The publisher uses the chosen property vocabulary as the basis to describe and register instances of that class. This means that the publisher associates values with the properties and registers the instance. But what is the vocabulary for the values? All parties may agree that something has the property color and on the meaning of that property, but if the publisher uses only primary colors and the subscriber's search criterion asks for the color pink, the latter will never find a match for items the first had catalogued. How does a client's requested value relate to a provider's published values? Do they agree on the vocabulary? Do they agree on the mechanism to mediate vocabulary mismatches?

- The publisher chooses a property vocabulary and creates instance descriptions by associating values. One can infer the properties the publisher considers important by which properties s/he chooses to populate, assuming values are not necessarily assigned for all possible properties. But what of the consumer's priorities when assigning search criteria? If the consumer assigns relative importance, how does the search engine trade off among different combinations of matches across the consumer's search criteria, and how are missing attribute values handled?

7.2 Uncertainty and Semantic Mapping

Figure 11 uses the example of a Delivery Control System to illustrate the semantic mapping challenges that are implicit in the above description. In the diagram, the Delivery Company (a service provider) has built a local vocabulary (Vocab α) based on a vocabulary available for Delivery and Shipping services (Vocab θ). The type of a truck in Vocab θ is defined as Property P and was copied unmodified from Vocab θ to Vocab α as property P1. User A, who is interested in the Delivery service, also based his vocabulary (Vocab β) on Vocab θ and has copied P with no changes to his own P2.

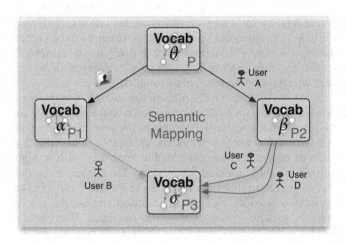

Fig. 11. Challenges of Synchronizing Vocabularies for the Publish-Find-Bind Exchange

Now given information about the Delivery Company and User A, how likely is that the same string (say "large container") in each vocabulary still refers to the same concept (and probably as meant by Vocab θ)?To address this question, it is necessary to keep track of characteristics such as the overall similarities of the concepts in the vocabularies, the origins of each concept, and other characteristics related to the problem context. A domain independent PR-OWL ontology would keep track of these characteristics, and thus be able to infer the likelihood that concepts with the same name have the same meaning.

In a more complex development of the situation depicted in Figure 11, assume Vocab σ exists as a vocabulary for Logistics. User B, who wants to use the Delivery service based on Vocab α, has its own system based on the Logistics Vocab σ, and so has developed a mapping from the Vocab α to Vocab σ. In Vocab σ, the type of a truck is indicated by property P3. Two other Users, C and D, have their own systems based on the Logistics Vocab σ, but are also customers of User A. To communicate with User A, Users C and D have each built mappings from User A system's Vocab β to their respective system's use of Vocab σ. With three different mappings linking two ontologies that were based on the same upper ontology to a fourth ontology (the Logistics ontology), what is the likelihood that each of them has the same intended meaning for the truck type, and that it matches the intended meaning of the original upper ontology's property P? If we wish to infer mappings between Vocab α and

Vocab β and have User B's mapping from Vocab α to Vocab σ, is the mapping of User C or of User D from Vocab β to Vocab σ more likely to reflect the needs in mapping between Vocab α and Vocab β, or does an appropriate combination of the User C and the User D mappings provide an optimum result? Again, a domain independent PR-OWL ontology can be used to keep track of such mappings, considering parameters such as the expertise of each user to help grading the likelihood of each mapping to be closer to the original and more aligned with the eventual need.

7.3 Uncertainty and Service Composition

Beyond publish-find-bind for a single service, the vision is to provide services at the appropriate granularity, combining atomic services into more complex tasks. For example, suppose a supplier needs to find the dimensions and weight limits for cargo containers for future shipments of items it produces. In today's integration paradigm, the supplier would need to query specific shipping agents directly, and might need to develop special-purpose software interfaces to support interactions with individual shipping agents. In the envisioned architecture, the supplier would invoke a service that (i) searches a UDDI registry for shipping agents; (ii) queries each for its respective restrictions; (iii) compares with the supplier's requirements; and (iv) selects a shipper that meets the requirements.

This simple scenario does not include other actions that must be included in such a transaction. For example, security will be needed to authenticate the supplier to the shipping agent and the shipping agent to the supplier. Other actions may be required to establish that each party is authorized to engage in business with the other. The interaction itself may require a guaranteed level of service that would fall into the realm of reliable messaging to guarantee delivery. Additionally, the response from the shipping agent could optionally include video showing details of container packing and handling, and these would not be appropriate to send if the supplier is using a low bandwidth communications link.

Security, reliable messaging, and results dissemination are examples of general-purpose services that could be combined with services for specific business functions, thus freeing the business service from the need to create and maintain all supporting services. All of these services will have associated service descriptions so that someone composing a robust service combination can identify the appropriate services and the process by which these will work together to provide the higher-level functionality. That said, what are the uncertainties in identifying the correct services and combining these to form a consistent package? Is uncertainty even a relevant concept, or is it a black-and-white issue of whether the pieces fit or not? When trying to decide among several services that appear to satisfy aspects of the same needed function, does the ability to reason under uncertainty come into play in identifying the component services to use and how to combine these?

The above questions do not have simple, universally valid answers. Undoubtedly, there will be problems for which deterministic implementations of SOA elements will suffice to build viable solutions. Nevertheless, there are issues that cannot be satisfactorily solved without a principled representation of uncertainty. Probabilistic ontology languages such as PR-OWL can fulfill this requirement.

7.4 Ontology Federation to Support SOA

Providing a detailed account of how to use PO languages to build standards for SOA elements, or even examples of (say) service descriptions with probabilistic elements would require detailed explanation that goes beyond the limits of this paper. Thus, as a means to explore another possible use of POs in a SOA environment, we now present a possible framework using a federation of ontologies (common and probabilistic) for tackling the problem of semantic mapping among concepts used in Web Services (WS) descriptions within a WS repository.

Figure 12 shows a simplified scheme for SOA using probabilistic semantic mapping. As a means to illustrate this scheme, we will devise fictitious examples involving Web Service providers within the geospatial reasoning domain. In this scheme, a service consumer or provider that conveys semantic information (ontology that it abides to, metadata about its requests, parameters, etc.) is called a SOA node Level 1, whereas a SOA node that has no semantic awareness is called a SOA node Level 0.

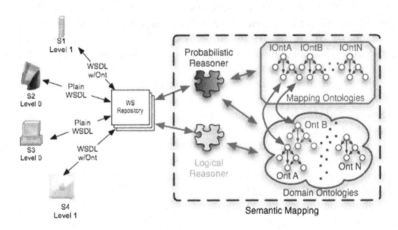

Fig. 12. Probabilistic Semantic Mapping for Web Services

In our first use case, S1 needs to generate a travel plan and requests a service for assessing the possibility of flooding in a given region due to recent heavy rains. Being a Level 1 client, S1 sends its request with embedded data about the ontology it references and other semantic information regarding its request (e.g. coordinate system used, expected QoS, etc.). The WS repository, which itself uses an ontology, finds S4, another Level 1 client using the same ontology as S1. This ontology is the PR-OWL ontology "OntB", which represents a probabilistic model of the geospatial domain and has the ability to perform a probabilistic assessment of the requested information. In this case, the request was probabilistic, but the uncertainty involved was related to the service itself (a probabilistic query on a uncertainty-laden domain), and not to the service exchanging process. In other words, the exchange was completed using the logical reasoner alone, since there was a perfect matching in terms of ontologies (both S1 and S4 abide to the same PR-OWL ontology) and the parameters of the requested service, and thus no probabilistic mapping was necessary. (Yet, note that S1's query made use of OntB's ability to represent uncertainty about the geospatial domain.)

In a variation of the previous case, let's suppose that no perfect match between the request and the available providers is found. In this case, the probabilistic reasoner accesses the WS repository to search for the most suitable service given the parameters of S1's request. During that process, it analyses the mapping ontologies related to "OntB" (the ontology referenced by S1) and the domain ontologies related to the services it deemed promising to fit S1's request. In the end, an ordered list of possible providers is built, and the best possible answers will be returned to S1. This simple example shows that there might be many combinations of the use of logical and probabilistic reasoners and ontologies to match the needs of a specific request.

8 Conclusion

This paper describes Probabilistic Ontologies as an initial step towards a coherent, comprehensive probabilistic framework for the Semantic Web. It also demonstrates how the use of such a framework can bring Semantic Web power to bear on the discovery and use of Web Services in a service oriented environment. In order to better convey the framework, we provided an explanation of the major concepts behind it, such as MEBN logic, probabilistic ontologies, and the PR-OWL language, and lay the groundwork for a more comprehensive effort focused on representing uncertainty in the Semantic Web.

A PR-OWL ontology editor that facilitates the creation of probabilistic ontologies built on MFrags and MTheories was also presented. It automates many of the steps in the ontology building, greatly facilitating the process of writing probabilistic ontologies. The automation includes defining MFrags to represent sets of related hypotheses, consistency checking and other tasks that demand unnecessary awareness of the inner workings of the present solution.

Finally, we discussed various aspects of SOA that would be enhanced by a means to represent and reason over uncertainty. We provided examples that demonstrate the benefits of probabilistic ontologies to enable semantic negotiation among independently developed but related vocabularies and to assist in composing complex solutions from services providing elementary functionality.

References

1. Berners-Lee, T., Fischetti, M.: Weaving the Web: the original design and ultimate destiny of the World Wide Web by its inventor, 1st edn., vol. ix, p. 246. HarperCollins Publishers, New York (2000)
2. Patel-Schneider, P.F., Hayes, P., Horrocks, I.: OWL Web ontology language - Semantics and abstract syntax. In: W3C Recommendation. 2004, World Wide Web Consortium, Boston, MA, W3C Recommendation (2004)
3. Baader, F., et al. (eds.): The Description Logic Handbook: Theory, Implementation and Applications, 1st edn., p. 574. Cambridge University Press, Cambridge (2003)
4. Korb, K.B., Nicholson, A.E.: Bayesian Artificial Intelligence. Series in Computer Science and Data, p. 392. Chapman & Hall/CRC (2003)
5. Laskey, K.B.: MEBN: A Language for First-Order Bayesian Knowledge Bases. Artificial Intelligence 172(2-3) (2007)

6. Laskey, K.B., Costa, P.C.G.: Of Klingons and Starships: Bayesian Logic for the 23rd Century. In: Uncertainty in Artificial Intelligence: Proceedings of the Twenty-first Conference. AUAI Press, Edinburgh (2005)

7. Heckerman, D., Mamdani, A., Wellman, M.P.: Real-world applications of Bayesian networks. Communications of the ACM 38(3), 24–68 (1995)

8. Ding, Z., Peng, Y.: A probabilistic extension to ontology language OWL. In: 37th Annual Hawaii International Conference on System Sciences (HICSS 2004), Big Island, Hawaii (2004)

9. Gu, T., Keng, P.H., Qing, Z.D.: A Bayesian approach for dealing with uncertainty contexts. In: Second International Conference on Pervasive Computing. Austrian Computer Society, Vienna (2004)

10. Calvanese, D., De Giacomo, G.: Expressive Description Logics, in The Description Logic Handbook: Theory. In: Baader, F., et al. (eds.) Implementations and Applications, pp. 184–225. Cambridge University Press, Cambridge (2003)

11. Pool, M., Aikin, J.: KEEPER: and Protégé: An elicitation environment for Bayesian inference tools. In: Workshop on Protégé and Reasoning held at the Seventh International Protégé Conference, Bethesda, MD, USA (2004)

12. Giugno, R., Lukasiewicz, T.: P-SHOQ(D): A probabilistic extension of SHOQ(D) for probabilistic ontologies in the Semantic Web. In: Flesca, S., Greco, S., Leone, N., Ianni, G. (eds.) JELIA 2002. LNCS, vol. 2424, pp. 86–97. Springer, Heidelberg (2002)

13. Schum, D.A.: Evidential Foundations of Probabilistic Reasoning. Wiley, New York (1994)

14. Shafer, G.: The Construction of Probability Arguments. Boston University Law Review 66(3-4), 799–816 (1986)

15. Costa, P.C.G.: Bayesian Semantics for the Semantic Web. In: Department of Systems Engineering and Operations Research, p. 312. George Mason University, Fairfax (2005), http://www.pr-owl.org

16. Daconta, M.C., Obrst, L.J., Smith, K.T.: The Sematic Web: A guide to the future of XML, Web services, and knowledge management, p. 312. Wiley Publishing, Inc., Indianapolis (2003)

17. Noy, N.F., Rector, A.: Defining N-ary relations on the Semantic Web: Use with individuals. In: W3C Working Draft. World Wide Web Consortium, Boston (2004)

18. Carvalho, R.N., Santos, L.L., Ladeira, M., Costa, P.C.G.: A Tool for Plausible Reasoning in the Semantic Web using MEBN. In: Proceedings of the Seventh International Conference on Intelligent Systems Design and Applications, pp. 381–386. IEEE Press, Los Alamitos (2007)

19. Carvalho, R.N., Ladeira, M., Santos, L.L., Matsumoto, S., Costa, P.C.G.: NBBayes-MEBN: Comments on Implementing a Probabilistic Ontology Tool. In: IADIS International Conference - Applied Computing 2008, Algarve, Portugal, April 10-13 (accepted, 2008)

20. Costa, P.C.G., Ladeira, M., Carvalho, R.N., Laskey, K.B., Santos, L.L., Matsumoto, S.: A First-Order Bayesian Tool for Probabilistic Ontologies. In: 21st International Florida Artificial Intelligence Research Society Conference (FLAIRS-21), Coconut Grove, Florida, USA, May 15-17 (to appear, 2008)

21. Mitra, N.: SOAP Version 1.2 Part 0: Primer, W3C Recommendation 24 June (2003), http://www.w3.org/TR/soap12-part0/

22. Chinnici, R., Moreau, J., Ryman, A., Weerawarana, S.: Web Services Description Language (WSDL) Version 2.0 Part 1: Core Language, W3C Candidate Recommendation, 27 March (2006), http://www.w3.org/TR/wsdl20

23. Paolucci, M., Kawamura, T., Payne, T., Sycara, K.: Importing the Semantic Web in UDDI. In: Pidduck, A.B., Mylopoulos, J., Woo, C.C., Ozsu, M.T. (eds.) CAiSE 2002. LNCS, vol. 2348. Springer, Heidelberg (2002)
24. Martin, D. (ed.): OWL-S: Semantic Markup for Web Services, http://www.daml.org/services/owl-s/1.1/overview/
25. Domingue, J., Lausen, H., Fensel, D.: (Chairs) ESSI Web Services Modeling Ontology Working Group, http://www.wsmo.org
26. Battle, S., et al.: Semantic Web Services Framework (SWSF) Overview (2005), http://www.daml.org/services/swsf/1.0/overview/
27. W3C Semantic Annotations for WSDL Working Group, (2006), http://www.w3.org/2002/ws/sawsdl
28. Mackensie, C.M., Laskey, K.J., McCabe, F., Brown, P.F., Metz, R.: Reference Model for Service Oriented Architecture 1.0. Committee Specification 1 (2006) (July 19, 2006), http://www.oasis-open.org/committees/download.php/19361/soa-rm-cs.pdf

Discovery and Uncertainty in Semantic Web Services

Francisco Martín-Recuerda[1] and Dave Robertson[2]

[1] Information Management Group (IMG), University of Manchester,
2.114 Kilburn Building, Oxford Road, Manchester, M13 9PL, UK
fmartin-recuerda@cs.man.ac.uk
[2] University of Edinburgh, School of Informatics, Centre for Intelligent Systems and their applications, Informatics Forum, Crichton Street, Edinburgh, EH89AB, Scotland, UK
dr@inf.ed.ac.uk

Abstract. Although Semantic Web service discovery has been extensively studied in the literature ([1], [2], [3] and [4]), we are far from achieving an effective, complete and automated discovery process. Using the Incidence Calculus [5], a truth-functional probabilistic calculus, and a lightweight brokering mechanism [6], this article explores the suitability of integrating probabilistic reasoning in Semantic Web services environments. We show how the combination of relaxation of the matching process and evaluation of Web service capabilities based on previous performances of Web service providers enables new possibilities in service discovery.

Keywords: Web services, Semantic Web services, discovery, broker, F-X, capability, probability, Incidence Calculus.

1 Introduction

Middleware is the "glue" that facilitates and manages the interaction between applications across heterogeneous computing platforms. Web services is a middleware infrastructure that provides descriptions of certain capabilities of an application (software component) and allow its remote execution using Internet protocols. To reduce manual efforts during the location, combination and use of Web services, machine processable semantics has been added to them creating Semantic Web services [7].

Web service discovery ([1], [8], [2] and [3]) is the act of locating Web services that meet certain functional criteria. Service requesters (clients) usually specify their wishes using a goal (a functional description of objectives that clients want to achieve using Web services). Service providers publish Web services capabilities (functional descriptions of a Web service) on Matchmakers and/or Brokers (service registry). Brokers like Matchmakers [9] are intermediate systems between clients and service providers that store web service capabilities and interfaces (description of how the functionality of the Web service is achieved), and locate Web services which capabilities match client's goals. Brokers also manage the interaction between clients and selected web services.

An exact match between a goal and a required Web service can be sometimes difficult to get. So, the relaxation of the matching conditions has been suggested to

P.C.G. da Costa et al. (Eds.): URSW 2005-2007, LNAI 5327, pp. 108–123, 2008.

improve Web service discovery [3]. Roughly speaking, the relaxation of the matching process between a goal and web services capabilities has been based on the following set of matching notions [4]: (i) exact-match, a goal and matched Web service capabilities are the same; (ii) plug-in-match, a goal is subsumed by matched Web service capabilities; (iii) subsume-match, matched Web service capabilities are subsumed by a goal; (iv) intersection-match, a goal and matched Web service capabilities have some elements in common; and (v) disjoint-match, a goal and matched Web service capabilities does not follow any of the previous definitions. Although matching notions relax the selection of target web services, in a future scenario in which thousands of services can potentially fulfil (or partially fulfil) the objectives described in a goal, a fine-grained classification of matching notions may be necessary for improving the degree of automation of the discovery process. One possible approach is to identify a degree of matching inside of each matching notion; other possibility, described in this paper, is to introduce an extra parameter that qualified a selected collection of Web services. Thus, if we found one thousand web services that follow an intersection-match pattern, we can filter which are the most promising web services for the goal requested based on a selected parameter. To do this, it would be useful to have a mechanism that collects valuable information about chosen Web services. Brokers can be a good choice for keeping a record of the quality of Web services, because brokers can analyze which Web services have been frequently available and which ones have been successfully used during a client request.

Identifying the "most promising" (or the "best possible") Web services based on their quality introduces a significant degree of uncertainty that requires a specific formalism to handle it. The Incidence Calculus [5] is a truth-functional probabilistic calculus in which the probabilities of composite formulae are computed from intersections and unions of the sets of worlds for which the atomic formulae hold true. Incidence Calculus can be easily integrated with other logic formalisms like propositional logic and logic programs that provide the foundations of Semantic Web services frameworks (e.g. OWL-S[1], WSMO[2] and Meteor-S[3]).

For testing purposes, we have used F-X [6], a modular formal knowledge management system developed at University of Edinburgh that includes a broker called F-Broker. The language used in F-Broker for describing Semantic Web services have common roots with WSMO (both follows the main principles of UPML [10]), and can deal with WSMO/OWL-S ontologies and Web services that fall into DLP fragment [11]. We will show in this paper how we have extended F-Broker to deal with relaxed matching notions, how this new version of F-Broker can filter Semantic Web services based on their quality, and how Incidence Calculus can be nicely integrated to deal with the uncertainty that quality measurements introduced.

The paper is structured as follows: section 2 introduces process F-X system. In section 3 is explained how we enhanced F-Broker using Incidence Calculus. Section 4 provides a short review of other improvements for F-Broker. Related work on probabilistic logic in the Semantic Web is described in section 5. Finally, conclusions and future work are included in section 6.

---

[1] http://www.daml.org/services/owl-s/

[2] http://www.wsmo.org/

[3] http://lsdis.cs.uga.edu/projects/meteor-s/

2 F-X, a Formal Management System

The **F-X system** [6] aims to provide a general modular and extendable architecture for formal management systems, spanning the entire knowledge management lifecycle (knowledge acquisition, transformation and publication). The F-X prototype (see figure 1) includes six different components: F-Comp (component representation language), F-Broker (broker component), F-Bus (communication component), F-Env (component for ontological envelope checker), F-Life (lifecycle manager component), and F-Pub (knowledge publication component). The design goal of **F-Comp** is to provide a simple language with a reduced set of primitives, but rich enough to describe all aspects of the design and interaction (communication) of any distributed collection of components capable of expressing knowledge in some form. **F-Broker** is the automated broker mechanism that stores the capabilities of knowledge components (i.e. problem-solving methods or Semantic Web services), identifies the assemblies of knowledge components appropriate for a given task, and manages (coordinates) interactions of selected knowledge components. F-Comp and F-Broker have been designed for describing and coordinating problem-solving methods and Semantic Web services [12]. Thus, F-X has become a vehicle for testing several aspects in the design and implementation of Semantic Web services. The interaction between knowledge components is handled by **F-Bus,** a compact communication system for knowledge components. F-Bus follows main principles of Linda communication style in which knowledge components publish and read in an asynchronous manner tuples (currently Prolog facts) in a tuple-space. **F-Life** is another component of F-X that defines an abstract calculus for modelling lifecycles of knowledge acquisition, transformation and publishing. F-Life provides tool support for building more specialized forms of lifecycles and for analyzing existing lifecycles. **F-Env** provides a compact meta-interpretation mechanism for ontological constraint checking. Finally, **F-Pub** includes tool support for synthesis of Web pages and Web sites from formally expressed knowledge.

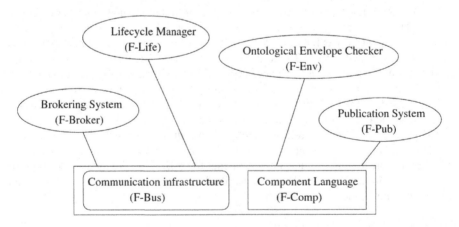

Fig. 1. F-X Architecture [6]

After this brief overview of F-X (more information in [6]), we will present in more detail two components (F-Comp and F-Broker) that are relevant to understand our work on service discovery using Incidence Calculus.

2.1   F-Comp, Component Representation Language

Inspired by previous work on coordination of distributed agents [13] and efforts on the *Unified Problem-Solving Method* (UPML [10]), **F-Comp** is a language that can represent *ontologies, domain descriptions, problem-solving methods,* and *bridges,* but unlike UPML, it is not able to represent tasks (although it can be possible to reintroduce if it is needed). An **ontology definition** in F-Comp specifies its signature (set of terms may be constraining which syntactic structures) and its axioms (set of formal definitions of each term of the signature). To characterize the domain knowledge available to solve certain problems, F-Comp includes **domain definitions**. The essential components of the domain description are: the name of the ontologies used; the properties of the knowledge expressed in the domain model; and the domain knowledge itself. A key element of F-Comp is a **problem-solving method**. It defines the reasoning process used to solve a concrete problem. The essential components of a problem-solving method are: the name of the ontology it uses; the capabilities which it provides; and the problem-solving mechanism delivering these capabilities. [12] reformulates the F-Comp form for problem-solving method into a close form of the DAML-S service profile and specification that is not compatible with the original F-Comp. Thus, using F-Comp we can model simple Semantic Web service capabilities that can be tested by F-Broker. Knowledge components might be specified using different ontologies. A **bridge** (also called **correspondence**) defines a translation between terms of different ontologies. The essential components of a bridge are: the name of the ontology being translated from; the name of the ontology being translated to; the renamings and mappings describing the translation; and the constraints applying to translated terms. For simplicity, **task descriptions** (similar to the notion of *goal* in Semantic Web services) are not supported by F-Comp. Essentially, task descriptions are problem-solving methods which do not commit to a specific method, and in F-X only existing problem-solving methods can advertise their capabilities.

2.2   Describing Capabilities Using F-Comp

The capability language included in F-Comp was designed to describe key information which can be obtained from a knowledge component without stipulating how is derived. F-Comp capability language is defined following horn clause notation [14]. The intuition behind the use of horn clauses for capabilities is that we have four reforms of capability, C, each of which is implemented within the expression cap(K, C), denoting that the agent named K can deliver capability, C in at least one instance or, if not, will signal failure. Valid options for C are [6]:

- A **unit goal** of the form P($A_1, ..., A_n$), where P is a predicate name and $A_1, ..., A_n$ are its arguments.
- A **conjunctive goal** of the form (C1∧...∧Cm), where each C_i is a unit goal or a set expression.

—   A **set expression** of the form setof(X,C,S) , where C is either a unit goal or
    a conjunctive goal; X is a tuple of variables appearing in C; and S is a set of in-
    stances of those tuples which satisfy C.
—   A **conditional goal** of the form $C_c \leftarrow C_p$, where C_c is a unit goal which the agent,
    K, will attempt to satisfy (but will not guarantee to satisfy) if the condition, C_p , is
    satisfied. C_p is either a unit goal or a conjunctive goal.

In addition, the capability language of F-Comp is able to describe **partial capabilities** which are capabilities that require additional information from another knowledge
component, mediated by the broker. A partial capability is defined by the expression
p_cap(K; C; E), which is identical to our original capability description but with
an additional argument, E, containing the capability required from another knowledge
component.

As we mentioned above, there is a version of F-Comp that provides a slightly different version of the capability description language, closer to DAML-S service profile. A detailed description of this revised version of F-Comp can be found in [12].

2.3 F-Broker, a Brokering Mechanism

The purpose of a broker is to find, for a given task (goal) posed by a client, the ways
in which knowledge components (i.e. problem-solving method, agent or Semantic
Web services) which have advertised their capabilities might be contacted in order to
satisfy that task. F-Broker requires that each knowledge component advertises first its
capabilities simply by sending each capability to F-Broker. For the given task, the
broker constructs from its capability descriptions its internal description, which is
called "*brokerage structure*" [6] of how the task might be complete based on those
capabilities. It then translates its brokerage structure into a sequence of communication acts (*performative statements* in KQML [15]) describing the messages which it
thinks should enable the task to be satisfied by requesting appropriate knowledge
components to discharge their capabilities. In the final stage, the performative information generated by the broker is used to establish an appropriate flow of messages
between the selected knowledge components in order to complete the task requested.

The brokering mechanism can be divided in three different elements: a formal way
of representing brokerage structures; a method for constructing brokerage structures;
and an algorithm for translating brokerage structures into message sequences.

A brokerable structure in F-Broker has the form c(K, C), where K is the name of
the knowledge component which should be able to deliver the capability and C is a
description of the sources of the capability. C can be in any of the following
forms [6]:

—   A capability available directly from K.
—   A term of the form c(K, dq(Q,QC)), where Q is a capability obtainable from
    K conditional on its other capabilities and QC describes how these capabilities are
    obtained.
—   A term of the form c(K, pdq(Q,QC,QP)), where Q is a capability obtainable
    from K conditional on its other capabilities and on capabilities external to K, and
    QC and QP describe how these internal and external capabilities (respectively) are
    obtained.

– A term of the form c(conj, co(CQ1,CQ2)), where CQ1 and CQ2 are two capability structures which must jointly be satisfied.
– A term of the form c(K, cn(Q, G, c(K1,Q1))), where K1 is the name of an agent different from K which allows capability structure Q to be delivered in combination with capability structure Q1 provided that the bridge constraints given by G are satisfiable.

We can now describe a method for constructing brokerage structures of the form given above using the capability and bridge definitions which reside in the brokering system. Notice that this does not involve any additional interaction with the individual knowledge components (the computation can be done entirely within the broker). We describe the algorithm below as a logic program because this is compact, precise and declarative but the mechanism itself could be implemented in a procedural language. The algorithm proceeds by cases corresponding to each of the forms of brokerage structure given above. The dq structure is obtained from a conditional cap definition; the pdq structure from a p_cap definition; the co structure from two capability structures; and the cn structure via bridge. In all cases where we introduce a new capability into our structure we must demonstrate that it too is obtainable from our definitions – hence the recursive use of broker in the algorithm. For partial capabilities (pdq structures) we need the same form of brokering but with the constraint that the external capability required by the agent comes from some other source. The easy way to describe this is simply to replicate the broker algorithm but with an additional argument (Kn in e broker below) that records the original agent name and prevents it being used to satisfy the external capability goal. The full definition of our broker is then as follows [6]:

```
% Capabilities of K
broker(Q,c(K,Q))
   ←cap(K,Q).
broker(Q, c(K, dq(Q,QC)))
   ←cap(K, (Q←C)) ∧
      broker(C,QC).
broker(Q, c(K1,pdq(Q,QC,QP)))
   ←p_cap(K1, (Q←C), P) ∧
      broker(C,QC) ∧
      e_broker(P,K1,QP).
broker((Q1,Q2), c(conj,
co(CQ1,CQ2)))
   ←broker(Q1,CQ1) ∧
      broker(Q2,CQ2).
broker(Q2, c(K2, cn(Q2, G,
c(K1,BQ))))
   ←corr(K1,Q1,K2,Q2,G) ∧
      Broker(Q1, c(K1,BQ)).
```

```
% External capabilities of K
e_broker(Q, Kn, c(K,Q))
   ←cap(K,Q) ∧ not(K=Kn).
e_broker(Q, Kn, c(K, dq(Q,QC)))
   ←cap(K, (Q←C)) ∧ not(K=Kn) ∧
      broker(C,QC).
e_broker(Q, Kn, c(K1,
pdq(Q,QC,QP)))
   ←p_cap(K1, (Q←C), P) ∧
      not(K1=Kn) ∧ broker(C,QC) ∧
      e_broker(P,K1,QP).
e_broker((Q1,Q2), Kn, c(conj,
co(CQ1, CQ2)))
   ←e_broker(Q1,Kn,CQ1) ∧
      e_broker(Q2,Kn,CQ2).
e_broker(Q2, Kn, c(Kn, cn(Q2, G,
c(K1,BQ))))
   ←corr(K1,Q1,Kn,Q2,G) ∧
      broker(Q1, c(K1,BQ)).
```

The brokerage structures and the brokering method described above do not prescribe the sequence in which it should be transmitted messages to their knowledge components which F-Broker is coordinating. How F-Broker establishes an appropriate sequence of messages depends of the conventions being used for message passing. F-Broker can support several message passing conventions (similarly to DAML-S/OWL-S and WSMO that provide several groundings including WSDL/SOAP grounding). Following previous work on KQML [15] (**Knowledge Query and Manipulation Language** is a language and protocol for communication among software agents and knowledge-based systems), F-Broker includes a simple version of KQML which uses three communication acts ("*performatives*" in the terminology of KQML) that are transmitted sequentially [6]:

~ ask(K; C) denoting that we are asking the agent named K to discharge the competence C. We must obtain a response to this message with an instance for C before proceeding with the rest of the sequence.

~ tell(K; C) denoting that we are informing the agent named K that a competence which it required externally can be discharged by a correspondence to another agent. We must obtain a response from K indicating that it accepts the information before proceeding with the rest of the sequence.

~ test(G) denoting that whatever system is sending the messages should attempt to satisfy the constraint, G, before sending any further messages in the sequence.

We now need an algorithm for translating the brokerage structures of the previous section into message sequences which conform to our message passing conventions. We describe this below in the style of a *Definite Clause Grammar* (DCG) the grammar is used to generate the sequence of terminal symbols, corresponding to performatives, by unpacking the brokerage structure. We assume in the definitions below that the DCG rules are mutually exclusive, so there is only one possible rule for each form of brokerage subterm. It is readily implemented in Prolog but could be implemented in other languages [6]:

```
assemble(c(S; dq(Q; QC))) ) fdependent queries(QC;DQ)g;
assemble(QC);
[ask(S; (Q  DQ))]
assemble(c(S1; pdq(Q; QC;QP ))) ) fdependent queries(QP;DQ)g;
assemble(QC);
assemble(QP );
[ask(S1; (Q  DQ))]
assemble(c(conj; co(CQ1;CQ2))) ) assemble(CQ1);
assemble(CQ2)
assemble(c(S; cn(Q;C; CQ))) ) assemble(CQ);
[test(C); tell(S; Q)]
assemble(c(S; Q)) ) [ask(S; Q)]
```

2.4 Travel Agency Example, Writing Capabilities in F-Broker

The example of this section describes a simplified version of the well-known *Virtual Travel Agency* (VTA) scenario. Clients can ask VTA for flight information or require the booking of selected flights. VTA stores capabilities of airline companies that are interested on advertising flights and selling the flight-tickets. Booking a flight requires payment service and passport verification. The former is done by credit-card financial entities, and the later is done by the police. We specify the capabilities of each airline company, credit-card company and the police in the following way:

```
% Financial capabilities

capability(financial_vs,, pay_order(Client_ID,
Client_Card_Number, Purchase_Order, Cost, Currency,
PaymentMethod)).

...

capability(financial_ms,,      pay_order(...)).

capability(financial_amex,,    pay_order(...)).

% Passport control capabilities

capability(police, has_passport (Client_ID, Nationality)).

...

% Airline capabilities

capability(airline_ib,   check_flight(Origin, Destination,
DepartureDate, ArrivalDate, NPassengers, Cost, Currency,
Flight_Number)).

capability(airline_ib,   book_seats (Origin, Destination,
DepartureDate, ArrivalDate, NPassengers, Cost, Currency,
Flight_Number)).

p_capability(airline_ib,(buy_flight_tickets (Origin,
Destination, DepartureDate, ArrivalDate, Cost, Currency,
Flight_Number, Purchase_Order, Client_ID, Nationality,
Client_Card_Number, Confirmation_Number) :-
    (check_flight(Origin, Destination, DepartureDate,
    ArrivalDate, NPassengers, Cost, Currency, Flight_Number),
    book_seats (Origin, Destination, DepartureDate,
    ArrivalDate, NPassengers, Cost, Currency,
    Flight_Number)), (pay_order(Client_ID,
    Client_Card_Number, Purchase_Order, Cost, Currency,
    PaymentMethod), has_passport (Client_ID, Nationality))).

capability(airline_aa,   check_flight(...)).

capability(airline_aa,   book_seats (...)).

p_capability(airline_aa,(buy_flight_tickets (...))).

...

capability(airline_ba,   check_flight(...)).

capability(airline_ba,   book_seats (...)).

p_capability(airline_ba,(buy_flight_tickets (...))).
```

If a potential client would like to buy flight-tickets, the brokerage structure generated by F-Broker is described next:

```
c (airline_ba, pdq (buy_flight_tickets (...),

    c (conj, co (c (airline_ba, check_flight(...)),

               c (airline_ba, book_seats (...)))),
```

```
c (airline_ba, cn ( c (financial_ms, pay_order(...))),
c (airline_ba, cn ( c (police, has_passport (...))))
))
```

The assembled message sequences corresponding to the brokerage structure described above would have the following form:

```
ask (airline_ba,    check_flight(...)),
ask (airline_ba,    book_seats (...)),
ask (police,        has_passport (...)),
ask (financial_ms,  pay_order(...)),
ask (airline_ba,    buy_flight_tickets (...):-
                    pay_order(...)),
                    has_passport (...)))
```

3   Quality-Based Service Selection Using Incidence Calculus

When a service requester (client) submits a goal (service request) to a broker, the broker has to find which Web service can fulfil the goal based on the service capabilities that have been stored in the broker's database. It is possible that the broker might find several candidates that can provide the service requested. Instead of choosing randomly the appropriate Web service, we have developed a simple but apparently effective technique for selecting Web services based on evidence of prior performance. Because brokers coordinate the interactions between service requester and service providers, brokers can track successful attempts of satisfying service requests. Thus, when a broker has several providers that meet the requirements of a new client's goal, the broker can select the Web service with better prior performance (number of previous occasions in which the service successfully attempt a service request).

Inferring the best possible Web service based on previous performance introduces a significant degree of uncertainty. To deal with this uncertainty, we use the Incidence Calculus [5] for our probabilistic calculations. Thus, F-Broker is able to select the Web service that maximises the probability of a successful outcome. To better understand what Incidence Calculus is and how it can be used in F-Broker, we provide a detailed description next in this section.

3.1   Incidence Calculus

Bundy [5] demonstrated that purely numeric probabilistic formalism can derive into contradictory results during the calculation of an uncertainty measure of complex formula. The key result of his analysis is that in general $P(A \wedge B) \neq P(A) * P(B)$.

Incidence Calculus [5] reviews the notions of probability theory and introduces an important novelty: *"the probability of a sentence is based on a sample space of elements. Each element defines a situation in a possible world where a sentence can be true or false. The sample space, T, contains an exhaustive and disjoint set of elements that for computational reasons should be finite"*.

The incidence of a sentence A, i(A), is the subset of W in which sentence A is true. The dependence or independence of two sentences, A and B, is defined by the amount of common points of the result of the intersection between their incidences, $i(A) \cap i(B)$.

The axioms of Incidence Calculus [5] associate a set of theoretic function with each connective, propositional constant and quantifier of Predicate (Propositional) Logic so that the incidence of a complex sentence can be calculated from the incidences of its sub-sentences. The probabilities of composite formulae are computed from intersections and unions of the sets of worlds for which the atomic formulae hold true. Bundy called the resulting system *Predicate (Propositional) Incidence Logic* [5]:

```
i(T) =        {}                      i(⊥)   = {}

i(A) =        i(A)                     i(¬A)  = i(T)\i(A)

i(A∧B) =      i(A)∩i(B)               i(A∨B) = i(A)∪i(B)

i(A→B) =      i(¬A∨B) = (i(T)\ i(A))∪i(B)
```

Thus, probabilities are calculated in the following way [5]:

```
P(T) =        |i(T)|  = 1              P(⊥)=  |i(⊥)| = 0

P(A) =        |i(A)| / |i(T)|          P(¬A)= 1-|i(A)| / |i(T)|

P(A∧B)  =     |i(A)∩i(B)| / |i(T)|

P(A∨B)  =     (|i(A) ∪i(B)| - |i(A)∩i(B)|) / |i(T)|

P(A|B)  =     |i(A)∩i(B)| / | i(B)|
```

As an illustration, consider the following set of incidences describing the weather of a given week adopted from [5]:

Suppose there are two propositions, P={rainy, windy} and seven possible worlds, T ={sunday, monday, tuesday, wednesday, thursday, friday, saturday}. Suppose that each possible world is equally probable (i.e. 1/7), and we learn that rainy is true in four possible worlds (*friday, saturday, sunday* and *monday*) and windy is true in three possible worlds (*monday, wednesday* and *friday*). Therefore, we can derivate the following incidence sets [5]:

```
i(rainy) = {friday, saturday, sunday, monday}

i(windy)= {monday,wednesday, friday}

i(windy∧rainy)= {monday, friday}
```

Moreover, we can calculate their probabilities in the following way:

```
P(rainy)  = |i(rainy)| / |i(T)|=4/7

P(windy)  = |i(windy)| / |i(T)|=3/7

P(windy∧rainy)= | i(windy)∩i(rainy)| / |i(T)|=2/7
```

3.2 Travelling Example Revised

In F-Broker, each time that a service request is received counts as one incidence. If the service request has been successfully attended by selected Web services, the capabilities involved during the creation of the brokerage structure are modified by the broker adding one incidence to the associated list of incidences. Initially, the incident database is empty, and the broker selects services at random. As more data is collected, a threshold is reached, at which point the broker begins to use the probabilities for selecting promising Web services.

The incidence database is composed of a proposition with all the incidences registered by the broker and an entrance for each atomic capability with the associated list of incidence (one incident for each successful used of the capability). The revised version of the *virtual travelling agency* scenario then is as follows:

```
i(all_requests, [1,2,3,...,20]).

i(capability(airline_aa, check_flight(...)), [2,3,9,10,11]).

i(capability(airline_ba, check_flight(...)), [1,4,5]).

i(capability(airline_ib, check_flight(...)),
    [6,7,8,12,13,14,15,16,17,20]).

...
```

In this example, we can observe that the first service has been successfully used in five different occasions. So, we can predict the goodness of the associated service by applying Incidence Calculus in the following way:

```
P(capability(airline_aa, check_flight(...))) =
|{2,3,9,10,11}|/|{1, ...,20}|
```

We can also observed that for a given request that can be satisfied by any of the Web services represented using the capabilities listed above, **capability**(airline_ib, **check_flight**(...)) is the most promising one. Further, by intersecting various sets of incidences, Incidence Calculus also allows F-Broker to compute the possible success of a group of capabilities. Let us examine the prior performance of the Web services associated with the partial capability "buy_flight_tickets" of the service provider "airline_ib". We ran F-Broker several times and we obtained the following incidence data:

```
i(all_requests, [1,2,3,...,20]).

...

i(capability(financial_amex,, pay_order(...)),
    [1,2,3,6,7,8,16,20]).

i(capability(police, has_passport (...)),
    [1,2,5,6,7,8,10,16,17]).

...

i(capability(airline_ib, check_flight(...)),
    [6,7,8,12,13,14,15,16,17,20]).
```

```
i(capability(airline_ib, book_seats(...)),
    [6,7,8,16,17,20]).
p_capability(airline_ib, (buy_flight_tickets (...) :-
    (check_flight(...),    book_seats (...)),
    (pay_order(...), has_passport (...)))).
...
```

There is no set of incidences associated with the partial capability "buy_flight_tickets", because its probability of success is computed using the incidences associated with the capabilities "check_flight", "book_seats", "pay_order", and "has_passport". The first two capabilities are associated to "airline_ib" and their joint distribution is calculated as the intersection of their incidences. The last two capabilities are dependent upon services other than "airline_ib" - their joint probability is conditional on the joint probability of "check_flight" and "book_seats". Thus, the probability of the Web service associated with "buy_flight_tickets" can be calculated as follows:

```
P(capability(airline_ib, buy_flight_tickets(...)))) =
   P(capability(airline_ib, check_flight(...)) ∧
     capability(airline_ib, book_seats(...)) |
     capability(financial_amex, pay_order(...)) ∧
     capability(police, has_passport (...))) =
   |{6,7,8,12,13,14,15,16,17,20} ∩ {6,7,8,16,17,20} ∩
   {1,2,3,6,7,8,16,20} ∩ {1,2,5,6,7,8,10,16,17}| /
   |{1,2,3,6,7,8,16,20} ∩ {1,2,5,6,7,8,10,16,17}|
```

Given that probabilities can be calculated from the dependencies established by the capability definitions, it is not necessary to modify the broker algorithm described in the previous section. F-Broker will try to "instantiate" first the capabilities with the highest number of incidences. The probability of the final brokerage structure can be calculated later, using the information about dependencies between capabilities. Using this technique can substantially improve performance over random selection of Web services which can individual meet the requirements. In addition, the use of Incidence Calculus does not degrade the performance of F-Broker.

3.3 Discussion

F-Broker has been designed for interaction between agents in architectures for which it is straightforward to gather data on the success or failure of an agent each time it has attempted to satisfy a goal. In this architecture, the enactment of interactions (which is complementary to the specification of agent capabilities) is controlled by specifications of the interaction process (analogous to the process model assumed by OWL-S). Each agent that wishes to become involved in an interaction must actively subscribe to one of the roles in that interaction; then when all the roles of an interaction are subscribed the corresponding agent group commit to the interaction before engaging in it. The fact that F-Broker connects agents through subscription to explicit roles and interactions means that it potentially can supply accurate statistics on agent performance in enacting capabilities.

The incidences gathered by F-Broker are considered independent. We must make that assumption because we have no means of knowing whether or not they have dependences "behind the scenes", and we are not concerned if they do because all we are interested in is whether agents have demonstrated their abilities to offer the capabilities they said they could offer.

The fact that we associate incidences to each atomic capability, and that we calculate probabilities following the rules defined by the Predicate Incidence Logic [5] should not confuse the reader about the nature of Incidence Calculus as a mechanism for handling uncertainty. In the scenario presented above, "counting incidences" is a strength, for two reasons: it allows us to use data on success/failure simply; and directly via Incidence Calculus, it allows us to distinguish incidences which avoids the problems in classical probability that stimulated Incidence Calculus in the first place [5].

An additional element for discussion is the way that we estimate the probability of complex capabilities. For complex capabilities, we derive an estimate of their probability by performing calculations over "primitive" incidences associated with atomic capabilities. We do not associate incidences directly with complex capabilities because we cannot know which agents to blame when a complex interaction has failed. For instance, if a complex capability, "Y", is defined by "A and B" where "A" and "B" are atomic, Incidence Calculus allows us to estimate probabilities for complex capabilities in a way that distinguishes the cases where A and B interact (because they were in the same incident) from those in which they do not.

We note here two significant problems that seem to be intrinsic to the technique that we presented above: the problem that F-Broker relies on the honest evaluation reported by the clients that have sent their goals; and the problem of quickly re-adapting when the performance of Web services changes. Since individual client services are responsible for the assigning of success metrics to goal satisfaction, there is scope for clients with unusual criteria or malicious intent to corrupt the database. We also note that F-Broker is not able to handle the changes that the environment undergoes in specific periods of time. For instance, the provider of a service with a large set of incidences (successful provider) might fail temporarily (no service then being available). Any request by clients that asks for this service will still be processed by the broker and the answer will include the service that the provider cannot supply. After many requests this could be remedied by another service outperforming the record of the unavailable service, but before then, the current broker will choose an unavailable service first.

4   Additional Improvements for F-Broker

We briefly present in this section two relevant modifications of F-Broker implemented in [16] for improving its ability of dealing with Semantic Web services. The first enhancement allows F-Broker to load many DAML-S Web services. This functionality is useful for increasing the range of testing data. The second improvement, also implemented in [16], extends the notion of matching in F-Broker which is able to find services that partially meet the requirements of a given goal.

4.1 Loading DAML-S Services

To test F-Broker with examples of Semantic Web Services implemented by institutions outside of University of Edinburgh, we developed a new module for loading DAML-S Service Profile descriptions. At the time of this work, DAML-S was the most important proposal for describing Semantic Web services, so it was a natural choice to focus on DAML-S. One of the main difficulties was to find a translation from DL logical statements into Prolog statements.

Description Logic Programs (DLP, [11]) is an expressive fragment of the intersection of Description Logics (DL) [17] and Logic Programs (LP) [14]. An important result of the development of this formalism is DLP-fusion, a bidirectional translation of premises and inferences from DLP fragment of DL into LP, and vice versa from DLP fragment of LP into DL. Our implementation of DLP-Fusion was good enough to load slightly adapted versions of DAML-S Service Profile descriptions from DAML[4], Mindswap[5] and Carnegie-Mellon[6].

4.2 Extending the Notion of Matching in F-Broker

The main motivation behind the use of Incidence Calculus for selecting Web services was to facilitate the work of the broker when "exact" matching is not the only notion of matching available. For testing purposes, F-Broker was modified to support *plug-in, subsume,* and *intersection* match. We implemented a new predicate, `matching_notion (Q1, Q2, Nmatch)`, that evaluates the amount of parameters in common between two capabilities, Q1 and Q2, and the predicate returns the kind of matching founded. The broker algorithm presented in previous section was modified to accommodate the new developed predicate. For instance, for a simple capability C directly available from Web service K, c(K, C), the brokerage predicate is:

```
brokerable(Q, c(S,Q,Nmatch)) :-
    capability(S, Q1),
    matchingnotion(Q1,Q,Nmatch),
    Nmatch<>"disjoint".
```

Supporting several notions of matching increases the amount of Web services that meet (perhaps partially) the requirements for a given goal. The Incidence Calculus was added to help F-Broker during the process of selecting the most promising Web services that match the requirements posted by a client.

5 Related Work

The use of probabilistic logic in the context of the Semantic Web has not been explored in detail. Even the inventor of the Semantic Web, Sir Tim Berners-Lee, claimed during the dev day lunchtime session at WWW2004 conference[7] that the Semantic Web stack does not need a representation of uncertainty. The first serious attempt to incorporate probabilistic reasoning in the Semantic Web was done with

---

[4] http:// www .daml.org/services/examples.html
[5] http://www.mindswap.org/2002/services/
[6] http://www. daml.ri.cmu.edu/ont/TaskModeler/TMont-index.html# Request Realtor1
[7] http://esw.w3.org/mt/esw/archives/000055.html

P-SHOQ[18]. This work has been recently revised, and as a result, the first implementation of P-SHOQ has been released. PRONTO [19] is an extension of Pellet [20] that enables probabilistic knowledge representation and reasoning in OWL ontologies. To the best of our knowledge, the use of Incidence Calculus for improving Web service discovering and composition was proposed first in [16]. Later, [21] incorporates the use of Incidence Calculus in an advance version of F-Broker that includes a lightweight coordination calculus (LCC) [22], a method for specifying agent interaction protocols. [21] provides a detailed analysis of performance of the use of Incidence Calculus for selecting web services and proves empirically the benefits of the use of Incidence Calculus for Web service discovery.

6 Conclusions and Future Work

The use of Incidence Calculus for improving service discovering is an excellent motivating scenario for encouraging the integration of probabilistic logic in Semantic Web service technology. Uncertainty is present in functional aspects of Web Services like discovery, composition, interoperation, mediation, monitoring and compensation. In this paper, we focused only in discovery, and in [16], composition is also studied.

Incidence Calculus was an excellent choice because its simplicity, rigor and compatibility with other classical logic formalisms. F-Broker provides an excellent test platform for the evaluation of Incidence Calculus in semantic web services. Although simple, F-Broker provides all basic functionality of a broker and allows the composition of web services capabilities and the execution of services based on an elementary vocabulary inspired in KQML. The code is compact and new extensions can be easily included. F-Broker assumes a service choreography architecture in which agent success/failure in interactions can accurately be recorded. An example of this sort of architecture has been developed by the OpenKnowledge project (www.openk.org). The Openknowledge kernel system (downloadable and available open source) allows peers on an arbitrarily large peer-to-peer network to interact with one another without any pre-established global agreements or knowledge of who to interact with or how interactions will proceed. This provides a concrete example of the sort of architecture assumed by F-Broker.

Future work will concentrate in the migration of the test platform to more realistic scenarios and the evaluation of other probabilistic logic formalism that combines logic programming with description logics.

Acknowledgements

We would like to thanks our anonymous reviewers for useful comments and feedback.

References

1. Gonzalez-Castillo, J., Trastour, D., Bartolini, C.: Description logics for matchmaking of services. In: KI 2001 Workshop on Applications of Description Logics (2001)
2. Li, L., Horrocks, I.: A Software Framework for Matchmaking Based on Semantic Web Technology. In: Proceedings of the Twelfth International World Wide Web Conference (WWW 2003), Budapest, Hungary (2003)

3. Paolucci, M., Kawamura, T., Payne, T., Sycara, K.: Semantic Matching of Web Service Capabilities. In: Horrocks, I., Hendler, J. (eds.) ISWC 2002. LNCS, vol. 2342, pp. 333–347. Springer, Heidelberg (2002)
4. Keller, U., Lara, R., Polleres, A. (eds.): WSMO Web Service Discovery. Technical report, DERI (2004), http://www.wsmo.org/TR/d5/d5.1/v0.1/
5. Bundy, A.: Incidence Calculus. In: Encyclopedia of Artificial Intelligence, pp. 663–668 (1992)
6. Robertson, D.: F-X: A Formal Knowledge Management System. Technical report, University of Edinburgh (unpublished, 2001)
7. Haas, H., Brown, A.: Web Services Glossary (2004), http://www.w3.org/TR/ws-gloss/
8. Sycara, K., Widoff, S., Klusch, M., Lu, J.: LARKS: Dynamic Matchmaking Among Heterogeneous Software Agents in Cyberspace. Autonomous Agents and Multi-Agent Systems, 173–203 (2002)
9. Decker, K., Sycara, K.: Middle-Agents for the Internet. In: Proceedings of the 15th International Joint Conference on Artificial Intelligence ICJCAI 1997, Nagoya, Japan (1997)
10. Fensel, D., Benjamins, V.R., Motta, E., Wielinga, B.J.: UPML: A Framework for Knowledge System Reuse. In: Proceedings of the 16th International Joint Conference on AI (IJCAI 1999), pp. 16–23 (1999)
11. Grosof, B.N., Horrocks, I., Volz, R., Decker, S.: Description Logic Programs: Combining Logic Programs with Description Logics. In: Proceedings of the Twelfth International World Wide Web Conference (WWW 2003), pp. 48–57 (2003)
12. Potter, S.: Broker description document. Technical report, University of Edinburgh (2003)
13. Robertson, D., Silva, F., Vasconcelos, W., de Melo, A.: A lightweight capability communication mechanism. In: Logananthara, R., Palm, G., Ali, M. (eds.) IEA/AIE 2000. LNCS, vol. 1821, pp. 660–670. Springer, Heidelberg (2000)
14. Lloyd, J.W.: Foundations of Logic Programming, 2nd edn. Springer series in symbolic computation. Springer, New York (1987)
15. Finin, T., Labrou, Y., Mayfield, J.: KQML as a Agent Communication Language. Sofware Agents, J.M. Bredshaw. AAAI Press/MIT Press (1997)
16. Martín-Recuerda, F.: Dealing with Uncertainty in Semantic Web Services. MSc Thesis. University of Edinburgh (2003)
17. Baader, F., Calvanese, D., McGuinness, D., Nardi, D., Patel-Schneider, P.F.: The Description Logic Handbook: Theory, Implementation, and Applications. Cambridge University Press, Cambridge (2003)
18. Lukasiewicz, T., Giugno, R.: P-SHOQ (Dn): A Probabilistic Extension of SHOQ(Dn) for Probabilistic Ontologies in the Semantic Web. Technical Report Nr. 1843-02-06, Institut für Informations systeme, Technische Universität Wien (2002)
19. Klinov, P., Parsia, B., Mazlack, M.J.: Pronto - a non-monotonic probabilistic description logic reasoner. OWLED 2008 DC (2008), http://www.webont.org/owled/2008dc/index.html
20. Sirin, E., Parsia, B., Cuenca Grau, B., Kalyanpur, A., Katz, Y.: Pellet: A practical OWL-DL reasoner. Journal of Web Semantics 5(2) (2007)
21. Lambert, D., Robertson, D.: Matchmaking and Brokering Multi-Party Interactions Using Historical Performance Data. In: Proceedings of fourth International Joint Conference on Autonomous Agents and Multi Agent Systems, Utrecht (2005)
22. Robertson, D.: A Lightweight Method for Coordination of Agent oriented Web Services. In: Proceedings of the 2004 AAAI Spring Symposium on Semantic Web Services, California, USA (2004)

An Approach to Description Logic with Support for Propositional Attitudes and Belief Fusion*

Matthias Nickles[1] and Ruth Cobos[2]

[1] Department of Computer Science, University of Bath,
Bath, BA2 7AY, United Kingdom
m.l.nickles@cs.bath.ac.uk
[2] Departamento de Ingeniería Informática, Universidad Autónoma de Madrid,
28049 - Madrid, Spain
ruth.cobos@uam.es

Abstract. In the (Semantic) Web, the existence or producibility of certain, consensually agreed or authoritative knowledge cannot be assumed, and criteria to judge the trustability and reputation of knowledge sources may not be given. These issues give rise to formalizations of web information which factor in heterogeneous and possibly inconsistent assertions and intentions, and make such heterogeneity *explicit* and manageable for reasoning mechanisms. Such approaches can provide valuable meta-knowledge in contemporary application fields, like open or distributed ontologies, social software, ranking and recommender systems, and domains with a high amount of controversies, such as politics and culture.

As an approach to this, we introduce a lean formalism for the Semantic Web which allows for the explicit representation of controversial individual and group opinions and goals by means of so-called *social contexts*, and optionally for the probabilistic belief merging of uncertain or conflicting statements.

Doing so, our approach generalizes concepts such as provenance annotation and voting in the context of ontologies and other kinds of Semantic Web knowledge.

Keywords: Semantic Web, OWL, Knowledge Integration, Context Logic, Voting, Provenance Annotation.

1 Introduction

Information found in open environments like the web can usually not be treated as objective, certain knowledge directly, and also not as truthful beliefs (due to the mental opaqueness of the autonomous information sources). Only a few approaches to the semantic modeling of what could be called subjective opinions, ostensible beliefs or "public assertions", which are neither truthful beliefs nor

---

* This work is a revised and extended version of a paper published in the Proceedings of the Second Workshop on Uncertainty Reasoning for the Semantic Web (URSW-06), 2006.

P.C.G. da Costa et al. (Eds.): URSW 2005-2007, LNAI 5327, pp. 124–142, 2008.
© Springer-Verlag Berlin Heidelberg 2008

objective knowledge, exist so far [11,12]. In contrast, most prevalent formal approaches to knowledge representation and reasoning for the Web handle logical inconsistencies and information source controversies mostly as something which should be avoided or filtered.

Against that, we argue that making (meta-)knowledge about the social, heterogeneous and controversial nature of web information *explicit* can be extremely useful - e.g., in order to gain a picture of the opinion landscape in controversial domains such as politics, for subsequent decision making and conflict resolution, for the acquisition and ranking of information from multiple, possibly dissent sources, and not at last for tasks like the learning whom (not) to trust. Such knowledge is especially crucial in domains with a strong viewpoint competition and difficult or impossible consensus finding like politics, product assessment and culture, and in current and forthcoming Semantic Web applications which support explicitly or implicitly people interaction, like (semantic) blogging, discussion forums, collaborative tagging and folksonomies, and in social computing in general. Approaching this issue, this work presents a lean approach to the formal representation of semantical heterogeneity by means of *social contexts* and the probabilistic weighting and fusion of inconsistent opinions.

The remainder of this paper is structured as follows: the following section defines the two most important concepts underlying our approach, namely social contexts and *social ontologies*. Section 3 introduces a formal, C-OWL based framework for the modeling of social contexts, and Section 4 shows how the formerly presented formal framework can be extended in order to allow for the fusion and probabilistic weighting of competing statements. Section 5 concludes with a discussion of related works.

2   Integration of Divergent Viewpoints and Intentions Using Social Contexts

In the following, we describe the main concepts underlying our approach. First we introduce a so-called *social ontology* of social entities and structures. This ontology is then used to obtain a certain type of logical contexts (called *social contexts*) which allow for the modularization of (ordinary) ontologies w.r.t. the addressee-dependent propositional attitudes of actors or organizations towards the axioms and facts in these ontologies.

A more in-depth exploration of these concepts can be found in [22].

2.1   Social Ontologies

Technically, our approach is based on implementing an interrelationship of a *social ontology* for the description of social concepts and individuals (like persons, agents and organizations, and maybe their relationships) on the one hand, and a set of possibly controversial or uncertain statements (*opinions*) on the other hand. Instances of the social ontology represent the knowledge sources which contribute these opinions. Special terms which are assembled using names from the social ontology then identify social contexts for the contextualization and optionally the fusion of semantically heterogeneous statements. The social ontology can thus be seen as a meta-ontology which is used to provide elements which are used to annotate facts and axioms of other ontologies (the ontologies which contains the opinions). The contextualization itself (independent of the social ontology) corresponds to the context-driven partitioning of a knowledge space, analogously to the approach presented in [2,4].

There is no canonical social ontology to be used with our approach. Basically any ontology could be used as long as it provides concepts, roles and instances for the modeling of the interacting agents and social groups, such as "Author", "Publisher" or "Reader", or, most basic, "Actor". We believe that information sources shall be seen as active, autonomous and - most important - communicating (i.e., *social*) actors, as well as the recipients of the information. A mere conceptualization of the (Semantic) Web as a kind of huge distributed document or knowledge base containing passive information fragments would be highly inadequate [23]. We see the Semantic Web rather as a place where actively pursued opinions and intentions will either compete against or strengthen each other interactively [24]. This viewpoint is independent from the concrete ways such interaction is technically performed (directly or indirectly, synchronously or asynchronously...).

The following example ontology fragment will do for the purpose of this work:

Definition 1: Social ontology SO (example)

$Actor(person_1), Actor(person_2), Actor(person_3)$

...

$Communication(com_1), Communication(com_2), Communication(com_3),$
$Communication(com_4)$

...

$Source(com_1, person_2), Addressee(com_1, person_3)$
$Source(com_2, person_1), Addressee(com_2, person_2)$

...

$Content(com_1, \text{"}a\_reified\_statement\text{"})$

$DegreeOfCertainty(com_1, 0.75)$
$DegreeOfCertainty(com_7, 0)$

..

$SocialGroup(group_1), SocialGroup(group_2)$

...

$hasMember(group_1, person_1), hasMember(group_2, person_1)$
$Actor(group_1)$

...

$Actor(organization_1)$

...

$Source(com_4, group_1), Addressee(com_4, organization_1)$

...

$CA(assertion), CA(publicBelief), CA(publicIntention)$
$Attitude(com_1, publicBelief), Attitude(com_2, assertion),$
$Attitude(com_3, publicIntention)$

...

$Aggregation(fusedPublicBelief)$

At this, *Actor* is the category of the participating actors, whereby these can be any kind of information sources or addressees, like persons, organizations, documents, web services, as well as the holder of a so-called public intention or goal (cf. below). *Communication* is the category of elementary communication acts, described by the properties *Source*, *Addressee*, *Attitude* and *Content* (the uttered statement or intention). A full-fledged approach would add further properties such as a time-stamp, but for many applications it will not be required to make *SO* explicit at all.

Information sources and addresses can be the roles of any kind of actors, not only individual persons. E.g., a social group or an organization such as a company can also act as a source. Social groups are modeled extensionally as sets, whereas organizations are legal entities. At this, it is very important to see that in our framework, opinions and public intentions uttered by a certain group or organization can be modeled fully independently from the opinions and intentions of its members and subgroups. I.e., a social group as a whole could exhibit opinion p, whereas each individual group member exhibits $\neg p$ simultaneously. Of course, in reality the opinions of group members influence the opinion of the group, by way of judgment aggregation [27]. But we think that no single particular way of group opinion settlement should be statically fixed. Instead, we will later introduce a special aggregation operator (informally denoted as *fusedPublicBelief* in *SO*) in order to model the quasi-democratic emergence of group opinions from individual opinions. But again, this is only one possibility: likewise, our framework allows to, e.g., model the case that a group always communicates the opinions of some dedicated opinion leader (dictatorship). It is also not necessarily the case that a social group as a whole forms a single actor at all.

At a first glance, it might seem that on the Semantic Web, the addressee of information is always the general public and thus a fine grained modeling of communication addressees would not be required. This is untrue at least for two reasons: firstly, Semantic Web technologies are also useful in environments where the set of recipients of some information is limited, such as in closed web communities. Secondly, even if some information is in principle visible to everybody, it is nevertheless usually targeted at some specific audience (although it might be difficult to obtain this kind of meta knowledge).

In this work we support the modeling of three public propositional attitudes: *assertion*, *publicBelief*, and *publicIntention*, all subsumed in the ontology under *CA* ("Communication Attitude").

assertion means that a certain statement is ostensibly believed *and* that the speaker (author) has the ostensible intention to make the addressee(-s) adopt the same attitude towards the respective statement also (e.g., "This product is the best-buy!"). This corresponds more or less to the communication act semantics which we have introduced in [11,5,12], and to Grice's conceptualization of speech acts as communications of intentions. *publicBelief* means here more or less the same as *assertion*, but in distinction from the latter *publicBelief* is a passive stance and does not necessarily comprise the person's intention to make the addressees approve the respective statement but merely that a person agrees with some statement (but note that it is not possible to communicate an information p without the implicit assertion that p is indeed an information...). We could likewise have called *publicBelief* *belief* instead, but avoid the latter in order to be able to distinguish between mental (truthful) beliefs and opinions.

Both *publicBelief* and *assertion*s are sometimes called "opinions" in this work. The pragmatic status of *publicBelief*, being a kind of "weak assertion", is somewhat unclear and mainly introduced for compatibility reasons w.r.t. [13], and we believe that *assertion* is sufficient to model most cases of information dissemination on the (Semantic) Web.

publicIntention finally is the communication attitude of ostensibly intending that a statement shall become true (i.e., an intention or goal of the actor to change the world appropriately). The attitude of *requesting* something from another actor is a subtype of *publicIntention*. As a simplification, we consider the attitude of *denial* as identical with the positive attitude towards the negation of the denied statement. This would perhaps be too simple for the modeling of inter-human dialogs, but should do in the context of the less dynamic information exchange on the web. These attitudes should be sufficient to represent most information, publishing and desiring acts on the internet.

assertion, *publicBelief* and *publicIntention* are no propositional attitudes in the usual mentalistic sense but *public* propositional attitudes, as they do not need to correspond to any sincere (i.e., mental) beliefs or intentions of the actors. Instead, they are possibly insincere *communication* or *social* attitudes - stances taken on statements in the course of social interaction. As a consequence, they can not be treated like their mental counterparts. E.g., an actor might hold the opinion ϕ towards addressee one and at the same time $\neg\phi$ informing addressee two (while believing neither ϕ nor $\neg\phi$ privately). As another example, opinions could even be bought, in contrast to sincere beliefs: it is known that opinions uttered in, e.g., web blogs have sometimes been payed for by advertising agencies. Even more, *all* information on the web is "just" opinion, simply due to the absence of a commonly accepted truth assessment authority.

fusedPublicBelief will be described later. It is used in place of communication attitudes, but it actually stands for the merging of opinions by some observer.

2.2   Social Contexts

Contexts (aka *microtheories*) have been widely used in AI since the early nineties, originally intended by McCarthy as a replacement of modal logic. [1,2] propose a context operator $ist(context, statement)$ which denotes that *statement* is true ("ist") within *context*. Building upon general approaches to contexts (specifically

[2,4]), and earlier works on social reification [24], we will use the notation of "context" to express formally that certain statements are being publicly asserted (informed about, ostensibly intended to become true, denied...) on the web by some information-*Source*(s), optionally facing some specific *Addressee*(s). The latter implies that our use of the term "public" optionally comprises "limited publics" in form of closed social groups also. Thus, such *social contexts* model the *social semantics* of the contextualized information. Here, the term "social semantics" has a twofold meaning itself: firstly, it refers to the pragmatic effects of the *communicative function* information publication on the web has - essentially, our contexts correspond to kinds of speech acts which express the particular attitudes web authors have towards statements. Although "propositional attitude" is traditionally a psychological concept, we use this term here for attitudes reported communicatively.

Secondly, the semantics is social in the sense that a fusion context can denote the meaning of a certain statement ascribed by *multiple* actors using some aggregation rule, e.g., the degree of truth assigned via consensus finding or voting, or other kinds of social choice among statements [27].

Defined as conceptualizations of domains, formal ontologies are usually associated with consensual and relatively stable and abstract knowledge. Contexts in contrast provide a powerful concept underlying approaches which aim at coping with the distributiveness and heterogeneity of environments by means of localizing information. This dichotomy of ontologies on the one hand and contexts on the other has been recognized already, but only since recently, the synergies of both concepts are being systematically explored.

Social contexts are special contexts which are used for the social contextualization of statements, i.e., their purpose is to express the social (= communicative) meaning of statements in a scenario like the web, with multiple synchronously or asynchronously communicating information providers and addressees. The major task now is thus to define a type of logical context which allows to model the communicated attitudes associated with information on the web.

The idea is to use parts of the descriptions of individual elementary communications as defined in *SO* as identifiers of contexts. That is, we maintain two ontologies: first *SO*, and second a dynamic context ontology, with context identifiers created from certain instances of *SO*. But for some applications, it will be sufficient to actually create and maintain only the latter ontology, whereas *SO* is given only implicitly in form of the context identifiers.

Definition 2: Social contexts

A *social context* is defined as a pair (id, c), with id being either a term which identifies communications in *SO*, or a *fusion context identifier* as specified below. c is the set of mutually consistent description logic statements (see the following section) which corresponds to the set of contents $\{c : Content(com_i, c)\}$ of all communications com_i which share the respective partial description id. id is called the *context identifier*. A "partial description" of a communication means the description of the communication in terms of the properties *Source*,

Addressee and *Attitude*. I.e., it comprises all role assertions for this communication, excluding those for the role *Content* (which flows into c instead). Thus, social contextualization essentially puts statements into the same context iff the communications which contain these statement as their content share the same properties speaker, hearer, and attitude. In some sense, this "un-reifies" the reified statements within SO in order to obtain contextualized logical statements, and reifies other parts of SO in order to obtain context identifiers.

We use the following syntax for (non-fusion) context identifiers:

$$source \xrightarrow{attitude} addresse$$

This term is obtained from a SO fragment

$$Source(com, source), Addressee(com, addressee), Attitude(com, attitude)$$

for a certain com with $Communication(com)$. We also allow for context identifiers with sets of actors in place of the source and/or the addressee (curly brackets omitted):

$$source_1, ..., source_n \xrightarrow{attitude} addresse_1, ..., addressee_n$$

But note that social groups like $source_1, ..., source_n$ can still only occur in the source role in (non-fusion) context identifiers if they *act as a group* as a source or a addressee.

As an abbreviation, we define $source_1, ..., source_n \xrightarrow{attitude} = source_1, ..., source_n \xrightarrow{attitude} Actor$, with $Actor$ being the extension of $Actor$ in SO. I.e., the communication is here addressed to the group of *all* potential addressees like it is the case with information found on an ordinary public web page. If the sources, addressees and the attitude are unspecified, for both sources and addressees the extension of $Actor$ is assumed, and *publicBelief* as the attitude.

At this, it is important to see that - like in real life - a certain source can hold mutually inconsistent attitudes even towards different members or subgroups of $Actor$ at the same time (but not towards the same addressee).

Fusion context identifiers will be used later in order to merge possibly inconsistent opinions uttered by multiple sources which do not necessarily form a social group with role *Source*. The syntax of fusion context identifiers is

$$source_1, ..., source_n \xrightarrow{fusedPublicBelief} addressee \cdot$$

or in case *addressee* is a social group alternatively:

$$source_1, ..., source_n \xrightarrow{fusedPublicBelief} addresse_1, ..., addressee_n \cdot$$

A question in this regard is how the information required in order to create social contexts (i.e., information source, addressee(-s), attitude) can be obtained.

Basically, the answer is analogous to the answer to the question where other Semantic Web data such as RDF or OWL documents shall come from: they need to be manually created or automatically generated. Other somewhat applicable analogies are the process of quotation, referencing, the provision of *named graphs* [20] and provenance annotation (but note that named graphs and all kinds of annotation are significantly weaker concepts compared to logical contexts). For example, authors could provide social contexts with their own statements on the web. Other knowledge workers or ontology creators could use social contexts in order to integrate statements provided by different people. As long as the authors of these statements are known (or at least URIs), at least the most simple kinds of social context identifiers can be easily generated. In contrast to techniques such as ontology mapping or trust assessment, social contextualization, if seen as a technical approach to quotation, is a simpler means to create correct and mutually consistent statements from inconsistent or dubious source statements (but of course it might require the recursive application of social contextualization...). Although social contexts only "wrap" the general problem of limited trustability on the web, they can be useful in order to integrate information on the fly, especially if no trust information is available. This functionality is shared with *RDF reification*, but the use of the long established context logic and its Semantic Web versions such as C-OWL appears to be a cleaner and better researched approach.

3   A Description Logic with Support for Social Contexts

We introduce now a formal language based on C-OWL [4] for the representation of ontologies with social contexts.

We settle on the $\mathcal{SHOIN}(D)$ description logic (over data types D), because ontology entailment in the current quasi-standard OWL DL can be reduced to $\mathcal{SHOIN}(D)$ knowledge base satisfiability [16]. Since we don't make use of any special features of this specific description language, our approach could trivially be adapted to any other description language or OWL variant, RDF(S), rule languages, or first-order logic.

Definition 3: $\mathcal{SHOIN}(D)$-ontologies

The context-free grammar of $\mathcal{SHOIN}(D)$ concepts C is as follows. Please find detailed information about the syntax and semantics of $\mathcal{SHOIN}(D)$ in [16,17].

$$C \rightarrow A|\neg C|C_1 \sqcap C_2|C_1 \sqcup C_2|\exists R.C|\forall R.C$$
$$| \geq nS| \leq nS|\{a_1, ..., a_n\}| \geq nT| \leq nT|\exists T_1, ..., T_n.D|\forall T_1, ..., T_n.D$$
$$D \rightarrow d|\{c_1, ..., c_n\}.$$

At this, C denote *concepts*, A denote *atomic concepts*, R denote *abstract roles* or *inverse roles* of abstract roles (R^-), S denote abstract *simple roles* [16], the

T_i denote *concrete roles*, d denotes a concrete *domain predicate*, and the a_i / c_i denote abstract / concrete *individuals*.

A $\mathcal{SHOIN}(D) - ontology$ (or *knowledge base*) is then a finite, non-empty set of TBox axioms and ABox axioms ("facts") $C_1 \sqsubseteq C_2$ (inclusion of concepts), $Trans(R)$ (transitivity), $R_1 \sqsubseteq R_2$, $T_1 \sqsubseteq T_2$ (role inclusion for abstract respectively concrete roles), $C(a)$ (concept assertion), $R(a, b)$ (role assertion), $a = b$ (equality of individuals), and $a \neq b$ (inequality of individuals). Concept equality can be expressed via mutual inclusion, i.e., $C_1 \sqsubseteq C_2, C_2 \sqsubseteq C_1$. Spelling out the semantics of $\mathcal{SHOIN}(D)$ is not required within the scope of this work, it can be found in [16].

Definition 4: SOC-OWL

Introducing ontologies and at the same time description logic knowledge bases with social contexts, we define *SOC-OWL* (*Social-Context-OWL* or simply "Social OWL") similarly to C-OWL [4]. While the syntax of SOC-OWL can be seen as a defined subset of the syntax of C-OWL, and SOC-OWL essentially shares with C-OWL the interpretation of concepts, individuals and roles, SOC-OWL satisfiability is constrained by meta-axioms (cf. 3.2) which go beyond C-OWL and put SOC-OWL somewhat close to BDI-style modal logics [11].

Essentially, SOC-OWL adds a kind of "S-Box" ("social box", i.e., social contexts) to a formal ontology language. In contrast to the mere *annotation* of axioms or facts with provenance information or other meta data, these contexts provide separate (but bridgeable) spheres of reasoning.

In the next section, the language P-SOC-OWL will be introduced, which also allows for uncertainty reasoning.

A SOC-OWL ontology parameterized with a social ontology SO is a finite, non-empty set $O = \{(id, s) : id \in Id, s \in AF\} \cup AF^i \cup B$, with AF being the set of all $\mathcal{SHOIN}(D)$ TBox and ABox axioms, AF^i being such axioms but with concepts, individuals and roles directly indexed with social contexts (i.e., $AF^i = \{(id_i, C_h) \sqsubseteq (id_j, C_k), (id_i, a_h) = (id_j, a_k), \dots : id_i, id_j \in Id\}$), and B being a set of *bridge rules* (see 3.1). A *social context* within O is a pair $(id, \{s : (id, s) \in O\})$.

Id is the set of all social context identifiers according to the social ontology SO (cf. Definition 1). The s within (id, s) are called *inner statements* which are said to "be *true* (or *intended* in case of *publicIntention*) within the respective context".

Examples (with multiple facts/axioms per row and (id, a) written as $id\ a$):

$InfluentialPainter(FrankFrazetta)$ $\qquad\qquad$ $InfluentialPainter \sqsubseteq Painter$

$tina \xrightarrow[]{assertion} tim,tom\ InnovativeArtist(FrankFrazetta)$

$tim,tom \xrightarrow[]{assertion} tina\ (\neg InnovativeArtist)(FrankFrazetta)$

$tim,tom \xrightarrow[]{assertion} tina\ TrashArtist(FrankFrazetta)$

$tom \xrightarrow[]{assertion} (\neg InnovativeArtist)(FrankFrazetta)$

$ControversialWikipediaArticle \sqsubseteq WikipediaArticle$

$NeutralWikipediaArticle \sqsubseteq WikipediaArticle$

$\overset{assertion}{tina}WikipediaArticle \sqsubseteq NeutralWikipediaArticle$

$ControversialWikipediaArticle(ArticleAboutFrankFrazetta)$

$\overset{assertion}{tim,tom \longrightarrow tina}(\neg NeutralWikipediaArticle)(ArticleAboutFrankFrazetta)$

This SOC-OWL ontology (modeling as a whole a sort of neutral point of view, like taken by an ideal Wikipedia article) expresses that the information sources Tim and Tom hold the opinion towards Tina that the painter Frank Frazetta is not an innovative artist but a trash artist, while Tina does allegedly believe that the opposite is true. But there is consensus of the whole group that Frazetta is an influential painter. Furthermore, Tina believes that all Wikipedia articles present a neutral point of view.

Notice that without explicit further constraints, bridge rules or meta-axioms, different social contexts are logically fully separated. Also, using only the above ontology it could *not* be inferred that $\overset{publicBelief}{tina \longrightarrow tim} InfluentialPainter$ (*FrankFrazetta*), because *InfluentialPainter*(*FrankFrazetta*) as an abbreviation of

$$\overset{publicBelief}{tina,tim,tom \longrightarrow tina,tim,tom} InfluentialPainter(FrankFrazetta)$$

in the example above is uttered/addressed *exactly* by/to the social group of all participants and not by/to any subgroup or individual. Consensus is always bound to a concrete social group and does not necessarily propagate to social subgroups. This principle allows to model the realistic case that someone conforms with some group opinion, but states some inconsistent opinion towards other groups (even a subgroup of the former group). Of course the co-presence of two or more inconsistent inner statements which indicate that a certain actor is insincere (as it would be the case with $\overset{assertion}{tina \longrightarrow tim}(\neg C)(x)$ and $\overset{assertion}{tina \longrightarrow tom}C(x)$ were contained within the same SOC-OWL ontology, which would be perfectly legal) could usually not be acquired directly from the web, since such actors would likely exhibit inconsistent opinions using different nicknames. Instead, some social reasoning or social data mining techniques would be required to obtain such SOC-OWL knowledge.

Obviously, each SOC-OWL statement (*contextId, statement*) corresponds to the "classic" [1,2] context logic statement *ist*(*context, statement*). But unfortunately, this "real" *ist* operator could not simply be made a first-class citizen of our language (which would allow for the nesting of context expressions), at least not without the need for a considerably more complicated semantics. As a further serious restriction compared to real context logic, it is not possible to relate contextualized statements freely with logical connectives like in $ist(c_1, s_x) \lor ist(c_2, s_y) \rightarrow ist(c_1, s_z)$.

Instead of these features, we allow for *bridge rules* and meta-axioms in order to interrelate social contexts.

The core idea underlying the following semantics of SOC-OWL is to group the axioms according to their social contexts, and to give each context its own interpretation function and domain within the model-based semantics, corresponding to the approach presented in [4]. In addition, we will provide meta-axioms (constraints) and bridge rules in order to state the relationships among the various communication attitudes (somewhat similarly to modal logic axiom schemes such as the well-known KD45 axioms of modal belief logic), and to allow for the interrelation of different attitudes, even across different contexts. E.g., we would like to express that a communication attitude such as $\underset{tina \longrightarrow tim,tom}{assertion}(\neg TrashArtist)(FrankFrazetta)$ implies (intuitively) $\underset{tina}{publicIntention}\left(\underset{tim,tom \longrightarrow tina}{publicBelief}(\neg TrashArtist)(FrankFrazetta)\right)$, i.e., that Tina not only expresses her ostensible beliefs, but also ostensibly intends that others adopt her opinion.

Definition 5: Interpretation of SOC-OWL

A SOC-OWL *interpretation* is a pair $(I, \{e_{i,j}\}_{i,j \in Id})$ with $I = \{I_{id}\}$ being a set of *local interpretations* I_{id}, with each $I_{id} = \langle \triangle^{I_{id}}, (.)^{I_{id}} \rangle$, $id \in Id$. $e_{i,j} \subseteq \triangle^{I_i} \times \triangle^{I_j}$ is a relation of two *local domains* $\triangle^{I_{id}}$ ($e_{i,j}$ is required for the definition of bridge rules in B (Definition 4) as explained later in 3.1). $(.)^{I_{id}}$ maps individuals, concepts and roles to elements (respectively subsets or the products thereof) of the domain $\triangle^{I_{id}}$.

To make use of this interpretation, contextualized statements of SOC-OWL impose a grouping of the concepts, roles and individuals within the inner statements into sets C_{id}, R_{id} and c_{id} [4]. This is done in order to "localize" the names of concepts, individuals and roles, i.e., to attach to them the respective local interpretation function I_{id} corresponding to the social context denoted by $id \in Id$: concretely, the sets C_{id}, R_{id} and c_{id} are defined inductively by assigning the concepts, individuals and role names appearing within the *statement* part of each SOC-OWL axiom/fact $(context_{Id}, statement)$ to the respective set C_{id}, c_{id} or R_{id}. With this, the interpretation of concepts, individuals etc. is as follows:

$C^{I_{id}} =$ any subset of $\triangle^{I_{id}}$ for $C \in C_{id}$
$(C_1 \sqcap C_2)^{I_{id}} = C_1^{I_{id}} \cap C_2^{I_{id}}$ for $C_1, C_2 \in C_{id}$
$(C_1 \sqcup C_2)^{I_{id}} = C_1^{I_{id}} \cup C_2^{I_{id}}$ for $C_1, C_2 \in C_{id}$
$(\neg C)^{I_{id}} = \triangle^{I_{id}} \setminus C^{I_{id}}$ for $C \in C_{id}$
$(\exists R.C)^{I_{id}} = \{x \in \triangle^{I_{id}} : \exists y : (x,y) \in R^{I_{id}} \wedge y \in C^{I_{id}}$ for $C \in C_{id}, R \in R_{id}$
$(\forall R.C)^{I_{id}} = \{x \in \triangle^{I_{id}} : \forall y : (x,y) \in R^{I_{id}} \rightarrow y \in C^{I_{id}}$ for $C \in C_{id}, R \in R_{id}$
$c^{I_{id}} =$ any element of $\triangle^{I_{id}}$, for $c \in c_{id}$
(Interpretation of concrete roles T analogously)

Satisfiability and Decidability

Given a SOC-OWL interpretation I, I is said to *satisfy* a (contextualized) statement ϕ ($I \models \phi$) if there exists an $id \in Id$ such that $I_{id} \models \phi$, with $I_{id} \in I$. A SOC-OWL ontology is then said to be "satisfied" if I satisfies each statement within the ontology (or statement set) and the ontology observes the meta-axioms listed below.

$I_{id} \models (id, C_1 \sqsubseteq C_2)$ iff $C_1^{I_{id}} \subseteq C_2^{I_{id}}$, $I_{id} \models (id, R_1 \sqsubseteq R_2)$ iff $R_1^{I_{id}} \subseteq R_2^{I_{id}}$, $I_{id} \models (id, C(a))$ iff $a^{I_{id}} \in C^{I_{id}}$ etc., i.e., as in the semantics of $\mathcal{SHOIN}(D)$, but with socially indexed interpretations.

With this configuration, the inherited semantics and decidability of $\mathcal{SHOIN}(D)$ remain unaffected in SOC-OWL "within" each context, since the new interpretation function simply decomposes the domain and the set of concepts etc. into local "interpretation modules" corresponding to the contexts.

3.1 Bridge Rules and Cross-Context Mappings

According to Definition 4, a SOC-OWL ontology can optionally comprise bridge rules [4] B and various stronger relationships AF^i among classes, individuals and roles from different contexts. As an example, consider $(context_i, x) \overset{\equiv}{\longrightarrow} (context_j, y)$ in B, with x, y being concepts, individuals or roles.

Informally, such a bridge rule states that the x and y denote corresponding elements even though they belong to different contexts $context_i, context_j$.

With, e.g., $(\overset{assertion}{tina}, FrankFrazetta) \overset{\equiv}{\longrightarrow} (\overset{assertion}{tim,tom}, FrankFrazetta)$ the interpretations of the "two Frank Frazettas" would abstractly refer to the same object. Analogously, $\overset{\sqsubseteq}{\longrightarrow}$ and $\overset{\perp}{\longrightarrow}$ state that the first concept is more specific than the second, or that both concepts are disjoint, respectively. These relationships are given by the relation $e_{i,j}$ (Definition 5).

Formally: $I \models (context_i, x) \overset{\equiv}{\longrightarrow} (context_j, y)$ iff $e_{i,j}(x^{I_i}) = y^{I_j})$ (resp. $e_{i,j}(x^{I_i}) \subseteq y^{I_j}$ and $e_{i,j}(x^{I_i}) \cap y^{I_j} = \emptyset$).

Please find details (which are out of the scope of this work) and analogously defined further bridge rules in [4]. Also, reasoning in the presence of bridge rules follows that with C-OWL.

A much stronger kind of relationship is stated by the syntax constructs where a concept, individual or role is directly indexed with a social context, as, e.g., in $(context_i, x) = (context_j, y)$, with x, y being concepts, individuals or roles.

Formally: $I \models (context_i, x) = (context_j, y)$ iff $x^{I_i} = y^{I_j}$ (analogously for $\sqsubseteq$ etc).

3.2 Meta-axioms

We state now some constraints, which will later be extended w.r.t. a different formal language with meta-axiom (PMA5). All so-called meta-axioms are in fact either entailment rules (which could not be formulated using SOC-OWL axiom schemes because the language is not expressive enough), or they put constraints regarding its integrity on an ontology which is sliced into social contexts. Although a practical reasoner could possibly take advantage of the latter kind of meta axioms (since these exclude certain constellations such as inconsistent contexts), they don't demand special reasoning procedures.

Actively asserting an opinion implies in our framework the intention of the source that the addressee(-s) adopt the asserted statement. With nested social contexts, we could formalize this using $\overset{assertion}{s_1,\ldots,s_n \longrightarrow a_1,\ldots,a_m} \varphi \rightarrow$ $(\overset{publicIntention}{s_1,\ldots,s_n \longrightarrow a_1,\ldots,a_m}(\overset{publicBelief}{a_1,\ldots,a_m \longrightarrow s_1,\ldots,s_n} \varphi)$. But this "strong" and problematic nesting is not possible in our language.

The next meta-axiom simply demands that assertions include the attitude of informing the addressee:

(MA1) $\overset{assertion}{s_1,\ldots,s_n \longrightarrow a_1,\ldots,a_m} \varphi \rightarrow \overset{publicBelief}{s_1,\ldots,s_n \longrightarrow a_1,\ldots,a_m} \varphi$

In this work, we do not provide a full meta-theory corresponding to the KD(45) axioms of (e.g.) modal Belief-Desire-Intention logics (but see [5,11]). Instead, we only demand that the inner statements of each context are mutually consistent (basic rationality):

(MA2) Each set a of statements such that for a specific *context* all $(context, a_i), a_i \in a$ are axioms of the same SOC-OWL ontology, is satisfiable (ensuring the consistency of one's opinions).

Furthermore, we demand - in accordance with many BDI-style logics - that the approval/assertion contexts of a certain actor on the one hand and his intention context on the other do not overlap addressing the same set of addressees, i.e., an actor does not (ostensibly) intent what he (ostensibly) believes to be the case already:

(MA3) For each a such that $(\overset{publicIntention}{s_1,\ldots,s_n \longrightarrow a_1,\ldots,a_n}, a)$ is part of an SOC-OWL ontology o, no axiom/fact $(\overset{publicBelief}{s_1,\ldots,s_n \longrightarrow a_1,\ldots,a_n}, b)$, $b \vdash a$, is part of o (analogously for assertions).

The following constraints are *not* demanded, but could be helpful in application domains were mutual opinion consistency of subgroups is desired (we use $\bigwedge$ to abbreviate a set of SOC-OWL statements).

(MAx1) $(\overset{attitude}{s_1,\ldots,s_n \longrightarrow a_1,\ldots,a_n} \varphi) \leftrightarrow \bigwedge_{s \in 2^{\{s_1,\ldots,s_n\}} - \{\emptyset\}} \overset{attitude}{s \longrightarrow a_1,\ldots,a_n} \varphi$

(MAx2) $(\overset{attitude}{s_1,\ldots,s_n \longrightarrow a_1,\ldots,a_n} \varphi) \leftrightarrow \bigwedge_{a \in 2^{\{a_1,\ldots,a_n\}} - \{\emptyset\}} \overset{attitude}{s_1,\ldots,s_n \longrightarrow a} \varphi$

But we can safely aggregate seemingly consented information in a separated fusion context:

(MA4) $\bigwedge_{s \in \{s_1,\ldots,s_n\}} (I_{\overset{publicBelief}{s \longrightarrow a_1,\ldots,a_n}} \models \varphi) \rightarrow (I_{\overset{fusedPublicBelief}{s_1,\ldots,s_n \longrightarrow a_1,\ldots,a_n}} \models \varphi)$ (analogously for assertions). In general, such group opinions induce a ranking of multiple statements with the respective rank corresponding to the size of the biggest group which supports the statement (this can be used, e.g., for a majority voting on mutually inconsistent statements).

4   Social Rating and Social Aggregation of Subjective Assertions

Building upon social contexts, the following extension of the previously presented logical framework is optional. It makes use of uncertainty reasoning and techniques from belief merging. They allow for i) the representation of gradual strengths of *uncertain opinions* held by individuals (corresponding to subjective probabilities) and social groups, and ii) the probabilistic *fusion* of semantically heterogeneous opinions held by different actors (basically by means of voting).

This feature is also useful in case traditional techniques to ontology integration fail, e.g., if the resulting merged ontology shall be accepted by all sources, but a consensus about the merging with traditional techniques to ontology mapping and alignment could not be found, or if the complexity of a high amount of heterogeneous information needs to be reduced by means of stochastic generalization. Probabilistic fusion is furthermore helpful in case statements shall be socially ranked, i.e., put in an order according to the amount of their respective social acceptance. In contrast to heuristical or surfer-behavior-related ways of information ranking or "knowledge ranking" such as those accomplished by most web search engines, the following approach is based on semantic opinion pooling [15].

In [10], the probabilistic extension $P-\mathcal{SHOQ}(D)$ of the $\mathcal{SHOQ}(D)$ description logic has been introduced. $\mathcal{SHOQ}(D)$ is very similar to $\mathcal{SHOIN}(D)$ and thus OWL DL, but does not have inverse roles, and is not restricted to unqualified number restrictions [16]. [10] shows that reasoning with $P-\mathcal{SHOQ}(D)$ is - maybe surprisingly - decidable. Instead of $P-\mathcal{SHOQ}(D)$, other probabilistic approaches to Semantic Web and ontology languages could likely also be used as a basis for our approach, e.g., [7]. $P-\mathcal{SHOQ}(D)$ is now used to define a probabilistic variant of SOC-OWL.

Definition 6: P-SOC-OWL

A *P-SOC-OWL* ontology is defined to be a finite subset of $\{((p_l, p_u], id, a_i)\} \cup \{(id, a_i)\} \cup \{a_i\} \cup AF^i \cup B$, with $p_l, p_u \in [0, 1], id \in Id, a_i \in AF$, AF being the set of all well-formed $\mathcal{SHOQ}(D)$ ontology axioms, and B and AF^i as in the previous section.

The syntax of $\mathcal{SHOQ}(D)$ can be obtained from that of $\mathcal{SHOIN}(D)$ by excluding inverse roles.

The $[p_l, p_u]$ are probability intervals. Non-interval probabilities p are syntactical abbreviations of $[p, p]$. If a probability is omitted, 1 is assumed.

Definition 7: Semantics of P-SOC-OWL

The semantics of a P-SOC-OWL ontology is given as a family of $P-\mathcal{SHOQ}(D)$ interpretations, each interpretation corresponding to a certain social context.

Formally, a P-SOC-OWL interpretation is a pair $(PI, \{e_{i,j}\}_{i,j \in Id})$ with $PI = \{(PI_{id}, \mu_{id}) : id \in Id\}$ being a set of *local probabilistic interpretations* (each denoted as Pr_{id}), each corresponding to a probabilistic interpretation of $P-\mathcal{SHOQ}(D)$ and a social context with identifier id.

$\mu_{id} : \Delta^{I_{id}} \rightarrow [0,1]$ is a subjective probability function, and the $\Delta^{I_{id}}$ are the domains. The relation $e_{i,j}$ (required to state bridge rules) is defined analogously to SOC-OWL. When restricted to a certain context (using the respective interpretation), reasoning in P-SOC-OWL remains decidable, since "within" this context, no bridge rules or meta-axioms need to be observed and thus P-SOC-OWL behaves in this case just like $P-\mathcal{SHOQ}(D)$.Individualistically assigned probabilities are constrained by the axioms of probability.

Example:

$[0.5, 0.8]: \overset{assertion}{\underset{tim,tom \longrightarrow tina}{}} TrashArtist(FrankFrazetta)$

$0.7: \overset{assertion}{\underset{tina}{}} InnovativeArtist(FrankFrazetta)$

$0.9: \overset{assertion}{\underset{tim}{}} InnovativeArtist(FrankFrazetta)$

This P-SOC-OWL ontology expresses inter alia that Tim and Tom (as a group, but not necessarily separately) hold the opinion that with some probability in $[0.5, 0.8]$, Frank Frazetta is a trash artist, while Tina does (publicly) believe he is an innovative artist with strength 0.7, and Tim believes so with strength 0.9 (i.e., his private opinion disagrees with the public group opinion of him and Tom).

In order to allow for a consistent fusion of opinions, we demand the following fusion meta-axiom, which effectively states how the probabilities of *social fusion contexts* are calculated. A social fusion context is a social context with more than one opinion source and a probability which pools the probabilities which subsets of the group assign to the respective statement. This allows to specify group opinions even if group members or subgroups do knowingly not agree with respect to this assertion. In this regard, we propose two versions of interpretation rules:

(PMA5') $(\bigwedge_{s_i \in \{s_1,...,s_n\}} (Pr_{publicBelief} \overset{}{\underset{s_i \longrightarrow addressees}{}} \models \varphi[p_i, p_i])) \rightarrow (Pr_{publicBelief} \overset{}{\underset{s_1,...,s_n \longrightarrow addressees}{}} \models$ $\varphi[p, p])$ with $p = pool^{poolingType}((p_1, ..., p_n), extraKnowledge)$. At this, $Pr_{id} \models \varphi[l, u]$ attests φ a probability within $[l, u]$ in context id, and $extraKnowledge$ is any knowledge the pooling function might utilize in addition to the p_i (see below for examples). (Analogously for the attitude *assertion*.)

A problem with (PMA5') is that it can lead to unsatisfiability (due to inconsistencies) in case the derived probability p is different than a probability assigned explicitly by this group of people - a group of agents is free to assign *any* truth value or probability to any statement, using *any* social choice procedure. A simple workaround is to use a new kind of context with aggregating "attitude" *fusedPublicBelief*, which is actually no speaker attitude of course, but a belief merging operator used by the observer who fuses opinions.

Another possibility would be to introduce some kind of defeasible logic or priority reasoning which gives priority to explicitly assigned probabilities.

(PMA5) $(\bigwedge_{s_i \in \{s_1,\ldots,s_n\}} (Pr_{publicBelief \atop s_i \longrightarrow addressees} \models \varphi[p_i, p_i])) \rightarrow (Pr_{fusedPublicBelief \atop s_1,\ldots,s_n \longrightarrow addressees} \models$
$\varphi[p,p])$ (remainder as PMA5').

As for $pool^{poolingType}$, there are several possibilities: in the most simple case of "democratic" Bayesian aggregation given the absence of any opinion leader or so-called "supra-Bayesian" [15], we define $pool^{avg}((p_1, \ldots, p_n), \emptyset) = \frac{\sum p_i}{n}$, i.e., $pool^{avg}$ averages over heterogeneous opinions. Using this aggregation operator, we could infer the following:

$$0.8 : {}^{fusedPublicBelief}_{tina,tim} InnovativeArtist(FrankFrazetta).$$

Social aggregation operators are traditionally studied in the field of *Bayesian belief aggregation* [15,3].

The most common fusion operator extends $pool^{avg}$ with expert weights (e.g., stemming from factors such as the opinion holder's trustability or reputation, or social power degrees of the information sources):

$pool^{LinOP}((p_1, \ldots, p_n), (weight_1, \ldots, weight_n)) = \sum weight_i p_i$, with $\sum weight_i = 1$. Also quite often, a geometric mean is used:
$pool^{LogOP}((p_1, \ldots, p_n), (weight_1, \ldots, weight_n)) = \kappa \prod_{i=1}^{n} p_i^{weight_i}$ (κ for normalization).

It is noteworthy that the operators given above do not deal with the problem of *ignorance* directly (e.g., by taking into account the evidence the information sources have obtained, as in Dempster-Shafer theory). But such ignorance could be modeled using the $weight_i$ of $pool^{LinOP}$ and $pool^{LogOP}$, and possibly using probability intervals instead of single probabilities. In case opinions with probability intervals $[p_i^l, p_i^u]$ shall be fused, the described fusion operators need to be accordingly applied to the interval boundaries.

One application of such rating in form of aggregated or individual probabilities is to take the probabilities (respectively, the mean values of the bounds for each interval) in order to impose an order (*ranking*) of the axioms of an ontology (TBox as well as ABox), so that inner statements can be directly ranked regard their degree of assumed social acceptance. The following is an example for how such a top-k list of socially preferred statements looks like.

$0.8 : {}^{fusedPublicBelief}_{voters} statement_1$ (highest social rating)
$[0.5, 0.8] : {}^{fusedPublicBelief}_{voters} statement_2$
...
$0.2 : {}^{fusedPublicBelief}_{voters} statement_3$ (lowest social rating)

Again, such a ranking can also be easily used to transform inconsistent ordinary ontologies into consistent ontologies by a voting on the statements of the inconsistent ontology: in case there are inner statements which are mutually inconsistent, a ranking can be used to obtain a consistent ordinary (i.e., OWL DL) ontology by removing from each smallest inconsistent subset of inner statements the statements with the lowest rating until all remaining elements of each subset are mutually consistent.

What could also be generated quite easily are rankings w.r.t. of the degrees of certainty assigned to the same statement by different voters or groups of voters:

$$0.8: \, ^{publicBelief}_{actor_1} \, statement_1$$
$$[0.5, 0.8]: \, ^{publicBelief}_{group_3} \, statement_1$$
$$0.4: \, ^{fusedPublicBelief}_{actor_1, actor_4} \, statement_1$$
...
$$0.1: \, ^{publicBelief}_{actor_2} \, statement_1$$

5 Related Works and Conclusion

The goal of this work is to provide a social semantics of possibly contradictory assertions on the web, i.e., to state their amount of social support, their communicative emergence and dissemination, and the consensus or dissent they give rise to. Doing so, we settle on the "opinion level" where neither true beliefs are visible (due to the mental opaqueness of the information sources) nor criteria for the selection of useful knowledge or semantic mappings from/among heterogenous information exist initially. This is both in contrast to the traditional aim of information integration and evolution for the determination of some consistent, reliable "truth" obtained from contributions of multiple sources as in traditional *multiagent belief representation and revision* (e.g., [21] - although this direction has still much in common with ours) and approaches to ontology alignment, merging and mapping.

Apart from the research field of knowledge and belief integration, the storage of heterogeneous information from multiple sources also has some tradition in the fields of *data warehousing* and *federated databases*, and view-generation for distributed and enterprise database systems [9], whereby such approaches do not take a social or communication-oriented perspective. *Opinions* are treated in the area of the (non-semantic) web (e.g., *opinion mining* in natural language documents) and in (informal) knowledge management (e.g., *KnowCat* [14]). The assignment of provenance information is mostly based on *tagging* and *punning* techniques, or makes use of the semantically problematic reification facility found in RDF. Meta knowledge modeling and reification techniques for the purpose of adding certain "slots" for provenance and statement identification data, and other useful meta information to Semantic Web languages can be found in [20,25,25]. These approaches, with *named graphs* [20] being currently the most popular representative, leave the original semantics of the underlying language more or less untouched and "merely" annotate traditional language constructs with some optional meta-information. In contrast, our approach aims at a truly social semantics and language.

[6] provides an approach to the grouping of RDF statements using contexts (including contexts for provenance and speech act performatives). Another related approach focusing on contexts including contexts for the aggregation of

RDF graphs, was presented in [2], and [4] provides a general formal account of contexts for OWL ontologies. Independently from web-related approaches, contexts have been widely used for the modeling of distributed knowledge and federated databases, see, e.g., [18,19].

To further explore and work out the new "social" perspective on uncertain information on the web modeled using contexts certainly constitutes a long-term scientific and practical endeavor of considerable complexity, with this work hopefully being a useful starting point.

Acknowledgements. This work was partially funded by the German National Research Foundation DFG (Br609/13-1, research project "Open Ontologies and Open Knowledge Bases") and by the Spanish National Plan of R+D, project no. TSI2005-08225-C07-06.

References

1. McCarthy, J.L.: Notes on formalizing context. In: IJCAI, pp. 555–562 (1993)
2. Guha, R.V., McCool, R., Fikes, R.: Contexts for the Semantic Web. In: McIlraith, S.A., Plexousakis, D., van Harmelen, F. (eds.) ISWC 2004. LNCS, vol. 3298. Springer, Heidelberg (2004)
3. Richardson, M., Domingos, P.: Building Large Knowledge Bases by Mass Collaboration. Technical Report UW-TR-03-02-04, Dept. of CSE, University of Washington (2003)
4. Bouquet, P., Giunchiglia, F., van Harmelen, F., Serafini, L., Stuckenschmidt, H.: C-OWL: Contextualizing Ontologies. In: Fensel, D., Sycara, K.P., Mylopoulos, J. (eds.) ISWC 2003. LNCS, vol. 2870. Springer, Heidelberg (2003)
5. Nickles, M., Fischer, F., Weiss, G.: Communication Attitudes: A Formal Approach to Ostensible Intentions, and Individual and Group Opinions. In: Procs. of the 3rd Intl. Workshop on Logic and Communication in Multiagent Systems, LCMAS 2005 (2005)
6. Klyne, G.: Contexts for RDF Information Modelling (2000), ll http://www.ninebynine.org/RDFNotes/RDFContexts.html
7. Costa, P., Laskey, K.B., Laskey, K.J.: PR-OWL: A Bayesian Framework for the Semantic Web. In: Procs. First Workshop on Uncertainty Reasoning for the Semantic Web, URSW 2005 (2005)
8. Froehner, T., Nickles, M., Weiss, G.: Towards Modeling the Social Layer of Emergent Knowledge Using Open Ontologies. In: Proceedings of The ECAI 2004 Workshop on Agent-Mediated Knowledge Management, AMKM 2004 (2004)
9. Ullmann, J.: Information Integration Using Logical Views. In: Proc. 6th Int'l Conference on Database Theory. Springer, Heidelberg (1997)
10. Giugno, R., Lukasiewicz, T.: P-SHOQ(d): A Probabilistic Extension of SHOQ(d) for Probabilistic Ontologies in the Semantic Web. In: Flesca, S., Greco, S., Leone, N., Ianni, G. (eds.) JELIA 2002. LNCS, vol. 2424, pp. 86–97. Springer, Heidelberg (2002)
11. Fischer, F., Nickles, M.: Computational Opinions. In: Procs. of the 17th European Conference on Artificial Intelligence, ECAI 2006 (to appear, 2006)
12. Gaudou, B., Herzig, A., Longin, D., Nickles, M.: A New Semantics for the FIPA Agent Communication Language based on Social Attitudes. In: Procs. of the 17th European Conference on Artificial Intelligence, ECAI 2006 (to appear, 2006)

13. Nickles, M.: Modeling Social Attitudes on the Web. In: Cruz, I., Decker, S., Allemang, D., Preist, C., Schwabe, D., Mika, P., Uschold, M., Aroyo, L.M. (eds.) ISWC 2006. LNCS, vol. 4273. Springer, Heidelberg (2006)

14. Cobos, R.: Mechanisms for the Crystallisation of Knowledge, a Proposal Using a Collaborative System. Ph.D. thesis. Universidad Autonoma de Madrid (2003)

15. Cooke, R.M.: Experts in Uncertainty: Opinion and Subjective Probability in Science. Oxford University Press, Oxford (1991)

16. Horrocks, I., Patel-Schneider, P.F.: Reducing OWL entailment to Description Logic Satisfiability. Journal of Web Semantics 1(4) (2004)

17. Haase, P., Motik, B.: A Mapping System for the Integration of OWL-DL Ontologies. In: Proceedings of the First International Workshop on Interoperability of Heterogeneous Information Systems. ACM Press, New York (2005)

18. Bonifacio, M., Bouquet, P., Cuel, R.: Knowledge Nodes: The Building Blocks of a Distributed Approach to Knowledge Management. Journal for Universal Computer Science 8(6) (2002)

19. Farquhar, A., Dappert, A., Fikes, R.,, W.: Pratt. Integrating Information Sources using Context Logic. In: Procs. of the AAAI Spring Symposium on Information Gathering from Distributed Heterogeneous Environments (1995)

20. Carroll, J., Bizer, C., Hayes, P., Stickler, P.: Named Graphs, Provenance and Trust. In: Procs. of the 14th International World Wide Web Conference (2005)

21. Dragoni, A., Giorgini, P.: Revisining Beliefs Received from Multiple Sources. In: Roth, H., Williams, M. (eds.) Frontiers in Belief Revision, pp. 431–444. Kluwer Academic Publisher, Dordrecht (2001)

22. Nickles, M.: Social Acquisition of Ontologies from Communication Processes. Applied Ontology 2(3-4) (2007)

23. Singh, M.P.: The Pragmatic Web: Preliminary thoughts. In: Proceedings of the NSF-OntoWeb Workshop on Database and Information Systems Research for Semantic Web and Enterprises, pp. 82–90 (2002)

24. Nickles, M., Weiss, G.: A Framework for the Social Description of Resources in Open Environments. In: Klusch, M., Omicini, A., Ossowski, S., Laamanen, H. (eds.) CIA 2003. LNCS, vol. 2782. Springer, Heidelberg (2003)

25. Tran, D.T., Haase, P., Motik, B., Grau, B.C., Horrocks, I.: Metalevel Information in Ontology-Based Applications. In: Procs. of the 23th AAAI Conference on Artificial Intelligence, AAAI 2008 (2008)

26. Schueler, B., Sizov, S., Staab, S., Tran, D.T.: Querying for Meta Knowledge. In: Procs. of the 17th International Conference on the World Wide Web, WWW 2008 (2008)

27. Pigozzi, G., Hartmann, S.: Judgment aggregation and the problem of truth-tracking. In: Procs. of the 11th Conference on Theoretical Aspects of Rationality and Knowledge, TARK 2007 (2007)

Using the Dempster-Shafer Theory of Evidence to Resolve ABox Inconsistencies

Andriy Nikolov, Victoria Uren, Enrico Motta, and Anne de Roeck

Knowledge Media Institute, The Open University, Milton Keynes, UK
{a.nikolov,v.s.uren,e.motta,a.deroeck}@open.ac.uk

Abstract. Automated ontology population using information extraction algorithms can produce inconsistent knowledge bases. Confidence values assigned by the extraction algorithms may serve as evidence in helping to repair inconsistencies. The Dempster-Shafer theory of evidence is a formalism, which allows appropriate interpretation of extractors' confidence values. This chapter presents an algorithm for translating the subontologies containing conflicts into belief propagation networks and repairing conflicts based on the Dempster-Shafer plausibility.

1 Introduction

The task of ontology population considers creation of concept and property assertions for a given ontological schema. One of the approaches for ontology population considers using automatic information extraction algorithms to annotate natural language data already available on the Web [1,2,3]. Automatic information extraction algorithms do not produce 100% correct output, which may lead to inconsistency of the whole knowledge base produced in this way. Errors can be introduced by human editors. Also information extracted from different sources can be genuinely contradictory. Finally, when information from different sources is fused together the identity problem has to be resolved: identical individuals referring to the same real-world entities must be linked or merged. Automatic matching algorithms produce further errors, which lead to knowledge base inconsistencies. So when performing knowledge fusion (integration of semantic data from different sources) it is important to resolve such inconsistencies automatically or provide the user with a ranking of conflicting options estimating how likely each statement is to be wrong. Extraction algorithms can often estimate the reliability of their output by attaching confidence values to produced statements [4]. Uncertain reasoning using these confidence values can help to evaluate the plausibility of statements and rank the conflicting options. Most of the ongoing research in the field of applying uncertain reasoning to the Semantic Web focuses on fuzzy logic and probabilistic approaches. Fuzzy logic was designed to deal with representation of vagueness and imprecision. This interpretation is not relevant for the problem occurring during population of crisp OWL knowledge bases, where we need to assess the likelihood for a statement to be true or false. The probabilistic approach is more appropriate for dealing with such problems.

P.C.G. da Costa et al. (Eds.): URSW 2005-2007, LNAI 5327, pp. 143–160, 2008.

However, as stated in [5], axioms of probability theory are implied by seven properties of belief measures. One of them is *completeness*, which states that "a degree of belief can be assigned to any well-defined proposition". However, this property cannot be ensured when dealing with confidence degrees assigned by extractors, because they do not always carry information about the probability of a statement being *false*. The Dempster-Shafer theory of evidence [6] presents a formalism that helps to overcome this problem. It allows belief measurements to be assigned to sets of propositions, thus specifying explicitly degrees of ignorance. In this paper, we describe an algorithm for resolving conflicts using the Dempster-Shafer belief propagation approach.

The paper is organized as follows: in the section 2 we discuss approaches to inconsistency resolution and uncertainty management. Section 3 briefly outlines the basics of the Dempster-Shafer uncertainty representation. Section 4 presents our algorithm of inconsistency resolution using belief propagation. Section 5 describes the results obtained in our experiments with a test dataset. Finally, section 6 summarizes our contribution and discusses proposed directions for the future work.

2   Related Work

There are several studies dealing with inconsistency handling in OWL ontologies, among others [7] and [8]. The general algorithm for the task of repairing inconsistent ontologies consists of two steps:

- Ontology diagnosis: finding sets of axioms, which contribute to inconsistency;
- Repairing inconsistencies: changing/removing the axioms most likely to be erroneous.

Choosing the axioms for change and removal is a non-trivial task. Existing algorithms working with crisp ontologies (e.g., [8]) utilize such criteria as syntactic relevance (how often each entity is referenced in the ontology), impact (the influence of removal of the axiom on the ontology should be minimized) and provenance (reliability of the source of the axiom). The last criterion is especially interesting for the automatic ontology population scenario since extraction algorithms do not extract information with 100% accuracy. A study described in [9] specifies an algorithm which utilizes the confidence value assigned by the extraction algorithm. The strategy employed by the authors was to order the axioms according to their confidence and add them incrementally, starting from the most certain one. If adding the axiom led to inconsistency of the ontology then a minimal inconsistent subontology was determined and the axiom with the lowest confidence was removed from it. A disadvantage of such a technique is that it does not take into account the impact of an axiom: e.g., when an axiom violates several restrictions, it does not increase its chances to be removed. Also it does not consider the influence of redundancy: if the same statement was extracted from several sources, this should increase its reliability. Using uncertain reasoning would provide a more sound approach to rank potentially erroneous statements and resolve inconsistencies.

In the Semantic Web domain the studies on uncertain reasoning are mostly focused on two formalisms: probability theory and fuzzy logic. Existing implementations of fuzzy description logic [10,11] are based on the notion of fuzzy set representing a vague concept. The uncertainty value in this context denotes a membership function $\mu_F(x)$ which specifies the degree to which an object x belongs to a fuzzy class F. Probabilistic adaptations of OWL-DL include Bayes OWL[12] and PR-OWL [13]. However, as we discuss below, both of these formalisms do not fully reflect the properties of the problems we are dealing with in the fusion scenario.

In [5] a framework for choosing an appropriate uncertainty handling formalism was presented. The framework is based on the following seven properties of belief measurements:

1. *Clarity*: Propositions should be well-defined.
2. *Scalar continuity*: A single real number is both necessary and sufficient for representing a degree of belief.
3. *Completeness*: A degree of belief can be assigned to any well-defined proposition.
4. *Context dependency*: The belief assigned to a proposition can depend on the belief in other propositions.
5. *Hypothetical conditioning*: There exists some function that allows the belief in a conjunction of propositions to be calculated from the belief in one proposition and the belief in the other proposition given that the first proposition is true.
6. *Complementarity*: The belief in the negation of a proposition is a monotonically decreasing function of the belief in the proposition itself.
7. *Consistency*: There will be equal belief in propositions that have the same truth value.

It was proven that accepting all seven properties logically necessitates the axioms of probability theory. Alternative formalisms allow weakening of some properties. Fuzzy logic deals with the case when the *clarity* property does not hold, i.e., when concepts and relations are vague. Such an interpretation differs from the one we are dealing with in the fusion scenario, where the ontology TBox contains crisp concepts and properties. Confidence value attached to a type assertion *ClassA(Individual1)* denotes a degree of belief that the statement is true in the real world rather than the degree of inclusion of the entity *Individual1* into a fuzzy concept *ClassA*. This makes fuzzy interpretation inappropriate for our case.

Probabilistic interpretation of the extraction algorithm's confidence may lead to a potential problem. If we interpret the confidence value c attached to a statement returned by an extraction algorithm as a Bayesian probability value p, we, at the same time, introduce a belief that the statement is false with a probability $1 - p$. However, the confidence of an extraction algorithm reflects only the belief that the document supports the statement and does not itself reflect the probability of a statement being false in the real world. Also while statistical extraction algorithms [14] are able to assign a degree of probability to each extracted statement, rule-based algorithms [15,16] can only assign the same confidence value to all statements extracted by the same rule based on the rule's performance on some evaluation set. Any extraction produced by a rule with a low confidence value in this case will serve as a negative evidence rather than simply lack of evidence. This issue becomes more important if the reliability of sources is included in the analysis: it is hard to assign the conditional probability of a statement being false given that the document supports it. It means that the *completeness* property does not always hold.

The Dempster-Shafer theory of evidence [6] allows weakening of the completeness property. Belief can be assigned to sets of alternative options rather than only to atomic elements. In the case of binary logic, it means that the degree of ignorance can be explicitly represented by assigning a non-zero belief to the set {true;false}. On the other hand, it still allows the Bayesian interpretation of confidence to be used, when it is appropriate (in this case the belief assigned to the set {true;false} is set to 0). This paper presents an algorithm for resolving inconsistencies by translating the inconsistency-preserving subset of an ontology into the Dempster-Shafer belief network and choosing the axioms to remove based on their plausibility. We are not aware of other studies adapting the Dempster-Shafer approach to the Semantic Web domain.

Alternative approaches to uncertainty representation, which were not applied so far to ontological modelling, include probability intervals [17] and higher-order probability [18]. However, the first of these approaches uses min and max operators for aggregation, which makes it hard to represent cumulative evidence, and the second focuses on resolving different kinds of problems (namely expressing probability estimations of other probability estimations). There are also other approaches to belief fusion in the Semantic Web (e.g., [19] and [20]). These studies deal with social issues of representing trust and provenance in a distributed knowledge base and focus on the problem of establishing the certainty of statements asserted by other people. These approaches, however, do not focus on resolving the inconsistencies and just deal with direct conflicts (i.e., statement A is true vs statement A is false). They do not take into account ontological inference and mutual influence of statements in the knowledge base. In this way, they can be considered complementary to ours.

3   The Dempster-Shafer Belief Theory

Dempster-Shafer theory of evidence differs from the Bayesian probability theory as it allows assigning beliefs not only to atomic elements but to sets of elements as well. The base of the Dempster's belief distribution is the frame of discernment (Ω) - an exhaustive set of mutually exclusive alternatives. A belief distribution function (also called mass function or belief potential) $m(A)$ represents the influence of a piece of evidence on subsets of Ω and has the following constraints:

- $m(\oslash) = 0$ and
- $\sum_{A \subseteq \Omega} m(A) = 1$

$m(A)$ defines the amount of belief assigned to the subset A. When $m(A) > 0$, A is referred to as a focal element. If each focal element A contains only a single element, the mass function is reduced to be a probability distribution. Mass also can be assigned to the whole set of Ω. This represents the uncertainty of the piece of evidence about which of the elements in Ω is true. In our case each mass function is defined on a set of variables $D = \{x_1, ..., x_n\}$ called the domain of m. Each variable is boolean and represents an assertion in the knowledge base. For a single variable we can get degree of support $Sup(x) = m(\{true\})$ and degree of plausibility $Pl(x) = m(\{true\}) + m(\{true; false\})$. Plausibility specifies how likely it is that the statement is false. Based on plausibility it is possible to select from a set of statements the one to be removed.

4   Description of the Algorithm

Our algorithm consists of four steps:

1. Inconsistency detection.
   At this stage a subontology is selected containing all axioms contributing to an inconsistency.
2. Constructing a belief network.
   At this stage the subontology found at the previous step is translated into a belief network.
3. Assigning mass distributions.
   At this stage mass distribution functions are assigned to nodes.
4. Belief propagation.
   At this stage uncertainties are propagated through the network and the confidence degrees of ABox statements are updated.

4.1   Illustrating Example

In order to illustrate our algorithm, we use an example from the banking domain. Supposedly, we have an ontology describing credit card applications, which defines two disjoint classes of applicants: reliable and risky. In order to be reliable, an applicant has to have UK citizenship and evidence that (s)he was never bankrupt in the past. For example, the TBox contains the following axioms:

T1: $RiskyApplicant \sqsubseteq CreditCardApplicant$
T2: $ReliableApplicant \sqsubseteq CreditCardApplicant$
T3: $RiskyApplicant \sqsubseteq \neg ReliableApplicant$
T4: $ReliableApplicant \equiv \exists wasBankrupt.False \sqcap \exists hasCitizenship.UK$
T5: $\top \sqsubseteq\, \le 1\ wasBankrupt$ ($wasBankrupt$ is functional)
The ABox contains the following axioms (with attached confidence values):
A1: $RiskyApplicant(Ind1)$: 0.7
A2: $wasBankrupt(Ind1, False)$: 0.6
A3: $hasCitizenship(Ind1, UK)$: 0.4
A4: $wasBankrupt(Ind1, True)$: 0.5

As given, the ontology is inconsistent: the individual *Ind1* is forced to belong to both classes *RiskyApplicant* and *ReliableApplicant*, which are disjoint, and the functional property *wasBankrupt* has two different values. If we choose to remove the axioms with the lowest confidence values, it will require removing A3 and A4. However, inconsistency can also be repaired by removing a single statement A2. The fact that A2 leads to the violation of two ontological constraints should increase the likelihood it is wrong.

4.2   Inconsistency Detection

The task of the inconsistency detection step is to retrieve all minimal inconsistent subontologies (MISO) of the ontology and combine them. As defined in [7], an ontology O' is a minimal inconsistent subontology of an ontology O, if $O' \subseteq O$ and O' is inconsistent and for all O'' such that $O'' \subset O' \subseteq O$, O'' is consistent. OWL reasoner Pellet [8] is able to return the MISO for the first encountered inconsistency in the ontology. To calculate all MISO $O'_1, ..., O'_n$ in the ontology we employ Reiter's hitting set tree algorithm [21]. The algorithm is an adaptation of the breadth-first tree search aimed at finding all diagnoses: sets of statements which, if removed, transfer the knowledge base into a consistent one. Each edge in the tree represents an axiom potentially included into diagnosis. Each node represents a MISO returned by the reasoner if all axioms contained in the path from the root of the tree to the node are deleted. The algorithm provides a guidance for building and pruning of the tree, which optimizes the diagnosis process. After all conflict sets were identified, the next step involves constructing belief networks from each set. If for two subontologies $O'_i \cap O'_j \neq \emptyset$ then these two subontologies are replaced with $O' = O'_i \cup O'_j$.

For our illustrating example, the conflict detection algorithm is able to identify two conflict sets in this ontology: the first, consisting of {T3, T4, A1, A2, A3} (individual *Ind1* belongs to two disjoint classes), and the second {T5, A2, A4} (individual *Ind1* has two instantiations of a functional property). The statement A2 belongs to both sets and therefore the sets are merged.

4.3   Constructing Belief Networks

The networks for propagation of Dempster-Shafer belief functions (also called valuation networks) were described in [22]. By definition, the valuation network is an undirected graph represented as a 5-tuple: $\{\Psi, \{\Omega_X\}_{X \in \Psi}, \{\tau_1, ..., \tau_n\}, \downarrow, \otimes\}$, where Ψ is a set of variables, $\{\Omega_X\}_{X \in \Psi}$ is a collection of state spaces, $\{\tau_1, ..., \tau_n\}$ is a collection of valuations (belief potentials of nodes), $\downarrow$ is a marginalization operator and $\otimes$ is a combination operator. In our case Ψ consists of ABox assertions, every $\{\Omega_X\}_{X \in \Psi} = \{0; 1\}$ and $\{\tau_1, ..., \tau_n\}$ are created using rules described below. The operators are used for propagation of beliefs and are described in the following subsections. The network contains two kinds of nodes: variable nodes corresponding to explicit or inferred ABox assertions and valuation nodes representing TBox axioms. Variable nodes contain only one variable, while valuation nodes contain several variables.

Translation of an inconsistent subontology into a belief propagation network is performed using a set of rules (Table 1). Each rule translates a specific OWL-DL construct into a set of network nodes and links between them. Rules 1 and 2 directly translate each ABox statement into a variable node. Other rules process TBox axioms and create two kinds of nodes: one valuation node to represent the TBox axiom and one or more variable nodes to represent inferred statements. Such rules only fire if the network already contains variable nodes for ABox axioms, which are necessary to make the inference. For example, a rule processing the class equivalence axiom (Rule 4) is interpreted as the following:

"If there is a node N_1 representing the type assertion $I \in X$ and an *owl:equivalentClass* axiom $X \equiv Y$, then:

– Create a node N_2 representing the assertion $I \in Y$;
– Create a node N_3 representing the axiom $X \sqsubseteq Y$;
– Create links between N_1 and N_3 and between N_3 and N_2."

A particularly interesting case is the representation of the *owl:sameAs* axiom. In the fusion scenario this axiom represents both a schema-level rule allowing inferencing new statements and a data-level assertion, which has its own confidence (e.g., produced by a matching algorithm). Thus, each *owl:sameAs* axiom in the knowledge base triggers creation of both variable and valuation nodes. If a rule requires creating a node, which already exists in the network, then the existing node is used.

Applying the rules described above to our illustrating example (rules 1, 2, 5, 6, 7, 10, 21) will result in the following network (Fig. 1).

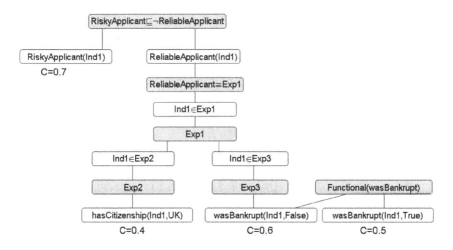

Fig. 1. Belief network example ($Exp1 = \exists \ wasBankrupt.False \sqcap \exists \ hasCitizenship.UK$, $Exp2 = \exists \ hasCitizenship.UK$, $Exp3 = \exists \ wasBankrupt.False$)

Table 1. Belief network construction rules

N	Pre-conditions	Nodes to create	Links to create
1	$I \in X$	$N_1 : I \in X$	
2	$R(I_1, I_2)$	$N_2 : R(I_1, I_2)$	
3	$I_1 = I_2$	$N_3 : I_1 = I_2$ (variable)	
4	$N_1 : I \in X, X \sqsubseteq Y$	$N_2 : I \in Y, N_3 : X \sqsubseteq Y$	$(N_1,N_3),(N_3,N_2)$
5	$N_1 : I \in X, X \equiv Y$	$N_2 : I \in Y, N_3 : X \equiv Y$	$(N_1,N_3),(N_3,N_2)$
6	$N_1 : I \in X, X \sqsubseteq \neg Y$	$N_2 : I \in Y, N_3 : X \sqsubseteq \neg Y$	$(N_1,N_3),(N_3,N_2)$
7	$N_1 : I \in X, X \sqcap Y$	$N_2 : I \in X \sqcap Y, N_3 : X \sqcap Y, N_4 : I \in Y$	$(N_1,N_3),(N_4,N_3), (N_3,N_2)$
8	$N_1 : I \in X, X \sqcup Y$	$N_2 : I \in X \sqcup Y, N_3 : X \sqcup Y, N_4 : I \in Y$	$(N_1,N_3),(N_4,N_3), (N_3,N_2)$
9	$N_1 : I \in X, \neg X$	$N_2 : I \in \neg X, N_3 : \neg X$	$(N_1,N_3),(N_3,N_2)$
10	$\top \sqsubseteq\leq 1R, N_1 : R(I,o_1), N_2 : R(I,o_2)$	$N_3 : \top \sqsubseteq\leq 1R$	$(N_1,N_3),(N_2,N_3)$
11	$\top \sqsubseteq\leq 1R^-, N_1 : R(I_2,I_1), N_2 : R(I_3,I_1)$	$N_3 : \top \sqsubseteq\leq 1R^-$	$(N_1,N_3),(N_2,N_3)$
12	$R \equiv R^-, N_1 : R(I_1,I_2)$	$N_2 : R \equiv R^-, N_3 : R(I_2,I_1)$	$(N_1,N_2),(N_2,N_3)$
13	$R \equiv Q, N1 : R(I_1,I_2)$	$N_2 : R \equiv Q, N_3 : Q(I_1,I_2)$	$(N_1,N_2),(N_2,N_3)$
14	$R \sqsubseteq Q, N_1 : R(I_1,I_2)$	$N_2 : R \sqsubseteq Q, N_3 : Q(I_1,I_2)$	$(N_1,N_2),(N_2,N_3)$
15	$R \equiv Q^-, N1 : R(I_1,I_2)$	$N_2 : R \equiv Q^-, N_3 : Q(I_2,I_1)$	$(N_1,N_2),(N_2,N_3)$
16	$Trans(R), N_1 : R(I_1,I_2), N_2 : R(I_2,I_3)$	$N_3 : Trans(R), N_4 : R(I_1,I_3)$	$(N_1,N_3),(N_2,N_3), (N_3,N_4)$
17	$\leq 1.R, N_1 : R(I_1,o_1), N_2 : R(I_1,o_2)$	$N_3 :\leq 1.R, N_4 : I \in\leq 1.R$	$(N_1,N_3),(N_2,N_3), (N_3,N_4)$
18	$\geq 1.R, N_1 : R(I_1,o_1), N_2 : R(I_1,o_2)$	$N_3 :\geq 1.R, N_4 : I \in\geq 1.R$	$(N_1,N_3),(N_2,N_3)$
19	$= 1.R, N_1 : R(I_1,o_1), N_2 : R(I_1,o_2)$	$N_3 : I \in= 1.R$	$(N_1,N_3),(N_2,N_3)$
20	$\forall R.X, N_1 : R(I_1,I_2), N_2 : I_2 \in X$	$N_3 : \forall R.X, N_4 : I_1 \in \forall R.X$	$(N_1,N_3),(N_2,N_3), (N_3,N_4)$
21	$\exists R.X, N_1 : R(I_1,I_2), N_2 : I_2 \in X$	$N_3 : \exists R.X, N_4 : I_1 \in \exists R.X$	$(N_1,N_3),(N_2,N_3), (N_3,N_4)$
22	$\exists R.\top \sqsubseteq X, N_1 : R(I_1,I_2), N_2 : I_1 \in X$	$N_3 : \exists R.\top \sqsubseteq X$	$(N_1,N_3),(N_2,N_3)$
23	$\top \sqsubseteq \forall R.X, N_1 : R(I_1,I_2), N_2 : I_2 \in X$	$N_3 : \top \sqsubseteq \forall R.X$	$(N_1,N_3),(N_2,N_3)$
24	$N_1 : I_1 = I_2$ (variable), $N_2 : I_2 \in X$	$N_3 : I_1 = I_2$ (valuation), $N_4 : I_1 \in X$	$(N_1,N_3),(N_2,N_3), (N_3, N_4)$
25	$N_1 : I_1 = I_2$ (variable), $N_2 : R(I_2,o_1)$	$N_3 : I_1 = I_2$ (valuation), $N_4 : R(I_1,o_1)$	$(N_1,N_3),(N_2,N_3), (N_3, N_4)$

4.4   Assigning Mass Distributions

After the nodes were combined into the network, the next step is to assign the mass distribution functions to the nodes. There are two kinds of variable nodes: (i) nodes representing statements supported by the evidence and (ii) nodes representing inferred statements. Initial mass distribution for the nodes of the first type is assigned based on their extracted confidence. If a statement was extracted with a confidence degree c, it is assigned the following mass distribution: $m(True) = c, m(True; False) = 1 - c$. It is possible that the same statement is extracted from several sources. In this case, multiple pieces of evidence have to be combined using Dempster's rule of combination.

Nodes created artificially during network construction are only used for propagation of beliefs from their neighbours and do not contain their own mass assignment. Valuation nodes specify the TBox axioms and are used to propagate beliefs through the network. For the crisp OWL ontologies only mass assignments of 0 and 1 are possible. The principle for assigning masses is to assign the mass of 1 to the set of all combinations of variable sets allowed by the corresponding axiom. Table 2 shows the mass assignment functions for OWL-DL T-Box axioms [1].

In our example, we assign distributions based on the extractor's confidence values to the variable nodes representing extracted statements: A1:(m(1)=0.7, m({0;1})=0.3), A2: (m(1)=0.6, m({0;1})=0.4), A3: (m(1)=0.4, m({0;1})=0.6), A4: (m(1)=0.5, m({0;1})=0.5). The valuation nodes obtain their distributions according to the rules specified in the Table 2: T3 (rule 3), T4 (rules 2, 4, 18) and T5 (rule 7).

4.5   Belief Propagation

The axioms for belief propagation were formulated in [23]. The basic operators for belief potentials are marginalization $\downarrow$ and combination $\otimes$. Marginalization takes a mass distribution function m on domain D and produces a new mass distribution on domain $C \subseteq D$.

$$m^{\downarrow C}(X) = \sum_{Y \downarrow C = X} m(Y)$$

For instance, if we have the function m defined on domain $\{x, y\}$ as $m(\{0; 0\}) = 0.2$, $m(\{0; 1\}) = 0.35$, $m(\{1; 0\}) = 0.3$, $m(\{1; 1\}) = 0.15$ and we want to find a marginalization on domain $\{y\}$, we will get $m(0) = 0.2 + 0.3 = 0.5$ and $m(1) = 0.35 + 0.15 = 0.5$. The combination operator is represented by the Dempster's rule of combination:

$$m_1 \otimes m_2(X) = \frac{\sum_{X_1 \cap X_2 = X} m_1(X_1) m_2(X_2)}{1 - \sum_{X_1 \cap X_2 = \emptyset} m_1(X_1) m_2(X_2)}$$

Belief propagation is performed by passing messages between nodes according to the following rules:

---

[1] For nodes allowing multiple operands (e.g., intersection or cardinality) only the case of two operands is given. If the node allows more than two children, then number of variables and the distribution function is adjusted to represent the restriction correctly.

Table 2. Belief distribution functions for valuation nodes

N	Node type	Variables	Mass distribution
1	$X \sqsubseteq Y$	$I \in X, I \in Y$	m({0;0}, {0;1}, {1;1})=1
2	$X \equiv Y$	$I \in X, I \in Y$	m({0;0},{1;1})=1
3	$X \sqsubseteq \neg Y$	$I \in X, I \in Y$	m({0;0},{0;1},{1;0})=1
4	$X \sqcap Y$	$I \in X, I \in Y, I \in X \sqcap Y$	m({0;0;0},{0;1;0},{1;0;0},{1;1;1})=1
5	$X \sqcup Y$	$I \in X, I \in Y, I \in X \sqcup Y$	m({0;0;0},{0;1;1},{1;0;1},{1;1;1})=1
6	$\neg X$	$I \in X, I \in \neg X$	m({0;1},{1;0})=1
7	$\top \sqsubseteq \leq 1R$	$R(I,o_1), R(I,o_2)$	m({0;0},{0;1},{1;0})=1
8	$\top \sqsubseteq \leq 1R^-$	$R(I_2,I_1), R(I_3,I_1)$	m({0;0},{0;1},{1;0})=1
9	$R \equiv R^-$	$R(I_1,I_2), R(I_2,I_1)$	m({0;0},{1;1})=1
10	$R \equiv Q$	$R(I_1,I_2), Q(I_1,I_2)$	m({0;0},{1;1})=1
11	$R \sqsubseteq Q$	$R(I_1,I_2), Q(I_1,I_2)$	m({0;0},{0;1},{1;1})=1
12	$R \equiv Q^-$	$R(I_1,I_2), Q(I_2,I_1)$	m({0;0},{1;1})=1
13	$Trans(R)$	$R(I_1,I_2), R(I_2,I_3), R(I_1,I_3)$	m({0;0;0},{0;0;1},{0;1;0},{0;1;1},{1;0;0},{1;0;1},{1;1;1})=1
14	$\leq 1.R$	$R(I_1,o_1), R(I_1,o_2), I_1 \in \leq 1.R$	m({0;0;1},{0;1;1},{1;0;1},{1;1;0})=1
15	$\geq 1.R$	$R(I_1,o_1), R(I_1,o_2), I_1 \in \geq 1.R$	m({0;0;0},{0;1;1},{1;0;1},{1;1;1})=1
16	$= 1.R$	$R(I_1,o_1), R(I_1,o_2), I_1 \in = 1.R$	m({0;0;0},{0;1;1},{1;0;1},{1;1;0})=1
17	$\forall R.X$	$R(I_1,I_2), I_2 \in X, I_1 \in \forall R.X$	m({0;0;1},{0;1;1},{1;0;0},{1;1;1})=1
18	$\exists R.X$	$R(I_1,I_2), I_2 \in X, I_1 \in \exists R.X$	m({0;0;1},{0;1;1},{1;0;0},{1;1;1})=1
19	$\exists R.\top \sqsubseteq X$	$R(I_1,I_2), I_1 \in X$	m({0;0}, {0;1}, {1;1})=1
20	$\top \sqsubseteq \forall R.X$	$R(I_1,I_2), I_2 \in X$	m({0;0}, {0;1}, {1;1})=1
21	$I_1 = I_2$	$I_1 = I_2, I_2 \in X, I_1 \in X$	m({0;0;0}, {0;0;1}, {0;1;0}, {0;1;1}, {1;0;0}, {1;1;1})=1
22	$I_1 = I_2$	$I_1 = I_2, R(I_2,o_1), R(I_1,o_1)$	m({0;0;0}, {0;0;1}, {0;1;0}, {0;1;1}, {1;0;0}, {1;1;1})=1

1. Each node sends a message to its inward neighbour (towards the root of the tree). If $\mu^{A \to B}$ is a message from A to B, $N(A)$ is a set of neigbours of A and the potential of A is m_A, then the message is specified as a combination of messages from all neighbours except B and the potential of A:

$$\mu^{A \to B} = (\otimes \{\mu^{X \to A} | X \in (N(A) - \{B\}) \otimes m_A\})^{\downarrow A \cap B}$$

2. After a node A has received a message from all its neighbors, it combines all messages with its own potential and reports the result as its marginal.

As the message-passing algorithm assumes that the graph is a tree, it is necessary to eliminate loops. All valuation nodes constituting the loop are replaced by a single node with the mass distribution equal to the combination of mass distributions of its constituents. The marginals obtained after propagation for the nodes corresponding to initial ABox assertions will reflect updated mass distributions. After the propagation we can remove the statement with the lowest plausibility from each of the MISO found at the diagnosis stage.

Calculating the beliefs for our example gives the following Dempster-Shafer plausibility values for ABox statements: Pl(A1)=0.94, Pl(A2)=0.58, Pl(A3)=0.8,

Pl(A4)=0.65. In order to make the ontology consistent it is sufficient to remove from both conflict sets an axiom with the lowest plausibility value (A2). In this example, we can see how the results using Dempster-Shafer belief propagation differ from the Bayesian interpretation. Bayesian probabilities, in this case, are calculated in the same way as Dempster-Shafer support values. If we use confidence values as probabilities and propagate them using the same valuation network we will obtain the results: P(A1)=0.66, P(A2)=0.35, P(A3)=0.32 and P(A4)=0.33. In this scenario, we would remove A3 and A4 because of the negative belief bias. Also we can see that all three statements A2, A3 and A4 will be considered wrong in such a scenario (resulting probability is less than 0.5). The Dempster-Shafer approach provides more flexibility by making it possible to reason about both support ("harsh" queries) and plausibility ("lenient" queries).

5   Evaluation

In order to test the approach we performed experiments with publicly available datasets describing the domain of scientific citations. Our datasets were structured according to the SWETO-DBLP ontology[2] (Fig. 2), and contained instances of two types: *opus:Article* (journal articles) and *opus: Article_in_Proceedings* (conference and workshop papers).

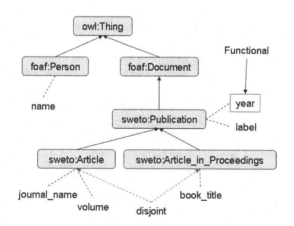

Fig. 2. Class hierarchy in the SWETO-DBLP ontology

We used two kinds of restrictions: classes *opus:Article* and *opus:Article_in_Proceedings* were disjoint and the property *opus:year* (the year of publication) was functional. Three different datasets were used:

---

[2] http://lsdis.cs.uga.edu/projects/semdis/swetodblp/august2007/opus_august2007.rdf

1. AKT EPrints archive[3]. This is a small dataset containing information about papers produced within the AKT research project.
2. Rexa dataset. This dataset was extracted from the Rexa search server[4] constructed in the University of Massachusets using automatic IE algorithms.
3. SWETO DBLP dataset. A well-known publicly available dataset listing publications from the computer science domain.

We know that the AKT EPrints archive was created by the authors themselves, who entered the data about their publications manually. The Rexa dataset was extracted using automatic IE algorithms (the authors reported extraction accuracy of 0.97 and coreferencing accuracy "in the 90s"[5].) However, sometimes the information was incorrectly reported in the sources (e.g., when it was extracted from a citation in a third-party publication), which lowers the actual quality of the data. The DBLP dataset was primarily constructed using the data reported in the proceedings and journal contents, which makes it a more reliable source.

5.1   Experimental Setup and Results

We performed the a priori mass assignment in the following way. First, we ranked the datasources according to our confidence in their quality (DBLP>Rexa> EPrints). As a rough clue to rank the sources by their quality we used the coreference quality of the papers' authors. We considered the case when the same author was referred to in the same dataset using different labels as an error and calculated the percentage of correct individuals (for EPrints this percentage was 0.46, for Rexa 0.63 and for DBLP 0.93). Then, we treated the class assignments as more reliable than datatype property assignments because the IE algorithms used by Rexa and the HTML wrappers sometimes made errors by assigning the wrong property (e.g., venue instead of year) or by assigning the borders of the value incorrectly (e.g., dropping part of the paper's title). Finally, for the Rexa dataset we had additional information: each paper record had a number of citations indicating the number of sources referring to the paper. We estimated the dependency between the reliability of records and the number of citations by randomly selecting a subset of paper records and manually counting the number of "spurious" records, which contained some obvious error (e.g., like assigning the name of the conference as paper title). We randomly selected 400 records for each value of the *hasCitations* property from 0 to 5 and counted the number of spurious records. If the total number of papers for some interval was lower than 400, then we selected all available records in the interval. Based on these reliability assignments, we adjusted the reliability of datatype property assignments for the Rexa dataset. This led us to assign belief masses to the statements from each source as shown in the Table 3. Of course, such confidence estimation was subjective, but we cannot expect it to be precise in most real-life scenarios, unless the complete gold standard data is available in advance.

---

[3] http://eprints.aktors.org/

[4] http://www.rexa.info/

[5] http://www.rexa.info/faq

Table 3. Initial belief mass assignment

Dataset	Class assertions	Datatype assertions
DBLP	0.99	0.95
Rexa	0.95	0.81 (<2 citations) 0.855 (>2 citations)
EPrints	0.9	0.85

Table 4. Evaluation results

No	Matching algorithm	Total clusters	Matching precision (before)	Conflicts found	Accuracy	Matching precision (after)
	EPrints/Rexa					
1	Jaro-Winkler	88	0.95	9	0.61	0.97
2	L2 Jaro-Winkler	96	0.88	13	0.73	0.92
3	Jaro-Winkler (mutated dataset)	88	0.95	55	0.92	0.95
	EPrints/DBLP					
4	Jaro-Winkler	122	0.92	12	0.83	0.99
5	L2 Jaro-Winkler	217	0.39	110	0.9	0.84
6	Jaro-Winkler (mutated dataset)	122	0.92	84	0.91	0.92
	Rexa/DBLP					
7	Jaro-Winkler	305	0.9	21	0.73	0.94
8	L2 Jaro-Winkler	425	0.55	149	0.87	0.82
9	Jaro-Winkler (mutated dataset)	305	0.9	213	0.94	0.9

We ran matching algorithms determining identical individuals for each pair of datasets. Their results were evaluated using precision/recall measures. To each *owl:sameAs* statement produced by a matching algorithm we assigned a support belief mass based on the precision of the algorithm (Table 4, column 4). We made tests with two kinds of string similarity algorithms[24]: Jaro-Winkler similarity directly applied to the papers' titles (rows 1, 4 and 7) and L2 Jaro-Winkler similarity, when both compared values are tokenized, each pair of tokens is compared using the standard Jaro-Winkler measure and the maximal total score is selected. The L2 measure can catch the cases when part of the label is missing (e.g., only the last part of a paper title was recognized), which slightly increases the recall, but its precision is usually significantly lower (rows 2, 5, 8).

Finally, in order to test the algorithm in a situation where the quality of one of the data sources is low, we had to introduce noise into our datasets. We did it in the following way: for one of the datasets in the pair (the smaller one, EPrints or Rexa depending on the case) we randomly mutated 40% of *rdf:type* assertions (changing from *Article* to *Article_in_Proceedings* and vice versa) and *opus:year* assertions (by + or -1). The support belief mass of all statements in the dataset was proportionally reduced: the values in the rows 2 and 3 in the Table 3 were multiplied by 0.6. We measured the quality of inconsistency resolution by comparing the resulting ranking produced after the belief propagation with the "correct" conflict resolution. A conflict was considered correctly resolved if the genuinely incorrect statements got the lowest belief after propagation. In cases when a conflict set contained several incorrect statements and only some of them were correctly recognized, we assigned a reduced score to such a conflict (e.g., 0.33 if only one statement out of three was properly recognized as wrong). Incorrect statements, which received low support, but were plausible, counted as 0.5 of a correct answer. The results we obtained are given in the Table 4. The clusters shown in the third column are sets of mapped individuals. Each cluster represents a group of up to 3 individuals mutually linked by an *owl:sameAs* relation.

5.2   Discussion and Examples

Since in our use case the conflicts occurred because of incorrect data, applying high-precision matching algorithms to the original datasets resulted in very small numbers of conflicts. Thus, the results obtained in such experiments (rows 1, 2, 4, 7) only illustrate common cases rather then provide reliable quantitative evaluation. As was expected, the algorithm's performance was better in "trivial" cases when the wrong statement had a priori support significantly lower than the statements, with which it was in conflict. This was the most frequent pattern in the experiments with a low-precision matching algorithms (rows 5, 8) and artificially distorted datasets (rows 3, 6, 9). In more complex cases, if the conflict set contained a correct statement with a lower support than the actual wrong statement, the algorithm was still able to resolve the conflict correctly if additional evidence was available. One typical cause of such a conflict was the situation when the same authors first presented a paper in a conference and after that published its extended version in a journal. For instance, a belief network built for such a case is shown in the Fig. 3a. While each conflict separately would be resolved by removing the assertions related to *Ind2* (*In_Proc(Ind2)* and *year(Ind2, 2005)* in the Fig. 3a), cumulative evidence allowed the algorithm to recognize the actual incorrect *sameAs* link (*Ind1=Ind2*). A similar situation occurred when one instance (*Ind1*) was considered similar to two others (*Ind2* and *Ind3*), but only one of the *sameAs* links (e.g., *Ind1=Ind2*) led to the inconsistency (e.g., disjoint axiom violation). In that case the existence of the correct *sameAs* link (*Ind1=Ind3*) increased the support of the corresponding class assertion and again caused the wrong link (*Ind1=Ind2*) to be removed. As would be expected, in cases when the wrong statement was considered a priori more reliable than the conflicting ones and the evidence was not sufficient, the algorithm made a mistake. For instance, the conflict in Fig. 3a was resolved wrongly when the dataset containing *Ind2* was artificially distorted. Although the statements involved in the conflict were not affected, the initial support of the *Ind2* assertions was significantly lower (0.51 instead of 0.9), which was insufficient to break the link. The capabilities of the Dempster-Shafer representation were important in cases when the a priori support of some statements was low. For instance, using L2 Jaro-Winkler similarity for the EPrints/DBLP datasets achieved a very low precision (0.39). In such cases the plausibility allows us to distinguish the

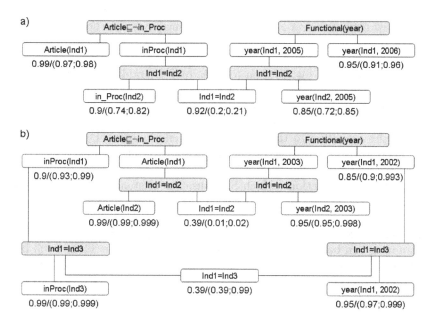

Fig. 3. Examples of belief networks constructed during the experimental testing. The numbers show the support before propagation and support and plausibility after propagation. a) Incorrect *sameAs* mapping violates two restrictions. b) Influence of the Dempster-Shafer plausibility: correct *sameAs* relation has low support but high plausibility because it does not contribute to inconsistency.

cases when a statement is considered unreliable because of insufficient evidence from the cases when there is sufficient evidence against it. For instance, Fig. 3b shows such a case. A record in the EPrints dataset describing a conference paper was linked to two different papers in the DBLP dataset. One of the links was wrong. After belief propagation the support values of both links were still below 0.5. However, the evidence against the wrong link was significantly stronger, so its plausibility was low (0.02) while the plausibility of the correct link remained high (0.99).

6   Conclusion and Future Work

In this paper, we described how the Dempster-Shafer theory of evidence can be used for dealing with ABox-level inconsistencies produced by inaccurate information extraction and human errors and reported the experiments we performed with publicly available datasets. The experiments have shown that in the majority of cases the algorithm was able to resolve inconsistencies that occurred when fusing data from several sources. Belief networks allowed related pieces of evidence to be utilized and the Dempster-Shafer mass distribution allowed more fine-grained ranking of statements than the probabilistic representation.

However, a feature of the belief propagation algorithm was its sensitivity to the initial mass distribution: a small initial difference between the belief masses of conflicting statements significantly increased after the beliefs were propagated. This can be a limitation in cases when initial mass distribution significantly differs from the actual distribution (especially if the ranking is wrong).

At the moment we are working on the extension of the framework in order to improve corefencing quality by propagating positive evidence as well as negative. As the most interesting direction for the future work we consider potential application of the algorithm to the multi-ontology environment. Often the datasets to be fused are structured according to different ontologies. Automatic ontology schema-matching algorithms such as those reported in [25] also can introduce errors. The restrictions, which can cause inconsistencies, are required to be correctly translated between the ontologies. Resolving such issues is a research challenge. Finally, it would be interesting to investigate whether the capabilities of the Dempster-Shafer uncertainty representation (e.g., explicit representation of ignorance) can be utilized for knowledge modelling at the TBox level. In [26] it was shown that the Dempster-Shafer approach may lead to problems when it is used to represent uncertainty of inferencing rules (i.e., TBox-level) rather than just of pieces of evidence (ABox assertions). These problems occur if the ontology contains contradictory pieces of knowledge, and are caused by the fact that the Dempster-Shafer approach does not distinguish pieces of evidence regarding specific individuals from generic rules applicable to all individuals. It will be interesting to investigate whether these problems can be avoided when modelling description logic axioms.

Acknowledgements

This work was funded by the X-Media project (www.x-media-project.org) sponsored by the European Commission as part of the Information Society Technologies (IST) programme under EC grant number IST-FP6-026978.

References

1. Welty, C., Murdock, W.: Towards knowledge acquisition from information extraction. In: 5th International Semantic Web Conference, Athens, GA, USA, November 5-9, 2006, pp. 709–722 (2006)
2. Popov, B.: KIM - a semantic platform for information extraction and retrieval. Natural language engineering 10, 375–392 (2004)
3. d'Amato, C., Fanizzi, N., Esposito, F.: Query answering and ontology population: an inductive approach. In: Bechhofer, S., Hauswirth, M., Hoffmann, J., Koubarakis, M. (eds.) ESWC 2008. LNCS, vol. 5021, pp. 288–302. Springer, Heidelberg (2008)
4. Iria, J.: Relation extraction for mining the Semantic Web. In: Dagstuhl Seminar on Machine Learning for the Semantic Web, Dagstuhl, Germany (2005)

5. Horvitz, E.J., Heckerman, D.E., Langlotz, C.P.: A framework for comparing alternative formalisms for plausible reasoning. In: AAAI 1986, pp. 210–214 (1986)
6. Shafer, G.: A mathematical theory of evidence. Princeton University Press, Princeton (1976)
7. Haase, P., van Harmelen, F., Huang, Z., Stuckenschmidt, H., Sure, Y.: A framework for handling inconsistency in changing ontologies. In: Gil, Y., Motta, E., Benjamins, V.R., Musen, M.A. (eds.) ISWC 2005. LNCS, vol. 3729, pp. 353–367. Springer, Heidelberg (2005)
8. Kalyanpur, A., Parsia, B., Sirin, E., Grau, B.C.: Repairing unsatisfiable concepts in OWL ontologies. In: Sure, Y., Domingue, J. (eds.) ESWC 2006. LNCS, vol. 4011, pp. 170–184. Springer, Heidelberg (2006)
9. Haase, P., Volker, J.: Ontology learning and reasoning - dealing with uncertainty and inconsistency. In: International Workshop on Uncertainty Reasoning for the Semantic Web (URSW), pp. 45–55 (2005)
10. Stoilos, G., Stamou, G., Tzouvaras, V., Pan, J.Z., Horrocks, I.: Fuzzy OWL: Uncertainty and the semantic web. In: International Workshop of OWL: Experiences and Directions, ISWC 2005 (2005)
11. Straccia, U.: Towards a fuzzy description logic for the Semantic Web (preliminary report). In: Gómez-Pérez, A., Euzenat, J. (eds.) ESWC 2005. LNCS, vol. 3532, pp. 167–181. Springer, Heidelberg (2005)
12. Ding, Z., Peng, Y.: A probabilistic extension to ontology language OWL. In: 37th Hawaii International Conference On System Sciences, HICSS-37 (2004)
13. da Costa, P.C.G., Laskey, K.B., Laskey, K.J.: PR-OWL: A Bayesian ontology language for the semantic web. In: Workshop on Uncertainty Reasoning for the Semantic Web, ISWC 2005 (2005)
14. Li, Y., Bontcheva, K., Cunningham, H.: SVM based learning system for information extraction. Deterministic and Statistical Methods in Machine Learning, 319–339 (2005)
15. Ciravegna, F., Wilks, Y.: Designing Adaptive Information Extraction for the Semantic Web in Amilcare. In: Annotation for the Semantic Web (2003)
16. Zhu, J., Uren, V., Motta, E.: ESpotter: Adaptive named entity recognition for web browsing. In: Althoff, K.-D., Dengel, A.R., Bergmann, R., Nick, M., Roth-Berghofer, T.R. (eds.) WM 2005. LNCS, vol. 3782. Springer, Heidelberg (2005)
17. de Campos, L.M., Huete, J.F., Moral, S.: Uncertainty management using probability intervals. In: Bouchon-Meunier, B., Yager, R.R., Zadeh, L.A. (eds.) IPMU 1994. LNCS, vol. 945, pp. 190–199. Springer, Heidelberg (1995)
18. Gaifman, H.: A theory of higher order probabilities. In: TARK 1986: Proceedings of the 1986 conference on Theoretical aspects of reasoning about knowledge, pp. 275–292. Morgan Kaufmann Publishers Inc., San Francisco (1986)
19. Richardson, M., Agrawal, R., Domingos, P.: Trust management for the Semantic Web. In: Fensel, D., Sycara, K.P., Mylopoulos, J. (eds.) ISWC 2003. LNCS, vol. 2870, pp. 351–368. Springer, Heidelberg (2003)
20. Nickles, M.: Social acquisition of ontologies from communication processes. Journal of Applied Ontology 2(3-4), 373–397 (2007)
21. Reiter, R.: A theory of diagnosis from first principles. Artificial Intelligence 32(1), 57–95 (1987)
22. Shenoy, P.P.: Valuation-based systems: a framework for managing uncertainty in expert systems. In: Fuzzy logic for the management of uncertainty, pp. 83–104. John Wiley & Sons, Inc., New York (1992)

23. Shenoy, P.P., Shafer, G.: Axioms for probability and belief-function propagation. In: Readings in uncertain reasoning, pp. 575–610. Morgan Kaufmann Publishers Inc., San Francisco (1990)
24. Cohen, W.W., Ravikumar, P., Fienberg, S.E.: A comparison of string metrics for matching names and records. In: KDD Workshop on Data Cleaning and Object Consolidation (2003)
25. Euzenat, J., Shvaiko, P.: Ontology matching. Springer, Heidelberg (2007)
26. Pearl, J.: Bayesian and belief-function formalisms for evidential reasoning: a conceptual analysis. In: Readings in uncertain reasoning, pp. 575–610. Morgan Kaufmann Publishers Inc., San Francisco (1990)

An Ontology-Based Bayesian Network Approach for Representing Uncertainty in Clinical Practice Guidelines

Hai-Tao Zheng, Bo-Yeong Kang, and Hong-Gee Kim*

Biomedical Knowledge Engineering Laboratory, Seoul National University
28 Yeongeon-dong, Jongro-gu, Seoul, Korea
hgkim@snu.ac.kr

Abstract. Clinical Practice Guidelines (CPGs) play an important role in improving quality of care and patient outcomes. Although several machine-readable representations of practice guidelines have been implemented with semantic web technologies, there is no implementation to represent uncertainty in activity graphs in clinical practice guidelines. In this paper, we explore a Bayesian Network(BN) approach for representing the uncertainty in CPGs based on ontologies. Using this representation, we can evaluate the effect of an activity on the whole clinical process, which can help doctors judge the risk of uncertainty for other activities when making a decision. A variable elimination algorithm is applied to implement the BN inference and a validation of an aspirin therapy scenario for diabetic patients is proposed.

1 Introduction

Clinical Practice Guidelines (CPGs) play an important role in improving quality of care and patient outcomes; therefore, the task of clinical guideline-sharing across different medical institutions is a prerequisite to many EMR (Electronic Medical Record) applications including medical data retrieval [18], medical knowledge management [7], and clinical decision support systems (CDSSs) [13]. To facilitate clinical guideline-sharing, GLIF (GuideLine Interchange Format) and SAGE (Standards-based Sharable Active Guideline Environment) have been the focus of extensive research [12]. GLIF is a semantic web based standard for representing clinical guidelines [15] and SAGE is an interoperable guideline execution engine, which encodes the content of the clinical guideline to an ontology representation, and executes the ontology through the functions of a CIS (clinical information system) [17].

Most previous approaches using GLIF and SAGE are designed to proceed from one step to the next only if there is no uncertain data in the former step [13]. However, this expectation is unrealistic in practice. For example, a guideline, which requires risk factors for heart disease to be assessed, needs to proceed

---

* Corresponding author.

P.C.G. da Costa et al. (Eds.): URSW 2005-2007, LNAI 5327, pp. 161–173, 2008.

even when the information about this item is uncertain. In the clinical process, uncertain data can be (1) data stemming from unreliable sources (e.g., a patient can not remember the results of his/her last glucose test); (2) data not obtainable (e.g., no historical data on familial diabetes); and (3) data not yet collected (e.g., levels of serum glucose today) [14]. If data represented in CPGs is uncertain, the activities that handle these uncertain data become uncertain as well. For instance, in CDSS systems, when using the diabetes clinical guideline, it is necessary to obtain a family history to evaluate the risk of insulin therapy. However, in real hospital environments, clinicians cannot easily obtain all the necessary data for their health care activities. Based on these issues, the goal of this paper is to construct an approach to represent the uncertainty in CPGs and help doctors judge the risk of these uncertainties in the clinical process. Uncertainty in CPGs means that the activity graphs composing the CPGs contain uncertain activities.

As a model for uncertainty, Bayesian Networks (BNs) occupy a prominent position in many medical decision making processes and statistical inference [11,3,2]. However, there have been few reports applying BNs to the representation of uncertainty in CPGs. Therefore, to address this issue, we propose an ontology-based representation of uncertainty in CPGs by using BNs.

In this paper, we first introduce BNs, then we describe the use of BNs for the medical domain, and review previous work on applying semantic web technology to model CPGs in section 2; Section 3 elaborates the mechanism of encoding uncertainty into a CPG ontology; Section 4 describes a scenario validation based on BN inference; Section 5 discusses the conclusions and future work.

2   Background and Related Work

2.1   Bayesian Network

There are several models that are used to represent uncertainty, such as fuzzy-logic and BNs. Generally, a BN of n variables consists of a DAG (Directed Acyclic Graph) of n nodes and a number of arcs. Nodes X_i in a DAG correspond to random variables, and directed arcs between two nodes represent direct causal or influential relations from one variable to the other. The uncertainty of the causal relationship is represented locally by the CPT (Conditional Probability Table). $P(X_i|pa(X_i))$ associated with each node X_i, where $pa(X_i)$ is the parent set of X_i. Under the conditional independence assumption, the joint probability distribution of $\mathbf{X} = (X_1, X_2, ..., X_n)$ can be factored out as a product of the CPTs in the network, namely, the chain rule of BN: $P(\mathbf{X}) = \prod_i P(X_i|pa(X_i))$. With the joint probability distribution, BNs theoretically support any probabilistic inference in the joint space. Besides the probabilistic reasoning provided by BNs themselves, we are attracted to BNs in this work for the structural similarity between the DAG of a BN and activity graphs of CPGs: both of them are directed graphs, and direct correspondence exists between many nodes and arcs in the two graphs. Moreover, BNs can be utilized to represent the uncertainty visually, to provide inference effectively and to facilitate human understanding of CPGs.

Motivated by these advantages, we apply BNs to represent the uncertainty in CPGs.

Considering the advantages of BNs, we apply BNs to represent the uncertainty in CPGs.

2.2   Bayesian Networks for the Medical Domain

Because BNs occupy a prominent position as a model for uncertainty in decision making and statistical inference, they have been applied to many medical decision support systems [11,3,2]. Atoui [3] adopted a decision making solution based on a BN that he trained to predict the risk of cardiovascular events (infarction, stroke, or cardiovascular death) based on a set of demographic and clinical data. Aronsky [2] presented a BN for the diagnosis of community-acquired pneumonia and he showed that BNs are an appropriate method for detecting pneumonia patients with high accuracy. With respect to clinical guidelines, Mani [11] proposed BNs for the induction of decision tables and generated a guideline based on these tables. However, although these methods focus on predicting some features or risk of disease using BN inference, there has been no implementation to represent the uncertainty in activity graphs in CPGs. These methods do not provide the probabilities of target activities based on uncertainty reasoning, which is the focus of our approach.

2.3   Semantic Web for Clinical Practice Guidelines

A representational form of clinical guideline knowledge, which promotes completeness and minimizes inconsistency and redundancy, is essential if we want to implement and share guidelines for computer-based applications. Semantic Web technology offers such sharable and manageable methodology for modeling CPGs. GLIF [15] and SAGE [17] are two good examples. For creation and maintenance of implementable clinical guideline specifications, an architecture is presented in [8]. This architecture includes components such as a rules engine, an OWL-based classification engine and a data repository storing patient data. Moreover, approaches for modeling clinical guidelines are discussed and they show that guideline maintenance is tractable when a CPG is represented in an ontology. Here, we apply an ontology to represent the uncertainty in CPGs because it is more extensible and maintainable than other methods such as relational databases.

3   Encoding Uncertainty into a CPG Ontology

Figure 1 depicts the overall procedure of the proposed method. Firstly, the original CPG is encoded into an ontology model that contains uncertainty features using BNs. For this, we propose a formal model of a CPG Ontology to represent uncertainty and an algorithm to construct the CPTs (Conditional Probability Tables) of the BN. The CPG ontology can be shared and utilized in different

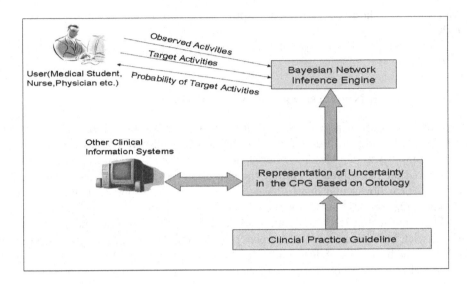

Fig. 1. The framework

clinical information systems. Then, when a user provides his/her observed evidence in the clinical process, the BN inference engine will load the CPG ontology as a BN and mark the nodes that are observed by the user in the BN. Based on the observed evidence, the BN inference engine can reason out the probabilities of target activities asked by the user. Given the reasoning results, the user can judge the risk of unobserved activities and make a further decision.

3.1   Clinical Practice Guideline Ontology

CPGs typically include multiple recommendation sets represented as an activity graph that show the recommended activities during a clinical process [4]. An activity graph describes the relationship between activities in the recommendation set as a process model. In this article, we use a single recommendation set in the SAGE diabetes CPG [1], which is an activity graph of aspirin therapy for diabetic patients, to illustrate how we represent the uncertainty in CPGs based on the ontology (Fig. 2). Typically, an activity graph contains three kinds of activities: context activities, decision activities, and action activities. Each activity graph segment within a guideline begins with a context activity node that serves as a control point in guideline execution by specifying the clinical context for that segment. A decision activity node in the SAGE guideline model represents clinical decision logic by listing alternatives (typically subsequent action activity nodes), and specifying the criteria that need to be met to reach those nodes. An action activity node encapsulates a set of work items that must be performed by either a computer system or persons.

To represent activities in CPGs, we create the activity class that represents all the nodes in an activity graph as shown in Figure 3. Because there are three kinds of activities, we construct an action class, a context class, and a decision

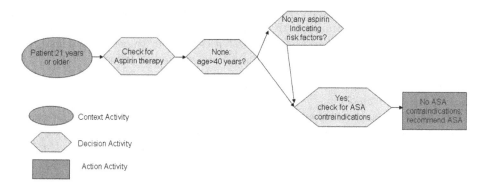

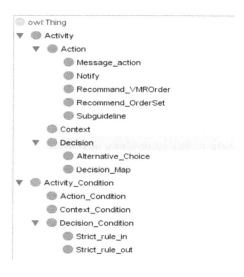

Fig. 2. Clinical practice guideline of aspirin therapy for diabetic patients(ASA means aspirin therapy)

class as sub classes of the activity class in the ontology. In CPGs, activities may include internal conditions that restrict their execution. For example, for the decision activity "Yes;check for ASA(aspirin therapy) contraindications" (Fig. 2), there are many internal conditions to make sure the ASA contraindications will be checked correctly, such as checking family history, checking for the hypertensive disorder, etc. We encode these activity internal conditions as an activity condition classes in the ontology (Fig. 3).

A CPG Ontology with uncertainty features is defined as follows:

Definition 1. (CPG Ontology) CPG Ontology $O := \{C, I, Ps, cinst\}$, with an activity class set C, an activity instance set I, a property set Ps, and an activity class instantiation function $cinst : C \rightarrow 2^I$.

Fig. 3. Classes representation for clinical practice guideline

In CPG ontology, the activity instance set I represents the set of real activities that belong to corresponding activity classes. The property set Ps is proposed to represent the different attributes of activities in order to encode the features of the BN into ontology. The property set Ps is defined as follows:

Definition 2. *(Properties for uncertainty representation) Property Set Ps :=* {*cause, hasCondition, hasState, isObserved, hasPriorProValue, hasCondiProValue*}, *has a property function cause : I → I, a property function hasCondition : I → I, a property function hasState : I → Boolean, a property function isObserved: I → Boolean, a property function hasPriorProValue: I → Float, and a property function hasCondiProValue: I → Float.*

In CPGs, if the criteria associated with an activity node are satisfied, it will be successfully executed, which will cause the execution of subsequent nodes in the activity graph. Therefore, the relationship between activities is defined as a *cause* relationship. For example, in Figure 2, the context activity "Patient 21 years or older" causes the decision activity "Check for Aspirin therapy". To represent this relationship in the ontology, we construct the property *cause* whose domain and range both are the activity class and the activity condition class. The *hasCondition* property is proposed as an inverse property of the *cause* property, which describes the "parent" activities of an activity that *cause* its execution. For example, the decision activity "Check for aspirin therapy" has the property *hasCondition* with value "Patient 21 years or older" activity that causes its execution. With the *hasCondition* property, users can easily figure out all of the conditions that cause the execution of any activity. The *cause* property plays the role of "directed arc" and all the activity instances play the role of "node" in the DAGs of BNs. Another property, the *hasState* property, which has a boolean value range, is denoted as the state of the activity instance; the *isObserved* property shows if the activity instances have been observed or not.

Prior probability and conditional probability are two important features that represent the uncertainty level of nodes in BNs. To encode prior probability and conditional probability of activity instances into the ontology respectively, *hasPriorProValue* property and *hasCondiProValue* property are employed. Let A, B be the instances of the activity class representing two concrete activities. We interpret $P(A = a)$ as the prior probability that a value a is a state of instance A and $P(B = b|A = a)$ as the conditional probability that when A has state a, B has state b. For example, when A is activity "Patient 21 years or older", B is activity "Check for Aspirin therapy", $P(A = true) = 0.5$ can be expressed in the ontology as follows:

```
<Context rdf:ID="Patient_21_yo_or_older">
    <hasPriorProValue
    rdf:datatype="http://www.w3.org/2001/XMLSchema#float"
    >0.5</hasPriorProValue>
    <hasState
    rdf:datatype="http://www.w3.org/2001/XMLSchema#boolean"
    >true</hasState>
```

```
    <cause rdf:resource="#Check_for_Aspirin_therapy"/>
</Context>
```

The conditional probability $P(B = true|A = true) = 1.0$ can be expressed in the ontology as follows:

```
<Decision rdf:ID="Check_for_Aspirin_therapy">
    <hasCondition>
        <Context rdf:ID="Patient_21_yo_or_older">
        <hasState rdf:datatype="http://www.w3.org/2001/XMLSchema#boolean"
        >true</hasState>
        </Context>
    </hasCondition>
    <hasState rdf:datatype="http://www.w3.org/2001/XMLSchema#boolean"
    >true</hasState>
    <hasCondiProValue rdf:datatype="http://www.w3.org/2001/XMLSchema#float"
    >1.0</hasCondiProValue>
    <cause rdf:resource="#Check_fo_age_older_than_40"/>
</Decision>
```

3.2   Construction of Conditional Probability Tables

In this section, we introduce an algorithm used to construct the CPTs of nodes in BNs. After creating the properties to represent the uncertainty in the ontology, the CPTs must be constructed for BNs, because BN inference is based on the CPTs of each node in the BNs. Because of the features of CPGs, we initialize the CPTs based on noisy OR-gate model. To demonstrate this idea more clearly, we use a subsect of clinical practice guideline of primary care clinic visit in SAGE diabetes CPGS as an example (Fig. 4). This example includes two context activities, "physician accesses the patient record" and "physician accesses the patient record", and four action activities and five action activities, "aspirin therapy", "Retrieve diabetes lab data items and calculate items due", "Out-of-goal notifications via inbox", "Check today's bp and issue appropriate alerts", and "Retrieve consult-related information and calculate items due". For simplicity, let P_1 stand for "physician accesses the patient record", P_2 stand for "physician accesses the patient record", A stand for "aspirin therapy", R stand for "Retrieve diabetes lab data items and calculate items due", O stand for "Out-of-goal notifications via inbox", C stand for "Check today's bp and issue appropriate alerts", and E stand for "Retrieve consult-related information and calculate items due".

First, we assign prior probabilities to activities, i.e, the activities have prior probabilities when they have no "parents" in the BN. In our example, the activity is P_1. Since every activity in our example has two states, true and false, 0.5 is assign to P_1 as the prior probabilities (Fig. 4). Second, when an activity causes a set of activities, each activity in the set is true if and only if the causing activity is true. In the example, activities A, R, C, and E have state true if and only if P_1's state is true. Similarly, activity O has state true if and only if R's state is true. The conditional probabilities of activities A, R, C, E, and O are shown in Figure 4. Third, when an activity is caused by a set of activities, the activity

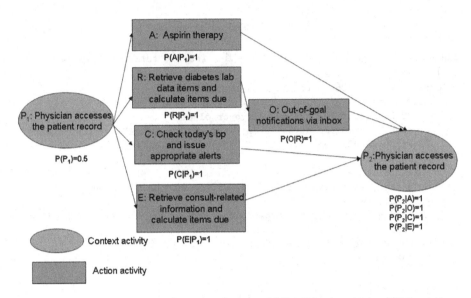

Fig. 4. Some activities in clinical practice guideline of primary care clinic visit

has state true if and only if one activity in the set of activities has state true. In the example, activity P_2 has state true if and only if one of activities A, O, C, and R has state true. The conditional probability of activity P_2 is $P(P2|A) = 1$, $P(P2|O) = 1$, $P(P2|C) = 1$, and $P(P2|E) = 1$ (Fig. 4).

After initializing the CPTs of the BN, we update the conditional probabilities based on clinical cases. If a population of patients with similar characteristics conducts a sequence of activities, the conditional probabilities are updated based on the prior records. The records correspond to how many patients conduct an activity after they finish the causing activity. For example, in Figure 4, if ten patients conduct activity R and seven of them conduct activity O consequently, the conditional probability $P(O|R)$ is updated as 0.7.

Finally, we encoded the BNs into ontologies that represent the uncertainty in CPGs, namely, representing with ontologies the activity graphs containing uncertain activities. When a BN inference engine loads this ontology, the ontology will be converted into a BN. The instances of the activity class and the activity condition class are translated into the nodes of the BN. The conditional probability tables of nodes are also converted from the properties of those instances in the CPG ontology accordingly. In the BN, an arc is drawn between nodes if the corresponding two activity instances are related by the *cause* property, the direction originating from the activity instance that has the *cause* property.

4   A Scenario Validation Based on Bayesian Network Inference

We apply the variable elimination algorithm [9,5] to perform BN inference. To verify the feasibility of our approach, a scenario of aspirin therapy for a diabetic patient is proposed. Based on this scenario, we applied our ontology-based BN approach to represent the uncertainty in CPGs and carried out BN inference based on this BN.

4.1 Bayesian Network Inference

There are many algorithms that manipulate BNs to produce posterior values [16,10]. The variable elimination algorithm [9,5] and the bucket elimination algorithm [6] are focused on algebraic operations. Since algebraic schemes like variable and bucket elimination compute marginal probability values for a given set of variables that is suitable for inference on observed evidence, we apply the variable elimination algorithm to implement the BN inference on the uncertainty of CPGs.

We assume all random variables have a finite number of possible values. The set of variables are denoted in bold; for instance, $\mathbf{X}$. The set of all variables that belong to $\mathbf{X}$ but do not belong to $\mathbf{Y}$ is indicated by $\mathbf{X} \backslash \mathbf{Y}$. The expression $\sum_{\mathbf{X}} f(\mathbf{X}, \mathbf{Y})$ indicates the sum of the function $f(\mathbf{X}, \mathbf{Y})$ is taken for all variables in $\mathbf{X}$. Denoted by $P(X)$ is the *probability density of* X: $P(x)$ is the probability measure of the event $\{X = x\}$. Denoted by $P(X|Y)$ is the probability density of X conditional on values of Y.

The semantics of the BN model are determined by the Markov condition: Every variable is independent of its non-descendants and non-parents given its parents. This condition leads to a unique joint probability density:

$$P(\mathbf{X}) = \prod_i (P(X_i|pa(X_i))).$$ (1)

where $pa(X_i)$ is denoted as the parent set of X_i.

Given a BN, the event E denotes the *observed evidence* in the network. Denoted by $\mathbf{X}_E$ is the set of observed variables. Inferences with BNs usually involve the calculation of the posterior marginal for a set of query variables $\mathbf{X}_q$. The posterior of $\mathbf{X}_q$ given E is:

$$P(\mathbf{X}_q|E) = \frac{P(\mathbf{X}_q, E)}{P(E)} = \frac{\sum_{\mathbf{X} \backslash \{\mathbf{X}_q, \mathbf{X}_E\}} P(\mathbf{X})}{\sum_{\mathbf{X} \backslash \mathbf{X}_E} P(\mathbf{X})}$$ (2)

The variable elimination algorithm can be described as follows:

1. Generate an ordering for the N requisite, non-observed, non-query variables.
2. Place all network densities in a pool of densities.
3. For i from 1 to N:
   (a) Create a data structure B_i, called a *bucket*, containing the variable, called the *bucket variable*; all densities that contain the bucket variable are called the *bucket densities*.
   (b) Multiply the densities in B_i. Store the resulting unnormalized density in B_i; the density is called B_i's *cluster*.

(c) Take the sum of X_i from B_i's cluster. Store the resulting unnormalized density in B_i's; the density is called B_i's *separator*.

(d) Place the bucket separator in the density pool.

4. At the end of the process, collect the densities that contain the query variables in a bucket B_q. Multiply the densities in B_q together and normalize the result.

The detail of variable elimination algorithm can be found in [9,5].

4.2 A Validation of an Aspirin Therapy Scenario for Diabetic Patients

We demonstrate the validity of our approach by applying an experiment to the CPG of aspirin therapy for diabetic patients (Fig. 2). Let us consider a scenario:

Scenario 1. *A user(medical student, nurse or physician etc.) is trying to apply aspirin therapy for a diabetic patient using the diabetes CPG. When he/she tries to check the aspirin risk factors, he/she can get some observed evidence, such as observations of hypertensive disorder, tobacco user finding, hyperlipidemia, and myocardial infarction. In this case, the user wants to evaluate target activities that he is concerned about in this CPG. In this way, he/she hopes the results can help him understand the effect of the observed evidence on the target activities during the whole clinical process.*

In the scenario, the CPG of aspirin therapy for diabetic patients is used. Since there are some uncertain activities in the activity graph in this CPG, we can apply our ontology-based BN approach to represent this uncertainty. Details are described in Section 3. As a result, figure 5 shows the ontology-based BN representing the uncertainty in the CPG of aspirin therapy for diabetic patients.

After loading the ontology-based BN, the BN inference engine can process the uncertainty inference when the user provides his/her observed evidence, such as observations of hypertensive disorder, tobacco user findings, hyperlipidemia, and myocardial infarctions in this scenario (Fig. 5). If the user queries the target activities, the BN inference engine will output the probability of their successful execution.

For example, after the user has obtained the observed evidence of some aspirin risk factors, he/she wants to know the probability of activity "No ASA (aspirin therapy) contraindications; recommend ASA" to help him/her to judge whether or not his/her observations of aspirin risk factors are adequate. In the BN inference engine, since the activity instance "presence of problem hypertensive disorder" is observed, its property *isObserved* is set as *true* and the property *hasState* is set as *false*. Similarly, the activities instances "presence of problem myocardial infarction", "presence of tobacco user finding", and "presence of problem hyperlipidemia" are also set in the same manner. After initializing the CPTs in this BN, Equation 2 (Section 4.1) is applied to calculate the probability of activity instance "No ASA contraindications; recommend ASA" :

$$P(\mathbf{X}_q|E) = \frac{P(\mathbf{X}_q, E)}{P(E)} = \frac{\sum_{\mathbf{X} \setminus \{\mathbf{X}_q, \mathbf{X}_E\}} P(\mathbf{X})}{\sum_{\mathbf{X} \setminus \mathbf{X}_E} P(\mathbf{X})} = 0.775$$

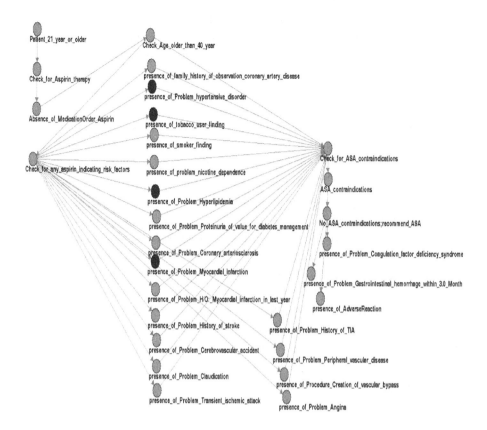

Fig. 5. An ontology-based Bayesian network of aspirin therapy for diabetic patients derived from figure 2 (blue nodes are the observed ones)

where $\mathbf{X}_q =\{$ "No ASA contraindications; recommend ASA" $\}$, and $E=\{$ "presence of problem hypertensive disorder" $= false,$ "presence of problem myocardial infarction" $= false$, "presence of tobacco user finding" $= false$, "presence of problem hyperlipidemia" $= false$ $\}$.

In another case, when the user wants to get the degree of uncertainty for the activity instance "presence of problem coagulation factor deficiency syndrome", he/she can query this target activity instance based on the observed evidence E. Through BN inference, we can obtain:

$$P(\mathbf{X}_q|E) = \frac{P(\mathbf{X}_q, E)}{P(E)} = 0.6425$$

where $\mathbf{X}_q=\{$ "presence of problem coagulation factor deficiency syndrome" $\}$ and E is the same as the above case.

The results in the two cases show high probabilities for the target activities, which suggest the user can make a decision to go ahead based on the observed evidence. When we consulted ten medical experts with this scenario, eight of them agreed with these results. We believe that the experimental results that the majority of medical experts agree with show the feasibility of our approach.

5 Conclusion and Future Work

In this paper, we contributed an ontology-based BN approach to represent the uncertainty in CPGs. With this uncertainty representation in ontology, computers can: (1) calculate the uncertainty of target activities in CPGs; (2) remind users of missing important data or event items, which should be observed in the clinical process; (3) simulate the clinical process under uncertain situations, which can be applied to e-learning systems in medical schools.

In the future, we are planning to combine our approach with a real CIS environment and apply uncertain clinical data to our application. A more comprehensive evaluation based on real clinical data should also be carried out.

Acknowledgements

This study was supported by the IT R&D program of MIC/IITA. [2005-S-083-02, Development of Semantic Service Agent Technology for the Next Generation Web].

References

1. Sage diabetes guideline,
   `http://sage.wherever.org/cpgs/diabetes/diabetes_html/phtml.html`
2. Aronsky, D., Haug, P.J.: Diagnosing community-acquired pneumonia with a bayesian network. In: Proc. AMIA Symp., pp. 632–636 (1998)
3. Atoui, H., Fayn, J., Gueyffier, F., Rubel, P.: Cardiovascular risk stratification in decision support systems:a probabilistic approach. application to phealth. Computers in Cardiology 33, 281–284 (2006)
4. Campbell, J., Tu, S., Mansfield, J., Boyer, J., McClay, J., Parker, C., Ram, P., Scheitel, S., McDonald, K.: The sage guideline model:a knowledge representation framework for encoding interoperable cpgs. Stanford Medical Informatics Report SMI-2003-0962 (2003)
5. Cozman, F.G.: Generalizing variable elimination in bayesian networks. In: Workshop on Probabilistic Reasoning in Artificial Intelligence, pp. 27–32 (2000)
6. Dechter, R.: Bucket elimination: A unifying framework for probabilistic inference. In: Jordan, M.I. (ed.) Learning in Graphical Models, pp. 75–104. MIT Press, Cambridge (1999)
7. Hripcsak, G.: Writing arden syntax medical logic modules. Comput. Biol. Med. 24, 331–363 (1994)
8. Kashyap, V., Morales, A., Hongsermeier, T.: Creation and maintenance of implementable clinical guideline specifications. In: Gil, Y., Motta, E., Benjamins, V.R., Musen, M.A. (eds.) ISWC 2005. LNCS, vol. 3729. Springer, Heidelberg (2005)
9. Zhang, N.L., Poole, D.: Exploiting causal independence in bayesian network inference. Artificial Intelligence Research, 301–328 (1996)

10. Lauritzen, S.L., Spiegelhalter, D.J.: Local computations with probabilities on graphical structures and their application to expert systems. Journal of the Royal Statistical SocietyB 50(2), 157–224 (1988)
11. Mani, S., Pazzani, M.J.: Guideline generation from data by induction of decision tables using a bayesian network framework. JAMIA supplement, 518–522 (1998)
12. Morris, A.H.: Developing and implementing computerized protocols for standardization of clinical decisions. Annal of Internal Medicine 132(5), 373–383 (2000)
13. Musen, M., Tu, S., Das, A., Shahar, Y.: Eon:a component-based approach to automation of protocol-directedtherapy. J. Am. Med. Inform. Assoc. 2, 367–388 (1996)
14. Ohno-Machado, L.: Representing uncertainty in clinical practice guidelines. In: An Invitational Workshop: Towards Representations for Sharable Guidelines (March 2000)
15. Ohno-Machado, L., Murphy, S.N., Oliver, D.E., Greenes, R.A., Barnett, G.O.: The guideline interchange format: A model for representing guidelines. Journal of the Americal Medical Informatics Association 5(4), 357–372 (1998)
16. Pearl, J.: Probabilistic Reasoning in Intelligent Systems: Networks of Plausible Inference. Morgan Kaufmann, SanMateo (1988)
17. Ram, P., Berg, D., Tu, S., Mansfield, G., Ye, Q., Abarbanel, R., Beard, N.: Executing clinical practice guidelines using the sage execution engine. Medinfo., 251–255 (2004)
18. Stoufflet, P., Ohno-Machado, L., Deibel, S., Lee, D., Greenes, R.: Geode-cm:a state-transition framework for clinical management. In: 20th Annu. Symp. Comput. Appl. Med. Care, p. 924 (1996)

A Crisp Representation for Fuzzy $\mathcal{SHOIN}$ with Fuzzy Nominals and General Concept Inclusions

Fernando Bobillo, Miguel Delgado, and Juan Gómez-Romero

Department of Computer Science and Artificial Intelligence, University of Granada
C. Periodista Daniel Saucedo Aranda, 18071 Granada, Spain
Ph.: +34 958243194, Fax: +34 958243317
fbobillo@decsai.ugr.es, mdelgado@ugr.es, jgomez@decsai.ugr.es

Abstract. Fuzzy Description Logics are a family of logics which allow the representation of (and the reasoning with) structured knowledge affected by imprecision and vagueness. They were born to overcome the limitations of classical Description Logics when dealing with such kind of knowledge, but they bring out some new challenges, requiring an appropriate fuzzy language to be agreed and needing practical and highly optimized implementations of the reasoning algorithms. In the current paper we face these problems by presenting a reasoning preserving procedure to obtain a crisp representation for a fuzzy extension of the Description Logic $\mathcal{SHOIN}$, which makes possible to reuse a crisp representation language as well as currently available reasoners, which have demonstrated a very good performance in practice. As additional contributions, we define the syntax and semantics of a novel fuzzy version of the nominal construct and allow to reason within fuzzy general concept inclusions.

1 Introduction

Ontologies [1] are a core element in the layered architecture of the Semantic Web [2]. Description Logics (DLs for short) [3] are a family of logics for representing structured knowledge. The name of each logic is composed by some labels which identify the constructs of the logic. DLs have been proved to be very useful as ontology languages [4].

As it has been widely pointed out, classical ontologies and DLs are not appropriate to handle imprecise and vague knowledge and since imprecision and vagueness are inherent to a lot of real-world application domains, there is a need for the Semantic Web to provide means to manage them. A well studied solution is to extend DLs with fuzzy sets theory [5], producing fuzzy DLs [6].

Nowadays, the World Wide Web Consortium (W3C) standard for ontology representation is OWL Web Ontology Language [7], a language comprising three sublanguages of increasing expressive power: OWL Lite, OWL DL and OWL Full, being OWL DL the most used level and nearly equivalent to the DL $\mathcal{SHOIN}(\mathcal{D})$ [8].

In order to deal with uncertain knowledge, OWL may be extended to a fuzzy DL-based language e.g. FuzzyOWL [9], with the drawback that the large number

P.C.G. da Costa et al. (Eds.): URSW 2005-2007, LNAI 5327, pp. 174–188, 2008.

of resources available (e.g. editors, reasoners or ontologies to be imported) should be adapted. Furthermore, reasoning within expressive DLs has a very high worst-case complexity (e.g. NExpTime in $\mathcal{SHOIN}$) and, consequently, there exists a significant gap between the design of a decision procedure and the achievement of a practical implementation [10]. Actually, some of the OWL DL reasoners used in practice do not support full $\mathcal{SHOIN}(\mathcal{D})$. For instance, Racer [11] does not support nominals.

Regarding fuzzy DLs, there does not exist any implemented reasoner for fuzzy $\mathcal{SHOIN}$. A reasoner for fuzzy $\mathcal{SHIF}(\mathcal{D})$ has been recently developed (fuzzyDL [12]), but its efficiency is still to be investigated. Moreover, the experience with crisp DLs ([10]) induces us to think that developing highly optimized implementations will be a hard task where ad-hoc mechanisms should be used for every particular fuzzy DL.

An alternative way to obtain fuzzy ontologies facing these two problems is to represent fuzzy DLs using crisp DLs and to reduce reasoning within fuzzy DLs to reasoning within crisp DLs [13,14,15,16]. This way it would be possible to translate them automatically into a crisp ontology language (e.g. OWL) and to use currently available reasoners (e.g. Pellet [17]).

On the other hand, current fuzzy DLs still present some limitations which we think that should be overcome. Some works on fuzzy DLs deal with nominals (named individuals) but they choose not to fuzzify the nominal construct arguing that a fuzzy singleton set does not represent any real concept world [18,9]. Hence, only crisp concepts can be defined extensively, as nominals either have to fully belong to them or not. Besides, the most used semantics for general concept inclusions (GCIs) is based on the Zadeh's set inclusion (a fuzzy set C is included in a fuzzy set D iff $\forall x, \mu_C(x) \leq \mu_D(x)$) and hence, it becomes a yes-no question. Although fuzzy GCIs, which allow to restrict the value of a GCI, have been proposed [18], current reasoning algorithms do not allow them.

Our work provides the following contributions. Firstly, we propose a different definition of fuzzy $\mathcal{SHOIN}$, including a fuzzy nominal construct and fuzzy GCIs. Secondly, we reduce reasoning in $f_{KD}\mathcal{SHOIN}$ to reasoning in $\mathcal{SHOIN}$, extending [13]. To the very best of our knowledge, this is the first reasoning algorithm dealing with such kind of fuzzy GCIs.

The present paper is organized as follows. The following section reviews some background on DLs and fuzzy logic. Next, in Section 3 we describe our fuzzy extension of $\mathcal{SHOIN}$. We have not considered (for the moment) fuzzy datatypes since OWL does not allow to define customised datatypes. Then, Section 4 shows how to reduce it into crisp $\mathcal{SHOIN}$. Finally, in Section 5 we set out some conclusions and ideas for future work.

2   Preliminaries

This section recalls some basic notions on DLs (defining the DL which will be treated along this paper, $\mathcal{SHOIN}$) and fuzzy set theory. The confident reader may choose to skip this part and pass directly to Section 3.

2.1   The Description Logic $\mathcal{SHOIN}$

Syntax. $\mathcal{SHOIN}$ assumes three alphabets of symbols, for concepts, roles and individuals. The *concepts of the language* (denoted C or D) can be built inductively from atomic concepts (A), roles (R), top concept $\top$, bottom concept $\bot$, named individuals (o_i) and simple roles $(S)^1$ according to the following syntax rule (where n, m are natural numbers, $n \geq 0, m > 0$):

$$
\begin{aligned}
C, D \rightarrow \quad & A \mid && \text{(atomic concept)} \\
& \top \mid && \text{(top concept)} \\
& \bot \mid && \text{(bottom concept)} \\
& C \sqcap D \mid && \text{(concept conjunction)} \\
& C \sqcup D \mid && \text{(concept disjunction)} \\
& \neg C \mid && \text{(concept negation)} \\
& \forall R.C \mid && \text{(universal quantification)} \\
& \exists R.C \mid && \text{(full existential quantification)} \\
& \{o_1, \dots, o_m\} \mid && \text{(nominals)} \\
& (\geq n \, S) \mid && \text{(at-least unqualified number restriction)} \\
& (\leq n \, S) && \text{(at-most unqualified number restriction)}
\end{aligned}
$$

If R_A is an atomic role, *complex roles* are built using this syntax rule:

$$
\begin{aligned}
R \rightarrow \quad & R_A \mid && \text{(atomic role)} \\
& R^- && \text{(inverse role)}
\end{aligned}
$$

A Knowledge Base (KB) comprises two parts: the intensional knowledge, i.e. general knowledge about the application domain (a Terminological Box or TBox $\mathcal{T}$ and a Role Box or RBox $\mathcal{R}$), and the extensional knowledge, i.e. particular knowledge about some specific situation (an Assertional Box or ABox $\mathcal{A}$ with statements about individuals).

An *ABox* consists of a finite set of assertions about individuals (denoted a and b), which can be one of the following types:

- Concept assertions $a : C$ (meaning that a is an instance of C).
- Role assertion $(a, b) : R$ ((a, b) is an instance of R).
- Individual assertion $a \neq b$ (a and b are different individuals).
- Individual assertion $a = b$ (a and b refer to the same individual).

A *TBox* consists of a finite set of terminological axioms about concepts, of the following types:

- General concept inclusions (GCI) $C \sqsubseteq D$ (C is more specific than D), where *general* means that they can refer to any concept of the language.
- Concept definitions $C \equiv D$ (C and D are equivalent), or abbreviation of the pair of axioms $C \sqsubseteq D$ and $D \sqsubseteq C$.

---

[1]   A simple role is a role with no transitive sub-roles. A role R is a sub-role of R' if $R \sqsubseteq^* R'$, where $\sqsubseteq^*$ is the reflexive-transitive closure of the inclusion relation $\sqsubseteq$, which will be defined below.

A *RBox* consists of a finite set of role axioms of the following types:

- role inclusions of the form $R \sqsubseteq R'$ (R is more specific than R'),
- role definitions $R \equiv R'$, a short hand for both $R \sqsubseteq R'$ and $R' \sqsubseteq R$,
- transitive role axioms $trans(R)$ (R is transitive).

Semantics. An interpretation $\mathcal{I}$ is a pair $(\Delta^{\mathcal{I}}, \cdot^{\mathcal{I}})$ consisting of a non empty set $\Delta^{\mathcal{I}}$ (the interpretation domain) and an interpretation function $\cdot^{\mathcal{I}}$ mapping every individual onto an element of $\Delta^{\mathcal{I}}$, every atomic concept A onto a set $A^{\mathcal{I}} \subseteq \Delta^{\mathcal{I}}$ and every atomic role R onto a binary relation $R^{\mathcal{I}} \subseteq \Delta^{\mathcal{I}} \times \Delta^{\mathcal{I}}$.

The interpretation is extended to complex concepts by the following inductive definitions (where $\sharp X$ denotes the cardinality of the set X):

$$\top^{\mathcal{I}} = \Delta^{\mathcal{I}}$$
$$\bot^{\mathcal{I}} = \emptyset$$
$$(C \sqcap D)^{\mathcal{I}} = C^{\mathcal{I}} \cap D^{\mathcal{I}}$$
$$(C \sqcup D)^{\mathcal{I}} = C^{\mathcal{I}} \cup D^{\mathcal{I}}$$
$$(\neg C)^{\mathcal{I}} = \Delta^{\mathcal{I}} \setminus C^{\mathcal{I}}$$
$$(\forall R.C)^{\mathcal{I}} = \{x \mid \forall y, (x, y) \notin R^{\mathcal{I}} \text{ or } y \in C^{\mathcal{I}}\}$$
$$(\exists R.C)^{\mathcal{I}} = \{x \mid \exists y, (x, y) \in R^{\mathcal{I}} \text{ and } y \in C^{\mathcal{I}}\}$$
$$\{o_1, \ldots, o_m\}^{\mathcal{I}} = \{o_1^{\mathcal{I}}, \ldots, o_m^{\mathcal{I}}\}$$
$$(\geq n\ S)^{\mathcal{I}} = \{x \mid \sharp\{y \mid (x, y) \in S^{\mathcal{I}}\} \geq n\}$$
$$(\leq n\ S)^{\mathcal{I}} = \{x \mid \sharp\{y \mid (x, y) \in S^{\mathcal{I}}\} \leq n\}$$
$$(R^-)^{\mathcal{I}} = \{(y, x) \in \Delta^{\mathcal{I}} \times \Delta^{\mathcal{I}} \mid (x, y) \in R^{\mathcal{I}}\}$$

An interpretation $\mathcal{I}$ satisfies (is a model of):

- An assertion $a : C$ iff $a^{\mathcal{I}} \in C^{\mathcal{I}}$.
- An assertion $(a, b) : R$ iff $(a, b)^{\mathcal{I}} \in R^{\mathcal{I}}$.
- An assertion $\langle a \neq b \rangle$ iff $a^{\mathcal{I}} \neq b^{\mathcal{I}}$.
- An assertion $\langle a = b \rangle$ iff $a^{\mathcal{I}} = b^{\mathcal{I}}$.
- A GCI $C \sqsubseteq D$ iff $C^{\mathcal{I}} \subseteq D^{\mathcal{I}}$.
- A concept definition $C \equiv D$ iff $C^{\mathcal{I}} = D^{\mathcal{I}}$.
- A role inclusion $R \sqsubseteq R'$ iff $R^{\mathcal{I}} \subseteq R'^{\mathcal{I}}$.
- A role definition $R \equiv R'$ iff $R^{\mathcal{I}} = R'^{\mathcal{I}}$.
- A transitive role axiom $trans(R)$ iff $(R)^{\mathcal{I}}$ is transitive.
- An ABox $\mathcal{A}$ (resp. a TBox $\mathcal{T}$, a RBox $\mathcal{R}$) iff $\mathcal{I}$ satisfies each element in $\mathcal{A}$ (resp. $\mathcal{T}$, $\mathcal{R}$).
- A KB $K = \langle \mathcal{A}, \mathcal{T}, \mathcal{R} \rangle$ iff it satisfies all $\mathcal{A}$, $\mathcal{T}$ and $\mathcal{R}$.

A DL not only stores axioms and assertions, but also offers some reasoning services, such as KB satisfiability, concept satisfiability, subsumption or instance checking. However, if a DL is closed under negation, then all the basic reasoning services are reducible to KB satisfiability [19].

2.2 Fuzzy Set Theory

Fuzzy set theory and fuzzy logic were proposed by L. Zadeh [5] to manage imprecise and vague knowledge. While in classical set theory elements either belong to a set or not, in fuzzy set theory elements can belong to a set to some degree. More formally, let X be a set of elements called the reference set. A fuzzy subset A of X is defined by a membership function $\mu_A(x)$, or simply $A(x)$, which assigns any $x \in X$ to a value in the interval of real numbers between 0 and 1. As in the classical case, 0 means no-membership and 1 full membership, but now a value between 0 and 1 represents the extent to which x can be considered as an element of X. If the reference set is finite ($X = \{x_1, \ldots, x_n\}$), the membership function can be expressed using the notation $A = \{\mu_A(x_1)/x_1, \ldots \mu_A(x_n)/x_n\}$.

For every $\alpha \in [0,1]$, the α-cut of a fuzzy set A is defined as the (crisp) set such that its elements belong to A with degree at least α, i.e. $\{x \mid \mu_A(x) \geq \alpha\}$. Similarly, the *strict α-cut* is defined as $\{x \mid \mu_A(x) > \alpha\}$.

All crisp set operations are extended to fuzzy sets. The intersection, union, complement and implication set operations are performed by a t-norm function $\otimes$, a t-conorm function $\oplus$, a negation function $\ominus$ and an implication function $\Rightarrow$, respectively. For a complete definition of these functions as well as their properties, we refer the reader to [20,21].

There are commonly two types of fuzzy implications used. The first class is *S-implications*, which are defined by the operation $\alpha \Rightarrow \beta = (\ominus\alpha) \oplus \beta$ and can be seen as a fuzzy extension of the crisp proposition $a \rightarrow b = \neg a \vee b$. The second class is *R-implications* (residuum-based implications), which are defined as $\alpha \Rightarrow \beta = \sup\{\gamma \in [0,1] \mid (\alpha \otimes \gamma) \leq \beta\}$ and can be used to define a fuzzy complement as $\ominus a = a \Rightarrow 0$. They always verify that $\alpha \Rightarrow \beta = 1$ iff $\alpha \leq \beta$. Product and Gödel implications are R-implications, the implication of the Zadeh family which is called Kleene-Dienes (KD) is an S-implication and the Łukasiewicz implication belongs to both types.

A fuzzy implication specifies a family of fuzzy operators. If it is an S-implication this notation also specifies the fuzzy complement and t-conorm, while if it is an R-implication then we also know the t-norm and the fuzzy complement. The missing operators are usually defined using duality of the t-norms and the t-conorms. Table 1 shows the most important families of fuzzy operators: Zadeh, Łukasiewicz, Gödel and Product.

Table 1. Popular families of fuzzy operators

Family	t-norm $\alpha \otimes \beta$	t-conorm $\alpha \oplus \beta$	negation $\ominus\alpha$	implication $\alpha \Rightarrow \beta$
Zadeh	$\min\{\alpha, \beta\}$	$\max\{\alpha, \beta\}$	$1 - \alpha$	$\max\{1 - \alpha, \beta\}$
Łukasiewicz	$\max\{\alpha + \beta - 1, 0\}$	$\min\{\alpha + \beta, 1\}$	$1 - \alpha$	$\min\{1 - \alpha + \beta, 1\}$
Gödel	$\min\{\alpha, \beta\}$	$\max\{\alpha, \beta\}$	$\begin{cases} 1, & \alpha = 0 \\ 0, & \alpha > 0 \end{cases}$	$\begin{cases} 1 & \alpha \leq \beta \\ \beta, & \alpha > \beta \end{cases}$
Product	$\alpha \cdot \beta$	$\alpha + \beta - \alpha \cdot \beta$	$\begin{cases} 1, & \alpha = 0 \\ 0, & \alpha > 0 \end{cases}$	$\begin{cases} 1 & \alpha \leq \beta \\ \beta/\alpha, & \alpha > \beta \end{cases}$

3   Fuzzy $\mathcal{SHOIN}$

In this section we define fuzzy $\mathcal{SHOIN}$, which extends $\mathcal{SHOIN}$ to the fuzzy case by letting (i) concepts denote fuzzy sets of individuals and (ii) roles denote fuzzy binary relations between individuals.

Syntax. Our logic is similar to [18,9], adding fuzzy nominals and fuzzy GCIs.

Fuzzy $\mathcal{SHOIN}$ assumes three alphabets of symbols, for concepts, roles and individuals. The *complex concepts* (denoted C or D) can be built inductively from atomic concepts (A), roles (R), top concept $\top$, bottom concept $\bot$, named individuals (o_i) and simple roles (S) according to the following syntax rule (where n, m are natural numbers, $n \geq 0, m > 0, \alpha_i \in [0,1]$):

$$C, D \to A \mid \top \mid \bot \mid C \sqcap D \mid C \sqcup D \mid \neg C \mid \forall R.C \mid \exists R.C \mid$$
$$\{\alpha_1/o_1, \ldots, \alpha_m/o_m\} \mid\; \geq m\, S \mid\; \leq n\, S$$

Note that the only difference is the presence of *fuzzy nominals*.

If R_A is an atomic role, *complex roles* are built using this syntax rule:

$$R \to R_A \mid R^-$$

We do not impose unique name assumption, i.e. two nominals might refer to the same individual.

A fuzzy Knowledge Base (fKB) comprises two parts: the intensional knowledge, i.e. general knowledge about the application domain (a fuzzy Terminological Box or TBox $\mathcal{T}$ and a fuzzy Role Box or RBox $\mathcal{R}$), and the extensional knowledge, i.e. particular knowledge about some specific situation (a fuzzy Assertional Box or ABox $\mathcal{A}$ with statements about individuals).

A *fuzzy ABox* $\mathcal{A}$ consists of a finite set of fuzzy assertions, which can be individual assertions or constraints on the truth value of a concept or role assertion. An individual assertion is either an inequality of individuals $\langle a \neq b \rangle$ or an equality of individuals $\langle a = b \rangle$ (they are necessary since we do not impose unique name assumption). Note that individual assertions are considered to be crisp, since the equality and inequality of individuals has always been considered crisp in the fuzzy DL literature [18].

A constraint on the truth value of a concept or role assertion is an expression of the form $\langle \Psi \geq \alpha \rangle$, $\langle \Psi > \beta \rangle$, $\langle \Phi \leq \beta \rangle$, $\langle \Phi < \alpha \rangle$, where Ψ is an assertion of the form $a : C$ or $(a, b) : R$, Φ is an assertion of the form $a : C$, $\alpha \in (0,1]$ and $\beta \in [0,1)$. Note that fuzzy assertions of the form $\langle (a, b) : R \leq \beta \rangle, \langle (a, b) : R < \alpha \rangle$ are not allowed. In fact, as we will see in Section 4, if these role assertions were allowed we would need some additional role constructs (role conjunction, role disjunction, bottom role and top role) which are not allowed in crisp $\mathcal{SHOIN}$.

A *fuzzy TBox* $\mathcal{T}$ consists of a finite set of fuzzy terminological axioms. A fuzzy terminological axiom is either a fuzzy GCI or a concept definition. A fuzzy GCI constrains the truth value of a GCI i.e. it is an expression of the form $\langle \Omega \geq \alpha \rangle$, $\langle \Omega > \beta \rangle$, $\langle \Omega \leq \beta \rangle$ or $\langle \Omega < \alpha \rangle$, where Ω is a GCI of the form $C \sqsubseteq D$, $\alpha \in (0,1]$ and $\beta \in [0,1)$. We think that concept definitions should not be fuzzified, so $C \equiv D$ is an abbreviation of the pair of axioms $\langle C \sqsubseteq D \geq 1 \rangle$ and $\langle D \sqsubseteq C \geq 1 \rangle$.

A *fuzzy RBox* $\mathcal{R}$ consists of a finite set of fuzzy role axioms. A fuzzy role axiom is either a fuzzy role inclusion $R \sqsubseteq R'$, a fuzzy role definition $R \equiv R'$ (a short hand for both $R \sqsubseteq R'$ and $R' \sqsubseteq R$) or a transitive role axiom $trans(R)$.

Semantics. A fuzzy interpretation $\mathcal{I}$ is a pair $(\Delta^{\mathcal{I}}, \cdot^{\mathcal{I}})$ consisting of a non empty set $\Delta^{\mathcal{I}}$ (the interpretation domain) and a fuzzy interpretation function $\cdot^{\mathcal{I}}$

mapping every individual onto an element of $\Delta^{\mathcal{I}}$, every concept C onto a function $C^{\mathcal{I}} : \Delta^{\mathcal{I}} \to [0,1]$ and every role R onto a function $R^{\mathcal{I}} : \Delta^{\mathcal{I}} \times \Delta^{\mathcal{I}} \to [0,1]$. $C^{\mathcal{I}}$ (resp. $R^{\mathcal{I}}$) is interpreted as the membership degree function of the fuzzy concept C (resp. fuzzy rol R) w.r.t. $\mathcal{I}$. $C^{\mathcal{I}}(a)$ (resp. $R^{\mathcal{I}}(a,b)$) gives us the degree of being the individual a an element of the fuzzy concept C (resp. the degree of being (a,b) an element of the fuzzy role R) under the fuzzy interpretation $\mathcal{I}$. The fuzzy interpretation function is extended to complex concepts and roles as:

$$
\begin{aligned}
\top^{\mathcal{I}}(x) &= 1 \\
\bot^{\mathcal{I}}(x) &= 0 \\
(C \sqcap D)^{\mathcal{I}}(x) &= C^{\mathcal{I}}(x) \otimes D^{\mathcal{I}}(x) \\
(C \sqcup D)^{\mathcal{I}}(x) &= C^{\mathcal{I}}(x) \oplus D^{\mathcal{I}}(x) \\
(\neg C)^{\mathcal{I}}(x) &= \ominus C^{\mathcal{I}}(x) \\
(\forall R.C)^{\mathcal{I}}(x) &= \inf_{y \in \Delta^{\mathcal{I}}}\{R^{\mathcal{I}}(x,y) \Rightarrow C^{\mathcal{I}}(y)\} \\
(\exists R.C)^{\mathcal{I}}(x) &= \sup_{y \in \Delta^{\mathcal{I}}}\{R^{\mathcal{I}}(x,y) \otimes C^{\mathcal{I}}(y)\} \\
\{\alpha_1/o_1, \ldots, \alpha_m/o_m\}(x) &= \sup_{i \mid x = o_i^{\mathcal{I}}} \alpha_i \\
(\geq m\ S)^{\mathcal{I}}(x) &= \sup_{y_1, \ldots, y_m \in \Delta^{\mathcal{I}}} \otimes_{i=1}^{m} S^{\mathcal{I}}(x, y_i) \bigotimes (\otimes_{j<k}\{y_j \neq y_k\}) \\
(\leq n\ S)^{\mathcal{I}}(x) &= \inf_{y_1, \ldots, y_{n+1} \in \Delta^{\mathcal{I}}} \otimes_{i=1}^{n+1} S^{\mathcal{I}}(x, y_i) \Rightarrow (\oplus_{j<k}\{y_j = y_k\}) \\
(R^-)^{\mathcal{I}}(x, y) &= R^{\mathcal{I}}(y, x)
\end{aligned}
$$

We will shortly justify our decision of fuzzifying the nominal construct by showing an example.

Example 1. Suppose we want to represent the concept of country where German is a widely spoken language. Previous approaches allow to represent a fuzzy disjunction of nominals $C \equiv \{germany\} \sqcup \{austria\} \sqcup \{switzerland\}$. Since the semantics of the nominal construct is crisp ($\{o_i\}^{\mathcal{I}}(x) = 1$ if $x = o_i^{\mathcal{I}}$ or 0 otherwise), it forces *switzerland* to fully belong to the concept or not, despite of German-speaking community of Switzerland represents only about two thirds of the total population of the country. On the contrary, following our approach we are able to define $C \equiv \{1/germany, 1/austria, 0.67/switzerland\}$.

Let us comment the semantics of the fuzzy nominals $\{\alpha_1/o_1, \ldots, \alpha_m/o_m\}^{\mathcal{I}}(x)$ $= \sup_{i \mid x = o_i^{\mathcal{I}}} \alpha_i$. Since we are not imposing unique name assumption, it is possible that $x = o_i^{\mathcal{I}}$ for more than one o_i. Then, we take the supremum over the membership degrees α_i associated to these named individuals o_i. In the previous example, if x can be interpreted as *germany* and *switzerland*, we take the supremum (maximum) over 1 and 0.67. And, of course, if $\forall i \in \{1, \ldots, m\}, x \neq o_i^{\mathcal{I}}$, then $\{\alpha_1/o_1, \ldots, \alpha_m/o_m\}^{\mathcal{I}}(x) = \sup \emptyset = 0$.

Note that previous approaches consider nominals to be crisp singletons arguing that they do not represent real-life concepts [18,9]. In these approaches it is possible to represent a fuzzy disjunction of crisp singletons. However, we consider fuzzy nominals as proper fuzzy sets, which do represent real-life concepts. It is easy to see that our definition generalizes the previous definition for the nominal construct, as $\{o_1\} \sqcup \cdots \sqcup \{o_m\}$ is equivalent to $\{1/o_1, \ldots, 1/o_m\}$.

A fuzzy interpretation $\mathcal{I}$ satisfies (is a model of):

- A fuzzy assertion $\langle a : C \geq \alpha \rangle$ iff $C^{\mathcal{I}}(a^{\mathcal{I}}) \geq \alpha$. Similar definitions can be given for $> \beta$, $\leq \beta$ and $< \alpha$.
- A fuzzy assertion $\langle (a, b) : R \geq \alpha \rangle$ iff $R^{\mathcal{I}}(a^{\mathcal{I}}, b^{\mathcal{I}}) \geq \alpha$. Similar definitions can be given for $> \beta$, $\leq \beta$ and $< \alpha$.
- An assertion $\langle a \neq b \rangle$ iff $a^{\mathcal{I}} \neq b^{\mathcal{I}}$ (resp. $\langle a = b \rangle$ iff $a^{\mathcal{I}} = b^{\mathcal{I}}$).
- A fuzzy GCI $\langle C \sqsubseteq D \geq \alpha \rangle$ iff $\inf_{x \in \Delta^{\mathcal{I}}} \{ C^{\mathcal{I}}(x) \Rightarrow D^{\mathcal{I}}(x) \} \geq \alpha$. Similar definitions can be given for $> \beta$, $\leq \beta$ and $< \alpha$.
- A concept definition $C \equiv D$ iff $C^{\mathcal{I}} = D^{\mathcal{I}}$.
- A role inclusion axiom $R \sqsubseteq R'$ iff iff $\forall x, y \in \Delta^{\mathcal{I}}, R^{\mathcal{I}}(x, y) \leq (R')^{\mathcal{I}}(x, y)$.
- A role definition axiom $R \equiv R'$ iff $R^{\mathcal{I}} = R'^{\mathcal{I}}$.
- An axiom $trans(R)$ iff $\forall x, y \in \Delta^{\mathcal{I}}, R^{\mathcal{I}}(x, y) \geq \sup_{z \in \Delta^{\mathcal{I}}} R^{\mathcal{I}}(x, z) \otimes R^{\mathcal{I}}(z, y)$.
- A fKB $\langle \mathcal{A}, \mathcal{T}, \mathcal{R} \rangle$ iff it satisfies each element in $\mathcal{A}$, $\mathcal{T}$ and $\mathcal{R}$.

The definition of fuzzy GCIs allows concept subsumption to hold to a certain degree in $[0, 1]$. This does not hold for role inclusion axioms, which leads to a certain asymmetry in the expressivity. While this is not too elegant, it is a restriction imposed by the choice of the implication function, which would require the subjacent DL to have negated roles and role disjunction. However, for a higher practical utility, we have preferred to restrict ourselves to $\mathcal{SHOIN}$, closer to the DL underlying OWL DL.

Similarly as in the crisp case, in fuzzy DLs most reasoning services are reducible to fKB satisfiability [22], so here in after we will only consider this task.

Some logical properties. The following lemma shows that our definition of fuzzy $\mathcal{SHOIN}$ is a sound extension of crisp $\mathcal{SHOIN}$:

Lemma 1. *Fuzzy interpretations coincide with crisp interpretations if we restrict to the membership degrees of 0 and 1 [9].*

Here in after we will concentrate on $f_{KD}SHOIN$, restricting ourselves to the Zadeh family of fuzzy operators. For instance, in the semantics of the at-least unqualified number restriction, $\otimes_{i<j} \{ y_i \neq y_j \}$ means that there must exist n distinct elements of the domain. The choice of the t-norm and the t-conorm will be justified in Section 4.

On the other hand, in fuzzy DLs it is very common to use the KD implication in the semantics of universal quantification, so for the sake of coherence we have chosen to use it in the semantics of fuzzy GCIs as well.

Similarly as in [23], $f_{KD}SHOIN$ allows some sort of modus ponens over concepts and roles, even with the new semantics of fuzzy GCIs:

Lemma 2. *For $\alpha, \beta, \gamma \in [0, 1]$, $\rhd \in \{\geq, >\}$ and $\alpha \not\rhd 1 - \beta$ ($+\geq \; = \; >, +> \; = \; \geq$), the following properties are verified:*

(i) $\langle a : C \rhd \alpha \rangle$ *and* $\langle C \sqsubseteq D \rhd \beta \rangle$ *imply* $\langle a : D \rhd \beta \rangle$.
(ii) $\langle (a, b) : R \rhd \gamma \rangle$ *and* $\langle R \sqsubseteq R' \rangle$ *imply* $\langle (a, b) : R' \rhd \gamma \rangle$.
(iii) $\langle (a, b) : R \rhd \alpha \rangle$ *and* $\langle a : \forall R.C \rhd \beta \rangle$ *imply* $\langle b : C \rhd \beta \rangle$.

Unfortunately, the use of the KD implication in the semantics of fuzzy GCIs brings about two counter-intuitive effects. Firstly, a concept does not fully subsume itself i.e. $C \sqsubseteq C \Rightarrow \inf_{a \in \Delta^{\mathcal{I}}} \max\{1 - C^{\mathcal{I}}(a), C^{\mathcal{I}}(a)\} \geq 0.5$. Secondly, crisp concept subsumption forces fuzzy concepts to be crisp i.e. $\langle C \sqsubseteq D \geq 1 \rangle \Rightarrow \inf_{a \in \Delta^{\mathcal{I}}} \max\{1 - C^{\mathcal{I}}(a), D^{\mathcal{I}}(a)\} \geq 1$ which is true iff for each element of the domain $D^{\mathcal{I}}(a) = 1$ or $1 - C^{\mathcal{I}}(a) \geq 1 \Rightarrow C^{\mathcal{I}}(a) = 0$. These problems point out the need of further investigation involving alternative fuzzy operators. For example, using an R-implication would fix the first problem; while using Łukasiewicz or Gödel implication would fix the second one.

4   A Crisp Representation for Fuzzy $\mathcal{SHOIN}$

In this section we show how to reduce a $f_{KD}\mathcal{SHOIN}$ fKB into a crisp Knowledge Base (KB). The procedure preserves reasoning, so existing $\mathcal{SHOIN}$ reasoners could be applied to the resulting KB. [13] presents a reasoning preserving transformation for $f_{KD}\mathcal{ALCH}$ into crisp $\mathcal{ALCH}$: firstly, some new atomic concepts and roles are defined, then some new axioms are added to preserve the semantics of the fKB and finally the ABox, the TBox and the RBox are mapped separately. Our reduction extends this work to $f_{KD}\mathcal{SHOIN}$. A slight difference is that our mapping of the TBox can introduce some new assertions about new individuals (not appearing in the initial fKB).

New Elements. Let A^{fK} and R^{fK} be the set of atomic concepts and atomic roles occurring in a fKB $fK = \langle \mathcal{A}, \mathcal{T}, \mathcal{R} \rangle$. In [13] it is shown that the set of the degrees which must be considered for any reasoning task is defined as $N^{fK} = X^{fK} \cup \{1 - \alpha \mid \alpha \in X^{fK}\}$, where X^{fK} is defined as follows:

$$X^{fK} = \{0, 0.5, 1\} \cup \{\alpha \mid \langle \Psi \geq \alpha \rangle \in \mathcal{A}\} \cup \{\beta \mid \langle \Psi > \beta \rangle \in \mathcal{A}\}$$
$$\cup \{1 - \beta \mid \langle \Phi \leq \beta \rangle \in \mathcal{A}\} \cup \{1 - \alpha \mid \langle \Phi < \alpha \rangle \in \mathcal{A}\}$$
$$\cup \{\alpha \mid \langle \Omega \geq \alpha \rangle \in \mathcal{T}\} \cup \{\beta \mid \langle \Omega > \beta \rangle \in \mathcal{T}\}$$
$$\cup \{1 - \beta \mid \langle \Omega \leq \beta \rangle \in \mathcal{T}\} \cup \{1 - \alpha \mid \langle \Omega < \alpha \rangle \in \mathcal{T}\}$$

This also holds in $f_{KD}\mathcal{SHOIN}$, because the fuzzy operators do not introduce new degrees, but note that it is no longer true when other fuzzy operators are considered. For example, the combination of the degrees 0.5 and 0.3 using product t-norm introduces the new degree $0.5 \cdot 0.3 = 0.15$. In that case, the process may calculate all possible degrees in $[0, 1]$ with a given precision, but further investigation is required. Without loss of generality, it can be assumed that $N^{fK} = \{\gamma_1, \ldots, \gamma_{|N^{fK}|}\}$ and $\gamma_i < \gamma_{i+1}, 1 \leq i \leq |N^{fK}| - 1$.

Now, for each $\alpha, \beta \in N^{fK}, \alpha \in (0, 1], \beta \in [0, 1)$, for each relation $\geq, >, \leq$, $<$, for each $A \in A^{fK}$ and for each $R \in R^{fK}$, four new atomic concepts $A_{\geq \alpha}, A_{> \beta}, A_{\leq \beta}, A_{< \alpha}$ and two new atomic roles $R_{\geq \alpha}, R_{> \beta}$ are introduced. $A_{\geq \alpha}$ represents the crisp set of individuals which are instance of A with degree higher or equal than α i.e the α-cut of A. The other new elements are defined in a similar way. Neither $A_{<0}, A_{>1}, R_{>1}$ are considered (they are always empty sets) nor $A_{\leq 1}, A_{\geq 0}, R_{\geq 0}$ (they are always equivalent to the top concept).

The semantics of these newly introduced atomic concepts and roles is preserved by some terminological and role axioms. For each $1 \leq i \leq |N^{fK}| - 1$, for each $2 \leq j \leq |N^{fK}|$, for each $A \in A^{fK}$ and for each $R \in R^{fK}$, $T(N^{fK})$ is the smallest terminology containing the following axioms:

$$
\begin{array}{ll}
A_{\geq \gamma_{i+1}} \sqsubseteq A_{> \gamma_i} & A_{> \gamma_i} \sqsubseteq A_{\geq \gamma_i} \\
A_{< \gamma_j} \sqsubseteq A_{\leq \gamma_j} & A_{\leq \gamma_i} \sqsubseteq A_{< \gamma_{i+1}} \\
A_{\geq \gamma_j} \sqcap A_{< \gamma_j} \sqsubseteq \bot & A_{> \gamma_i} \sqcap A_{\leq \gamma_i} \sqsubseteq \bot \\
\top \sqsubseteq A_{\geq \gamma_j} \sqcup A_{< \gamma_j} & \top \sqsubseteq A_{> \gamma_i} \sqcup A_{\leq \gamma_i}
\end{array}
$$

Similarly, $R(N^{fK})$ is the smallest terminology containing these two axioms:

$$
R_{\geq \gamma_{i+1}} \sqsubseteq R_{> \gamma_i} \qquad\qquad R_{> \gamma_i} \sqsubseteq R_{\geq \gamma_i}
$$

It is easy to see that allowing expressions of the type $\langle (a,b) : R \leq \beta \rangle, \langle (a,b) : R < \alpha \rangle$ would need additional role constructs (role conjunction, role disjunction, bottom role and top role) which are not part of $\mathcal{SHOIN}$.

Mapping the ABox. Let ρ be a mapping, inductively defined on the structure of concepts and roles as shown in Table 2. For instance, given a fuzzy concept C, $\rho(C, \geq \alpha)$ is a crisp set containing all the elements which belong to C with a degree greater or equal than α (the other cases are similar).

Fuzzy assertions are mapped into $\mathcal{SHOIN}$ assertions using a mapping σ. Let $\gamma \in N^{fK}, \bowtie \in \{\geq, <, \leq, >\}, \triangleright \in \{\geq, <\}, \sigma(\mathcal{A}) = \{\sigma(\Phi) \mid \Phi \in \mathcal{A}\}$, where $\sigma(\Phi)$ is defined as in the following table:

$$
\begin{array}{l}
\sigma(\langle a : C \bowtie \gamma \rangle) = \{a : \rho(C, \bowtie \gamma)\} \\
\sigma(\langle (a,b) : R \triangleright \gamma \rangle) = \{(a,b) : \rho(R, \triangleright \gamma)\} \\
\sigma(\langle a \neq b \rangle) = \{a \neq b\} \\
\sigma(\langle a = b \rangle) = \{a = b\}
\end{array}
$$

Example 2. Let us consider the reduction of an assertion of the form $\langle a : \forall R.C \geq \alpha \rangle$. If it is satisfied, there exists a fuzzy interpretation $\mathcal{I}$ such that $\inf_{y \in \Delta^{\mathcal{I}}} \max\{1 - R^{\mathcal{I}}(a^{\mathcal{I}}, y), C^{\mathcal{I}}(y)\} \geq \alpha$. For an arbitrary y, $R^{\mathcal{I}}(a^{\mathcal{I}}, y) \leq 1 - \alpha$ or $C^{\mathcal{I}}(y) \geq \alpha$ must hold. Hence, if $R^{\mathcal{I}}(a^{\mathcal{I}}, y) \leq 1 - \alpha$ is not satisfied (i.e., $R^{\mathcal{I}}(a^{\mathcal{I}}, y) > 1 - \alpha$), then we deduce that $C^{\mathcal{I}}(y) \geq \alpha$, which is the semantics of the crisp assertion $a : \forall \rho(R, > 1 - \alpha).\rho(C, \geq \alpha)$.

Mapping the TBox. $f_{KD}\mathcal{SHOIN}$ fuzzy terminological axioms to either (crisp) terminological axioms (for the cases $\geq$ or $>$) or assertions (for the cases $\leq$ and $<$). In the former case, we redefine $\kappa(fK, \mathcal{T})$ as $\kappa(fK, \mathcal{T}) = \bigcup_{\Omega \in \mathcal{T}} \kappa(\Omega)$, where $\Omega = \{\langle C \sqsubseteq D\{\geq, >\}\gamma \rangle\}$ and $\kappa(\Omega)$ is defined as:

$$
\begin{array}{l}
\kappa(\langle C \sqsubseteq D \geq \alpha \rangle) = \{\rho(C, > 1 - \alpha) \sqsubseteq \rho(D, \geq \alpha)\} \\
\kappa(\langle C \sqsubseteq D > \beta \rangle) = \{\rho(C, \geq 1 - \beta) \sqsubseteq \rho(D, > \beta)\}
\end{array}
$$

Table 2. Mapping ρ

x	y	$\rho(x,y)$
A	$\geq \gamma$	$A_{\geq \gamma}$ if $\gamma \neq 0$, $\top$ otherwise
A	$> \gamma$	$A_{> \gamma}$, if $\gamma \neq 1$, $\bot$ otherwise
A	$\leq \gamma$	$A_{\leq \gamma}$ if $\gamma \neq 0$, $\top$ otherwise
A	$< \gamma$	$A_{< \gamma}$, if $\gamma \neq 1$, $\bot$ otherwise
R	$\geq \gamma$	$R_{\geq \gamma}$ if $\gamma \neq 0$, $\top$ otherwise
R	$> \gamma$	$R_{> \gamma}$, if $\gamma \neq 1$, $\bot$ otherwise
$\top$	$\geq \gamma$	$\top$
$\top$	$> \gamma$	$\top$ if $\gamma \neq 1$, $\bot$ otherwise
$\top$	$\leq \gamma$	$\top$ if $\gamma = 1$, $\bot$ otherwise
$\top$	$< \gamma$	$\bot$
$\bot$	$\geq \gamma$	$\top$ if $\gamma = 0$, $\bot$ otherwise
$\bot$	$> \gamma$	$\bot$
$\bot$	$\leq \gamma$	$\top$
$\bot$	$< \gamma$	$\top$ if $\gamma \neq 0$, $\bot$ otherwise
$C \sqcap D$	$\{\geq, >\}\, \gamma$	$\rho(C, \{\geq, >\}\, \gamma) \sqcap \rho(D, \{\geq, >\}\, \gamma)$
$C \sqcap D$	$\{\leq, <\}\, \gamma$	$\rho(C, \{\leq, <\}\, \gamma) \sqcup \rho(D, \{\leq, <\}\, \gamma)$
$C \sqcup D$	$\{\geq, >\}\, \gamma$	$\rho(C, \{\geq, >\}\, \gamma) \sqcup \rho(D, \{\geq, >\}\, \gamma)$
$C \sqcup D$	$\{\leq, <\}\, \gamma$	$\rho(C, \{\leq, <\}\, \gamma \sqcap \rho(D, \{\leq, <\}\, \gamma)$
$\neg C$	$\{\geq, >\}\, \gamma$	$\rho(C, \{\leq, <\}\, 1 - \gamma)$
$\neg C$	$\{\leq, <\}\, \gamma$	$\rho(C, \{\geq, >\}\, 1 - \gamma)$
$\exists R.C$	$\{\geq, >\}\, \gamma$	$\exists \rho(R, \{\geq, >\}\, \gamma).\rho(C, \{\geq, >\}\, \gamma)$
$\exists R.C$	$\{\leq, <\}\, \gamma$	$\forall \rho(R, \{>, \geq\}\, \gamma).\rho(C, \{\leq, <\}\, \gamma)$
$\forall R.C$	$\{\geq, >\}\, \gamma$	$\forall \rho(R, \{>, \geq\}\, 1 - \gamma).\rho(C, \{\geq, >\}\, \gamma)$
$\forall R.C$	$\{\leq, <\}\, \gamma$	$\exists \rho(R, \{\geq, >\}\, 1 - \gamma).\rho(C, \{\leq, <\}\, \gamma)$
$\{\{\alpha_1/o_1, \ldots, \alpha_m/o_m\}\}$	$\bowtie \gamma$	$\{o_i \mid \alpha_i \bowtie \gamma, 1 \leq i \leq m\}$
$\geq m\, S$	$\{\geq, >\}\, \gamma$	$\geq m\, \rho(S, \{\geq, >\}\, \gamma)$
$\geq m\, S$	$\{\leq, <\}\, \gamma$	$\leq m-1\, \rho(S, \{>, \geq\}\, \gamma)$
$\leq n\, S$	$\{\geq, >\}\, \gamma$	$\leq n\, \rho(S, \{>, \geq\}\, 1 - \gamma)$
$\leq n\, S$	$\{\leq, <\}\, \gamma$	$\geq n+1\, \rho(S, \{\geq, >\}\, 1 - \gamma)$
R^-	$\bowtie \gamma$	$\rho(R, \bowtie \gamma)^-$

Example 3. Consider the reduction of a GCI $\langle C \sqsubseteq D \geq \alpha \rangle$. If it is satisfied, $\inf_{x \in \Delta^{\mathcal{I}}} C^{\mathcal{I}}(x) \Rightarrow D^{\mathcal{I}}(x) \geq \alpha$. As this is true for the infimum, an arbitrary $x \in \Delta^{\mathcal{I}}$ must satisfy $C^{\mathcal{I}}(x) \Rightarrow D^{\mathcal{I}}(x) \geq \alpha$. From the semantics of the KD implication, this is true if $\max\{1 - C^{\mathcal{I}}(x), D^{\mathcal{I}}(x)\} \geq \alpha$, which is true if $1 - C^{\mathcal{I}}(x) \geq \alpha \equiv C^{\mathcal{I}}(x) \leq 1 - \alpha$ or $D^{\mathcal{I}}(x) \geq \alpha$. Hence, if $C^{\mathcal{I}}(x) > 1 - \alpha$ we deduce that $D^{\mathcal{I}}(x) \geq \alpha$, which is the semantics of the crisp GCI $\rho(C, > 1 - \alpha) \sqsubseteq \rho(D, \geq \alpha)$.

In the latter case, new assertions are necessary since negated terminological axioms are not part of crisp $\mathcal{SHOIN}$. A new function $A(\mathcal{T})$ adds these new assertions to the ABox. $A(\mathcal{T}) = \bigcup_{\Xi \in \mathcal{T}} A(\Xi)$, where $\Xi = \{\langle C \sqsubseteq D\{\leq, <\}\gamma\rangle\}$ and $A(\Xi)$ is defined as follows (where x is a new individual):

$$A(\langle C \sqsubseteq D \leq \beta \rangle) = \{x : \rho(C, \geq 1 - \beta) \sqcap \rho(D, \leq \beta)\}$$
$$A(\langle C \sqsubseteq D < \alpha \rangle) = \{x : \rho(C, > 1 - \alpha) \sqcap \rho(D, < \alpha)\}$$

Example 4. Consider a GCI $\langle C \sqsubseteq D \leq \beta \rangle$. If it is satisfied, $\inf_{x \in \Delta^{\mathcal{I}}} C^{\mathcal{I}}(x) \Rightarrow D^{\mathcal{I}}(x) \leq \beta$. As this is true for the infimum, there exists some $x \in \Delta^{\mathcal{I}}$ which satisfies $C^{\mathcal{I}}(x) \Rightarrow D^{\mathcal{I}}(x) \leq \beta$. This is true if $\max\{1 - C^{\mathcal{I}}(x), D^{\mathcal{I}}(x)\} \leq \beta$, which is true if $1 - C^{\mathcal{I}}(x) \leq \beta \equiv C^{\mathcal{I}}(x) \geq 1 - \beta$ and $D^{\mathcal{I}}(x) \leq \beta$. Hence, for a new individual x, the crisp assertion $x : \rho(C, \geq 1 - \beta) \sqcap \rho(D, \leq \beta)$ holds.

Mapping the RBox. Role axioms are reduced using a function $\kappa(fK, \mathcal{R}) = \bigcup_{\Omega \in \mathcal{R}} \kappa(fK, \Omega)$, where $\kappa(fK, \Omega)$ is defined as:

$$\kappa(fK, R \sqsubseteq R') = \bigcup_{\gamma \in N^{fK}, \triangleright \in \{\geq, >\}} \{\rho(R, \triangleright\gamma) \sqsubseteq \rho(R', \triangleright\gamma)\}$$
$$\kappa(fK, trans(R)) = \bigcup_{\gamma \in N^{fK}, \triangleright \in \{\geq, >\}} \{trans(\rho(R, \triangleright\gamma))\}$$

Discussion. A fKB $fK = \langle \mathcal{A}, \mathcal{T}, \mathcal{R} \rangle$ is reduced into a KB $K(fK) = \langle \sigma(\mathcal{A}) \cup A(\mathcal{T}), T(N^{fK}) \cup \kappa(fK, \mathcal{T}), R(N^{fK}) \cup \kappa(fK, \mathcal{R}) \rangle$. The following important theorem shows that the reduction to a crisp DL preserves reasoning:

Theorem 1. *A $fK_D\mathcal{SHOIN}$ fKB fK is satisfiable iff $K(fK)$ is satisfiable.*

The complexity of our procedure is quadratic: the ABox is linear while the TBox and the RBox are quadratic. It is interesting to note that, while [13] reduces a fuzzy terminological axiom into a set of crisp terminological axioms, our semantics for fuzzy GCIs allows to reduce each axiom into either an axiom or an assertion. This reduction in the size of the TBox (although it is still quadratic) is very interesting since reasoning with GCIs is a source of computational complexity [24].

Example 5. To illustrate the reduction, let us present an example concerning interchange of medical knowledge. A known issue in health-care support is that consensus in the used vocabulary is required to achieve understanding among different physicians and systems. Medical taxonomies are an effort in this direction, as they provide a well-defined catalogue of codes to label diseases univocally. Two examples are ICD (for general medicine) and DSM-IV (for mental disorders), which identify prototypical clinical medical profiles with a name and a code. Medical taxonomies have been developed to be essentially crisp, so they can be transcribed almost directly to OWL. However, vagueness could be introduced at different levels of the taxonomy so that richer semantics would be represented:

- In order to associate diagnostic codes to patient electronic records, fuzzy assertions would be useful, allowing the knowledge base to contain statements such as "Patient001's Serotonin Level is quite low" or "Patient001's disease is likely to be an Obsessive-Compulsive Disorder".
- In the current version of DSM-IV, "Substance-Induced Anxiety Disorder" is defined only as a specialization of "Substance-Related Disorder". Using a fuzzy GCI, we may say that a concept subsumes another to some degree, being possible to assert that a "Substance-Induced Anxiety Disorder can be partially considered a Substance-Related Disorder", as well as an "Anxiety Disorder".

Hence, assume a fuzzy fKB representing the following knowledge:

- $\langle patient001 : \exists hasSerotoninLevel.HighLevel \leq 0.25 \rangle$
- $\langle patient001 : \exists hasDisease.ObsessiveCompulsiveDisorder \geq 0.75 \rangle$
- $\langle SubstanceInducedAnxietyDisorder \sqsubseteq AnxietyDisorder \geq 0.75 \rangle$

Firstly, we compute the number of degrees of truth to be considered: $X^{fK} = \{0, 0.5, 1, 0.75\}$, so $N^{fK} = \{0, 0.25, 0.5, 0.75, 1\}$.

Next, we create some new elements and some axioms preserving their semantics. The new axioms in $R(N^{fK})$, due to the new atomic concepts are:

$HighLevel_{\geq 1} \sqsubseteq HighLevel_{>0.75}, \quad AnxietyDisorder_{\geq 1} \sqsubseteq AnxietyDisorder_{>0.75},$
$HighLevel_{>0.75} \sqsubseteq HighLevel_{\geq 0.75}, \quad AnxietyDisorder_{>0.75} \sqsubseteq AnxietyDisorder_{\geq 0.75},$
$HighLevel_{\geq 0.75} \sqsubseteq HighLevel_{>0.5}, \quad AnxietyDisorder_{\geq 0.75} \sqsubseteq AnxietyDisorder_{>0.5},$
$HighLevel_{>0.5} \sqsubseteq HighLevel_{\geq 0.5}, \quad AnxietyDisorder_{>0.5} \sqsubseteq AnxietyDisorder_{\geq 0.5},$
$HighLevel_{\geq 0.5} \sqsubseteq HighLevel_{>0.25}, \quad AnxietyDisorder_{\geq 0.5} \sqsubseteq AnxietyDisorder_{>0.25},$
$HighLevel_{>0.25} \sqsubseteq HighLevel_{\geq 0.25}, \quad AnxietyDisorder_{>0.25} \sqsubseteq AnxietyDisorder_{\geq 0.25},$
$HighLevel_{\geq 0.25} \sqsubseteq HighLevel_{>0} \quad AnxietyDisorder_{\geq 0.25} \sqsubseteq AnxietyDisorder_{>0}$

(and analogously for $ObsessiveCompulsiveDisorder$ and $SubstanceInduced$ $AnxietyDisorder$).

Similarly , $R(N^{fK})$ contains the following axioms:

$hasSerotoninLevel_{\geq 1} \sqsubseteq hasSerotoninLevel_{>0.75}, \quad hasDisease_{\geq 1} \sqsubseteq hasDisease_{>0.75},$
$hasSerotoninLevel_{>0.75} \sqsubseteq hasSerotoninLevel_{\geq 0.75}, \quad hasDisease_{>0.75} \sqsubseteq hasDisease_{\geq 0.75},$
$hasSerotoninLevel_{\geq 0.75} \sqsubseteq hasSerotoninLevel_{>0.5}, \quad hasDisease_{\geq 0.75} \sqsubseteq hasDisease_{>0.5},$
$hasSerotoninLevel_{>0.5} \sqsubseteq hasSerotoninLevel_{\geq 0.5}, \quad hasDisease_{>0.5} \sqsubseteq hasDisease_{\geq 0.5},$
$hasSerotoninLevel_{\geq 0.5} \sqsubseteq hasSerotoninLevel_{>0.25}, \quad hasDisease_{\geq 0.5} \sqsubseteq hasDisease_{>0.25},$
$hasSerotoninLevel_{>0.25} \sqsubseteq hasSerotoninLevel_{\geq 0.25}, \quad hasDisease_{>0.25} \sqsubseteq hasDisease_{\geq 0.25},$
$hasSerotoninLevel_{\geq 0.25} \sqsubseteq hasSerotoninLevel_{>0} \quad hasDisease_{\geq 0.25} \sqsubseteq hasDisease_{>0}$

Finally, we map the axioms in the fKB:

- $\kappa(\langle patient001 : \exists hasSerotoninLevel.HighLevel \leq 0.25 \rangle) =$
  $patient001 : \forall hasSerotoninLevel_{>0.25}.HighLevel_{\leq 0.25}$
- $\kappa(\langle patient001 : \exists hasDisease.ObsessiveCompulsiveDisorder \geq 0.75 \rangle) =$
  $patient001 : \exists hasDisease_{\geq 0.75}.ObsessiveCompulsiveDisorder_{\geq 0.75}$
- $\kappa(\langle SubstanceInducedAnxietyDisorder \sqsubseteq AnxietyDisorder \rangle \geq 0.75) =$
  $SubstanceInducedAnxietyDisorder_{>0.25} \sqsubseteq AnxietyDisorder_{\geq 0.75}$

5 Conclusions and Future Work

This paper has presented an alternative approach to achieve fuzzy ontologies, reusing currently existing crisp ontology languages and reasoners. In particular, after having presented a sound fuzzy extension of $\mathcal{SHOIN}$ including fuzzy nominals (enabling to define fuzzy sets extensively) and fuzzy GCIs (allowing to constrain the truth value of a GCI), we have presented a reasoning preserving procedure (quadratic in complexity) to reduce a $f_{KD}\mathcal{SHOIN}$ fKB into a crisp one. The semantics of fuzzy GCIs reduces the size of the resulting TBox w.r.t. [13], but imposes some counter-intuitive effects.

The main direction for future work is to perform an empirical evaluation in order to validate the theoretical results. From a theoretical point view, we are considering different fuzzy operators to avoid the counter-intuitive effects of the KD implication. We also plan to include a crisp representation for fuzzy datatypes. Since OWL does not currently allow to define customised datatypes, it seems interesting to consider OWL Eu [25], a promising extension of OWL supporting them. Another interesting direction for future research is to consider the more expressive DL $\mathcal{SROIQ}$ [26] (providing some additional role constructs such as disjoint roles and negated role assertions) and which is the subjacent DL of OWL 1.1 [27], an extension of OWL which has been recently proposed. The additional expressivity would allow to overcome the asymmetry in the definitions of fuzzy concept and role inclusion axioms.

Acknowledgements

Fernando Bobillo holds a FPU scholarship from the Spanish Ministerio de Educación y Ciencia. Juan Gómez-Romero holds a scholarship from Consejera de Innovación, Ciencia y Empresa (Junta de Andalucía).

References

1. Gruber, T.R.: A translation approach to portable ontologies. Knowledge Acquisition 5(2), 199–220 (1993)
2. Berners-Lee, T., Hendler, J., Lassila, O.: The semantic web. Scientific American 284(5), 34–43 (2001)
3. Baader, F., Calvanese, D., McGuinness, D., Nardi, D., Patel-Schneider, P.F.: The Description Logic Handbook: Theory, Implementation, and Applications. Cambridge University Press, Cambridge (2003)
4. Baader, F., Horrocks, I., Sattler, U.: Description logics as ontology languages for the semantic web. In: Hutter, D., Stephan, W. (eds.) Mechanizing Mathematical Reasoning. LNCS (LNAI), vol. 2605, pp. 228–248. Springer, Heidelberg (2005)
5. Zadeh, L.A.: Fuzzy sets. Information and Control 8, 338–353 (1965)
6. Lukasiewicz, T., Straccia, U.: Managing uncertainty and vagueness in description logics for the Semantic Web. Journal of Web Semantics (to appear)
7. W3C: OWL Web Ontology Language Overview (2004),
   http://www.w3.org/TR/owl-features
8. Horrocks, I., Patel-Schneider, P.F.: Reducing OWL entailment to description logic satisfiability. Journal of Web Semantics 1(4), 345–357 (2004)
9. Stoilos, G., Stamou, G., Tzouvaras, V., Pan, J.Z., Horrocks, I.: Fuzzy OWL: Uncertainty and the semantic web. In: Proceedings of the International Workshop on OWL: Experience and Directions, OWLED 2005 (2005)
10. Sirin, E., Cuenca-Grau, B., Parsia, B.: From wine to water: Optimizing description logic reasoning for nominals. In: Proceedings of the 10th International Conference of Knowledge Representation and Reasoning (KR 2006), pp. 90–99 (2006)
11. Haarslev, V., Möller, R.: Racer system description. In: Goré, R.P., Leitsch, A., Nipkow, T. (eds.) IJCAR 2001. LNCS, vol. 2083, pp. 701–705. Springer, Heidelberg (2001)

12. Bobillo, F., Straccia, U.: fuzzyDL: An expressive fuzzy description logic reasoner. In: Proceedings of the 17th IEEE International Conference on Fuzzy Systems, FUZZ-IEEE 2008 (to appear, 2008)
13. Straccia, U.: Transforming fuzzy description logics into classical description logics. In: Alferes, J.J., Leite, J. (eds.) JELIA 2004. LNCS, vol. 3229, pp. 385–399. Springer, Heidelberg (2004)
14. Straccia, U.: Description logics over lattices. International Journal of Uncertainty, Fuzziness and Knowledge-Based Systems 14(1), 1–16 (2006)
15. Kang, D., Lu, J., Xu, B., Li, Y., He, Y.: Two reasoning methods for extended fuzzy $\mathcal{ALCH}$. In: Meersman, R., Tari, Z. (eds.) OTM 2005. LNCS, vol. 3761, pp. 1588–1595. Springer, Heidelberg (2005)
16. Li, Y., Xu, B., Lu, J., Kang, D.: Reasoning technique for extended fuzzy $\mathcal{ALCQ}$. In: Gavrilova, M.L., Gervasi, O., Kumar, V., Tan, C.J.K., Taniar, D., Laganá, A., Mun, Y., Choo, H. (eds.) ICCSA 2006. LNCS, vol. 3981, pp. 1179–1188. Springer, Heidelberg (2006)
17. Sirin, E., Parsia, B., Cuenca-Grau, B., Kalyanpur, A., Katz, Y.: Pellet: A practical OWL-DL reasoner. Journal of Web Semantics 5(2), 51–53 (2007)
18. Straccia, U.: A Fuzzy Description Logic for the Semantic Web. In: Fuzzy Logic and the Semantic Web. Capturing Intelligence, vol. 1, pp. 73–90. Elsevier, Amsterdam (2006)
19. Schaerf, A.: Reasoning with individuals in concept languages. Data & Knowledge Engineering 13(2), 141–176 (1994)
20. Hájek, P.: Metamathematics of Fuzzy Logic. Kluwer, Dordrecht (2001)
21. Klir, G.J., Yuan, B.: Fuzzy Sets and Fuzzy Logic: Theory and Applications. Prentice-Hall, Englewood Cliffs (1995)
22. Straccia, U.: Reasoning within fuzzy description logics. Journal of Artificial Intelligence Research 14, 137–166 (2001)
23. Straccia, U.: A fuzzy description logic. In: Proceedings of the 15th National Conference on Artificial Intelligence (AAAI 1998), pp. 594–599 (1998)
24. Nebel, B.: Terminological reasoning is inherently intractable. Artificial Intelligence 43(2), 235–249 (1990)
25. Pan, J.Z., Horrocks, I.: OWL-Eu: Adding customised datatypes into OWL. Journal of Web Semantics 4, 29–39 (2006)
26. Horrocks, I., Kutz, O., Sattler, U.: The even more irresistible $\mathcal{SROIQ}$. In: Proceedings of the 10th International Conference of Knowledge Representation and Reasoning (KR 2006), pp. 452–457 (2006)
27. W3C: OWL 1.1 Web Ontology Language Overview (2006), http://www.w3.org/Submission/owl11-overview/

Optimizing the Crisp Representation of the Fuzzy Description Logic $\mathcal{SROIQ}$

Fernando Bobillo, Miguel Delgado, and Juan Gómez-Romero

Department of Computer Science and Artificial Intelligence, University of Granada
C. Periodista Daniel Saucedo Aranda, 18071 Granada, Spain
Ph.: +34 958243194; Fax: +34 958243317
fbobillo@decsai.ugr.es, mdelgado@ugr.es, jgomez@decsai.ugr.es

Abstract. Classical ontologies are not suitable to represent imprecise nor uncertain pieces of information. Fuzzy Description Logics were born to manage the former type of knowledge, but they require an appropriate fuzzy language to be agreed and an important number of available resources to be adapted. This paper faces these problems by presenting a reasoning preserving procedure to obtain a crisp representation for a fuzzy extension of the logic $\mathcal{SROIQ}$ which uses Gödel implication in the semantics of fuzzy concept and role subsumption. This reduction allows crisp representation languages as well as currently available reasoners to be reused. Our procedure is optimized with respect to the related work, since it reduces the size of the resulting knowledge base, and is implemented in DELOREAN, the first reasoner that supports fuzzy OWL DL.

1 Introduction

Description Logics (DLs for short) [1] are a family of logics for representing structured knowledge. Each logic is denoted by using a string of capital letters which identify the constructors of the logic and therefore its complexity. DLs have proved to be very useful as ontology languages. For instance, $\mathcal{SROIQ}(\mathbf{D})$ [2] is the subjacent DL of OWL 1.1 [3], a recent extension of the standard language OWL which is its most likely immediate successor.

Nevertheless, it has been widely pointed out that classical ontologies are not appropriate to deal with imprecise and vague knowledge, which is inherent to several real-world domains. Since fuzzy logic is a suitable formalism to handle these types of knowledge, several fuzzy extensions of DLs can be found in the literature (see [4] for an overview).

Defining a fuzzy DL brings about that crisp standard languages are no longer appropriate, new fuzzy languages need to be used, and hence the large number of resources available need to be adapted to the new framework, requiring an important effort. An additional problem is that reasoning within (crisp) expressive DLs has a very high worst-case complexity (e.g. NExpTime in $\mathcal{SHOIN}$) and, consequently, there exists a significant gap between the design of a decision procedure and the achievement of a practical implementation [5].

An alternative is to represent fuzzy DLs using crisp DLs and to reduce reasoning within fuzzy DLs to reasoning within crisp ones. This has several advantages:

P.C.G. da Costa et al. (Eds.): URSW 2005-2007, LNAI 5327, pp. 189–206, 2008.
© Springer-Verlag Berlin Heidelberg 2008

- There is no need to agree on a new standard fuzzy language, but every developer could use its own language expressing fuzzy $\mathcal{SROIQ}$, as long as he implements the reduction that we describe.
- We can continue using standard languages with a lot of resources available. Although it would be desirable to assist the user in tasks such as fuzzy ontology editing, reducing the fuzzy ontology into a crisp one or fuzzy querying, once the reduction is performed, we may use the resources available for the crisp language.
- We may continue using existing crisp reasoners. We do not claim that reasoning will be more efficient, but this approach offers a workaround to support early reasoning in future fuzzy languages. In fact, nowadays there is no reasoner fully supporting a fuzzy extension of OWL 1.1.

Under this approach an immediate practical application of fuzzy ontologies is feasible, because of its tight relation with already existing languages and tools which have proved their validity.

Although there has been a relatively significant amount of works in extending DLs with fuzzy set theory ([4] is a good survey), the representation of them by using crisp description logics has not received such attention. The first efforts in this direction are due to U. Straccia, who considered fuzzy $\mathcal{ALCH}$ [6] and fuzzy $\mathcal{ALC}$ with truth values taken from an uncertainty lattice [7]. F. Bobillo et al. extended Straccia's work to $\mathcal{SHOIN}$, including fuzzy nominals and fuzzy General Concept Inclusions (GCIs) with a semantics given by Kleene-Dienes implication [8]. Finally, G. Stoilos et al. extended this work to a subset of $\mathcal{SROIN}$ [9].

The contributions of this work can be summarized as follows:

- We provide a full representation of fuzzy $\mathcal{SROIQ}$, differently from [9] which do not show how to reduce qualified cardinality restrictions, local reflexivity concepts in expressions of the form $\rho(\exists S.Self, \triangleleft\gamma)$ nor negative role assertions.
- [6,9] force GCIs and Role Inclusion Axioms (RIAs) to be either true or false, but we will allow them to be verified up to some degree by using Gödel implication in the semantics.
- We improve one of their starting points (the reduction presented in [6]) by reducing the number of new atomic elements and their corresponding axioms.
- We show how to optimize some important cases of GCIs, as well as irreflexive role axioms.
- We present DeLorean, our implementation of the reduction and the first implemented reasoner supporting fuzzy $\mathcal{SHOIN}$.

The remainder is organized as follows. Section 2 recalls some preliminaries on fuzzy set theory. Then, Section 3 describes a fuzzy extension of $\mathcal{SROIQ}$ and discusses some logical properties. Section 4 depicts a reduction into crisp $\mathcal{SROIQ}$, whereas Section 5 presents our implementation of the procedure. Finally, in Section 6 we set out some conclusions and ideas for future work.

2 Fuzzy Set Theory

Fuzzy set theory and fuzzy logic were proposed by L. Zadeh [10] to manage imprecise and vague knowledge. While in classical set theory elements either belong to a set or not, in fuzzy set theory elements can belong to a set to some degree. More formally, let X be a set of elements called the reference set. A fuzzy subset A of X is defined by a membership function $\mu_A(x)$, or simply $A(x)$, which assigns any $x \in X$ to a value in the interval of real numbers between 0 and 1. As in the classical case, 0 means no-membership and 1 full membership, but now a value between 0 and 1 represents the extent to which x can be considered as an element of X. If the reference set is finite ($X = \{x_1, \ldots, x_n\}$), the membership function can be expressed using the notation $A = \{\mu_A(x_1)/x_1, \ldots \mu_A(x_n)/x_n\}$.

For every $\alpha \in [0, 1]$, the α-cut of a fuzzy set A is defined as the (crisp) set such that its elements belong to A with degree at least α, i.e. $\{x \mid \mu_A(x) \geq \alpha\}$. Similarly, the strict α-cut is defined as $\{x \mid \mu_A(x) > \alpha\}$.

All crisp set operations are extended to fuzzy sets. The intersection, union, complement and implication set operations are performed by a t-norm function $\otimes$, a t-conorm function $\oplus$, a negation function $\ominus$ and an implication function $\Rightarrow$, respectively. For a complete definition of these functions as well as their properties, we refer the reader to [11,12].

Two types of fuzzy implications are commonly used. The first class is S-implications, which extend the crisp proposition $a \to b = \neg a \lor b$ to the fuzzy case and are defined by the operation $\alpha \Rightarrow \beta = (\ominus \alpha) \oplus \beta$. The second class is R-implications (residuum-based implications), which are defined as $\alpha \Rightarrow \beta = \sup\{\gamma \in [0, 1] \mid (\alpha \otimes \gamma) \leq \beta\}$ and can be used to define a fuzzy complement as $\ominus a = a \Rightarrow 0$. They always verify that $\alpha \Rightarrow \beta = 1$ iff $\alpha \leq \beta$. Product and Gödel implications are R-implications, the implication of the Zadeh family, which is called Kleene-Dienes (KD), is an S-implication and the Łukasiewicz implication belongs to both types.

A fuzzy implication specifies a family of fuzzy operators. If it is an S-implication this notation also specifies the fuzzy complement and t-conorm, while if it is an R-implication then we also know the t-norm and the fuzzy complement. The missing operators are usually defined by using duality of the t-norms and the t-conorms. Table 1 shows the most important families of fuzzy operators: Zadeh, Łukasiewicz, Gödel and Product.

Table 1. Popular families of fuzzy operators

Family	t-norm $\alpha \otimes \beta$	t-conorm $\alpha \oplus \beta$	negation $\ominus \alpha$	implication $\alpha \Rightarrow \beta$
Zadeh	$\min\{\alpha, \beta\}$	$\max\{\alpha, \beta\}$	$1 - \alpha$	$\max\{1 - \alpha, \beta\}$
Łukasiewicz	$\max\{\alpha + \beta - 1, 0\}$	$\min\{\alpha + \beta, 1\}$	$1 - \alpha$	$\min\{1 - \alpha + \beta, 1\}$
Gödel	$\min\{\alpha, \beta\}$	$\max\{\alpha, \beta\}$	$\begin{cases} 1, & \alpha = 0 \\ 0, & \alpha > 0 \end{cases}$	$\begin{cases} 1 & \alpha \leq \beta \\ \beta, & \alpha > \beta \end{cases}$
Product	$\alpha \cdot \beta$	$\alpha + \beta - \alpha \cdot \beta$	$\begin{cases} 1, & \alpha = 0 \\ 0, & \alpha > 0 \end{cases}$	$\begin{cases} 1 & \alpha \leq \beta \\ \beta/\alpha, & \alpha > \beta \end{cases}$

3 Fuzzy $\mathcal{SROIQ}$

In this section we define $f\mathcal{SROIQ}$, which extend $\mathcal{SROIQ}$ to the fuzzy case by letting *(i)* concepts denote fuzzy sets of individuals and *(ii)* roles denote fuzzy binary relations. Axioms are also extended to the fuzzy case and some of them hold to a degree.

The following definition extends [13,9] with fuzzy nominals [8] and the use of Gödel implication in the semantics of GCIs and RIAs.

In the rest of the paper we will assume $\bowtie \in \{\geq, <, \leq, >\}$, $\alpha \in (0, 1]$, $\beta \in [0, 1)$ and $\gamma \in [0, 1]$. The symmetric $\bowtie^-$ and the negation $\neg \bowtie$ of an operator $\bowtie$ are defined as:

$\bowtie$	$\bowtie^-$	$\neg \bowtie$
$\geq$	$\leq$	$<$
$>$	$<$	$\leq$
$\leq$	$\geq$	$>$
$<$	$>$	$\geq$

Syntax. $f\mathcal{SROIQ}$ assumes three alphabets of symbols, for concepts, roles and individuals.

Complex roles (denoted R) are built from atomic roles (R_A) and the universal role (U) as follows:

$$R \to R_A \mid \text{(atomic role)}$$
$$R^- \mid \text{(inverse role)}$$
$$U \quad \text{(universal role)}$$

Let n, m be natural numbers ($n \geq 0, m > 0$). The concepts (denoted C or D) of the language can be built inductively from atomic concepts (A), top concept $\top$, bottom concept $\bot$, named individuals (o_i), simple roles (S, which will be defined below), as follows:

$$C, D \to \qquad A \mid \text{(atomic concept)}$$
$$\top \mid \text{(top concept)}$$
$$\bot \mid \text{(bottom concept)}$$
$$C \sqcap D \mid \text{(concept conjunction)}$$
$$C \sqcup D \mid \text{(concept disjunction)}$$
$$\neg C \mid \text{(concept negation)}$$
$$\forall R.C \mid \text{(universal quantification)}$$
$$\exists R.C \mid \text{(existential quantification)}$$
$$\{\alpha_1/o_1, \ldots, \alpha_m/o_m\} \mid \text{(fuzzy nominals)}$$
$$(\geq m\ S.C) \mid \text{(at-least qualified number restriction)}$$
$$(\leq n\ S.C) \mid \text{(at-most qualified number restriction)}$$
$$\exists S.Self \quad \text{(local reflexivity)}$$

The only difference with the syntax of the crisp case are fuzzy nominals [8].

A fuzzy Knowledge Base (fKB) comprises a fuzzy ABox $\mathcal{A}$, a fuzzy Terminological Box (TBox) $\mathcal{T}$ and a fuzzy RBox $\mathcal{R}$.

An ABox consists of a finite set of *fuzzy assertions* about individuals:

- *concept assertions* $\langle a\!:\!C \geq \alpha \rangle, \langle a\!:\!C > \beta \rangle, \langle a\!:\!C \leq \beta \rangle$ or $\langle a\!:\!C < \alpha \rangle$, meaning that individual a is an instance of C with some degree,
- *role assertions* $\langle (a,b)\!:\!R \geq \alpha \rangle$, $\langle (a,b)\!:\!R > \beta \rangle$, $\langle (a,b)\!:\!R \leq \beta \rangle$ or $\langle (a,b)\!:\!R < \alpha \rangle$, meaning that (a,b) is an instance of R with some degree,
- *inequality assertions* $\langle a \neq b \rangle$,
- *equality assertions* $\langle a = b \rangle$.

A *fuzzy TBox* consists of *fuzzy GCIs*, which constrain the truth value of a GCI i.e., they are expressions of the form $\langle \Omega \geq \alpha \rangle$ or $\langle \Omega > \beta \rangle$, where $\Omega = C \sqsubseteq D$.

Let w be a role chain (a finite string of roles not including the universal role U). An RBox consists of a finite set of role axioms:

- *fuzzy role inclusion axioms (fuzzy RIAs)* $\langle w \sqsubseteq R \geq \alpha \rangle$ or $\langle w \sqsubseteq R > \beta \rangle$ for a role chain $w = R_1 R_2 \ldots R_n$ (meaning that the role chain w is more specific than R to some degree),
- *transitive role axioms* $trans(R)$,
- *disjoint role axioms* $dis(S_1, S_2)$,
- *reflexive role axioms* $ref(R)$,
- *irreflexive role axioms* $irr(S)$,
- *symmetric role axioms* $sym(R)$,
- *asymmetric role axioms* $asy(S)$.

A fuzzy axiom is *positive* (denoted $\langle \tau \rhd \alpha \rangle$) if it is of the form $\langle \tau \geq \alpha \rangle$ or $\langle \tau > \beta \rangle$, and *negative* (denoted $\langle \tau \lhd \alpha \rangle$) if it is of the form $\langle \tau \leq \beta \rangle$ or $\langle \tau < \alpha \rangle$. $\langle \tau = \alpha \rangle$ is equivalent to the pair of axioms $\langle \tau \geq \alpha \rangle$ and $\langle \tau \leq \alpha \rangle$. Of course, if the degree is omitted τ is interpreted as $\langle \tau \geq 1 \rangle$.

A strict partial order $\prec$ on a set A is an irreflexive and transitive relation on A. A strict partial order $\prec$ on the set of roles is called a regular order if it also satisfies $R_1 \prec R_2 \Leftrightarrow R_2^- \prec R_1$, for all roles R_1 and R_2.

As in the crisp case, role axioms cannot contain U and every RIA should be $\prec$-regular for a regular order $\prec$. A RIA $\langle w \sqsubseteq R \rhd \gamma \rangle$ is $\prec$-*regular* if $R = R_A$ and:

- $w = RR$, or
- $w = R^-$, or
- $w = S_1 \ldots S_n$ and $S_i \prec R$ for all $i = 1, \ldots, n$, or
- $w = R S_1 \ldots S_n$ and $S_i \prec R$ for all $i = 1, \ldots, n$, or
- $w = S_1 \ldots S_n R$ and $S_i \prec R$ for all $i = 1, \ldots, n$.

Simple roles are defined as follows:

- R_A is simple if does not occur on the right side of a RIA,
- R^- is simple if R is,
- if R occurs on the right side of a RIA, R is simple if, for each $\langle w \sqsubseteq R \rhd \gamma \rangle$, $w = S$ for a simple role S.

Notice that negative GCIs or RIAs are not allowed, because they correspond to negated GCIs and RIAs respectively, which are not part of crisp $\mathcal{SROIQ}$.

Semantics. A fuzzy interpretation $\mathcal{I}$ is a pair $(\Delta^{\mathcal{I}}, \cdot^{\mathcal{I}})$, where $\Delta^{\mathcal{I}}$ is a non empty set (the interpretation domain) and $\cdot^{\mathcal{I}}$ a fuzzy interpretation function mapping:

- every individual a onto an element $a^{\mathcal{I}}$ of $\Delta^{\mathcal{I}}$,
- every concept C onto a function $C^{\mathcal{I}} : \Delta^{\mathcal{I}} \to [0,1]$,
- every role R onto a function $R^{\mathcal{I}} : \Delta^{\mathcal{I}} \times \Delta^{\mathcal{I}} \to [0,1]$,

$C^{\mathcal{I}}$ (resp. $R^{\mathcal{I}}$) denotes the membership function of the fuzzy concept C (resp. fuzzy role R) w.r.t. $\mathcal{I}$. $C^{\mathcal{I}}(x)$ (resp. $R^{\mathcal{I}}(x,y)$) gives us the degree of being the individual x an element of the fuzzy concept C (resp. the degree of being (x,y) an element of the fuzzy role R) under the fuzzy interpretation $\mathcal{I}$. We do not impose unique name assumption, i.e., two nominals might refer to the same individual.

Given a t-norm $\otimes$, a t-conorm $\oplus$, a negation function $\ominus$ and an implication function $\Rightarrow$, the fuzzy interpretation function is extended to complex concepts and roles as follows:

$$\top^{\mathcal{I}}(x) = 1$$
$$\bot^{\mathcal{I}}(x) = 0$$
$$(C \sqcap D)^{\mathcal{I}}(x) = C^{\mathcal{I}}(x) \otimes D^{\mathcal{I}}(x)$$
$$(C \sqcup D)^{\mathcal{I}}(x) = C^{\mathcal{I}}(x) \oplus D^{\mathcal{I}}(x)$$
$$(\neg C)^{\mathcal{I}}(x) = \ominus C^{\mathcal{I}}(x)$$
$$(\forall R.C)^{\mathcal{I}}(x) = \inf_{y \in \Delta^{\mathcal{I}}} \{ R^{\mathcal{I}}(x,y) \Rightarrow C^{\mathcal{I}}(y) \}$$
$$(\exists R.C)^{\mathcal{I}}(x) = \sup_{y \in \Delta^{\mathcal{I}}} \{ R^{\mathcal{I}}(x,y) \otimes C^{\mathcal{I}}(y) \}$$
$$\{\alpha_1/o_1, \ldots, \alpha_m/o_m\}^{\mathcal{I}}(x) = \sup_{i \mid x = o_i^{\mathcal{I}}} \alpha_i$$
$$(\geq m\ S.C)^{\mathcal{I}}(x) = \sup_{y_1, \ldots, y_m \in \Delta^{\mathcal{I}}} [(\otimes_{i=1}^{n} \{ S^{\mathcal{I}}(x, y_i) \otimes C^{\mathcal{I}}(y_i) \}) \bigotimes (\otimes_{j<k} \{ y_j \neq y_k \})]$$
$$(\leq n\ S.C)^{\mathcal{I}}(x) = \inf_{y_1, \ldots, y_{n+1} \in \Delta^{\mathcal{I}}} [(\otimes_{i=1}^{n+1} \{ S^{\mathcal{I}}(x, y_i) \otimes C^{\mathcal{I}}(y_i) \}) \Rightarrow (\oplus_{j<k} \{ y_j = y_k \})]$$
$$(\exists S.Self)^{\mathcal{I}}(x) = S^{\mathcal{I}}(x,x)$$
$$(R^-)^{\mathcal{I}}(x,y) = R^{\mathcal{I}}(y,x)$$
$$U^{\mathcal{I}}(x,y) = 1$$

A fuzzy interpretation $\mathcal{I}$ satisfies (is a model of):

- $\langle a{:}C \bowtie \gamma \rangle$ iff $C^{\mathcal{I}}(a^{\mathcal{I}}) \bowtie \gamma$,
- $\langle (a,b){:}R \bowtie \gamma \rangle$ iff $R^{\mathcal{I}}(a^{\mathcal{I}}, b^{\mathcal{I}}) \bowtie \gamma$,
- $\langle a \neq b \rangle$ iff $a^{\mathcal{I}} \neq b^{\mathcal{I}}$,
- $\langle a = b \rangle$ iff $a^{\mathcal{I}} = b^{\mathcal{I}}$,
- $\langle C \sqsubseteq D \rhd \gamma \rangle$ iff $\inf_{x \in \Delta^{\mathcal{I}}} \{ C^{\mathcal{I}}(x) \Rightarrow D^{\mathcal{I}}(x) \} \rhd \gamma$,
- $\langle R_1 \ldots R_n \sqsubseteq R \rhd \gamma \rangle$ iff $\sup_{x_1 \ldots x_{n+1} \in \Delta^{\mathcal{I}}} \bigotimes [R_1^{\mathcal{I}}(x_1, x_2), \ldots, R_n^{\mathcal{I}}(x_n, x_{n+1})] \Rightarrow R^{\mathcal{I}}(x_1, x_{n+1}) \rhd \gamma$,
- $trans(R)$ iff $\forall x, y \in \Delta^{\mathcal{I}}, R^{\mathcal{I}}(x,y) \geq \sup_{z \in \Delta^{\mathcal{I}}} R^{\mathcal{I}}(x,z) \otimes R^{\mathcal{I}}(z,y)$,
- $dis(S_1, S_2)$ iff $\forall x, y \in \Delta^{\mathcal{I}}, S_1^{\mathcal{I}}(x,y) = 0$ or $S_2^{\mathcal{I}}(x,y) = 0$,
- $ref(R)$ iff $\forall x \in \Delta^{\mathcal{I}}, R^{\mathcal{I}}(x,x) = 1$,
- $irr(S)$ iff $\forall x \in \Delta^{\mathcal{I}}, S^{\mathcal{I}}(x,x) = 0$,
- $sym(R)$ iff $\forall x, y \in \Delta^{\mathcal{I}}, R^{\mathcal{I}}(x,y) = R^{\mathcal{I}}(y,x)$,
- $asy(S)$ iff $\forall x, y \in \Delta^{\mathcal{I}}$, if $S^{\mathcal{I}}(x,y) > 0$ then $S^{\mathcal{I}}(y,x) = 0$,
- a fKB iff it satisfies each element in $\mathcal{A}, \mathcal{T}$ and $\mathcal{R}$.

Notice that individual assertions are considered to be crisp, since the equality and inequality of individuals has always been considered crisp in the fuzzy DL literature [13,14].

In the rest of the paper we will only consider fKB satisfiability, since (as in the crisp case) most inference problems can be reduced to it [15].

Some logical properties. It can be easily shown that $f\mathcal{SROIQ}$ is a sound extension of crisp $\mathcal{SROIQ}$, because fuzzy interpretations coincide with crisp interpretations if we restrict the membership degrees to $\{0,1\}$.

In the fuzzy DLs literature, the notation $f_i\mathcal{DL}$ has been proposed [16], where i is the fuzzy implication function considered. Here in after we will concentrate on $f_{KD}\mathcal{SROIQ}$, restricting ourselves to the Zadeh family: minimum t-norm, maximum t-conorm, Łukasiewicz negation and KD implication, with the exception of GCIs and RIAs, where we will consider Gödel implication. This choice comes from the fact that KD implication specifies a t-norm, a t-conorm and a negation which make possible the reduction to a crisp KB, as we will see in Section 4 (other fuzzy operators are not suitable for a similar reduction).

However, the use of KD implication in the semantics of GCIs and RIAs brings about two counter-intuitive effects: *(i)* in general concepts (and roles) do not fully subsume themselves and *(ii)* crisp subsumption (holding to degree 1) forces some fuzzy concepts and roles to be interpreted as crisp [8].

Another common semantics which could be considered is the one based on Zadeh's set inclusion ($C \sqsubseteq D$ iff $\forall x \in \Delta^{\mathcal{I}}$, $C^{\mathcal{I}}(x) \leq D^{\mathcal{I}}(x)$) as in [15,17], but it forces the axioms to be either true or false. For example, under this semantics it is not possible that concept *Hotel* subsumes concept *Inn* with degree 0.5.

Gödel implication solves the afore-mentioned problems and is suitable for a classical representation as we will see in Section 4. Moreover, for GCIs of the form $\langle C \sqsubseteq D \geq 1 \rangle$, the semantics is equivalent to that of Zadeh's set inclusion.

Although in general Gödel implication provides better logical properties than KD, the latter allows for instance reasoning with *modus tolens*, since $C \sqsubseteq D \equiv \neg D \sqsubseteq \neg C$. In the rest of this paper we will allow these two implication functions in the semantics of the GCIs and RIAs of our language. We will write $\sqsubseteq$ to denote the use of the Gödel implication in the semantics, and $\sqsubseteq_{KD}$ to denote the use of the KD implication. Our approach is similar to [18], which proposes a representation language allowing three types of subsumption.

It would be possible to transform concept expressions into a semantically equivalent *Negation Normal Form* (NNF), which is obtained by pushing in the usual manner negation in front of atomic concepts, fuzzy nominals and local reflexivity concepts.

Irreflexive, transitive and symmetric role axioms are syntactic sugar for every R-implication (and consequently it can be assumed that they do not appear in fKBs) due to the following equivalences:

- $irr(S) \equiv \langle \top \sqsubseteq \neg\exists S.Self \geq 1 \rangle$,
- $trans(R) \equiv \langle RR \sqsubseteq R \geq 1 \rangle$,
- $sym(R) \equiv \langle R \sqsubseteq R^- \geq 1 \rangle$.

4   An Optimized Crisp Representation for Fuzzy $\mathcal{SROIQ}$

In this section we show how to reduce a $f_{KD}\mathcal{SROIQ}$ fKB into a crisp KB, similarly as in [6,8,9]. The procedure preserves reasoning, in such a way that existing $\mathcal{SROIQ}$ reasoners could be applied to the resulting KB.

Our reduction is optimized with respect to related work, in the sense that the number of generated axioms, and hence the size of the resulting crisp KB, is smaller here. The reduction is optimized but not necessarily optimal, since other optimizations could still be possible.

The basic idea is to create some new crisp concepts and roles, representing the α-cuts of the fuzzy concepts and relations, and to rely on them. Next, some new axioms are added to preserve their semantics and finally every axiom in the ABox, the TBox and the RBox is represented, independently from other axioms, using these new crisp elements.

4.1   Adding (an Optimized Number of) New Elements

Let A^{fK} and R^{fK} be the set of atomic concepts and atomic roles occurring in a fKB $fK = \langle fK_A, fK_T, fK_R \rangle$. In [6] it is shown that the set of the degrees which must be considered for any reasoning task is defined as $N^{fK} = X^{fK} \cup \{1 - \alpha \mid \alpha \in X^{fK}\}$, where $X^{fK} = \{0, 0.5, 1\} \cup \{\gamma \mid \langle \tau \bowtie \gamma \rangle \in fK\}$. This also holds in $fK_D\mathcal{SROIQ}$, because the fuzzy operators do not introduce new degrees, but note that it is no longer true when other fuzzy operators are considered. For example, the combination of the degrees 0.5 and 0.3 using product t-norm introduces the new degree $0.5 \cdot 0.3 = 0.15$. Without loss of generality, it can be assumed that $N^{fK} = \{\gamma_1, \ldots, \gamma_{|N^{fK}|}\}$ and $\gamma_i < \gamma_{i+1}, 1 \le i \le |N^{fK}| - 1$. It is easy to see that $\gamma_1 = 0$ and $\gamma_{|N^{fK}|} = 1$.

Example 1. In order to illustrate the reduction process, we will consider throughout this section a simple fuzzy KB fK, including the fuzzy relation *isCloseTo*. Closeness between individuals is usually a matter of degree, so we can expect this relation to appear for instance in Semantic Web ontologies dealing with geographical information. Assume that $fK = \{\langle sym(isCloseTo) \rangle, \langle (h_1, h_2) : isCloseTo \le 0.75 \rangle\}$. Firstly, the symmetric role axiom $\langle sym(isCloseTo) \rangle$ is represented using a fuzzy RIA, so $fK = \{\langle isCloseTo \sqsubseteq isCloseTo^- \ge 1 \rangle, \langle (h_1, h_2) : isCloseTo \le 0.75 \rangle\}$. Now, $X^{fK} = \{0, 0.5, 1\} \cup \{0.75\}$, so the set of degrees of truth which has to be considered is $N^{fK} = \{0, 0.25, 0.5, 0.75, 1\}$.

Now, for each $\alpha, \beta \in N^{fK}$ with $\alpha \in (0, 1]$ and $\beta \in [0, 1)$, for each $A \in A^{fK}$ and for each $R_A \in R^{fK}$, two new atomic concepts $A_{\ge \alpha}, A_{> \beta}$ and two new atomic roles $R_{\ge \alpha}, R_{> \beta}$ are introduced. $A_{\ge \alpha}$ represents the crisp set of individuals which are instance of A with degree higher or equal than α i.e. the α-cut of A. The other new elements are defined in a similar way. The atomic elements $A_{>1}, R_{>1}, A_{\ge 0}$ and $R_{\ge 0}$ are not considered because they are not necessary, due to the restrictions on the allowed degree of the axioms in the fKB (e.g. we do not allow GCIs of the form $C \sqsubseteq D \ge 0$).

The semantics of these newly introduced atomic concepts and roles is preserved by some terminological and role axioms. For each $1 \le i \le |N^{fK}| - 1, 2 \le j \le |N^{fK}| - 1$ and for each $A \in A^{fK}$, $T(N^{fK})$ is the smallest terminology containing these two axioms:

$$A_{\ge \gamma_{i+1}} \sqsubseteq A_{> \gamma_i}, A_{> \gamma_j} \sqsubseteq A_{\ge \gamma_j}$$

Similarly, for each $R_A \in R^{fK}$, $R(N^{fK})$ contains these axioms:

$$R_{\geq \gamma_{i+1}} \sqsubseteq R_{>\gamma_i}, R_{>\gamma_i} \sqsubseteq R_{\geq \gamma_i}$$

Example 2. Consider the fKB defined in Example 1. Fuzzy atomic role $isCloseTo$ introduces some new atomic concepts roles ($isCloseTo_{\geq 1}, isCloseTo_{>0.75}, isClo$-$seTo_{\geq 0.75}, isCloseTo_{>0.5}, isCloseTo_{\geq 0.5}, isCloseTo_{>0.25}, isCloseTo_{\geq 0.25}, isClo$ - $seTo_{>0}$), as well as some axioms preserving their semantics:

$$isCloseTo_{\geq 1} \sqsubseteq isCloseTo_{>0.75}, \quad isCloseTo_{>0.75} \sqsubseteq isCloseTo_{\geq 0.75},$$
$$isCloseTo_{\geq 0.75} \sqsubseteq isCloseTo_{>0.5}, \quad isCloseTo_{>0.5} \sqsubseteq isCloseTo_{\geq 0.5},$$
$$isCloseTo_{\geq 0.5} \sqsubseteq isCloseTo_{>0.25}, \quad isCloseTo_{>0.25} \sqsubseteq isCloseTo_{\geq 0.25},$$
$$isCloseTo_{\geq 0.25} \sqsubseteq isCloseTo_{>0}$$

4.2   Mapping Fuzzy Concepts, Roles and Axioms

Concept and role expressions are reduced using mapping ρ, as shown in the first part of Table 2. For instance, given a fuzzy concept C, $\rho(C, \geq \alpha)$ is a crisp set containing all the elements which belong to C with a degree greater or equal than α (the other cases are similar).

Example 3. The 0.4 cut of the fuzzy concept $\forall R.(C \sqcup (\leq 1\ R.\neg D))$ is computed as $\rho(\forall R.(C \sqcup (\leq 1\ R.\neg D)), \geq 0.4) = \forall \rho(R, > 0.6).\rho(C \sqcup (\leq 1\ R.\neg D), \geq 0.4) = \forall R_{>0.6}.\rho(C, \geq 0.4) \sqcup \rho(\leq 1\ R.\neg D, \geq 0.4) = \forall R_{>0.6}.C_{\geq 0.4} \sqcup (\leq 1\ \rho(R, > 0.6).\rho(\neg D, > 0.6) = \forall R_{>0.6}.C_{\geq 0.4} \sqcup (\leq 1\ R_{>0.6}.\rho(D, < 0.4) = \forall R_{>0.6}.C_{\geq 0.4} \sqcup (\leq 1\ R_{>0.6}.\neg D_{\geq 0.4})$.

In order to finish the reduction, we map the axioms in the ABox, TBox and RBox. Axioms are reduced as in the second part of Table 2, where σ maps fuzzy axioms into crisp assertions and κ maps fuzzy TBox (resp. RBox) axioms into crisp TBox (resp. RBox) axioms. Recall that we are assuming that irreflexive, transitive and symmetric role axioms do not appear in the RBox.

Our reduction of a fuzzy GCI $\langle C \sqsubseteq D \geq 1 \rangle$ is equivalent to the reduction of a GCI under a semantics based on Zadeh's set inclusion proposed in [6], although it introduces some unnecessary axioms: $C_{\geq 0} \sqsubseteq D_{\geq 0}$ and $C_{>1} \sqsubseteq D_{>1}$.

Observe that the reduction preserves simplicity of the roles and regularity of the RIAs. Note also that due to the restrictions in the definition of the fKB, some expressions cannot appear during the process:

- $\rho(R, \lhd\gamma)$ and $\rho(U, \lhd\gamma)$ can only appear in a (crisp) negated role assertion.
- $\rho(A, \geq 0), \rho(A, > 1), \rho(A, \leq 1)$ and $\rho(A, < 0)$ cannot appear due to the existing restrictions on the degree of the axioms in the fKB. The same also holds for $\top$, $\bot$ and R_A.

Example 4. The reduction of the axioms in the fKB defined in Example 1 is as follows:

- $\sigma(\langle (h_1, h_2) : isCloseTo \leq 0.75 \rangle) = (h_1, h_2) : \neg isCloseTo_{>0.75}$.

Table 2. Mapping of concept and role expressions, and reduction of the axioms

Fuzzy concepts	
$\rho(\top, \rhd\gamma)$	$\top$
$\rho(\top, \lhd\gamma)$	$\bot$
$\rho(\bot, \rhd\gamma)$	$\bot$
$\rho(\bot, \lhd\gamma)$	$\top$
$\rho(A, \rhd\gamma)$	$A_{\rhd\gamma}$
$\rho(A, \lhd\gamma)$	$\neg A_{\neg\lhd\gamma}$
$\rho(\neg C, \bowtie\gamma)$	$\rho(C, \bowtie^- 1-\gamma)$
$\rho(C\sqcap D, \rhd\gamma)$	$\rho(C,\rhd\gamma)\sqcap\rho(D,\rhd\gamma)$
$\rho(C\sqcap D, \lhd\gamma)$	$\rho(C,\lhd\gamma)\sqcup\rho(D,\lhd\gamma)$
$\rho(C\sqcup D, \rhd\gamma)$	$\rho(C,\rhd\gamma)\sqcup\rho(D,\rhd\gamma)$
$\rho(C\sqcup D, \lhd\gamma)$	$\rho(C,\lhd\gamma)\sqcap\rho(D,\lhd\gamma)$
$\rho(\exists R.C, \rhd\gamma)$	$\exists\rho(R,\rhd\gamma).\rho(C,\rhd\gamma)$
$\rho(\exists R.C, \lhd\gamma)$	$\forall\rho(R,\neg\lhd\gamma).\rho(C,\lhd\gamma)$
$\rho(\forall R.C, \{\geq,>\}\gamma)$	$\forall\rho(R,\{>,\geq\}1-\gamma).\rho(C,\{\geq,>\}\gamma)$
$\rho(\forall R.C, \lhd\gamma)$	$\exists\rho(R,\lhd^- 1-\gamma).\rho(C,\lhd\gamma)$
$\rho(\{\alpha_1/o_1,\ldots,\alpha_m/o_m\}, \bowtie\gamma)$	$\{o_i \mid \alpha_i\bowtie\gamma, 1\leq i\leq m\}$
$\rho(\geq m\, S.C, \rhd\gamma)$	$\geq m\,\rho(S,\rhd\gamma).\rho(C,\rhd\gamma)$
$\rho(\geq m\, S.C, \lhd\gamma)$	$\leq m-1\,\rho(S,\neg\lhd\gamma).\rho(C,\neg\lhd\gamma)$
$\rho(\leq n\, S.C, \{\geq,>\}\gamma)$	$\leq n\,\rho(S,\{>,\geq\}1-\gamma).\rho(C,\{>,\geq\}1-\gamma)$
$\rho(\leq n\, S.C, \lhd\gamma)$	$\geq n+1\,\rho(S,\lhd^- 1-\gamma).\rho(C,\lhd^- 1-\gamma)$
$\rho(\exists S.Self, \rhd\gamma)$	$\exists\rho(S,\rhd\gamma).Self$
$\rho(\exists S.Self, \lhd\gamma)$	$\neg\exists\rho(S,\neg\lhd\gamma).Self$
Fuzzy roles	
$\rho(R_A, \rhd\gamma)$	$R_{A\rhd\gamma}$
$\rho(R_A, \lhd\gamma)$	$\neg R_{A\neg\lhd\gamma}$
$\rho(R^-, \bowtie\gamma)$	$\rho(R,\bowtie\gamma)^-$
$\rho(U, \rhd\gamma)$	U
$\rho(U, \lhd\gamma)$	$\neg U$
Axioms	
$\sigma(\langle a{:}C\bowtie\gamma\rangle)$	$\{a{:}\rho(C,\bowtie\gamma)\}$
$\sigma(\langle(a,b){:}R\bowtie\gamma\rangle)$	$\{(a,b){:}\rho(R,\bowtie\gamma)\}$
$\sigma(\langle a\neq b\rangle)$	$\{a\neq b\}$
$\sigma(\langle a = b\rangle)$	$\{a = b\}$
$\kappa(C\sqsubseteq D\geq\alpha)$	$\bigcup_{\gamma\in N^{fK}\setminus\{0\}\,\mid\,\gamma\leq\alpha}\{\rho(C,\geq\gamma)\sqsubseteq\rho(D,\geq\gamma)\}$ $\bigcup_{\gamma\in N^{fK}\,\mid\,\gamma<\alpha}\{\rho(C,>\gamma)\sqsubseteq\rho(D,>\gamma)\}$
$\kappa(C\sqsubseteq D>\beta)$	$\kappa(C\sqsubseteq D\geq\beta)\cup\{\rho(C,>\beta)\sqsubseteq\rho(D,>\beta)\}$
$\kappa(C\sqsubseteq_{KD} D\geq\alpha)$	$\{\rho(C,>1-\alpha)\sqsubseteq\rho(D,\geq\alpha)\}$
$\kappa(C\sqsubseteq_{KD} D>\beta)$	$\{\rho(C,\geq 1-\beta)\sqsubseteq\rho(D,>\beta)\}$
$\kappa(\langle R_1\ldots R_n\sqsubseteq R\geq\alpha\rangle)$	$\bigcup_{\gamma\in N^{fK}\setminus\{0\}\,\mid\,\gamma\leq\alpha}\{\rho(R_1,\geq\gamma)\ldots\rho(R_n,\geq\gamma)\sqsubseteq\rho(R,\geq\gamma)\}$ $\bigcup_{\gamma\in N^{fK}\,\mid\,\gamma<\alpha}\{\rho(R_1,>\gamma)\ldots\rho(R_n,>\gamma)\sqsubseteq\rho(R,>\gamma)\}$
$\kappa(\langle R_1\ldots R_n\sqsubseteq R>\beta\rangle)$	$\kappa(\langle R_1\ldots R_n\sqsubseteq R\geq\beta\rangle)\cup$ $\{\rho(R_1,>\beta)\ldots\rho(R_n,>\beta)\sqsubseteq\rho(R,>\beta)\}$
$\kappa(\langle R_1\ldots R_n\sqsubseteq_{KD} R\geq\alpha\rangle)$	$\{\rho(R_1,>1-\alpha)\ldots\rho(R_n,>1-\alpha)\sqsubseteq\rho(R,\geq\alpha)\}$
$\kappa(\langle R_1\ldots R_n\sqsubseteq_{KD} R>\beta\rangle)$	$\{\rho(R_1,\geq 1-\beta)\ldots\rho(R_n,\geq 1-\beta)\sqsubseteq\rho(R,>\beta)\}$
$\kappa(dis(S_1,S_2))$	$\{dis(\rho(S_1,>0),\rho(S_2,>0))\}$
$\kappa(ref(R))$	$\{ref(\rho(R,\geq 1))\}$
$\kappa(asy(S))$	$\{asy(\rho(S,>0)\}$

- $\kappa(\langle isCloseTo \sqsubseteq isCloseTo^- \geq 1\rangle)$ is reduced into these axioms:

$$isCloseTo_{>0} \sqsubseteq isCloseTo^-_{>0}, \qquad isCloseTo_{\geq 0.25} \sqsubseteq isCloseTo^-_{\geq 0.25},$$
$$isCloseTo_{>0.25} \sqsubseteq isCloseTo^-_{>0.25}, \quad isCloseTo_{\geq 0.5} \sqsubseteq isCloseTo^-_{\geq 0.5},$$
$$isCloseTo_{>0.5} \sqsubseteq isCloseTo^-_{>0.5}, \qquad isCloseTo_{\geq 0.75} \sqsubseteq isCloseTo^-_{\geq 0.75},$$
$$isCloseTo_{>0.75} \sqsubseteq isCloseTo^-_{>0.75}, \quad isCloseTo_{\geq 1} \sqsubseteq isCloseTo^-_{\geq 1}$$

4.3 Properties of the Reduction

Summing up, a fKB $fK = \langle fK_A, fK_T, fK_R\rangle$ is reduced into a KB $\mathcal{K}(fK) = \langle \sigma(fK_A), T(N^{fK}) \cup \kappa(fK, fK_T), R(N^{fK}) \cup \kappa(fK, fK_R)\rangle$.
The following theorem shows that the reduction preserves reasoning:

Theorem 1. *A $f_{KD}\mathcal{SROIQ}$ fKB fK is satisfiable iff $\mathcal{K}(fK)$ is satisfiable.*

Complexity. It is easy to see that every fuzzy concept expression of depth k generates a crisp concept expression of depth k. Most of the axioms of the fuzzy KB generate one axiom in the crisp KB, but some of them (fuzzy GCIs and RIAs if Gödel implication is used in the semantics) generate several (at most $2 \cdot (|N^{\mathcal{K}}| - 1))$ axioms in the crisp KB.
$|\mathcal{K}(fK)|$ is $O(|fK|^2)$ i.e. the resulting KB is quadratic in size. The ABox is actually linear while the TBox and the RBox are both quadratic:

- $|N^{fK}|$ is linearly bounded by $|fK_A| + |fK_T| + |fK_R|$.
- $|\sigma(fK_A)| = |fK_A|$.
- $|T(N^{fK})| = (2 \cdot (|N^{fK}| - 1) - 1) \cdot |A^{fK}|$.
- $|\kappa(fK, T)| \leq 2 \cdot (|N^{fK}| - 1) \cdot |T|$.
- $|R(N^{fK})| = (2 \cdot (|N^{fK}| - 1) - 1) \cdot |R^{fK}|$.
- $|\kappa(fK, R)| \leq 2 \cdot (|N^{fK}| - 1) \cdot |R|$.

The resulting KB is quadratic because it depends on the number of relevant degrees $|N^{fK}|$. An immediate solution to obtain a KB which is linear in complexity is to fix the number of degrees which can appear in the knowledge base. From a practical point of view, in most of the applications it is sufficient to consider a small number of degrees, e.g. $\{0, 0.25, 0.5, 0.75, 1\}$.

Reusing the reduction. An interesting property of the procedure is that the reduction of an ontology can be reused when adding new axioms. In fact, for every new axiom τ, the reduction procedure generates only one new axiom or a (linear in size) set of axioms if τ does not introduce new atomic concepts nor new atomic roles and, in case τ is a fuzzy axiom, if it does not introduce a new degree of truth. Formally, given a fuzzy ontology fK and an axiom τ, the reduction of the extension of fK with τ, denoted $\mathcal{K}(fK \cup \tau)$ is equivalent to $\mathcal{K}(fK) \cup \mathcal{K}(\tau)$. Hence, this property is very useful when it is necessary to add a new axiom to an ontology in order to perform some reasoning task e.g. in ontology classification.
If τ introduces a new atomic concept, $T(N^{fK})$ needs to be recomputed. If τ introduces a new atomic role, $R(N^{fK})$ needs to be recomputed. If τ is a fuzzy

axiom that introduces a new degree of truth, X^{fK} changes. As a consequence, N^{fK} may change. If N^{fK} changes, we need to recompute: *(i)* $T(N^{fK})$, *(ii)* $R(N^{fK})$, *(iii)* the reduction of every fuzzy GCI in fK, and *(iv)* the reduction of every fuzzy RIA in fK.

4.4   Some Optimizations

Optimizing the number of new elements and axioms. Previous works use two more atomic concepts $A_{\leq\beta}, A_{<\alpha}$ and some additional axioms $(2 \leq k \leq |N^{fK}|)$ [6,8]:

$$A_{<\gamma_k} \sqsubseteq A_{\leq\gamma_k}, \qquad\qquad A_{\leq\gamma_i} \sqsubseteq A_{<\gamma_{i+1}}$$
$$A_{\geq\gamma_k} \sqcap A_{<\gamma_k} \sqsubseteq \bot, \qquad A_{>\gamma_i} \sqcap A_{\leq\gamma_i} \sqsubseteq \bot$$
$$\top \sqsubseteq A_{\geq\gamma_k} \sqcup A_{<\gamma_k}, \qquad \top \sqsubseteq A_{>\gamma_i} \sqcup A_{\leq\gamma_i}$$

However, we use $\neg A_{>\gamma_k}$ rather than $A_{\leq\gamma_k}$ and $\neg A_{\geq\gamma_k}$ instead of $A_{<\gamma_k}$, since the six axioms above follow immediately from the semantics of the crisp concepts as Proposition 1 shows:

Proposition 1. *If* $A_{\geq\gamma_{i+1}} \sqsubseteq A_{>\gamma_i}$ *and* $A_{>\gamma_k} \sqsubseteq A_{\geq\gamma_k}$ *hold, then the followings axioms are verified:*

(1) $\neg A_{\geq\gamma_k} \sqsubseteq \neg A_{>\gamma_k}$ *(2)* $\neg A_{>\gamma_i} \sqsubseteq \neg A_{\geq\gamma_{i+1}}$
(3) $A_{\geq\gamma_k} \sqcap \neg A_{\geq\gamma_k} \sqsubseteq \bot$ *(4)* $A_{>\gamma_i} \sqcap \neg A_{>\gamma_i} \sqsubseteq \bot$
(5) $\top \sqsubseteq A_{\geq\gamma_k} \sqcup \neg A_{\geq\gamma_k}$ *(6)* $\top \sqsubseteq A_{>\gamma_i} \sqcup \neg A_{>\gamma_i}$

Proof. (1) and (2) derive from the fact that in crisp DLs $A \sqsubseteq B \equiv \neg B \sqsubseteq \neg A$. (3) and (4) come from the law of contradiction $A \sqcap \neg A \sqsubseteq \bot$, while (5) and (6) derive from the law of excluded middle $\top \sqsubseteq A \sqcup \neg A$.   □

As a minor comment, those works also introduce unnecessarily a couple of elements $A_{\geq 0}$ and $R_{\geq 0}$, as well as the axioms $A_{>0} \sqsubseteq A_{\geq 0}$, $R_{>0} \sqsubseteq R_{\geq 0}$ [6,8].

Optimizing GCI reductions. GCI reductions can be optimized in several particular cases:

- $\langle C \sqsubseteq \top \bowtie \gamma \rangle$ and $\langle \bot \sqsubseteq D \bowtie \gamma \rangle$ are tautologies, so their reductions are unnecessary in the resulting KB.
- $\kappa(\top \sqsubseteq D \bowtie \gamma) = \top \sqsubseteq \rho(D, \bowtie \gamma)$. Note that this kind of axiom appears in role *range* axioms (C is the range of R iff $\top \sqsubseteq \forall R.C$ holds with degree 1) and role *domain* axioms (C is the domain of R iff $\top \sqsubseteq \forall R^-.C$ holds with degree 1).
- $\kappa(C \sqsubseteq \bot \bowtie \gamma) = \rho(C, > 0) \sqsubseteq \bot$. This appears when two concepts are *disjoint* i.e. C and D are disjoint iff $C \sqcap D \sqsubseteq \bot$ holds with degree 1.

Another optimization involving GCIs follows from the following observation. If the resulting TBox contains $A \sqsubseteq B$, $A \sqsubseteq C$ and $B \sqsubseteq C$, then $A \sqsubseteq C$ is unnecessary, since $\{A \sqsubseteq B, B \sqsubseteq C\} \models A \sqsubseteq C$. This is very useful in concept definitions involving the nominal constructor. For example, the reduction of the axiom

$$\kappa(C \sqsubseteq \{1/o_1, 0.5/o_2\}) = \{C_{>0} \sqsubseteq \{o_1, o_2\}, C_{\geq 0.5} \sqsubseteq \{o_1, o_2\}, C_{>0.5} \sqsubseteq \{o_1\}, C_{\geq 1} \sqsubseteq \{o_1\}\}$$

can be optimized as follows:

$$\kappa(C \sqsubseteq \{1/o_1, 0.5/o_2\}) = \{C_{>0} \sqsubseteq \{o_1, o_2\}, C_{>0.5} \sqsubseteq \{o_1\}\}$$

since the two unnecessary axioms trivially hold:

$$\{C_{\geq 0.5} \sqsubseteq C_{>0}, C_{>0} \sqsubseteq \{o_1, o_2\}\} \models C_{\geq 0.5} \sqsubseteq \{o_1, o_2\}$$
$$\{C_{\geq 1} \sqsubseteq C_{>0.5}, C_{>0.5} \sqsubseteq \{o_1\}\} \models C_{\geq 1} \sqsubseteq \{o_1\}$$

Optimizing irreflexive role axioms. For the sake of clarity, we are assuming in this paper that irreflexive role axioms do not appear in the RBox. Currently, an irreflexive role axiom is replaced with an equivalent RIA, which produces the following reduction (which is the same to the reduction of irreflexive role axioms proposed in [9])

$$\kappa(irr(R)) = \bigcup_{\gamma \in \mathcal{N}^{fK} \setminus \{0\}} irr(\rho(R, \geq \gamma)) \bigcup_{\gamma \in \mathcal{N}^{fK}} irr(\rho(R, > \gamma))$$

However, this reduction could be optimized to $\kappa(irr(R)) = irr(\rho(R, > 0))$. Proposition 2 shows that the other axioms follow immediately.

Proposition 2. *If $R_1 \sqsubseteq R_2$ and $irr(R_2)$, then it holds that $irr(R_1)$.*

Proof. Assume that $(x, y) \in R_1^{\mathcal{I}}$. Since $R_1 \sqsubseteq R_2$ is satisfied, then $(x, y) \in R_2^{\mathcal{I}}$. Since $irr(R_2)$, then it holds that $(y, x) \notin R_2^{\mathcal{I}}$. But the role inclusion implies that $(y, x) \notin R_1^{\mathcal{I}}$. For every pair of individuals, we have shown that $(x, y) \in R_1^{\mathcal{I}}$ implies $(y, x) \notin R_1^{\mathcal{I}}$. Hence, $irr(R_1)$ holds.   □

Allowing Crisp Concepts and Roles. It is easy to see that the complexity of the crisp representation is caused by fuzzy concepts and roles. Fortunately, in real applications not all concepts and roles will be fuzzy. Another optimization would be allowing to specify that a concept is crisp. For instance, suppose that A is a fuzzy concept. Then, we need $N^{fK} - 1$ concepts of the form $A_{\geq \alpha}$ and another $N^{fK} - 1$ concepts of the form $A_{>\beta}$ to represent it, as well as $2 \cdot (|N^{fK}| - 1) - 1$ axioms to preserve their semantics. On the other hand, if A is declared to be crisp, we just need one concept to represent it and no new axioms. The case for fuzzy roles is exactly the same.

5   Implementation: DeLorean

Our prototype implementation of the reduction process is called DELOREAN (DEscription LOgic REasoner with vAgueNess). It has been developed in Java with Jena API[1], the parser generator JavaCC[2], and using DIG 1.1 interface [19] to communicate with crisp DL reasoners. Since DIG interface does not yet support full $\mathcal{SROIQ}$, currently the logic supported is $f_{KD}\mathcal{SHOIN}$ (OWL DL).

---

[1] http://jena.sourceforge.net/
[2] https://javacc.dev.java.net

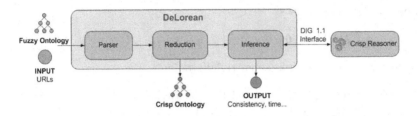

Fig. 1. Architecture of DELOREAN reasoner

Figure 1 illustrates the architecture of the system.

– The *Parser* reads an input file with a fuzzy ontology and translates it into an internal representation. As we have remarked in the Introduction, we could use any language to encode the fuzzy ontology, as long as the *Parser* can understand the representation and the reduction is properly implemented; consequently we will not get into details of our particular choice. For the moment, we do not allow to define crisp concepts and roles.
– In the next step, the *Reduction* module implements the procedure described in Section 4, building a Jena model from which an OWL file with an equivalent crisp ontology is created.
– Finally, the *Inference* module tests this ontology for consistency, using any crisp reasoner through the DIG interface.
– The *User interface* allows the user to introduce the inputs and shows the result of the reasoning and the elapsed time. (see Figure 2 for a screenshot).

We have carried out some experiments in order to evaluate our approach in terms of reasoning, that is, in order to check that the results of the reasoning tasks over the crisp ontology were the expected. The aim of this section is not to perform a full benchmark, which could be the topic of a forthcoming work. Nevertheless, we will show some performance examples to show that our approach is feasible and the increment of time for small ontologies when using a limited number of degrees of truth is acceptable. In any case, optimizations are crucial.

We considered the Koala ontology[3], a sample $\mathcal{ALCON}(\mathcal{D})$ ontology with 20 named classes, 15 anonymous classes, 4 object properties, 1 datatype property (which we have omitted) and 6 individuals. Regarding the axioms, it contains 6 concept assertions and 35 GCIs (15 proper GCIs, 5 concept equivalences which can be seen as 10 GCIs, 1 disjoint concept axiom, 4 domain axioms and 4 range axioms and 1 functional axiom).

We obtained a fuzzy version by extending its axioms with random (lower bound) degrees belonging to a variable set N^{fK}. For the moment, we have assumed that all of the fuzzy concepts and roles are fuzzy. Furthermore, in fuzzy GCIs and RIAs we always assume Gödel implication in the semantics (which introduces more axioms than Kleene-Dienes implication).

Then, we computed an equivalent crisp ontology in $\mathcal{ALCHON}(\mathcal{D})$ (since the reduction introduces role inclusion axioms). The resulting ontology has 6 concept

---

[3] http://http://protege.cim3.net/file/pub/ontologies/koala/koala.owl

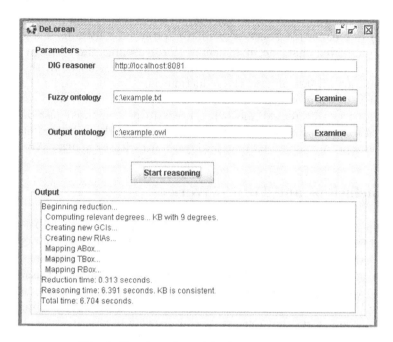

Fig. 2. User interface of DeLorean reasoner

assertions, $(2 \cdot (|N^{fK}| - 1) - 1) \cdot 4$ RIAs and at least $(2 \cdot (|N^{fK}| - 1) - 1) \cdot 20$ GCIs (added to keep the semantics of the new crisp elements). It also contains other GCIs added in the reduction of the original fuzzy GCIs. The number of axioms of this type depends on N^{fK} but also on the lower bound degree of every particular fuzzy GCI.

Once we obtained the crisp representation, reasoning was performed by using PELLET reasoner [20] through the DIG interface. Table 3 shows the influence of the number of degrees on the reduction time and on the time that requires a classification test over the resulting crisp ontology (the times are shown in seconds), together with some statistics about the resulting crisp ontology (the number of atomic concepts, atomic roles, concept assertions, GCIs and RIAs).

Table 3. Influence of the number of degrees in the performance of DeLorean

Number of degrees	crisp	3	5	7	9	11	21
Reduction time	-	0.062	0.079	0.125	0.141	0.172	0.312
Classification time	0.188	0.39	0.437	0.485	0.531	0.578	0.859
Atomic concepts	20	80	160	240	320	400	800
Atomic roles	4	16	32	48	64	80	160
Concept assertions	6	6	6	6	6	6	6
GCIs	35	152	318	484	650	816	1646
RIAs	0	12	28	44	60	76	156

Note that the result of the experimentation is different to that in a preliminary version of this paper [21], because we have optimized a little bit our reasoner. It can be observed that the increment in the reasoning time when the fuzzy ontology contains a small number of degrees can be assumed. In the current implementation of DELOREAN, the reduction time may be considered still high because it is just a prototype, but as already discussed in the previous section, the reduction can be reused and hence needs to be computed just once. It is interesting to mention that another implementation of the reduction has shown that the idea of the reduction fits well to reasoning with large fuzzy ABoxes [22].

6   Conclusions and Future Work

In this paper we have shown how to reduce a fuzzy extension of $\mathcal{SROIQ}$ with fuzzy GCIs and RIAs (under a novel semantics using Gödel implication) into $\mathcal{SROIQ}$. We have enhanced previous works by reducing the number of new elements and axioms. We have also presented DELOREAN, our implementation of this reduction procedure which is, to the very best of our knowledge, the first reasoner supporting fuzzy $\mathcal{SHOIN}$ (and hence and eventually fuzzy OWL DL). The very preliminary experimentation shows that our approach is feasible in practice when the number of truth degrees is small, even for our non-optimized prototype. This work means an important step towards the possibility of dealing with imprecise and vague knowledge in DLs, since it relies on existing languages and tools.

Future work could include the possibility of defining some concepts and roles to be interpreted as crisp, and the comparison of DELOREAN with other fuzzy DL reasoners (for instance FUZZYDL [23]), although they support different languages and features and, as far as we know, there does not exist any significant fuzzy knowledge base; the only one that we are aware of is a fuzzy extension of LUBM [24], but it is also a non expressive ontology (in fuzzy DL Lite). The reasoner is currently being extended to fuzzy $\mathcal{SROIQ}$ (and hence OWL 1.1) by using technologies other than Jena and DIG 1.1.

Acknowledgements

This research has been partially supported by the project TIN2006-15041-C04-01 (Ministerio de Educación y Ciencia). Fernando Bobillo holds a FPU scholarship from Ministerio de Educación y Ciencia. Juan Gómez-Romero holds a scholarship from Consejería de Innovación, Ciencia y Empresa (Junta de Andalucía).

References

1. Baader, F., Calvanese, D., McGuinness, D., Nardi, D., Patel-Schneider, P.F.: The Description Logic Handbook: Theory, Implementation, and Applications. Cambridge University Press, Cambridge (2003)

2. Horrocks, I., Kutz, O., Sattler, U.: The even more irresistible $\mathcal{SROIQ}$. In: Proceedings of the 10th International Conference of Knowledge Representation and Reasoning (KR 2006), pp. 452–457 (2006)
3. Patel-Schneider, P.F., Horrocks, I.: OWL 1.1 Web Ontology Language overview (2006), http://www.w3.org/Submission/owl11-overview/
4. Lukasiewicz, T., Straccia, U.: Managing uncertainty and vagueness in description logics for the semantic web. Journal of Web Semantics (to appear)
5. Sirin, E., Cuenca-Grau, B., Parsia, B.: From wine to water: Optimizing description logic reasoning for nominals. In: Proceedings of the 10th International Conference of Knowledge Representation and Reasoning (KR 2006), pp. 90–99 (2006)
6. Straccia, U.: Transforming fuzzy description logics into classical description logics. In: Alferes, J.J., Leite, J. (eds.) JELIA 2004. LNCS, vol. 3229, pp. 385–399. Springer, Heidelberg (2004)
7. Straccia, U.: Description logics over lattices. International Journal of Uncertainty, Fuzziness and Knowledge-Based Systems 14(1), 1–16 (2006)
8. Bobillo, F., Delgado, M., Gómez-Romero, J.: A crisp representation for fuzzy $\mathcal{SHOIN}$ with fuzzy nominals and general concept inclusions. In: da Costa, P.C.G., Laskey, K.B., Laskey, K.J., Fung, F., Pool, M. (eds.) Proceedings of the 2nd ISWC Workshop on Uncertainty Reasoning for the Semantic Web (URSW 2006). CEUR Workshop Proceedings, vol. 218 (2006)
9. Stoilos, G., Stamou, G.: Extending fuzzy description logics for the semantic web. In: Proceedings of the 3rd International Workshop on OWL: Experiences and Directions (OWLED 2007). CEUR Workshop Proceedings, vol. 258 (2007)
10. Zadeh, L.A.: Fuzzy sets. Information and Control 8, 338–353 (1965)
11. Hájek, P.: Metamathematics of Fuzzy Logic. Kluwer, Dordrecht (1998)
12. Klir, G.J., Yuan, B.: Fuzzy Sets and Fuzzy Logic: Theory and Applications. Prentice-Hall, Englewood Cliffs (1995)
13. Straccia, U.: A fuzzy description logic for the semantic web. In: Sanchez, E. (ed.) Fuzzy Logic and the Semantic Web. Capturing Intelligence, vol. 1, pp. 73–90. Elsevier Science, Amsterdam (2006)
14. Stoilos, G., Stamou, G., Pan, J.Z., Tzouvaras, V., Horrocks, I.: Reasoning with very expressive fuzzy description logics. Journal of Artificial Intelligence Research 30(8), 273–320 (2007)
15. Straccia, U.: Reasoning within fuzzy description logics. Journal of Artificial Intelligence Research 14, 137–166 (2001)
16. Stoilos, G., Stamou, G., Tzouvaras, V., Pan, J.Z., Horrocks, I.: Fuzzy OWL: Uncertainty and the semantic web. In: Proceedings of the 1st International Workshop on OWL: Experience and Directions (OWLED 2005). CEUR Workshop Proceedings, vol. 188 (2005)
17. Stoilos, G., Straccia, U., Stamou, G., Pan, J.Z.: General concept inclusions in fuzzy description logics. In: Proceedings of the 17th European Conference on Artificial Intelligence (ECAI 2006), pp. 457–461 (2006)
18. Ma, Y., Hitzler, P., Lin, Z.: Algorithms for paraconsistent reasoning with OWL. In: Franconi, E., Kifer, M., May, W. (eds.) ESWC 2007. LNCS, vol. 4519, pp. 399–413. Springer, Heidelberg (2007)
19. Bechhofer, S., Möller, R., Crowther, P.: The DIG description logic interface: DIG/1.1. In: Proceedings of the 16th International Workshop on Description Logics, DL 2003 (2003)
20. Sirin, E., Parsia, B., Cuenca-Grau, B., Kalyanpur, A., Katz, Y.: Pellet: A practical OWL-DL reasoner. Journal of Web Semantics 5(2), 51–53 (2007)

21. Bobillo, F., Delgado, M., Gómez-Romero, J.: Optimizing the crisp representation of the fuzzy description logic $\mathcal{SROIQ}$. In: Bobillo, F., et al. (eds.) Proceedings of the 3rd ISWC Workshop on Uncertainty Reasoning for the Semantic Web (URSW 2007). CEUR Workshop Proceedings, vol. 327 (2007)
22. Cimiano, P., Haase, P., Ji, Q., Mailis, T., Stamou, G.B., Stoilos, G., Tran, T., Tzouvaras, V.: Reasoning with large A-Boxes in fuzzy description logics using DL reasoners: An experimental valuation. In: Proceedings of the 1st Workshop on Advancing Reasoning on the Web: Scalability and Commonsense (ARea 2008). CEUR Workshop Proceedings, vol. 350 (2008)
23. Bobillo, F., Straccia, U.: fuzzyDL: An expressive fuzzy description logic reasoner. In: Proceedings of the 17th IEEE International Conference on Fuzzy Systems (FUZZ-IEEE 2008), pp. 923–930 (2008)
24. Pan, J.Z., Stamou, G., Stoilos, G., Thomas, E., Taylor, S.: Scalable querying service over fuzzy ontologies. In: Proceedings of the 17th International World Wide Web Conference, WWW 2008 (2008)

Uncertainty Issues and Algorithms in Automating Process Connecting Web and User

Alan Eckhardt[1], Tomáš Horváth[2], Dušan Maruščák[1], Róbert Novotný[2], and Peter Vojtáš[1]

[1] Charles University, Prague, 118 00 Czech Republic
{alan.eckhardt,peter.vojtas}@mff.cuni.cz
[2] P. J. Šafárik University, Košice, 04154 Slovakia
{tomas.horvath,robert.novotny}@upjs.sk

Abstract. We focus on replacing human processing web resources by automated processing. On an experimental system we identify uncertainty issues making this process difficult for automated processing and try to minimize human intervention. In particular we focus on uncertainty issues in a Web content mining system and a user preference mining system. We conclude with possible future development heading to an extension of OWL with uncertainty features.

Keyword: Uncertain reasoning, World Wide Web, web content mining, user profile mining.

1 Introduction

The amount of data accessible on Web is a great challenge for web search systems. Efficient utilization of these data (and information or knowledge hidden in them) can be a competitive advantage both for companies and individuals. Hence Web search systems form a part of different systems ranging from marketing systems, competitors and/or price tracking systems to private decision support systems. The main vision of Semantic web (see T. Berners-Lee, J. Hendler, O. Lassila. in [3]) is to automate some web search activities that a human is able to do personally. Using this automation of human search should speed up the search, access a wider range of resources and when necessary to soften our search criteria and optimize.

We quote the Uncertainty Reasoning for the World Wide Web (URW3) Incubator Group charter [23]: ...As work with semantics and services (on the Web) grows more ambitious, there is increasing appreciation of the need for principled approaches to representing and reasoning under uncertainty. In this Charter, the term "uncertainty" is intended to encompass a variety of forms of incomplete knowledge, including incompleteness, inconclusiveness, vagueness, ambiguity, and others. The term "uncertainty reasoning" is meant to denote the full range of methods designed for representing and reasoning with knowledge when Boolean truth values are unknown, unknowable, or inapplicable. Commonly applied approaches to uncertainty reasoning include probability theory,

P.C.G. da Costa et al. (Eds.): URSW 2005-2007, LNAI 5327, pp. 207–223, 2008.

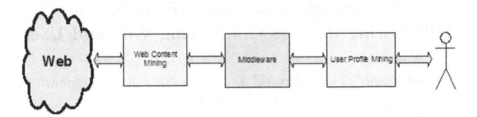

Fig. 1. Schema of an automated process connecting Web and User

Dempster-Shafer theory, fuzzy logic, and numerous other methodologies." In this paper we are using term "uncertainty" in this wider (generic) understanding and we would like to contribute to these efforts (for related discussion see [24]).

In this paper we concentrate especially to issues connected with replacing human abilities on the web by software. From this point of view, some sorts of uncertainty are not "human-to-machine-web" specific, like faulty sensors, input errors, data recorded statistically, medical diagnosis, weather prediction, gambling etc. These are difficult for human alone and also outside the web.

According to H. R. Turtle and W. B. Croft [18], uncertainty in information retrieval can be found especially in three areas: Firstly, there is the problem of the representation, annotation of a resource (service) and difficulties arise also when attempting to represent the degree to which a resource is relevant to the task. The second problem is the representation of the kind of information the user needs to retrieve or the action which he needs to perform (this is especially difficult since it typically changes during the session). Thirdly, it is necessary to match user needs to resource concepts. By our opinion, these areas of uncertainty apply also to our case, when replacing human activities on the web by software. Specific tasks connected to these three problems are depicted in Figure 1 and we will discuss them in this paper.

Our goal is to discuss uncertainty issues based on a system integrating the whole chain of tools from the Web to the user. The uncertainty problem here appears as a problem of two inductive procedures. Two types of data mining appearing in these systems will be discussed here. One is Web content mining and second is user profile (preference) mining. Middleware will do the matching part and query evaluation optimization.

1.1 Motivating Example

In this motivating example, assume users that are looking for a hotel in a certain region. Suppose that the amount of data about the accommodation possibilities is huge and they are distributed over several sites. Furthermore, the users have different preferences which are soft and difficult to express in a standard query language.

From the middleware point of view, there is no chance to evaluate user's query over all data. Therefore we have decided to use Fagin threshold algorithm [10], which can find best (top-k) answers without looking to all object. This algorithm works under following assumptions. First, we have approach to objects (in our case hotels) in different list ordered by user particular attribute ordering, equipped by a numerical score ranging from 0 to 1, e. g. $f_1(x) = \text{cheap}(x)$, $f_2(x) = \text{close}(x)$... Second, we have a combination function computing total fuzzy preference value of an object based on preference values of attributes, e. g. $(3 \cdot \text{cheap}(x) + \text{close}(x)/4)$.

In practical application we have to consider different users with possible different attribute ordering f_u^1, f_u^2 and combination function $@^u$. These represent the overall user preference $@^u(f_1^u, f_2^u)$ and the profile for this task. For user profile mining part there is a task to find these particular attribute orderings and the combination function (using ranking a sample of hotels).

On the web side of our system, very often information of vendors, companies or advertisement are presented using Web pages in structured layout containing data records. These serve for company presentation and mainly are assumed to be visited by a potential customer personally. Structured data objects are a very important type of information on the Web for systems dealing with competitor tracking, market intelligence or tracking of pricing information from sources like vendors.

We need to bring these data to our middleware. Due to the size of Web, the bottleneck is the degree of automation of data extraction. We have to balance the tradeoff between degree of automation of Web data extraction also from unvisited pages (usually with lower precision) and the amount of user (administrator) work, needed to train data extractor for special type of pages (increasing precision).

First restriction we make is we consider Web pages containing several structured data records. This is usually the case of Web pages of companies and vendors containing information on products and services and, in our case, hotels. Main problem is to extract data and especially attribute values to middleware.

Although we use a system which has modules in experimental implementation, we do not present this system here. Our main contributions are

- Identification of some uncertainty issues in web content mining system extracting attribute values from structured pages with several records
- Identification of some uncertainty issues in user profile model and use profile mining method
- Discussion of coupling of these systems via a middleware based on Fagin threshold algorithm like storage and querying. We point to uncertainty issues by inserting (**UNC**) in the appropriate place in the text.

2   Uncertainty in Web Content Mining

In this section we describe our experience with a system for information extraction from certain types of web pages and try to point out places where uncertainty occurred.

Using our motivation as a running example, imagine a user looking for a hotel in a certain location. A relevant page for a user searching for hotels can look similarly to that on Figure 2. The first goal for our middleware system

Fig. 2. A typical Web page containing several regions

is to automatically extract relevant data from various information sources and put them in a structured form (typically filling the attribute values of object instances), thus allowing the core components to perform either top-k search or other suitable object search methods.

Since most of the relevant data are presented on the web in the form of HTML pages, they are a suitable target for methods which operate on the semistructured input data. There are multiple existing approaches, which are generally semiautomatic – i. e. Lixto [1], Stalker [16] or that of Kushmerick et al. [15]. While these systems are reliable and can be used to cover pages for which the system was trained, pages that dynamically change not only the content, but also the structure, can require user's attention or even the repetition of training process. In cases when we have to collect data from a large number of structurally varying pages, these methods are not applicable or would require a substantial amount of effort.

Our solution is to concentrate on pages with several repeating records and search for similarities. There are many ways how to search for similar records in source tree (and also tree representations). System IEPAD [4] uses the Patricia tree (radix tree) representation and produces sometimes a bigger number of potential extraction rules choice of the right rule can be sometimes a problem and needs human intervention). This system is outperformed by the MDR system [5] which uses the HTML tree structure for search of repeating node sequences with same father. Nevertheless MDR (so IEPAD too) is searching for objects of interest in the whole web document. This is time consuming and as we have experienced, it surprisingly decreases precision. Moreover mentioned systems do not extract attribute values from data records.

It is clear that looking for attribute values depends on the domain. There are papers describing usage of ontologies for attribute value extraction [5]. These systems do not consider data records extraction.

In this paper we consider a system as a sequence of both data record extraction and attribute value selection, with possibility of ontology starting almost from scratch (e.g. user search key words).

The system will be described in several phases.

2.1 Data Regions Extraction

Identification of Data Regions is the first phase of our information extraction process. In the beginning we assume that data records, which are similar to each other, form a larger unit – a data region. Moreover, we expect that a data region contains at least two records.

At first, we build a Document Object Model (DOM) representation of the input document by using a JRex DOM parser (which besides other things resolves problems with invalid or malformed HTML). Besides the DOM tree we build an auxiliary data structure which simplifies the typical operations on this tree (like retrieving a subtree etc.). Searching for data regions starts with the pruning of the input tree, which reduces the complexity of the search process and improves the efficiency. It is obvious, that the desired data occur solely in the text nodes and therefore it is sufficient to consider only those subtrees which contain at least one text node. We perform a "colorization" of nodes. White nodes do not contain any text nodes in their subtrees. Grey-colored ones do not contain any text-nodes in their child nodes, but they can occur in the lower layers. Black nodes have a high probability of text-node occurence under their corresponding subtrees. The actual color is determined by the total text length in the subtree, corresponding HTML tag (since some HTML nodes by definition cannot contain children and therefore are implicitly white-colored) and the depth of the node in the DOM tree.

It is evident that only the gray and black-colored trees can contain the data regions. This can be an observation that leads us to the first uncertainty (**UNC1**) and a corresponding coefficient

$$unc_1 = \begin{cases} 0 & \text{for white notes} \\ 1 & \text{for gray nodes} \\ 2 & \text{for black nodes} \end{cases}$$

which can be used to annotate the extracted information, in support of (automated) agent usage of our sources.

To further reduce the search space, we have experimentally deduced that the fifth layer of the DOM tree is the suitable place to start the search.

The actual search algorithm is based on the breadth-first traversal of the DOM tree. For each non-white node we test, whether there is a repeating sequence of similar subtrees rooted on the nodes from the current layer. At first, we compare the subtrees rooted in the consecutive node (in the Figure 4: node 2 and 6, then node 6 and 10 etc.). Then we compare the consecutive node pairs (e. g. a union of subtrees from nodes 2 and 6 with a union of subtrees from nodes 10 and 12 etc.)

The similarity (**UNC2**) is determined according to the Levenshtein edit distance. We map the node names (i. e. HTML tags) to the letters of alphabet. Each subtree is mapped to a substring, which serves as a base for the comparison. The computed edit distance is normalized by the average length of both subtrees. Moreover, we have experimentally found the suitable threshold for the similarity equal to 0.2.

This gives another uncertainty coefficient

$$\text{unc}_2(0.2) = (p, r)$$

assigning precision and recall of detecting similar subtrees using this threshold.

The similar subtrees represent a candidate for the data region. However, we seek to find the largest possible data region. It means that we search for the node combination that covers the largest number of nodes. The breadth-first algorithm ensures that once a data region is discovered, it is not necessary to further examine its subnodes. It should be noted, that the algorithm can possibly find the data regions with data records out of scope of our particluar domain. Such regions will be resolved in the Attribute Identification phase.

The aforementioned pruning process serves yet another purpose. For example the data region can be represented by the repetition of elements TABLE, #text and BR. Since the last tag is only in the role of separator, it does not appear in the last record, which will be then incorrectly omitted.

2.2   Identification of Data Records

In the previous phase we have identified one or more data regions, where each of them contains multiple data records (represented by repeating node sequences). Each data record can be represented in two ways:

- A data region contains data records which are correspond to the contiguous HTML elements. Each data record then forms a proper subtree. This is an optimal case.

```
function BFSfindDR(LevelNodes) {
    NextLevelNodes = ∅;
    for each Node in LevelNodes do {
        if weight(Node) == 2 {
            regions := identDataRegions(normalized(Node.children));
            NextLevelNodes := NextLevelNodes ∪ (Node.Children not in regions)
        } else if weight(Node) == 1 {
            NextLevelNodes := NextLevelNodes ∪ Node.children
        }
    }
    if NextLevelNodes ≠ ∅ return regions ∪ BFDfindDR(LevelNodes)
    else return regions
}
```

Fig. 3. Pseudocode of Data Record Identification Phase

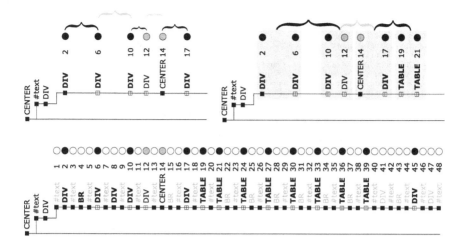

Fig. 4. A Sample DOM Model (left), a pruned DOM model with node pairs comparison (top right) and a pruned DOM model with node triples comparisons

- A data region contains data records which correspond to the HTML elements scattered among multiple subtrees (albeit forming a visually contiguous region). **(UNC3)**

The latter kind of regions can be resolved by two approaches. We can either syntactically analyse the data, tags and tables and estimate the number of data regions. Then we can try to pair up the individual data in the records. Alternatively we can postpone this solution to the Attribute Identification phase, where we use an ontology to match the attribute values to the corresponding objects.

2.3   Attribute Values Extraction from Master Page

In this phase we process the actual data records and parse them for the attribute values of the relevant objects **(UNC4)**. The basic idea is based on the traversal of the tree in the depth-first approach (to ensure the consistent order of the attributes), and the tokenization of the text nodes by the various separators.

The tokenized nodes are then matched against the extraction ontology, which determines the structure of data records, cardinality of attribute values etc. Generally, it allows us to specify the attribute name, regular expression patterns used for extraction, typical keywords associated with attribute, an enumeration of all attribute values (applicable in the case of attributes with fixed number of values) and maximum and minimum length of attribute.

It is natural that more elaborated ontology (more properties, detailed regular expressions, exact attribute labels) increases the success rate. Nondiscovered attributes can be found in the next phase, which deals with the extraction from the detail pages.

```
function findAttributes(page1, page2) {
    source1 = normalize(page1)
    dom1 = getDomTree(page1)
    source2 = normalize(page2)
    dom2 = getDomTree(page2)
    listOfDifferenceSpots = diff(source1, source2)
    listOfDifferenceSpots = listOfDifferenceSpots ∪ annotatedElements(dom1) ∪ annotatedElements(dom2)
    for each differenceSpot in listOfDifferenceSpots {
        element1 = findElementById(dom1)
        element2 = findElementById(dom1)
        attributeValue = applyOntology(extractText(element2))
        storeValue(attributeValue)
        annotateSpot(page1, element1)
        annotateSpot(page1, element2)
    }
}
```

Fig. 5. Pseudocode of Detail Pages Data Extraction

If there is a known number k of object attributes, we can determine the uncertainty coefficient as

$$\text{unc}_4 = \frac{n}{k},$$

where n is the number of discovered attributes.

2.4 Attribute Value Extraction from Detail Pages

The previous phase deals with attribute values extraction (**UNC4**). Inputs are single data records from the original page and extraction ontology, which contains and represents human effort inserted into our system. This ontology generally consists of multiple regular expressions, typical values, value intervals etc. which correspond to particular attributes.

As a complement to the master page extraction, we can easily discover the links to the detail pages (i. e. pages which contain information about a single

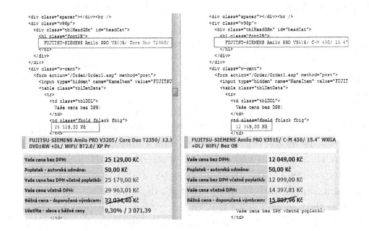

Fig. 6. A sample of difference spots

hotel), which usually contain the additional or better structured data. Moreover, a set of detail pages from a single site usually comes from a single template preprocessed by a template engine [26]. This template (invisible to us) can be seen as a skeleton of HTML page with variables which are on the server-side replaced with actual values. Comparing two detail pages by the diff algorithm [27] can yield the places of difference, which occur on three general occasions:

- structural differences between DOM subtrees. Corresponds to the situation when one subtree does not occur in the opposite part (typically when one product contains attribute, which does not occur in another one or to the redundant data on the page),
- the differences between HTML attributes — this represents differences between HTML attributes or their values. This kind of differences is of a lesser importances, since they contain only the hidden data (not displayed to the user)
- the differences between text nodes, which represents different values in HTML tag values.

By comparing a large number of web pages, we can discover the attribute values, which can be further refined by extraction ontology. The general overview of an algorithm is as follows:

1. retrieve a pair of two web pages and build their DOM trees.
2. normalize their HTML source trees and prepare them for the source code comparison
3. perform a diff algorithm comparison on the two normalized sources, retrieving a candidate difference spots. Categorize each difference spot into one of aforementioned three categories and if appropriate, retrieve the attribute values from the text nodes under the different nodes (**UNC4**).
4. retroactively annotate the input document with difference spot occurence metadata and statistics.
5. go to step 1.

The retroactive annotations prevents the obvious situation in which the attribute value is missed due to equal values on the particular place (a usual example is an input set of hotels from the same city. The same attribute value means skipped difference spot. However, annotated element can be seen as a forced difference spot which can cover such situation.)

The retroactive annotations are generated automatically. However, to improve and bootstrap the extraction process, one could manually preannotate a few web pages in a simple user interface.

Both value extraction approaches correspond to the epistemic uncertainty nature defined in the [24]. The extraction ontology imposes rather strict criteria on the required attribute and there in the cases in which the attribute was not extracted at all due to the limited knowledge of the tool. This is also the case of the source-comparing approach, in which the attribute could be incorrectly misspoted or there could be an data which do not correspond to any of the attributes.

Table 1. Optimal values for efficient extraction

Domain	K	s	l_{min}	l_{max}
Notebooks	1	0.1	800	100
Cars	1	0.2	300	80
Hotels	2	0.3	800	200

However, on the test data, we have reached a very high soundness / correctness, which was above 95%. In case of structured data we have experienced the only encountered misdiscovered data regions were those which contained nondomain data. However, as we have said, these data were eliminated in the Attribute Identication phase.

Moreover, we have experimentally found the optimal values for the most efficient extraction in the first two phases (see Table 1) in additional two domains. These values represent Levenshtein similarity (s), minimum and maximum length of attribute value text (l_{min} and l_{max}) and the maximum number of nodes used in the repeating sequences search (K).

When taking another review of the information extraction phases, we have identified possible sources of uncertainty in web information extraction and possible uncertainty coeficients unc_1, unc_2, unc_3 and unc_4. They can be used to annotate the extracted information, withour prescribing how a potential user of these informations will use them.

3 Middleware

3.1 Semantic Web Infrastructure

User preference mining is done locally and assumes the extracted data are stored in middleware. Extracted data have to be modeled on an OWA (Open World Assumption) model, and hence the traditional database models are not appropriate. Yet we are compatible with a semantic web infrastructure described in [20]. The storage is based on the ideas of Data Pile described in [2] in which a typical schema of record resembles a RDF statement with some statements about this statement (although the reification is not necessary in our case).

Resource	Attribute	Value	Extracted From	Extracted By	Using Ontology
$Hotel_1$	Price	V_1	URL1.html	$Tool_1$	O_1
$Hotel_1$	Distance	D_1	URL1.html	$Tool_1$	O_1

If a value of an attribute is missing, in our middleware system it means a missing record (thus implementing OWA). Note that we have records without any uncertainty degree attached, but this can be evaluated additionally the applications working on the data (e. g. it can be known that $Tool_1$ is highly reliable on extracting price, but less on distance).

To know what we are looking for and which attribute values to extract we need to know user interests. For middleware we moreover need to know the ordering of particular attributes and the combination function.

3.2 Using User Profiles

Another possibility is to create several user profiles, based on previous experiences with users coming to the web page. These profiles may be created as clusters of users or manually by an expert in the field (a hotel-keeper in our example). Manual creation is more suitable because we will know more details about user, but it is often impossible. Independent of the way profiles are created, we have ratings of objects associated with each profile, thus knowing the best and worst objects.

Suppose we have a set of user profiles $P_1, \ldots, P_k$ and we know the ideal hotel for each profile. We propose computing the distance d_i of user profile M_1 from each profile P_i in following way:

$$d_i = \frac{\sum_{j=1,\ldots,n} |\text{Rating}(\text{User}_1, o_j) - \text{Rating}(P_i, o_j)|}{n} \tag{1}$$

Equation (1) represents the average difference between the user's rating of an object o_j and profile's P_i's rating. The ideal hotel for the user can be computed as an average of ideal hotels for each profile P_i, weighted by the inverse of distance d_i (see (2)). The average is computed on attributes of hotels. Formally,

$$\text{IdealHotel}(\text{User}_1) = \frac{\sum_{i=1,\ldots,k} \text{IdealHotel}(P_i)/d_i}{\sum_{i=1,\ldots,k} 1/d_i} \tag{2}$$

Then, $\text{IdealHotel}(\text{User}_1)$ is the weighted centroid of profiles' best hotels. We can use this approach also for computing ideal values of attribute values:

$$\text{IdealPrice}(\text{User}_1) = \frac{\sum_{i=1,\ldots,k} \text{IdealPrice}(P_i)/d_i}{\sum_{i=1,\ldots,k} 1/d_i} \tag{3}$$

An example of data, user profiles' best hotel and user's best hotel is on Figure 7. User_1 is clearly closest to Profile 3.

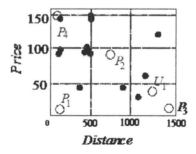

Fig. 7. Positions of best hotels for the user profiles and for the user

After the computation of the ideal hotel for the user, we will use it for computing ratings of remaining hotels. Disadvantage of this method is that one cannot use the Fagin threshold algorithm.

4   Uncertainty in User Preference Mining

In our meaning, user preferences are expressed in form of classification rules, where the values of attributes are assigned with *grades* corresponding to orderings of the domains of these attributes. The higher the grade the more appropriate (preferable) the value of an attribute is for the given user. This form of grading corresponds to *truth values* well-known in fuzzy community and thus the orderings correspond to fuzzy functions.

The combination function can be represented by a fuzzy aggregation function (see [10]). Fuzzy aggregation functions are monotone function of n variables ranging through the unit interval $[0, 1]$ of real numbers (in practical applications we use only a finite part of it).

Main assumption of our learning user preferences is that we have a (relatively small) sample of objects (hotels) evaluated by user (see smileys on the Figure 2). From this sample evaluation we would like to learn his/her preferences. The point is to use this learned user preference to retrieve top-k objects from a much bigger amount of data. Moreover, using user sample evaluation, we do not have to deal with the problem of matching the query language and document language. This rating is a form of QBE – query by example.

4.1   Learning Local Preferences

In [7] and [8] we have described several techniques of learning user's preferences of particular attributes (**UNC5**) represented by fuzzy functions f_1, f_2...on attribute domains. These techniques use regression methods. A problem occurs here. There can be potentially big number of hotels of one sort (e.g. cheap ones) but to detect user preference (cheap, medium or expensive) should not be influenced by the number of such hotels. Regression typically counts number of objects. We have introduced a special technique of discretization to get true user local preference (for details see [7] and [8]). We face the problem of outliers, here. Since, users often evaluate just a few objects, local preferences are learned from a small dataset. Thus, outliers have big impact to the final result of learning. The precision of local preference learning (e.g. regression) can be computed by a method, similar to the standard deviation, as follows:

$$\sigma = \sqrt{\frac{1}{n} \sum_{i=1}^{n} (f(x_i) - u(x_i))^2} \tag{4}$$

where n is the number of objects we learn local preferences from, $f(x_i)$ is the learned user's local preference for an attribute value x_i of an object i and $u(x_i)$ is the real user's local preference for an attribute value x_i of an object i.

This measure serves also as uncertainty level for (**UNC5**). It expresses how much the fuzzy function comply to user preferences.

We can simply put

$$unc_5 = \sigma$$

More complex case is in following section, where uncertainty for whole global preferences is studied. Similarly, other methods to computing the precision of local preference learning can be proposed.

Another approach not using regression is the following. The view of the whole domain of attribute *price* is in Figure 8. We can see that with increasing price, the rating is decreasing. This can be formalized (details are out of the scope of this paper) and we have experimented also with this possibility. These methods also gives local preference in the form of a fuzzy function (here small, cheap...) and hence are usable for Fagin Threshold algorithm.

4.2   Learning Combination Function

Second assumption of the Fagin's model [10] is to have a combination function @, which combines the particular attribute preference degrees f_1, f_2...(local preferences) to a overall score – $@(f_1, f_2 ...)$ – according to which the top-k answers will be computed.

There are several ways to learn (**UNC6**) the combination functions and several models. It is an instance of classification trees with monotonicity constraints (see [17], more references to ordinal classification are presented).

We learn the aggregation function by the method of Inductive Generalized Annotated Programming (IGAP) described in [13,14], nevertheless implemented only a fragment of IGAP method. The result of IGAP is a set of Generalized Annotated Program rules in which the combination function has a form of a function annotating the head of the rule – here the quality of hotel:

User1_hotel(H) good in degree at least $@(f_1(x), f_1(y), \ldots)$
    IF User1_hotel_price(x) good in degree at least $f1(x)$
                    AND
    User1_hotel_distance(y) good in degree at least $f_2(y)$
Note that these are rules of generalized annotated programs.

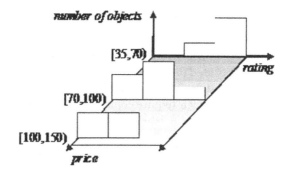

Fig. 8. Ratings for whole attribute domain

As stated in previous section, we have to learn global preferences from a small dataset. Thus, the learned rules have smaller prediction ability as if these are learned from the biggest dataset (training set). On the other hand, since, we learn global preferences using local preferences, the accuracy of rules strongly depends on the precision of learned local prefernces. Experiments show, that if local prefernces are learned correctly, IGAP learns rules with good prediction ability.

Our approach to measuring of the precision of our method to global preference learning is based on the following idea: Since, global preferences are used in the process of searching top-k objects, we are interested mainly in more precise global preferences for higher grades of classification. Thus higher precision for higher grades of classification is preferable.

We can express the precision of our method to global preference learning as follows:

$$P = \frac{\sum\limits_{i=2}^{n} i \cdot A_i}{\sum\limits_{i=2}^{n} i} \tag{5}$$

where i represents grades of classification (except the lowest grade, which global preferences are not learned for) and A_i is the accuracy of an ILP system ALEPH used in IGAP for the hypothesis learned for the grade i.

Moreover, we have to deal with a so called not-consistent user. When global preferences cannot be computed from the rated objects, the reason is often hidden in user's rating (mistake in rating, randomly rating user, etc.). The measure P may be interpreted as uncertainty level for **(UNC6)**. The accuracy A_i is the main source of uncertainty; it expresses how well the rules correspond with the real user preferences. P can be also interpreted as user consistency – if the user decides based on the attribute values consistently, P is much lower than if the user is largely inconsistent. This occurs when the significance of decision for the user is low.

Again, we can simply put

$$unc_6 = P$$

5   The Implementation and Experiments

Our Web content mining system has a modular implementation which allows additional modules to be incorporated (e. g. querying with preference-based querying). Communication between modules is based on the traditional Observer/Listener design pattern. All modules, which require communication with other ones, have to implement a Listener interface. All listeners are bound to the central Bus, which manages the communication between them. Each listener can specify a range of broadcasted and received events, which will be supported by it.

We proposed and implemented the middleware system for performing top-k queries over RDF data. As a Java library, our system can be used either on the server side, for example in a Web service, or on the client side. In both cases, it gathers information from local or Web data sources and combines them into one ordered list. To avoid reordering each time a user comes with different ordering, we have designed a general method using B+ trees to simulate arbitrary fuzzy ordering of a domain [6]. There are several implemented classes for standard user scoring functions, and TA and NRA algorithms. Detailed description of experiments is out of the scope of this paper. We can conclude that experiments have shown this solution is viable.

6   Conclusions and Future Work

Using an experimental implementation, in this paper we have identified several uncertainty challenges, when

(UNC1) identifying HTML nodes with relevant information in the sub-tree,
(UNC2) tuning similarity measures for discovery of similar tag subtrees,
(UNC3) identifying single data records in the HTML source,
(UNC4) extracting attribute values
(UNC5) learning user's preferences of particular attributes
(UNC6) learn the user preference combination function.

It is evident, that there can be found relations between certain uncertainties – for example unc_4 strongly influences unc_5 because when attribute values are determined wrong, the fuzzy function will also be wrong.

When unc_5 is high, unc_6 will be also high. High uncertainty about ordering of attribute domains will produce high uncertainty of their aggregation. Thanks to the degrees of unc_x we can quantify the overall degree of uncertainty and find where the process is mostly wrong. Corrections and parameter tuning will be applied at these places.

We have also experimented with some candidate solutions, which should enable annotation of resources by our uncertainty coefficients. However, we do not propose particular syntax for annotation. Moreover, it remains a problem, how should these be reflected in OWLU, a possible uncertainty modeling extension of OWL supporting machine processing of web task which is easy for humans. One can imagine an extension of OWL to OWLU where `owl:oneOf` extends to `owlU:typicallyOneOf` and `rdfs:subClassOf` extends to `owlU:typicallySubClassOf`. In our experiments we have used such flexible implementation when extracting attribute values.

Models and methods used in these experiments are compatible with our fuzzy $f\mathcal{EL}@$ description logic [19]. In this description logic only concepts are allowed to be fuzzy, roles remain crisp. Description logic $f\mathcal{EL}@$ can be embedded into two valued description logic with concrete domains and hence it is compatible with possible extension of OWL.

Acknowledgement. This work was supported in part by Czech projects 1ET 100300517 and 1ET 100300419 and Slovak projects VEGA 1/3129/06 and NA-ZOU.

References

1. Baumgartner, R., Flesca, S., Gottlob, G.: Visual Web Information Extraction. In: VLDB Conference (2001)
2. Bednárek, D., Obdržálek, D., Yaghob, J., Zavoral, F.: Data Integration Using DataPile Structure. In: Proceedings of the 9th East-European Conference on Advances in Databases and Information Systems, ADBIS 2005, Tallinn, pp. 178–188 (2005) ISBN 9985-59-545-9
3. Berners-Lee, T., Hendler, J., Lassila, O.: The Semantic Web. In: Scientific American Magazine (May 2001)
4. Chang, C.-H., Lui, S.-L.: IEPAD: Information extraction based on pattern discovery. In: WWW-10 (2001)
5. Liu, B., Grossman, R., Zhai, Y.: Mining Data Records in Web Pages. In: Procs. SIGKDD 2003, Washington, DC, USA, August 24-27 (2003)
6. Eckhardt, A., Pokorný, J., Vojtáš, P.: A system recommending top-k objects for multiple users preferences. In: 2007 IEEE Conference on Fuzzy Systems, pp. 1101–1106. IEEE, Los Alamitos (2007)
7. Eckhardt, A., Horváth, T., Vojtáš, P.: PHASES: A User Profile Learning Approach for Web Search. In: WI 2007 Web Intelligence Conference, Fremont, CA (November 2007) (accepted)
8. Eckhardt, A., Horváth, T., Vojtáš, P.: Learning different user profile annotated rules for fuzzy preference top-k querying. In: SUM 2007 Scalable Uncertainty Management Conference, Washington DC Area (October 2007) (accepted)
9. Embley, D.W., Campbell, D.M., Smith, R.D., Liddle, S.W.: Ontology-Based Extraction and Structuring of Information from Data-Rich Unstructured Documents. In: CIKM 1998, pp. 52–59 (1998)
10. Fagin, R., Lotem, A., Naor, M.: Optimal Aggregation Algorithms for Middleware. In: Proc. 20th ACM Symposium on Principles of Database Systems, pp. 102–113 (2001)
11. Galamboš, L.: Dynamization in IR Systems. In: Klopotek, M.A. (ed.) Proc. IIPWM 2004 - Intelligent Information Processing And Web Mining, pp. 297–310. Springer, Heidelberg (2004)
12. Galamboš, L.: Semi-automatic stemmer evaluation, ibid., 209–218
13. Gursky, P., Horvath, T., Novotny, R., Vanekova, V., Vojtas, P.: UPRE: User preference based search system. In: IEEE/WIC/ACM International Conference on Web Intelligence (WI 2006), pp. 841–844. IEEE, Los Alamitos (2006)
14. Horvath, T., Vojtas, P.: Ordinal Classification with Monotonicity Constraints. In: Perner, P. (ed.) ICDM 2006. LNCS, vol. 4065, pp. 217–225. Springer, Heidelberg (2006)
15. Kushmerick, N.: Wrapper induction: efficiency and expressiveness. Artificial Intelligence 118, 15–68 (2000)
16. Muslea, I., Minton, S., Knoblock, C.: A hierarchical approach to wrapper induction. In: Conf. on Autonomous Agents (1999)
17. Potharst, R., Feelders, A.J.: Classification trees for problems with monotonicity constraints. In: ACM SIGKDD Explorations Newsletter archive, vol. 4(1), pp. 1–10. ACM Press, New York (2002)
18. Turtle, H.R., Croft, W.B.: Uncertainty in Information Retrieval Systems. In: Proc. Second Workshop Uncertainty Management and Information Systems: From Needs to Solutions, Catalina, Calif., 1993 as quoted in S. Parsons. Current Approaches to Handling Imperfect Information in Data and Knowledge Bases. IEEE TKDE, vol. 8(3), pp. 353–372 (1996)

19. Vojtáš, P.: $\mathcal{EL}$ description logic with aggregation of user preference concepts. In: Duží, M., et al. (eds.) Information modeling and Knowledge Bases XVIII, pp. 154–165. IOS Press, Amsterdam (2007)
20. Yaghob, J., Zavoral, F.: Semantic Web Infrastructure using DataPile. In: Butz, C.J., et al. (eds.) Proc. 2006 IEEE/WIC/ACM International Conference on Web Intelligence and Intelligent Agent Technology, pp. 630–633. IEEE, Los Alamitos (2006)
21. JRex – The Java Browser Component, `http://jrex.mozdev.org/index.html`
22. org.w3c.dom.Document on, `http://www.w3.org`
23. Charter of W3C Uncertainty Reasoning for the World Wide Web Incubator Group, `http://www.w3.org/2005/Incubator/urw3/charter`
24. Wiki of W3C Uncertainty Reasoning for the World Wide Web XG Search, `http://www.w3.org/2005/Incubator/urw3/wiki/FrontPage`
25. `http://www.egothor.org/`
26. Smarty: Template Engine (April 2, 2008), `http://smarty.php.net`
27. DiffUtil (March 25, 2008), `http://www.gnu.org/software/diffutils/`

Granular Association Rules for Multiple Taxonomies: A Mass Assignment Approach

Trevor P. Martin[1,2], Yun Shen[1], and Ben Azvine[2]

[1] AI Group, University of Bristol, BS8 1TR UK
{Trevor.Martin,Yun.Shen}@bristol.ac.uk
[2] Intelligent Systems Lab, BT, Adastral Park, Ipswich IP5 3RE, UK
Ben.Azvine@bt.com

Abstract. The use of hierarchical taxonomies to organise information (or sets of objects) is a common approach for the semantic web and elsewhere, and is based on progressively finer granulations of objects. In many cases, seemingly crisp granulation disguises the fact that categories are based on loosely defined concepts that are better modelled by allowing graded membership. A related problem arises when different taxonomies are used, with different structures, as the integration process may also lead to fuzzy categories. Care is needed when information systems use fuzzy sets to model graded membership in categories - the fuzzy sets are not disjunctive possibility distributions, but must be interpreted conjunctively. We clarify this distinction and show how an extended mass assignment framework can be used to extract relations between fuzzy categories. These relations are association rules and are useful when integrating multiple information sources categorised according to different hierarchies. Our association rules do not suffer from problems associated with use of fuzzy cardinalities. Experimental results on discovering association rules in film databases and terrorism incident databases are demonstrated.

Keywords: Fuzzy, granules, association rules, hierarchies, mass assignments, semantic web, iPHI.

1 Introduction

The Semantic Web [1] provides general data and knowledge representation formats (e.g. RDF [2], OWL [3]) for integration and combination of data drawn from diverse sources, enabling these data to be shared and reused across application, enterprise and community boundaries. Ontologies play an essential role in achieving this vision and have become the target of extensive research over the last decade though there are still several major issues to be addressed.

Information granularity is considered as a key issue of knowledge representation in ontological structures [4]. Ontologies use taxonomies to define classes of objects and relations among them, enabling us to express a large number of relations among entities by assigning properties to classes and allowing subclasses

P.C.G. da Costa et al. (Eds.): URSW 2005-2007, LNAI 5327, pp. 224–243, 2008.

to inherit such properties. The use of taxonomic hierarchies to organise information and sets of objects into manageable chunks (granules) is widespread [5]. Granules were informally defined by Zadeh [6] as a way of decomposing a whole into parts, generally in a hierarchical way. We can regard a hierarchical categorisation as a series of progressively finer granulations, allowing us to represent problems at the appropriate level of granularity.

The idea of a taxonomy serves as an organisational principle for libraries, for document repositories, for corporate structure, for the grouping of species and very many other applications. It is therefore no surprise to note that the semantic web adopts hierarchical taxonomies as a fundamental structure, using the *subClassOf* construct. Although in principle the idea of a taxonomic hierarchy is crisply defined, in practice there is often a degree of arbitrariness in its definition. For example, we might divide the countries of the world by continent at the top level of a taxonomic hierarchy. However, continents do not have crisp definitions - Europe contains some definite members (e.g. France, Germany) but at the Eastern and South-Eastern border, the question of which countries belong / do not belong is less clear. Iceland is generally included in Europe despite being physically closer to Greenland (part of North America). Thus although the word "Europe" denotes a set of countries (i.e. it is a granule) and can be used as the basis for communication between humans, it does not have an unambiguous definition in terms of the elements that belong to the set. Different "authorities" adopt different definitions - the set of countries eligible to enter European football competitions differs from the set of countries eligible to enter the Eurovision song contest, for example.

Of course, mathematical and some legal taxonomic structures are generally very precisely defined - in axiomatic geometry, the class of polyhedra further subdivides into triangles, quadrilaterals, etc and triangles may be subdivided into equilateral, isosceles etc. Such definitions admit no uncertainty. Most information systems model the world in some way, and need to represent categories which correspond to the loosely defined classes used by humans in natural language. For example, a company may wish to divide adults into customers and non-customers, and then sub-divide these into high-value customers, dissatisfied customers, potential customers, etc. Such categories are not necessarily distinct (i.e. they may be a covering rather than a partition) but more importantly, membership in these categories is graded - customer X may be highly dissatisfied and about to find a new supplier whilst customer Y is only mildly dissatisfied. We argue that most hierarchical taxonomies involve graded or loosely defined categories, but the nature of computerised information systems means that a more-or-less arbitrary decision has to be made on borderline cases, giving the taxonomy the appearance of a crisp, well-defined hierarchy. This may not be a problem as long as a rigorous and consistent criterion for membership is used (e.g. a dissatisfied customer is defined as one who has made at least two calls complaining about service), but the lack of subjectivity in a definition is rare. The use of graded membership (fuzziness) in categories enhances their expressive power and usefulness.

The Semantic Web vision implies that conceptually overlapping ontologies will co-exist and interoperate [1]. In the real world, it is impossible to converge to a single unambiguous ontology that is acceptable to all knowledge engineers [7]. Ontologies vary in the level of detail and in the nature of their logical specification. In such a scenario the interoperability between ontologies will benefit from the ability to represent degrees of membership categories of target ontology, given information about its class membership in the source ontology.

For example, the category of "vintage wine" has a different (but objective) definition, depending on the country of origin. To a purist, vintage wines are made from grapes harvested in a single year - however, the European Union allows up to 5% of the grapes to be harvested in a different year, the USA allows 15% in some cases and 5% in others, while other countries such as Chile and South Africa may allow up to 25%. Thus even taking a simple (crisp) granulation of wines into vintage and non-vintage categories can lead to problems if we try to integrate different sources.

In this paper we describe a new method for calculating association rules to find correspondences between fuzzy granules in different hierarchies (with the same underlying universe). Knowing that a category in one taxonomy is identical, almost identical to, completely different from a category in another taxonomy is an important step in combining multiple taxonomies. We discuss the semantics of fuzzy sets when used to describe granules, and introduce a mass assignment-based method to rank association rules and show that the new method gives more satisfactory results than approaches based on fuzzy cardinalities. Ongoing work is focused on comparison of this approach to others (e.g. on ontology merging benchmarks), and with application to merging classified directory content.

2   Background

This work take place in the context of the iPHI system (intelligent Personal Hierarchies for Information) [8] which aims to combine and integrate multiple sources of information and to configure access to the information based on an individuals personal categories. We assume here that the underlying entities (instances) that are being categorised are known unambiguously - when integrating multiple sources, this is often not the case. We have outlined SOFT (the Structured Object Fusion Toolkit) elsewhere [9] as one solution to this problem.

2.1   Fuzzy Sets in Information Systems

Many authors (e.g. [10]) have proposed the use of fuzzy sets to model uncertain values in databases and other knowledge based applications . The standard interpretation of a fuzzy set in this context is as a *possibility distribution* [11,12] - that is to say it represents a single valued attribute that is not known exactly. For example we might use the fuzzy set *tall* to represent the height of a specific person or *low* to represent the value shown on a dice. The fuzzy sets *tall* and *low* admit a range of values, to a greater or lesser degree; the actual value is taken from the range. Knowing that a dice value val is $even$ restricts the possible values to $val = 2$ XOR $val = 4$ XOR $val = 6$ (where XOR is an exclusive or). If a fuzzy set on the same universe is defined as $low = \{1/1, 2/1, 3/0.4\}^1$ then knowing the value val is low restricts the possible values to $val = 1$ XOR $val = 2$ XOR $val = 3$ with corresponding memberships.

The conjunctive interpretation of a fuzzy set occurs when the attribute can have multiple values. For example, a person may be able to speak several languages; we could model this as a fuzzy set of languages, where membership would depend on the degree of fluency. This is formally a relation rather than a function on the underlying sets. If we write $speaks(John, \{Spanish/1, Portugese/0.8\})$ it is not clear whether this means John speaks Spanish AND Portugese or John speaks Spanish OR Portugese. Our position is to make a distinction between the conjunctive interpretation - modelled by a fuzzy relation - and the disjunctive interpretation - modelled by a possibility distribution. To emphasise the distinction, we use the notation

$$F(a) = \{x/\mu(x)|x \in U\}$$

to denote a single valued attribute F of some object a (i.e. a possibility distribution over a universe U) and

$$R(a) = [x/\chi(x)|x \in U]$$

to denote a multi-valued attribute (relation). Granules represent the latter case, since we have multiple values that satisfy the predicate to a greater or lesser degree.

2.2 Association Rules

In creating association rules within transaction databases (e.g. [13], see also [14] for a clear overview), the standard approach is to consider a table in which columns correspond to items and each row is a transaction. A column contains 1 if the item was bought, and 0 otherwise. The aim of association rule mining is to determine whether or not there are links between two disjoint subsets of items - for example, do customers generally buy biscuits and cheese when beer, lager and wine are bought?

Let X denote the set of items, so that any transaction can be represented as $tr \subseteq X$ and we have a multiset Tr of transactions. We must also specify two non-overlapping subsets of X, s and t. An association rule is of the form $S \Rightarrow T$ where S (respectively T) is the set of transactions containing the items s (respectively t). The rule is interpreted as stating that when the items in s appear in a transaction, it is likely that the items in t will also appear i.e. it is not an implication in the formal logical sense.

Most authors use two measures to assess the significance of association rules, although these measures can be misleading in some circumstances. The support

---

1 The notation 3/0.4 indicates that the element 3 has membership 0.4 in the fuzzy set.

of a rule is the fraction of transactions in which both s and t appear, and the confidence of a rule is an estimate (based on the samples) of the conditional probability of T given S

$$Support(S,T) = |S \cap T|$$

and

$$Conf(S,T) = \frac{|S \cap T|}{|S|}$$

where we operate on multisets rather than sets. Typically a threshold is chosen for the support, so that only frequently occurring sets of items s and t are considered; a second threshold filters out rules of low confidence.

Various approaches to fuzzifying association rules have been proposed e.g. [14,15,16]. The standard extension to the fuzzy case is to treat the (multi-) sets S, T as fuzzy and find the intersection and cardinality using a t-norm and sigma-count respectively.

$$Conf(S,T) = \frac{\sum_{x \in X} \mu_{S \cap T}(x)}{\sum_{x \in X} \mu_S(x)}$$

Clearly any fuzzy generalisation must reduce to the crisp case when memberships are restricted to $\{0,1\}$.

As pointed out by [14], using min and the sigma count for cardinality can be unsatisfactory because it does not distinguish between several tuples with low memberships and few tuples with high memberships - for example, given

$$S = [x_1/1, x_2/0.01, x_3/0.01, ..., x_{1000}/0.01]$$

and

$$T = [x_1/0.01, x_2/1, x_3/0.01, ..., x_{1000}/0.01]$$

leads to

$$Conf(S,T) = \frac{1000 \times 0.01}{1 + 999 \times 0.01} \approx 0.91$$

which is extremely high for two almost disjoint sets (this example originally appeared in [17]). Using a fuzzy cardinality (i.e. a fuzzy set over the possible cardinality values) is also potentially problematic since the result is a possibility distribution over rational numbers, and the extension principle [18] gives a wider bound than it should, due to neglect of interactions between the numerator and denominator in this expression. For example, given

$$S = [x1/1, x2/0.8]$$

$$T = [x1/1, x2/0.4]$$

the fuzzy cardinalities are $|S \cap T| = \{1/1, 2/0.4\}$, $|S| = \{1/1, 2/0.8\}$ leading to a confidence of $\{0.5/0.4, 1/1, 2/0.8\}$ which is clearly incorrect.

One approach to the problem is given in [19] where the fuzzy association rule is interpreted as a quantified sentence. The confidence of the fuzzy association rule $S \Rightarrow T$ in the set of fuzzy transactions T is the evaluation of quantified sentence "Q of Γ_S are Γ_T" where Γ refers to the (fuzzy) transactions containing S (respectively T) and Q refers to a quantifier. The confidence value in the above example is 0.01 by modelling the quantified sentence using fuzzy level sets.

In this paper, we propose the use of mass assignment theory in calculating the support and confidence of association rules between fuzzy categories.

The fuzziness in our approach arises because we allow partial membership in categories - for example, instead of looking for an association between biscuits and beer, we might look for an association between *alcoholic drinks* and *snack foods*. It is important to note that we are dealing with conjunctive fuzzy sets (monadic fuzzy relations) here. Mass assignment theory is normally applied to fuzzy sets representing possibility distributions and the operation of finding the conditional probability of one fuzzy sets given another is known as semantic unification [20]. This rests on the underlying assumption of a single valued attribute - a different approach is required to find the conditional probability when we are dealing with set-valued attributes.

2.3 Mass Assignments

A mass assignment [22] (see also [23]) is a distribution over a power set, representing disjunctive uncertainty about a value. For a universe U, m is defined as

$$m : P(U) \rightarrow [0, 1]$$
$$\sum_{X \subseteq U} m(X) = 1 \tag{1}$$

The mass assignment m_A is related to a fuzzy set (possibility distribution) A as follows.

Let μ_A be the membership function of A with range

$$Range(\mu_A) = \{\mu_A^1, \mu_A^2, \mu_A^3, ..., \mu_A^m\}$$

such that

$$\mu_A^1 > \mu_A^2 > ... > \mu_A^m$$

and A_i (also known as the focal elements) be the alpha-cuts [2] at these values

$$A_i = \{x | \mu_A(x) \geq \mu_A^i\}$$

---

[2] An alpha cut is a crisp set containing all elements with membership greater than or equal to the specified value of alpha. For details, see any introductory text on fuzzy sets e.g. Fuzzy Sets, Information and Uncertainty, Klir and Folger, Prentice Hall, 1988.

Then

$$m_A(A_i) = \mu_A^i - \mu_A^{i+1}$$

Given a fuzzy set A, the corresponding mass assignment can be written as

$$M(A) = \{A_i : m_A(A_i) | A_i \subseteq A\} \qquad (2)$$

where conventionally only the focal elements (non-zero masses) are listed in the mass assignment. The mass assignment represents a family of probability distributions on U, with the restrictions

$$p : U \to [0,1]$$

$$\sum_{x \in U} p(x) = 1$$

$$m(\{x\}) \le p(x) \le \sum_{x \in X} m(X) \qquad (3)$$

For example, if $X = \{a, b, c, d\}$ and A is the fuzzy set defined as

$$A = \{a/1, b/0.8, c/0.3, d/0.2\}$$

then

$$M(A) = \{\{a\} : 0.2, \{a, b\} : 0.5, \{a, b, c\} : 0.1, \{a, b, c, d\} : 0.2\}$$

In the example above, $p(a) = 0.4$, $p(b) = 0.3$, $p(c) = 0.1$, $p(d) = 0.2$ is a possible distribution, obtained by allocating the mass of 0.5 on the set $\{a, b\}$ to a (0.2) and b (0.3), and so on. We can also give a mass assignment definition of the cardinality of a fuzzy set as a distribution over integers

$$p(|A| = n) = \sum_{\substack{A_i \subseteq A \\ |A_i| = n}} m_A(A_i)$$

where $0 \le n \le |U|$.

In the example above, $p(|A| = 1) = 0.2$, $p(|A| = 2) = 0.5$, etc. Clearly in this framework, the cardinality of a fuzzy set can be left as a distribution over integer values, or an expected value can be produced from this distribution in the usual way. A similar definition of fuzzy cardinality was proposed by [24], also motivated by the problem of fuzzy association rules.

Baldwin introduced the least prejudiced distribution (LPD) which is a specific distribution satisfying (3) above but also obeying

$$LPD_A(x) = \sum_{x \in A_i} \frac{m(A_i)}{|A_i|} \qquad (4)$$

where $|A|$ indicates the cardinality of the set A and the summation is over all focal elements containing x.

Informally, wherever mass is associated with a non-singleton focal element, it is shared equally between the members of the set. Clearly a least prejudiced distribution is a restriction of the original assignment.

The steps from LPD to mass assignment and then to fuzzy set can be reversed, so that we can derive a unique fuzzy set for any frequency distribution on a finite universe, by assuming the relative frequencies are the least prejudiced distribution (proof in [25]).

If the relative frequencies are written

$$L_A = \{L_A(x_1), L_A(x_2), ..., L_A(x_n)\}$$

such that

$$L_A(x_1) > L_A(x_2) > ... > L_A(x_n)$$

then we can define

$$A_i = \{x | x \in U \wedge L_A(x) \geq L_A(x_i)\}$$

and the fuzzy set memberships are given by

$$\mu_A(x_i) = |A_i| \times L_A(x_i) + \sum_{j=i+1}^{n} (|A_j| - |A_{j-1}|) \times L_A(x_j)$$

3   Mass-Based Granular Association Analysis

3.1   Fuzzy Relations and Mass Assignments

A relation is a conjunctive set of ordered n-tuples i.e. it represents a conjunction of n ground clauses. For example, if U is the set of dice scores then we could define a predicate $differBy4or5$ on $U \times U$ as the set of pairs

$$[(1, 6), (1, 5), (2, 6), (5, 1), (6, 1), (6, 2)]$$

This is a conjunctive set in that each pair satisfies the predicate. In a similar way, a fuzzy relation represents a set of n-tuples that satisfy a predicate to a specified degree. Thus $differByLargeAmount$ could be represented by

$$[(1,\ 6)/1, (1,\ 5)/0.6, (2,\ 6)/0.6, (5,\ 1)/0.6, (6,\ 1)/1, (6,\ 2)/0.6]$$

3.2   Granular Association Analysis

Zadeh [6] pointed out that information granulation is inspired by the ways in which humans granulate information and reason with it. The core of information granulation is that information can be processed on different levels of abstraction in which data objects are organised into meaning granules [3] so as to convey a perception of information itself. Thus, it is necessary to study the criteria for

---

[3] We interpret granules as conjunctive fuzzy sets in this paper.

Table 1. Database of Sales Employees

name	sales	salary
a	100	1,000
b	80	400
c	50	800
d	20	700

deciding how "a clump of points (objects) are drawn together by indistinguisha-bility, similarity, proximity or functionality [6]". The formation of granules can also be interpreted from the point view of concept formation. In the logic theory [28], the intension of a concept consists of all properties that are valid for all those objects to which the concept applies. Similarly, the granule applies certain measure criteria/function to select common features of all the entities belonging to itself.

We consider two granules, represented as monadic fuzzy relations S and T on the same domain, and wish to calculate the degree of association between them. For example, consider a database of sales employees, salaries and sales figures. We can categorise employees according to whether their salaries are *high*, *medium* or *low* and also according to whether their sales figures are *good*, *moderate* or *poor*. A mining task might be to find out whether the *good* sales figures are achieved by the *highly paid* employees. For example, given the Table 1, according to our subjective perception, we might define the monadic fuzzy relations,

$$S = goodSales = [a/1, b/0.8, c/0.5, d/0.2]$$

and

$$T = highSalary = [a/1, b/0.4, c/0.8, d/0.7]$$

These represent sets of values (1-tuples) that all satisfy the related predicate to a degree. The confidence in an association rule can be calculated as follows: For a source granule

$$S = [x_1/\chi_S(x_1), x_2/\chi_S(x_2), ..., x_{|S|}/\chi_S(x_{|S|})]$$

and a target granule

$$T = [x_1/\chi_T(x_1), x_2/\chi_T(x_2), ..., x_{|T|}/\chi_T(x_{|T|})]$$

we can define the corresponding mass assignments as follows. Let the set of distinct memberships in S be,

$$\{\chi_S^{(1)}, \chi_S^{(2)}, ..., \chi_S^{(n_S)}\}$$

where

$$\chi_S^{(1)} > \chi_S^{(2)} > ... > \chi_S^{(n_S)}$$

and

$$n_S \leq |S|$$

Let

$$S_1 = \{[x | \chi_S(x) = \chi_S^{(1)}]\}$$

$$S_i = \{[x | \chi_S(x) \geq \chi_S^{(i)}]\} \cup S_{i-1}$$

where $1 < i \leq n_S$.

Then the mass assignment corresponding to S is

$$\{S_i : m_S(S_i)\}, \quad 1 \leq i \leq n_X$$

where

$$m_S(S_k) = \chi_S^{(k)} - \chi_S^{(k+1)} \quad (\chi_S^{(i)} = 0 \; if \; i > n_S)$$

For example, the source granule

$$S = [a/1, b/0.8, c/0.5, d/0.2]$$

Table 2. Maximum Confidence

		0.2	0.1		0.3			0.4			
		[a]	[a]	[ac]	[a]	[ac]	[acd]	[a]	[ac]	[acd]	[abcd]
0.2	[a]	0.2									
0.3	[a] [ab]		0.1								0.2
0.3	[a] [ab] [abc]					0.3					
0.2	[a] [ab] [abc] [abcd]							0.2			

$$Conf(S \rightarrow T) = \frac{0.2 \times 1 + 0.1 \times 1 + 0.2 \times 2 + 0.3 \times 1 + 0.2 \times 1}{0.2 \times 1 + 0.1 \times 1 + 0.2 \times 2 + 0.3 \times 1 + 0.2 \times 1}$$
$$= 1$$

it has the corresponding mass assignment

$$M_S = \{\{[a]\} : 0.2, \{[a], [a, b]\} : 0.3, \{[a], [a, b], [a, b, c]\} : 0.3, \{[a], [a, b], [a, b, c], [a, b, c, d]\} : 0.2\}$$

The mass assignment corresponds to a distribution on the power set of relations, and we can define the least prejudiced distribution in the same way as for the standard mass assignment. In the example above, we get the LPD of the source granule as a distribution on granules (monadic relations)

$$LPD_S = \{[a] : 0.5, [a, b] : 0.3, [a, b, c] : 0.15, [a, b, c, d] : 0.05\}$$

We can now calculate the confidence in the association between the granules S and T using mass assignment theory. In general, this will be an interval as we are free to move mass (consistently) between elements of each S_i and T_j [21].

For two mass assignments,

$$M_S = \{\{S_{p_i}\} : m_S(S_i)\}, \quad 1 \leq p_i \leq i \leq n_S$$

and

$$M_T = \{\{T_{q_j}\} : m_T(T_i)\}, \quad 1 \leq q_j \leq j \leq n_T$$

the composite mass assignment is

$$M_C = M_S \bigoplus M_T = \{X : m_C(X)\}$$

where m_C is the composite mass allocation function $C(i, j, S_{p_i}, T_{q_j})$ subject to

$$\sum_{j=1}^{n_T} \sum_{\substack{1 \leq q_j \leq j \\ 1 \leq p_i \leq i}} C(i, j, S_{p_i}, T_{q_j}) = m_S(S_i)$$

$$\sum_{i=1}^{n_S} \sum_{\substack{1 \leq q_j \leq j \\ 1 \leq p_i \leq i}} C(i, j, S_{p_i}, T_{q_j}) = m_T(T_i)$$

By this it follows Equation (5)

$$Conf(S \rightarrow T) = \frac{\displaystyle\sum_{i=1}^{n_S} \sum_{j=1}^{n_T} \sum_{\substack{1 \leq q_j \leq j \\ 1 \leq p_i \leq i}} C(i, j, S_{p_i}, T_{q_j}) \times |S_{p_i} \cap T_{q_j}|}{\displaystyle\sum_{i=1}^{n_S} \sum_{j=1}^{n_T} \sum_{\substack{1 \leq q_j \leq j \\ 1 \leq p_i \leq i}} C(i, j, S_{p_i}, T_{q_j}) \times |S_{p_i}|} \tag{5}$$

This can be visualised using a mass tableau (see [22]). Each row (column) represents a focal element of the mass assignment, and is split into sub-rows (sub-columns). The mass associated with a row (column) is shown at the far left (top) and can be distributed amongst the sub-rows (sub-columns). For example consider the granules

$$S = [a/1, b/0.8, c/0.5, d/0.2]$$

and

$$T = [a/1, b/0.4, c/0.8, d/0.7]$$

The rule confidence is given by equation (5).

Clearly the mass can be allocated in many ways, subject to the column constraints and it is not always straightforward to find the minimum and maximum confidences arising from different composite mass allocations. Two extreme examples are shown in Table 2 and Table 3. The head row represents the mass

Table 3. Minimum Confidence

		0.2	0.1		0.3			0.4			
		[a]	[a]	[ac]	[a]	[ac]	[acd]	[a]	[ac]	[acd]	[abcd]
0.2	[a]	0.2									
0.3	[a]		0.1					0.2			
	[ab]										
0.3	[a]				0.3						
	[ab]										
	[abc]										
0.2	[a]										
	[ab]										
	[abc]										
	[abcd]							0.2			

$$Conf(S \rightarrow T) = \frac{0.2 \times 1 + 0.1 \times 1 + 0.2 \times 1 + 0.3 \times 1 + 0.2 \times 1}{0.2 \times 1 + 0.1 \times 2 + 0.2 \times 2 + 0.3 \times 3 + 0.2 \times 4}$$
$$= 0.4$$

distribution of T and the first column represents the mass distribution of S. The confidence in the association rule between the two granules lies in the interval $[0.4, 1]$. In general there can be considerable computation involved in finding the maximum and minimum confidences for a rule. When ranking association rules it is preferable to have a single figure for confidence, rather than an interval that can lead to ambiguity in the ordering.

We can redistribute the mass according to the least prejudiced distribution i.e. split the mass in each row (column) equally between its sub-rows (sub-columns) and taking the product as the mass in each cell. In this case, the calculation is simplified by (a) combining rows (columns) with the same label and (b) re-ordering the summations. This enables us to calculate association confidences with roughly $O(n)$ complexity, rather than $O(n^4)$ where n is the number of focal elements in the source granule S. We get the confidence as,

$$Conf_{LPD}(S,T) = \frac{\sum_{i=1}^{n_S}\sum_{j=1}^{n_T} LPD_S(S_i) \times LPD_T(T_j) \times |S_i \cap T_j|}{\sum_{i=1}^{n_S} LPD_S(S_i) \times |S_i|} \tag{6}$$

(although the numerator in this expression is written as a double summation, it can be calculated in $O(n)$ time by stepping through the cells on the leading diagonal, due to the nested structure of the sets). If we choose the least prejudiced distribution and re-arrange sub-rows into single rows with the same label (also columns) we obtain the following intersections in Table 4.

Table 4. LPD Table

	[a]:0.45	[ac]:0.25	[acd]:0.2	[abcd]:0.1
[a]:0.5	[a]	[a]	[a]	[a]
[ab]:0.3	[a]	[a]	[a]	[ab]
[abc]:0.15	[a]	[ac]	[ac]	[abc]
[abcd]:0.05	[a]	[ac]	[acd]	[abcd]

$$
\begin{aligned}
Conf_{LPD}(S,T) = &(0.5 \times (0.45 + 0.25 + 0.2 + 0.1) \times 1 \\
&+0.3 \times (0.45 + 0.25 + 0.2) \times 1 + 0.3 \times 0.1 \times 2 \\
&+0.15 \times 0.45 \times 1 + 0.15 \times (0.25 + 0.2) \times 2 + 0.15 \times 0.1 \times 3 \\
&+0.05 \times 0.45 \times 1 + 0.05 \times 0.25 \times 2 + 0.05 \times 0.2 \times 3 + 0.05 \times 0.1 \times 4)/1.75 \\
= &\ 0.67
\end{aligned}
$$

This method gives a confidence of 0.67 - lying in the interval shown in Table 2 and Table 3. Using the least prejudiced distribution (LPD) allows us to replace the calculation in Equation (5) with straightforward calculations of the expected values of the cardinality of the source set and the intersection.

The example above gives a similar result to the cardinality-based method, but this is not always the case. For example if given,

$$S = [x_1/1, x_2/0.01, x_3/0.01, ..., x_{1000}/0.01]$$

and

$$T = [x_1/0.01, x_2/1, x_3/0.01, ..., x_{1000}/0.01]$$

then a fuzzy cardinality based approach gives a confidence $\frac{10}{10.99} \approx 0.91$ whereas we get a value of approximately 10^{-5}. Clearly this is a far more reasonable answer, as there are no elements with strong membership in both granules.

4   Experimental Results

4.1   Film Genre

We have carried out tests on the approach by finding associations between movie genres from different online sources. The two online movie databases IMDB (http://www.imdb.com) and Rotten Tomatoes (http://www.rottentomatoes.com) have been used in previous work [26] to test instance matching methods. We have used the SOFT [9] method to establish correspondence between the (roughly) 95, 000 movies in the databases. Within these two sources, movies are assigned to one or more genres and our task is to find strong associations between genres. In principle one would expect genres to be at different granularity (e.g. comedy could be sub-divided into slapstick, satire, situation comedy, etc). At this stage, there is no benchmark for comparison but the results are intuitively reasonable as shown in Table 5.

Table 5. Movie Genres Association Analysis

IMDB Genre	Tomato Genre	Association Value
Thriller	Suspense	0.351
	Drama	0.35
	Horror/Suspense	0.326
War	Drama	0.531
	Action	0.47
	Horror	0.319
Adult	Non-explicit	0.53
	Beauty/Fashion	0.375
	Education/General Interest	0.318

4.2 Terrorism Databases Analysis

We have carried out tests on the Worldwide Incidents Tracking System (WITS) and MIPT Terrorism Knowledge Base (TKB) data [4]. 15,900 incidents between January 2005 and January 2006 are analysed. We construct three granules modelling civilian casualties - *high casualty*, *medium casualty* and *low casualty* - according to the selection criteria shown in the Figure 1. We also construct the weapon granules, such as *Vehicle bomb*, *Explosive* and etc., by identifying predefined keywords. Fuzzy grammar [27] has been used to identify the weapons from the text fragments. The results are shown in Table 6. As we can see, "High casualty" has tight association with "Explosive" and "Vehicle bomb" while "Medium casualty" is normally connected with "Firearm".

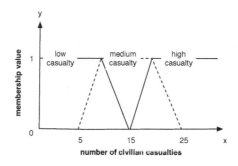

Fig. 1. Casualty Granules

---

[4] WITS data is available at http://wits.nctc.gov/Export.do. TKB ceased operation on 31st March 2008 and is now part of Global Terrorism Database (GTD) at http://www.start.umd.edu/data/gtd/

Table 6. Terrorism Data Association Analysis (I)

Casualty	Weapons	Association Value
High Casualty	Explosive	0.304
	Vehicle bomb	0.303
	Firearm	0.18
Medium Casualty	Firearm	0.316
	Explosive	0.224
	Vehicle bomb	0.165

Moreover, we also construct geography granules for the regions such as *near Iraq* and *south asia mainland* in which the partial membership of a country depends on the degree of political/military connection with a certain region. For example, Iran has the second largest membership in the "near Iraq" granule while "Afghanistan" holds the largest membership in the "south asia mainland" granule. Perpetrators can be modelled through a similar way to that of the products. The result is shown in Table 7. The confidence values are low because most of the perpetrators responsible for the attacks are "unknown" in the data sets.

Table 7. Terrorism Data Association Analysis (II)

Region	Perpetrators	Association Value
South Asia Mainland	Secular Political	0.311
Near Israel	Islamic Extremists (Sunni)	0.277
Near Iran	Islamic Extremists (Sunni)	0.093

Table 8. Terrorism Data Association Analysis (III)

Perpetrators	Weapons	Association Value
Islamic Extremist (Sunni)	Firearm	0.333
	Vehicle bomb	0.164
	Explosive	0.151
	Missile/Rocket	0.148
Islamic Extremist (Shia)	Firearm	0.793
	Missile/Rocket	0.103
Secular Political/Anarchist	Firearm	0.312
	Explosive	0.215

Table 9. Supermarket Transaction Database

Association Rule	Mass-Based Conf.	QS Conf.	Mass-Based Pos.	QS Pos.
$< P, L > \rightarrow < SF, H >$	0.639	0.744	1	1
$< P, M > \rightarrow < SF, M >$	0.248	0.444	2	2
$< P, M > \rightarrow < SF, H >$	0.195	0.317	3	3
$< P, M > \rightarrow < SF, L >$	0.139	0.179	4	5
$< P, L > \rightarrow < SF, M >$	**0.093**	**0.258**	5	4

Table 10. Fuzzy Text Transactions based on Harry Potter Book

Association Rule	Mass-Based Conf.	QS Conf.
$Dumbledore \rightarrow Harry$	0.753	0.808
$Hermione \rightarrow Ron$	**0.292**	**0.683**
$Harry \rightarrow Question$	**0.07**	**0.140**
$Harry \rightarrow Ron$	0.178	0.338
$Harry \rightarrow Hermione$	0.137	0.265

It is also interesting to identify the association between a specific group and the weapons. Though most of the perpetrators are classified as "Unknown", we focus on finding the strong association rules when the perpetrators are identified. The results are shown in Table 8. The example shows that the method can handle reasonable scale problems, and that results presented are intuitively reasonable but (in common with all association rules) not objectively verifiable.

4.3 Comparative Study

Since the quantified sentence (QS) approach [19] is using a similar multiset method, we also compare our mass-based association rules mining to QS on two datasets kindly supplied by Jose Maria Serrano Chica[5]. The first dataset contains 1,458 items from the supermarket transaction database. The price (P) and sales figures (SF) are respectively evaluated by three granules; High (H), Medium (M) and Low (L). The comparison result is show in Table 9. The result from mass-based association analysis is almost consistent with the quantified sentence approach except that quantified sentence approach ranks the rule $< P, L > \rightarrow < SF, M >$ higher than the rule $< P, M > \rightarrow < SF, L >$.

The second dataset contains 3,234 terms extracted by Chica [private communication] from parts of the Harry Potter series of books and is modelled as a fuzzy text transaction database according to [29]. Let us consider $T_D = \{d_1, ..., d_n\}$ as the set of transactions from the collection of documents D, and $I = \{t_1, ..., t_m\}$

---

[5] Datasets are supplied in the private communication.

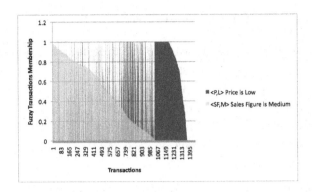

Fig. 2. Fuzzy Transactions of "<P, L>" (Price is Low) and "<SF, M>" (Sales Figure is Medium)

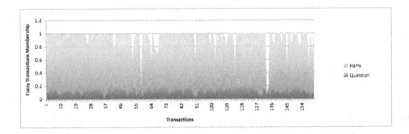

Fig. 3. Fuzzy Transactions of "Hermione" and "Ron"

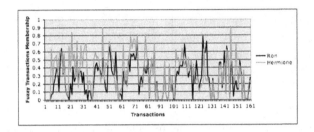

Fig. 4. Fuzzy Transactions of "Harry" and "Question"

as the text items obtained from all the representation documents $d_i \in D$ with their membership to the transaction expressed by $W_i = \{w_{i1}, ..., w_{im}\}$, where w_{ij} is obtained from a weight function such as frequency weighting scheme and the TF-IDF weighting scheme [29]. The fuzzy association rule $term_s \rightarrow term_t$ shall be interpreted as "This means that the appearance of $term_s$ in a document *suggests* the appearance of $term_t$". The experimental results from both mass-based association analysis and quantified sentence approach are shown in Table 10.

However, we argue that mass-based association analysis gives more reasonable values in cases of low overlap. Take the rule $< P, L > \rightarrow < SF, M >$ for example. The QS approach gives its confidence value 0.258. From Figure 2, we can see there are significant areas where the two fuzzy sets hardly overlap. The confidence value 0.093 from mass-based association method is more reasonable. Another result arising from low overlap is given in Figure 3. Although most of the items in the fuzzy set of "Hermione" have the membership lower than 0.6, the QS gives the confidence value 0.683 to the rule $Hermione \rightarrow Ron$. Similar issue can also be found in Figure 4, where QS gives the confidence value 0.140 to the rule $Harry \rightarrow Question$. However, there is no strong association between "Harry" and "Question" as the latter has a low membership across the whole range. Thus the association value 0.07 from the mass-based association method is better to describe this weak association.

5   Summary

We have described a new method for generating association rules between granules in different information hierarchies. These rules enable us to find related categories without leading to spurious relations suggested by association rules based on fuzzy cardinalities. Results were presented for discovery of links between film genres and terrorism incident databases in different classification hierarchies, giving intuitively reasonable associations. The new method is currently undergoing further tests, looking at benchmark instance-matching problems, finding associations between music genres and finding links between categories in different classified business directories.

Acknowledgments. This work was partly funded by BT and the Defence Technology Centre Data and Information Fusion. This is an extended version of a paper published at URSW07, and we would like to thank the reviewers of the original and extended papers for their careful reviews and helpful suggestions.

References

1. Berners-Lee, T., Hendler, J., Lassila, O.: The Semantic Web. Scientific American Magazine (2001)
2. Klyne, G., Carroll, J.J. (eds.): Resource Description Framework (RDF): Concepts and Abstract Syntax. W3C Recommendation, http://www.w3.org/TR/rdf-concepts/
3. Patel-Schneider, P.F., Hayes, P., Horrocks, I.: OWL Web Ontology Language Semantics and Abstract Syntax. W3C Recommendation, http://www.w3.org/TR/owl-semantics
4. Hüllermeier, E.: Fuzzy sets in machine learning and data mining. Applied Soft Computing (in press, 2008)
5. Yao, Y.Y., Zhong, N.: Granular Computing Using Information Tables. In: Lin, T.Y., Yao, Y.Y., Zadeh, L.A. (eds.) Data Mining, Rough Sets and Granular Computing, pp. 102–124. Physica-Verlag, Heidelberg (2002)

6. Zadeh, L.A.: Toward a theory of fuzzy information granulation and its centrality in human reasoning and fuzzy logic. Fuzzy Sets and Systems 90, 111–127 (1997)
7. Abulaish, M., Dey, L.: Interoperability among Distributed Overlapping Ontologies - A Fuzzy Ontology Framework. In: Web Intelligence (WI), pp. 397–403 (2006)
8. Martin, T.P., Azvine, B.: Acquisition of Soft Taxonomies for Intelligent Personal Hierarchies and the Soft Semantic Web. BT Technology Journal 21, 113–122 (2003)
9. Martin, T.P., Azvine, B.: Soft Integration of Information with Semantic Gaps. In: Sanchez, E. (ed.) Fuzzy Logic and the Semantic Web, pp. 150–180. Elsevier, Amsterdam (2005)
10. Bosc, P., Bouchon-Meunier, B.: Introduction: Databases and fuzziness. International Journal of Intelligent Systems 9, 419 (1994)
11. Zadeh, L.: Fuzzy Sets as the Basis for a Theory of Possibility. Fuzzy Sets and Systems 1, 3–28 (1978)
12. Dubois, D., Prade, H.: Possibility Theory, Probability Theory and Multiple-valued Logics: A Clarification. Annals of Mathematics and Artificial Intelligence 32, 35–66 (2001)
13. Agrawal, R., Srikant, R.: Fast Algorithms for Mining Association Rules in Large Databases. In: 20th Int. Conf. Very Large Data Bases (VLDB), pp. 487–499 (1994)
14. Dubois, D., Hüllermeier, E., Prade, H.: A systematic approach to the assessment of fuzzy association rules. Data Mining and Knowledge Discovery 13, 167–192 (2006)
15. Bosc, P., Pivert, O.: On Some Fuzzy Extensions of Association Rules. In: IFSA World Congress and 20th NAFIPS International Conference, vol. 2, pp. 1104–1109 (2001)
16. Kacprzyk, J., Zadrozny, S.: Linguistic Summarization of Data Sets Using Association Rules. In: 12th IEEE International Conference on Fuzzy Systems, vol. 1, pp. 702–707 (2003)
17. Martin-Bautista, M.J., Vila, M.A., Larsen, H.L., Sanchez, D.: Measuring Effectiveness in Fuzzy Information Retrieval. In: 4th International Conference on Flexible Query Answering Systems, pp. 396–402 (2000)
18. Zadeh, L.A.: The concept of a linguistic variable and its application to approximate reasoning, Parts 1- 3. Information Science 8, 9, 199–249, 301–357, 43–80 (1975)
19. Delgado, M., Marin, N., Sanchez, D., Vila, M.A.: Fuzzy Association Rules: General Model and Applications. IEEE transactions on Fuzzy Systems 11, 214–225 (2003)
20. Baldwin, J.F., Lawry, J., Martin, T.P.: Efficient Algorithms for Semantic Unification. In: Proc. Information Processing and the Management of Uncertainty (1996)
21. Baldwin, J.F., Martin, T.P.: Pilsworth, B.W.: FRIL - Fuzzy and Evidential Reasoning in AI. Research Studies Press (1995)
22. Baldwin, J.F.: The Management of Fuzzy and Probabilistic Uncertainties for Knowledge Based Systems. In: Shapiro, S.A. (ed.) Encyclopedia of AI, 2nd edn., pp. 528–537. John Wiley, Chichester (1992)
23. Dubois, D., Prade, H.: On Several Representations of an Uncertain Body of Evidence. In: Gupta, M.M., Sanchez, E. (eds.) Fuzzy Information and Decision Processes, pp. 167–181. North-Holland Publishing Company, Amsterdam (1982)
24. Delgado, M., Sanchez, D., Martin-Bautista, M.J., Vila, M.A.: A probabilistic definition of a nonconvex fuzzy cardinality. Fuzzy Sets and Systems 126, 177–190 (2002)
25. Baldwin, J.F., Lawry, J., Martin, T.P.: A Mass Assignment Theory of the Probability of Fuzzy Events. Fuzzy Sets and Systems 83, 353–367 (1996)

26. Martin, T.P., Shen, Y.: Improving access to multimedia using multi-source hierarchical meta-data. In: Detyniecki, M., Jose, J.M., Nürnberger, A., van Rijsbergen, C.J.'. (eds.) AMR 2005. LNCS, vol. 3877, pp. 266–278. Springer, Heidelberg (2006)
27. Martin, T.P., Azvine, B., Shen, Y.: Incremental Evolution of Fuzzy Grammar Fragments to Enhance Instance Matching and Text Mining. In: IEEE Transactions on Fuzzy Systems (in press, 2008)
28. Novák, D.: Intensional Theory of Granular Computing. Soft Computing 8, 281–290 (2003)
29. Martin-Bautista, M.J., Sanchez, D., Chamorro-Martinez, J., Serrano, J.M., Vila, M.A.: Mining Web Documents to Find Additional Query Terms Using Fuzzy Association Rules. Fuzzy Sets and Systems 148, 85–104 (2004)

A Fuzzy Semantics for the Resource Description Framework

Mauro Mazzieri and Aldo Franco Dragoni

Università Politecnica delle Marche
{m.mazzieri,a.f.dragoni}@univpm.it

Abstract. Semantic Web languages cannot currently represent vague or uncertain information. However, their crisp model-theoretic semantics can be extended to represent uncertainty in much the same way first-order logic was extended to fuzzy logic. We show how the interpretation of an RDF graph (or an RDF Schema ontology) can be a matter of values, addressing a common problem in real-life knowledge management. While unmodified RDF triples can be interpreted according to the new semantics, an extended syntax is needed in order to store fuzzy membership values within the statements. We give conditions an extended interpretation must meet to be a model of an extended graph. Reasoning in the resulting fuzzy languages can be implemented by current inferencers with minimal adaptations.

Keywords: Fuzzy Logic, Knowledge Representation, Semantic Web, RDF, RDF Schema.

1 Knowledge Representation on the Web

The Semantic Web is an extension of the current web in which information is given well-defined meaning[1] by the use of knowledge representation (KR) languages.

The KR languages used (RDF, RDF Schema and OWL) have the characteristics that make them useful on the web[2]:

- the elements of the domain are represented by URIs;
- there is no global coherence requirement, as local sources can make assertions independently without affecting each other's expressiveness.

Those languages have the ability to describe, albeit not formally, much more than their semantics can express. Their model theory captures only a formal notion of meaning, given by inference rules; the exact 'meaning' of a statement can depend on many factors, not all accessible to machine processing [3].

This feature can be exploited also to represent information from fields that require knowledge representation paradigms other than the FOL-like RDF Model Theory or the expressive Description Logics used by OWL. Amongst those paradigms there is fuzzy logic, to represent vague or ambiguous knowledge.

P.C.G. da Costa et al. (Eds.): URSW 2005-2007, LNAI 5327, pp. 244–261, 2008.

There is an extensive literature on fuzzy semantics for OWL. OWL is a standard semantic web language [4] based on description logics [5]. The original idea of generalizing term subsumption languages (ancestors of description logics) is from Yen [6]; Straccia defined a fuzzy version of the DL $\mathcal{ALC}$ [7] and then of the expressive logic $\mathcal{SHOIN}(\mathbf{D})$ [8]. This allowed the definition of fuzzy OWL by Stoilo et. al. [9]. Other works on fuzzy description logics are [10,11,12,13,14,15,16,17,18].

RDF is the basic semantic web language, with a non-standard model theory defined in [3]. The idea of a fuzzy extension of RDF model theory is from Mazzieri [19]. This paper is a revision and extension of the fuzzy RDF model theory as presented in [20]. A comparable approach is *Annotated RDF* by Udrea et. al. [21]; their approach is swallower (only a limited subset of RDF Schema is extended) but broader (annotation can be from any partially ordered set, fuzzy annotation is one of the possible scenarios).

Next sections are structured as follows. Section 2 is a summary of RDF and RDF Schema syntax and semantics. In Section 3 we define an extended interpretation for RDF graphs, an extended syntax and an interpretation for extended graphs. Section 4 defines an interpretation for fuzzy RDF Schema ontology language, with emphasis on some subtleties in the interpretation of language terms involving set inclusion. A proof-of-concept implementation of the extended syntax and semantics within an RDF/RDF Schema inferencer is described in Section 5.

2   RDF Model Theory

This is a short summary of RDF syntax and semantics, to make the paper self-contained. Normative references are [22] and [3].

2.1   Syntax

An RDF graph is a set of RDF *triples* (or *statements*). A triple consists of a subject, a predicate and an object. Subjects and objects are called *nodes* and can be a literal, an URI reference [23] or a blank node (a unique node with no name). Predicates are URI references.

2.2   Simple Interpretation

An interpretation assign meaning to a set of names in a vocabulary. An interpretation is "simple" if it does not assign special meanings to any particular set of names. A simple interpretation I of a vocabulary V defines an non empty set IR of resources, a set IP of properties, a mapping IEXT from IP to IR×IR, a mapping IS from URI references in V to IR ∪ IP, a mapping IL from typed literals in V into IR, and a distinguished subset LV of IR containing the literal values.

Properties, unlike first-order predicates, are not relations, but elements of the domain, which have an *extension* IEXT, which in turn is a relation on couples of elements. The trick of distinguishing a property as a domain element from its relational extension allows a property to occur in its own extension.

2.3 Denotations for Ground Graphs

A graph with no blank nodes is called *ground*. Truth or falsity of a ground RDF graph in an interpretation is given recursively from its syntactic constituents by a set of rules: literals correspond to themselves, typed literals are mapped by IL, URI references are mapped by IS. A ground triple is true if the subject **s**, the predicate **p** and the object **o** are in the vocabulary V of I, the interpretation of the property I(**p**) is in IP, and the couple $\langle I(\mathbf{s}), I(\mathbf{o}) \rangle$ is in IEXT(I(**p**)), otherwise is false. A ground graph is false (given an interpretation) if some triple of the graph is false, otherwise is true. Interpretations for which the graph is true are called *models*.

Blank nodes are interpreted as existential variables. A non-ground graph is true, given an interpretation I, if there is a mapping from blank nodes to IR for which the graph is true, otherwise is false.

2.4 Entailment

The key idea for RDF reasoning is *entailment*: a set of graphs S entails graph G if every model of every member of S is also a model of G.

Section 2 of RDF Semantics [3] shows many lemmas that apply to simple interpretations.

Lemma 1 (Empty Graph). *The empty set of triples is entailed by any graph, and does not entail any graph except itself.*

A *subgraph* of a graph is a subset of the triples of the graph.

Lemma 2 (Subgraph). *A graph entails all its subgraphs.*

A graph obtained by a graph G replacing some of its blank nodes with literals, blank nodes and URI references is an *instance* of G.

Lemma 3 (Instance). *A graph is entailed by any of its instances.*

The *merge* of graphs is different from a simple union since blank nodes are existentially quantified in the scope of a single graph.

Lemma 4 (Merging). *The merge of a set S of RDF graphs is entailed by S, and entails every member of S.*

Lemma 5 (Interpolation). *S entails a graph E if and only if a subgraph of S is an instance of E.*

An instance of a graph is *proper* if a blank node has been replaced by a name or two blank nodes have been mapped in the same node. A *lean* graph is a graph with no instance which is a proper subgraph of the graph.

Lemma 6 (Anonymity). *Suppose E is a lean graph and E' is a proper instance of E. Then E does not entail E'.*

Lemma 7 (Monotonicity). *Suppose S is a subgraph of S' and S entails E. Then S' entails E.*

Lemma 8 (Compactness). *If S entails E and E is a finite graph, then some finite subset S' of S entails E.*

Proofs for the lemmas are in [3]. Proofs of interpolation, anonymity, monotonicity and compactness lemma need a way to build a model of a graph using lexical items in the graph itself, the so called *Herbrand interpretation* [24].

2.5 RDF Interpretation

RDF-interpretations give particular meaning to URI references in the RDF vocabulary, made of URI in the $\mathtt{rdf:}$ namespace[1]. An RDF interpretation must satisfy all the axiomatic triples in Table 1 and the following extra conditions:

- $x \in IP$ if and only if $\langle x, \, I(\mathtt{rdf:Property})) \rangle \in IEXT(I(\mathtt{rdf:type}))$
- If $\mathtt{"xxx"\hat{}\hat{}rdf:XMLLiteral}$ is in V and $\mathtt{xxx}$ is a well-typed XML literal string, then
  - $IL(\mathtt{"xxx"\hat{}\hat{}rdf:XMLLiteral})$ is the XML value of $\mathtt{xxx}$;
  - $IL(\mathtt{"xxx"\hat{}\hat{}rdf:XMLLiteral}) \in LV$;
  - $\langle IL(\mathtt{"xxx"\hat{}\hat{}rdf:XMLLiteral}), \, I(\mathtt{rdf:XMLLiteral}) \rangle$ $\in IEXT(I(\mathtt{rdf:type}))$
- If $\mathtt{"xxx"\hat{}\hat{}rdf:XMLLiteral}$ is in V and $\mathtt{xxx}$ is an ill-typed XML literal string, then
  - $IL(\mathtt{"xxx"\hat{}\hat{}rdf:XMLLiteral}) \notin LV$;
  - $\langle IL(\mathtt{"xxx"\hat{}\hat{}rdf:XMLLiteral}), \, I(\mathtt{rdf:XMLLiteral}) \rangle$ $\notin IEXT(I(\mathtt{rdf:type}))$

Table 1. RDF axiomatic triples

```
rdf:type        rdf:type rdf:Property .
rdf:subject     rdf:type rdf:Property .
rdf:predicate   rdf:type rdf:Property .
rdf:object      rdf:type rdf:Property .
rdf:first       rdf:type rdf:Property .
rdf:rest        rdf:type rdf:Property .
rdf:value       rdf:type rdf:Property .
rdf:_1          rdf:type rdf:Property .
rdf:_2          rdf:type rdf:Property .
                  . . .
rdf:nil         rdf:type rdf:List      .
```

A consequence of the first condition is that, in RDF, IP is a subset of IR.

---

[1] $\mathtt{http://www.w3.org/1999/02/22\text{-}rdf\text{-}syntax\text{-}ns\#}$

2.6 RDF Schema Interpretation

RDF Schema has a larger vocabulary then RDF, composed of URIs in the `rdfs:` namespace[2]. The semantics is conveniently expressed in terms of classes: a *class* is a resource with a class extension ICEXT, which is a subset of IR. As a consequence of this definition, a class can have itself as a member. The relation between a class and a class member is given using the RDF vocabulary property `rdf:type`, and the set of all classes is IC.

An RDF Schema interpretation is an RDF interpretation which satisfies RDFS axiomatic triples of Table 2 and the following conditions:

- $x \in ICEXT(y)$ if and only if $\langle x, y \rangle \in IEXT(I(\text{rdf}:\text{type}))$
  - $IC = ICEXT(I(\text{rdfs}:\text{Class}))$
  - $IR = ICEXT(I(\text{rdfs}:\text{Resource}))$
  - $IL = ICEXT(I(\text{rdfs}:\text{Literal}))$
- If $\langle x, y \rangle \in IEXT(I(\text{rdfs}:\text{domain}))$ and $\langle u, v \rangle \in IEXT(x)$
  then $u \in ICEXT(y)$
- If $\langle x, y \rangle \in IEXT(I(\text{rdfs}:\text{range}))$ and $\langle u, v \rangle \in IEXT(x)$
  then $v \in ICEXT(y)$
- $IEXT(I(\text{rdfs}:\text{subPropertyOf}))$ is transitive and reflexive on IP
- If $\langle x, y \rangle \in IEXT(\text{rdfs}:\text{subPropertyOf})$ then $x, y \in IP$ and $IEXT(x) \subseteq IEXT(y)$
- If $x \in IC$ then $\langle x, I(\text{rdfs}:\text{Resource})\rangle \in IEXT(I(\text{rdfs}:\text{subClassOf}))$
- If $\langle x, y \rangle \in IEXT(I(\text{rdfs}:\text{subClassOf}))$ then $x, y \in IC$ and $ICEXT(x) \subseteq ICEXT(y)$
- $IEXT(I(\text{rdfs}:\text{subClassOf}))$ is transitive and reflexive on IC
- If $x \in ICEXT(I(\text{rdfs}:\text{ContainerMembershipProperty}))$
  then $\langle x, I(\text{rdfs}:\text{member})\rangle \in IEXT(I(\text{rdfs}:\text{subPropertyOf}))$
- If $x \in ICEXT(I(\text{rdfs}:\text{Datatype}))$
  then $\langle x, I(\text{rdfs}:\text{Literal})\rangle \in IEXT(I(\text{rdfs}:\text{subClassOf}))$

3 Fuzzy RDF

In this section we will introduce fuzzy RDF as follows. In 3.1 and 3.2 we will first show how a a fuzzy interpretation can be given to plain RDF graphs. Then, after introducing a syntax for fuzzy graphs in 3.3, we give an interpretation for them in 3.4, 3.5, and 3.6.

3.1 Simple Fuzzy Interpretation

The intuitive meaning of a statement is that a relation, given by the property, holds between the subject and the object. An uncertainty on a statement must thus be reflected by an uncertainty on the subsistence of the relation between subject and object.

---

[2] `http://www.w3.org/2000/01/rdf-schema#`

Table 2. RDF axiomatic statements

rdf:type	rdfs:domain	rdfs:Resource	.
rdfs:domain	rdfs:domain	rdf:Property	.
rdfs:range	rdfs:domain	rdf:Property	.
rdfs:subPropertyOf	rdfs:domain	rdf:Property	.
rdfs:subClassOf	rdfs:domain	rdfs:Class	.
rdf:subject	rdfs:domain	rdf:Statement	.
rdf:predicate	rdfs:domain	rdf:Statement	.
rdf:object	rdfs:domain	rdf:Statement	.
rdfs:member	rdfs:domain	rdfs:Resource	.
rdf:first	rdfs:domain	rdf:List	.
rdf:rest	rdfs:domain	rdf:List	.
rdfs:seeAlso	rdfs:domain	rdfs:Resource	.
rdfs:isDefinedBy	rdfs:domain	rdfs:Resource	.
rdfs:comment	rdfs:domain	rdfs:Resource	.
rdfs:label	rdfs:domain	rdfs:Resource	.
rdf:value rdfs:domain	rdfs:Resource	.	
rdf:type	rdfs:range	rdfs:Class	.
rdfs:domain	rdfs:range	rdfs:Class	.
rdfs:range	rdfs:range	rdfs:Class	.
rdfs:subPropertyOf	rdfs:range	rdf:Property	.
rdfs:subClassOf	rdfs:range	rdfs:Class	.
rdf:subject	rdfs:range	rdfs:Resource	.
rdf:predicate	rdfs:range	rdfs:Resource	.
rdf:object	rdfs:range	rdfs:Resource	.
rdfs:member	rdfs:range	rdfs:Resource	.
rdf:first	rdfs:range	rdfs:Resource	.
rdf:rest	rdfs:range	rdf:List	.
rdfs:seeAlso	rdfs:range	rdfs:Resource	.
rdfs:isDefinedBy	rdfs:range	rdfs:Resource	.
rdfs:comment	rdfs:range	rdfs:Literal	.
rdfs:label	rdfs:range	rdfs:Literal	.
rdf:value	rdfs:range	rdfs:Resource	.
rdf:Alt	rdfs:subClassOf	rdfs:Container	.
rdf:Bag	rdfs:subClassOf	rdfs:Container	.
rdf:Seq	rdfs:subClassOf	rdfs:Container	.
rdfs:ContainerMembershipProperty	rdfs:subClassOf	rdf:Property	.
rdfs:isDefinedBy	rdfs:subPropertyOf	rdfs:seeAlso	.
rdf:XMLLiteral	rdf:type	rdfs:Datatype	.
rdf:XMLLiteral	rdfs:subClassOf	rdfs:Literal	.
rdfs:Datatype	rdfs:subClassOf	rdfs:Class	.
rdf:_1	rdf:type	rdfs:ContainerMembershipProperty	.
rdf:_1	rdfs:domain	rdfs:Resource	.
rdf:_1	rdfs:range	rdfs:Resource	.
rdf:_2	rdf:type	rdfs:ContainerMembershipProperty	.
rdf:_2	rdfs:domain	rdfs:Resource	.
rdf:_2	rdfs:range	rdfs:Resource	.
. . .			

Starting from RDF syntax and model theory, a new theory is developed with the aim to make the minimal amount of changes needed; in the following, the emphasis will be on the part where it branches from RDF semantics.

In fuzzy RDF model theory, a couple ⟨subject, object⟩ has a membership degree to the extension of the predicate. Membership degrees will be taken, without loosing generality, as real numbers in the interval [0, 1]. Thus, the extension is not an ordinary set of couples anymore, but a fuzzy set of couples.

We have chosen not to make the mapping between vocabulary items and domain elements fuzzy. Instead, the membership of a resource to the domain is fuzzy. This is a step which poses some theoretical problems, in particular when we have to deal with properties in simple interpretations. In RDF interpretations, the property domain IP is a subset of the resource domain IR, so in fuzzy RDF interpretations would be enough to make IP a fuzzy subset of IR; in simple interpretations, however, there is no formal relation between IP and IR, so when the mapping IS from URI references to $IR \cup IP$ becomes fuzzy we need a further device. The chosen solution is to define a domain IDP for properties, so that IP is a fuzzy subset of IDP, and to modify the definition of IS to a mapping from URI references in V to $IR \cup IDP$. RDF interpretations will not need IDP, as IP can be shown to be a fuzzy subset of IR. The membership degree of a resource to IP is intuitively related to the use of the resource as a property; this would be clearer later.

Definition 1 (simple interpretation)

A simple fuzzy interpretation I_f of a vocabulary V is defined by:

1. *A non-empty set IR of resources, called the domain or universe of I_f.*
2. *A set IDP, called the property domain of I_f.*
3. *A fuzzy subset IP of IDP, called the set of properties of I_f.*
4. *A fuzzy mapping $IEXT$ from property domain IDP to the powerset of $IR \times IR$, i.e. the set of pairs $\langle x, y \rangle$ with $x, y \in IR$.*
5. *A mapping IS from URI references in V to $IR \cup IDP$.*
6. *A mapping IL from typed literals in V to IR.*
7. *A distinguished subset $LV \subseteq IR$, called the set of literal values, which contains all the plain literals of V.*

3.2 Fuzzy Denotations for Ground Graphs

The truth of a ground triple in fuzzy RDF is given by a value. The degree of satisfaction for a statement s p o. is the degree of membership of the couple, formed by the interpretation of the subject and the interpretation of the object, to the extension of the interpretation of the predicate.

We will use the abbreviated Zadeh's notation $A(x) = n$, instead of $\mu_A(x) = n$, to state that the membership degree of the element x to the set A is equal to n [25].

Denotation of ground graph from syntactic constituents is the same as in 2.3; only the condition of truth and falsity of a ground statement in the interpretation is affected. Truth values for statement under a fuzzy interpretation is given as follows:

Definition 2 (Semantic conditions for ground graphs). *If E is a ground triple s p o., and s, p and o are in the vocabulary V of a fuzzy interpretation I_f, then $I_f(E) = \min\{IP(I_f(p)), IEXT(I_f(p))(\langle I_f(s), I_f(o) \rangle)\}$, otherwise $I_f(E) = 0$.*

If E is a ground RDF graph, than $I_f(E) = \min_{E' \in E}\{I_f(E')\}$.

From the given definition of denotation of a statement follows that a resource p used as a property in some statement must belong to set of properties with a degree IP(p) greater then or equal each statement's value.

The truth of a graph is defined as the minimum of statements' truth values. This, as every semantics based on minima with respect of a population, can give unreasonable results when a single statements has atypically low truth value. For example, given an interpretation I_f, if there is a statement E' with interpretation much smaller than the interpretation of every other statement in the graph, the interpretation of the entire graph E is the same as the interpretation of E'.

This behavior reflects the fact that a chain (of inferences) is only as strong as its weakest link; or, as in classical philosophical logic, *pejorem semper sequitur conclusio partem* (the conclusion follows always from the weaker part). A semantics based on minima of function cannot be avoided in fuzzy logic, as any other t-norm would give values smaller then or equal to min[3].

3.3 Syntax for Fuzzy Graphs

So far we have seen how a plain graph can be given a fuzzy interpretation. However, it is not clear how membership values must be chosen, or at least how membership values can be given or represented.

To represent uncertainty at the triple level, the RDF syntax must be extended to add to the triple ⟨subject, predicate, object⟩ the membership value. This is not an extension from a 3-elements tuple to a 4-elements tuple as it may seem at a first glance; the added element has a syntactic nature different from the others: it is not an element of the domain of the discourse, but a property related to the formalism used by the language to represent uncertainty and vagueness.

To show the extended syntax we will modify the N-Triples [29] concrete syntax, a line-based, plain text format for encoding an RDF graph, used to show RDF test cases. in the original for a statement is **s p o.**, where **s**, **p** and **o** are respectively the subject, the predicate and the object of the statement. Our extended syntax add an optional prefix n:, where n is a decimal number representing the fuzzy truth-value of the triple. The use of decimal numbers instead of real numbers is only a limitation of the syntax and does not undermine the discussion. Modifications to N-Triples' EBNF are given in Table 3.

Table 3. Modification to N-Triples EBNF for Fuzzy N-Triples. Definitions of white space (ws), comment, end-of-line characters (eoln), subject, predicate and object are unchanged from [29] and not shown here.

fuzzyNtripleDoc	::= line*
line	::= ws* (comment | statement)? eoln
statement	::= (value ':' ws+)? subject ws+ predicate ws+ object ws* '.' ws*
value	::= 1 | 0.[0–9]+

The term *triple*, used in the EBNF for N-Triple, is replaced with the more generic term *statement*. Triple and statement are often used in semantic web literature as synonyms, but we prefer to use the latter to avoid confusion between a

---

[3] For any other operator $\triangle$ such that

$$x \triangle 1 = 1 \triangle x = x \quad \text{and}$$

$$x_1 \triangle y_1 \leq x_2 \triangle y_2 \text{ if } x_1 \leq x_2 \text{ and } y_1 \leq y_2 \quad \text{(monotonicity)}$$

it can be easily shown that $x \triangle y \leq \min\{x, y\}$. See [26] or any introductive text in fuzzy logic, such as [27] or [28].

plain RDF statement (made actually of three parts) and a fuzzy RDF statement (that, although still a triple semantically, is made up of four elements).

The fuzzy *value* is defined as optional. This way, the syntax is backward-compatible; the intended semantics is that a statement with the form s p o. is equivalent to the statement 1: s p o.. With such a (syntactic only) default, and the defined semantics, we could take an inference engine implementing fuzzy RDF, let it parse plain RDF statements, and get the same results of a conventional RDF inference engine. Furthermore, as it would be clear in the description of fuzzy RDF inference rules (section 5), even the complexity of the computation would be of the same order.

We will not give an abstract syntax, as in [22], nor a RDF/XML based syntax, as in [30]. All "physical" data (i.e., data transmitted between host or processes) is supposed to be encoded using plain RDF reified statements[4]. The extended syntax is to be used only to write examples and test cases.

3.4   Denotations for Fuzzy Ground Graphs

Denotation for statements and graphs, in the case of fuzzy graphs, is modified as follows:

Definition 3 (Semantic conditions for ground graphs). *If E is a ground triple s p o., and s, p and o are in the vocabulary V of a fuzzy interpretation I, then $I(E) \leq \min\{IP(I(p)), IEXT(I(p))(\langle I(s), I(o)\rangle)\}$, otherwise $I(E) = 0$. If E is a ground RDF graph, than $I(E) = \min_{E' \in E}\{I(E')\}$.*

Given this denotation, a statement n: s p o. intuitively corresponds to a restriction on interpretations imposing that membership of p to IP must be at least n, and membership of $\langle s, o \rangle$ to $IEXT(p)$ must be at least n[5]

---

[4] To encode a fuzzy statements in pain RDF, the value is given to the reification of the statement by a property. Let's call `fuzzyValue` the property denoting the fuzzy membership value of a statement. The encoding of n: s p o. would be:

```
_:xxx rdf:type rdf:Statement .
_:xxx rdf:subject s .
_:xxx rdf:predicate p .
_:xxx rdf:object o .
_:xxx fuzzyValue n .
```

In the example, `_:xxx` is a blank node and n a placeholder for a typed literal with a numerical value.

[5] In fact, an alternative (maybe clearer syntax) for statement n: s p o. would be

$$s\ p\ o\ \geq\ n .$$

At this point, why not introduce also the opposite statements, with the $\leq$ instead of $\geq$? However, the resulting mixing of semantic conditions would introduce a problem usually non relevant in plain RDF: *consistency*. While every RDF graph has at least a simple interpretation that is a model (the Herbrand interpretation), a graph made by the statements s p o $\geq$ n_1 . and s p o $\leq$ n_2 ., with $n_1 > n_2$, would have no models..

Property 1. A graph where the same statement appears more than once, with different values, is equivalent to a graph where the statement appears only once, with a value equal to the maximum of the membership degrees.

3.5   Simple Entailment among Fuzzy Graphs

The definition of simple entailment between RDF graphs is not affected. A set S of RDF graphs *(simply) entails* a graph E if every interpretation which satisfies every member of S also satisfies E.

Property 2. If a set S of fuzzy RDF graphs entails a statement **n:** **s p o.**, then S entails any statement **n':** **s p o.** with $n' \le n$.

Definition 4. *If S is a set of fuzzy RDF graphs and E is the plain statement* **s p o.**, *the* minimum value *of E (with respect to S) is the minimum value n such that S entails* **n:** **s p o.**.

Property 3. If E is the plain statement **s p o.** and n is the minumum value E with respect to a set S of fuzzy RDF graphs, then S entails any statement **n':** **s p o.** with $n' \le n$.

Lemmas that apply to simple interpretation (§3.5) retain their validity within fuzzy RDF Model Theory. Even the proofs from appendix A of [3], with minimal syntactical adjustments, are the same.

   Regarding the empty graph lemma, it is still valid is we keep the same definition of empty graph as in plain RDF:

Definition 5. *An* empty fuzzy graph *is a graph containing no statements.*

An empty fuzzy graph can not be defined as a graph with no not-zero-valued statements. Statements such as **0:** **s p o.** cannot be ignored, as the semantic requirements that **s**, **p** and **o** must belong to the graph's vocabulary still apply.

   Proofs of interpolation, anonymity, monotonicity and compactness lemma need to build an Herbrand interpretation. The fuzzy version of the Herbrand interpretation can be defined as in Def. 6.

Definition 6. *Given a nonempty fuzzy graph G, the (simple) fuzzy Herbrand interpretation of G, written* Herb(G), *is the interpretation defined as follows.*

- $LV_{\text{Herb}}(G)$ *is the set of all plain literals in G;*
- $IR_{\text{Herb}}(G)$ *is the set of all names and blank nodes which occur in subject or object position of statements in G;*
- $IDP_{\text{Herb}}(G)$ *is the set of URI references which occur in the property position of statements in G;*
- $IP_{\text{Herb}(G)}(\mathbf{p})$ *is the maximum of n for all statements in which $\mathbf{p}$ occur in property position;*
- $IEXT_{\text{Herb}(G)}(\langle \mathbf{s}, \mathbf{o} \rangle)$ *is the maximum n for all the statements* **n:** **s p o.** *in G*
- $IS_{\text{Herb}}(G)$ *and* $IL_{\text{Herb}}(G)$ *are both identity mappings on the appropriate parts of the vocabulary of G.*

3.6   Fuzzy RDF Interpretation

A fuzzy RDF interpretation must satisfy all the RDF axiomatic triples in Table 1 and the following conditions:

Definition 7 (Fuzzy RDF Semantic Conditions)

- $IP(x) = IEXT(I(\text{rdf}:\text{type}))(\langle x, I(\text{rdf}:\text{Property}))$
- If $"xxx"\hat{}\hat{}rdf\!:\!XMLLiteral$ is in V and xxx is a well-typed XML literal string, then
    - $IL("xxx"\hat{}\hat{}rdf\!:\!XMLLiteral)$ is the XML value of xxx;
    - $IL("xxx"\hat{}\hat{}rdf\!:\!XMLLiteral) \in LV$;
    - $IEXT(I(\text{rdf}:\text{type}))(\langle IL("xxx"\hat{}\hat{}rdf\!:\!XMLLiteral),$
      $I(\text{rdf}:\text{XMLLiteral}))) = 1$
- If $"xxx"\hat{}\hat{}rdf\!:\!XMLLiteral$ is in V and xxx is an ill-typed XML literal string, then
    - $IL("xxx"\hat{}\hat{}rdf\!:\!XMLLiteral) \notin LV$;
    - $IEXT(I(\text{rdf}:\text{type}))(\langle IL("xxx"\hat{}\hat{}rdf\!:\!XMLLiteral),$
      $I(\text{rdf}:\text{XMLLiteral}))) = 0$

As the first RDF semantic condition has the consequence that IP must be a subset of IR, there is no more need of IDP, and IP can be directly defined as a fuzzy subset of IR. Membership to class rdf:Property is defined as equivalent to membership to IP.

The second and third conditions amounts to seeing the well-formedness of an XML Literal as crisp truth-valued. As the algorithm to classify an XML literal as well-formed or not is deterministic, the classification is crisp.

By definition, axiomatic triples have a unit truth value. Given the (syntactic) convention that a triple s p o. is equivalent to the fuzzy statement 1: s p o., the set of fuzzy RDF axiomatic statements is the same as plain RDF (Table 1).

4   Fuzzy RDF Schema

The path from RDF Schema to fuzzy RDF Schema follows the same guidelines of the previous section: the main concern is the choice of the characteristic function to soften in order to capture a fuzzy semantic.

While in RDF the main construct is the *extension*, the RDF Schema semantics is stated in terms of *classes* [3].

As a class is a resource with a *class extension*, $ICEXT$, which represents a set of domain element, the definition of class relies on the definition of extension. If an extension is a set of couples, and a fuzzy extension is a fuzzy set of couples, fuzzy class extensions in RDF Schema are fuzzy sets of domain's elements.

4.1 RDF Schema Interpretation

An RDF Schema interpretation defines the domains for resources (IR), literals (IL) and literal values (LV) in terms of fuzzy classes, i.e. fuzzy subdomains of IR.

After showing the semantic conditions, we will try to explain the more problematic definitions (domains and ranges in §4.3, subproperties and subclasses in §4.4) in terms of a definition of *inclusion grade* between fuzzy subsets (§4.2).

A fuzzy RDF Schema interpretation is a fuzzy RDF interpretation which satisfies RDFS axiomatic triples of Table 2 and the following conditions:

Definition 8 (RDFS semantic conditions)

- $ICEXT(y)(x) = IEXT(I(\mathtt{rdf:type}))(\langle x,\ y\rangle)$
  - $IC = ICEXT(I(\mathtt{rdfs:Class}))$
  - $IR = ICEXT(I(\mathtt{rdfs:Resource}))$
  - $IL = ICEXT(I(\mathtt{rdfs:Literal}))$
- $\inf_{a\in IR}min\{1,\ 1\ -\ max_{b\in IR}IEXT(x)(\langle a,\ b\rangle)\ +\ ICEXT(y)(a)\} \geq$ $IEXT(I(\mathtt{rdfs:domain}))(\langle x,\ y\rangle)$
- $\inf_{b\in IR}min\{1,\ 1\ -\ max_{a\in IR}IEXT(x)(\langle a,\ b\rangle)\ +\ ICEXT(y)(b)\} \geq$ $IEXT(I(\mathtt{rdfs:range}))(\langle x,\ y\rangle)$
- $IEXT(I(\mathtt{rdfs:subPropertyOf}))$ *is transitive and reflexive on* IP
- If $IEXT(\mathtt{rdfs:subPropertyOf})(\langle x,\ y\rangle) = n$, then $IP(x) \geq n$, $IP(y) \geq n$, $\inf_{a,\ b\in IR}\ min\{1,\ 1\ -\ IEXT(x)(\langle a,\ b\rangle)\ +\ IEXT(y)(\langle a,\ b\rangle)\} \geq n$
- $IEXT(I(\mathtt{rdfs:subClassOf}))(\langle x,\ I(\mathtt{rdfs:Resource})\rangle) = IC(x)$
- If $IEXT(\mathtt{rdfs:subClassOf})(\langle x,\ y\rangle) = n$, then $IC(x) \geq n$, $IC(y) \geq n$, $\inf_{a\in IR}min(1,\ 1\ -\ ICEXT(x)(a) + ICEXT(y)(a)\} \geq n.$
- $IEXT(I(\mathit{rdfs:subClassOf}))$ *is transitive and reflexive on* IC
- $IEXT(I(\mathtt{rdfs:subPropertyOf}))(\langle x,\ I(\mathtt{rdfs:member})\rangle) = ICEXT(I(\mathtt{rdfs:ContainerMembershipProperty}))(x)$
- $ICEXT(I(\mathtt{rdfs:Datatype}))(x) = IEXT(I(\mathtt{rdfs:subClassOf}))(\langle x, I(\mathtt{rdfs:Literal})\rangle)$

As for fuzzy RDF, given the syntactic default, the set of fuzzy RDF Schema axiomatic statements is the same as plain RDF Schema (Table 2).

4.2 Inclusion Grades

To define the semantics of some RDF Schema vocabulary terms (**domain** and **range**, **subClassOf** and **subPropertyOf**), we need a relation of set inclusion between fuzzy sets that takes into account also the degree of the relation of inclusion itself. This relation must be transitive and reflexive.

Zadeh's definition of fuzzy subset [31][6] ($A \subseteq B \iff \forall x \in X\quad A(x) \leq B(x)$) is transitive and reflexive, but it is a rigid definition: either the set A is a subset of B, or not. What we need is instead a weaker fuzzy subset relation; a relation that reduces to the Zadeh's one when the inclusion is certain. It must also maintain the reflexivity and transitivity properties.

---

[6] Again, we use the abbreviation $A(x)$ for the membership function $\mu_A(x)$.

Dubois and Prade [32] define *weak inclusion* $\prec_\alpha$ as

$$A \prec_\alpha B \iff x \in (\overline{A} \cup B)_\alpha \; \forall x \in X \;,$$

where α is a parameter and $(\cdot)_\alpha$ is the α-cut[7]. However, this relation is transitive only for $\alpha > \frac{1}{2}$.

Other definitions of weak inclusion make use of *inclusion grades*. An inclusion grade $I(A, B)$ is a scalar measure of the inclusion of the set A in the set B. In general, $A \subseteq_\alpha B$ iff $I(A, B) \geq \alpha$, where $\subseteq_\alpha$ denote a weak inclusion with inclusion grade α.

We have chosen to use the inclusion grade defined in [32] as:

$$I(A, B) = \inf_{x \in X} \overline{(A \mid - \mid B)}(x)$$

where inf is the infimum and $\mid - \mid$ is the *bounded difference*[8].

When $A \subseteq B$, $I(A, B) = 1$ [32]. This inclusion grade could also be written as $I(A, B) = \inf_{x \in X}(1 - \max(0, \ A(x) - B(x))) = \inf_{x \in X}\min(1, \ (1 - A(x) + B(x)))$; in this form corresponds to Lukasiewicz implication operator, a base operator around whom other generalized implication operators are built [34].

It could be interesting to ask how much this definition differs from the condition for classical fuzzy subsets, i.e. $A(x) \leq B(x)$. If $A \subseteq B$, then $I(A, B) = 1$ and the inclusion holds for any grade.

Let's call $d(x)$ the difference $d(x) = A(x) - B(x)$, so that $1 - A(x) + B(x) = 1 - d(x)$. A requirement that inclusion grade between A and B must be at least n could be written as

$$\inf_{x \in X}\min\{1, 1 - d(x)\} \geq n \quad .$$

But $n \leq 1$, so the maximum value of $d(x)$ for which this condition holds is equal to $1 - n$. Finally, if n is the required inclusion, $1 - n$ can represent the lack of inclusion, and we can conclude that the maximum allowable difference between $A(x)$ and $B(x)$ is equal to the lack of inclusion allowed between A and B.

4.3   Domains and Ranges

In non-fuzzy RDF Schema, if a resource u is used as subject for a statement whose property is x, then u must belong to the domain of x. This condition could be stated in term of sets. First, we can define dom(p) (the "actual" domain) as the set of resources used as subject for some statement whose property is p:

$$\mathrm{dom}(p) = \{s : \langle s, o \rangle \in IEXT(p) \text{ for some } o \in IR\}$$

Thus the condition for domains is that the "actual" domain is a subset of the stated domain: if p has domain y, then $\mathrm{dom}(p) \subseteq ICEXT(y)$.

---

[7] The *α-cut* A_α of A is the set of all elements with a membership value to A greater than α, with $\alpha \in (0, 1]$ $A_\alpha = \{x \mid A(x) \geq \alpha\}$.

[8] $\forall x \in X, \quad (A \mid - \mid B)(x) = \max(0, \ A(x) - B(x))$ [33].

In fuzzy set theory, the definition of domain for a fuzzy relation R on $X \times Y$ is $\text{dom}(R)(x) = \sup_y R(x,y)$ [32], i.e. the least upper bound of $R(x,y)$ for all y. We can thus define the fuzzy set of resources used as subject for some statement whose property is p:

$$\text{dom}(p)(s) = \max_{o \in IR}\{IEXT(p)(\langle s,o \rangle)\}$$

In fuzzy RDF Schema a simple condition such as $\text{dom}(p) \subseteq ICEXT(y)$, i.e. the actual domain required to be fuzzy subset of the stated domain, is not enough. We have to deal both with a fuzzy notion of domain, and with a fuzzy assignment of a domain to a property. The resulting condition is the combination of the two requirements, using the aforementioned notion of inclusion grade.

The semantic condition for domains can thus be read as:

The inclusion grade of the domain of x into y must be greater then or equal to the membership of $\langle x,y \rangle$ into $IEXT(I(\mathtt{rdfs : domain}))$.

The condition for ranges is completely analogous. The "actual" range $\text{ran}(p)$ is the set of resources used as object for some statement whose property is p:

$$\text{ran}(p) = \{o : \langle s,o \rangle \in IEXT(p) \text{ for some } s \in IR\}$$

Given this definition, the semantic condition for ranges can be read as:

The inclusion grade of the range of x into y must be greater then or equal to the membership of $\langle x,y \rangle$ into $IEXT(I(\mathtt{rdfs : domain}))$.

4.4 Subproperties and Subclasses

To define the semantics of `subPropertyOf` and `subClassOf` we use again the previously defined inclusion grade between fuzzy sets. The condition on sub-properties corresponds to:

The inclusion grade of property x into property y must be greater then or equal to the membership of $\langle x,y \rangle$ into $IEXT(I(\mathtt{rdfs : subPropertyOf}))$.

Subproperties and subclasses and are fully analogous concepts in RDFS: the inclusion is between extensions for the former, between class extensions for the latter. The condition on subclasses is:

The inclusion grade of class x into class y must be greater then or equal to the membership of $\langle x,y \rangle$ into $IEXT(I(\mathtt{rdfs : subClassOf}))$.

5   Fuzzy RDF Entailment Rules

RDF model theory specification [3] suggests a set of entailment rules for RDF and RDF Schema. All the entailment rules are of the same form: add a statement to a graph when it contains triples conforming to a pattern. The rules are *valid*

Table 4. Fuzzy RDF inference rules

#	antecedents	consequent
1	iii: xxx aaa yyy	iii: aaa rdf:type rdf:Property
2.1	iii: xxx aaa yyy jjj: aaa rdfs:domain zzz	kkk: xxx rdf:type zzz where kkk = min(iii, jjj)
2.2	iii: aaa rdfs:domain zzz jjj: xxx aaa yyy	kkk: xxx rdf:type zzz where kkk = min(iii, jjj)
3.1	iii: xxx aaa uuu jjj: aaa rdfs:range zzz	kkk: uuu rdf:type zzz where kkk = min(iii, jjj)
3.2	iii: aaa rdfs:range zzz jjj: xxx aaa uuu	kkk: uuu rdf:type zzz where kkk=min(iii, jjj)
4a	iii: xxx aaa yyy	jjj: xxx rdf:type rdfs:Resource
4b	iii: xxx aaa uuu	iii: uuu rdf:type rdfs:Resource
5a.1	iii: aaa rdfs:subPropertyOf bbb jjj: bbb rdfs:subPropertyOf ccc	kkk: aaa rdfs:subPropertyOf ccc where kkk=min(iii, jjj)
5a.2	iii: bbb rdfs:subPropertyOf ccc jjj: aaa rdfs:subPropertyOf bbb	kkk: aaa rdfs:subPropertyOf ccc where kkk=min(iii, jjj)
5b	iii: xxx rdf:type rdf:Property	iii: xxx rdfs:subPropertyOf xxx *reflexivity of rdfs:subPropertyOf*
6.1	iii: xxx aaa yyy jjj: aaa rdfs:subPropertyOf bbb	kkk: xxx bbb yyy where kkk=min(iii, jjj)
6.2	iii: aaa rdfs:subPropertyOf bbb jjj: xxx aaa yyy	kkk: xxx bbb yyy where kkk=min(iii, jjj)
7a	iii: xxx rdf:type rdfs:Class	iii: xxx rdfs:subClassOf rdfs:Resource
7b	iii: xxx rdf:type rdfs:Class	iii: xxx rdfs:subClassOf xxx *reflexivity of rdfs:subClassOf*
8.1	iii: xxx rdfs:subClassOf yyy jjj: yyy rdfs:subClassOf zzz	kkk: xxx rdfs:subClassOf zzz where kkk=min(iii, jjj)
8.2	iii: yyy rdfs:subClassOf zzz jjj: xxx rdfs:subClassOf yyy	kkk: xxx rdfs:subClassOf zzz where kkk=min(iii, jjj)
9.1	iii: xxx rdfs:subClassOf yyy jjj: aaa rdf:type xxx	kkk: aaa rdf:type yyy where kkk=min(iii, jjj)
9.2	iii: aaa rdf:type xxx jjj: xxx rdfs:subClassOf yyy	kkk: aaa rdf:type yyy where kkk=min(iii, jjj)
10	iii: xxx rdf:type rdfs:ContainerMembershipProperty	iii: xxx rdfs:subPropertyOf rdfs:member
11	iii: xxx rdf:type rdfs:Datatype	jjj: xxx rdfs:subClassOf rdfs:Literal
X1	iii: xxx rdf:_* yyy	jjj: rdf:_* rdf:type rdfs:ContainerMembershipProperty *This is an extra rule for list membership* *properties (_1, _2, _3, ...). The RDF MT* *does not specify a production for this.*

in the sense that a graph entails any larger graph obtained adding statements according to such rules.

Each rule has only one or two antecedent statements and derives only one new inferred statement; either $P \vdash R$ or $P, Q \vdash R$. Given the way fuzzy RDF semantics is defined, the corresponding inference rules for fuzzy RDF are analogous; only the fuzzy truth values of inferred statements must be computed. The simplest possible choice that respect the semantics is:

- With rules as $P \vdash Q$, having only one antecedent, the membership value of the inferred consequent Q is taken to be the same of the antecedent P.
- With rules as $P, Q \vdash R$, the truth value of the inferred consequent R is the minimum between the membership values of P and Q.

Such inference patterns gives *maximal* membership values: the starting graph would entail any statement with the same consequent and a membership value lesser than or equal than the one given by the rules.

The resulting inference rules for RDF/RDFS are shown in Table 4.

The rules table is derived from the rules used by the Sesame[35] forward-chaining inferencer. Sesame is a generic architecture for storing and querying RDF and RDF Schema, that makes use of a forward-chaining inferencer to compute and store the closure of its knowledge base whenever a transaction adds data to the repository[36]. A simple proof-of-concept fuzzy RDF storage and inference tool was obtained modifying Sesame inferencer, making it compute the correct membership values for the inferred statements. The underlying statements storage was also extended to save statements' membership values.

As the number of required steps is the same, and additional computations amount at most to a single comparison of membership values, complexity of reasoning in fuzzy RDF as the same complexity order then reasoning in plain RDF.

There is no known proof that an inference engine implementing the rules in Table 4 would be complete. In fact, there is no such proof for plain RDF/S inference rules either.

Acknowledgements

This work was supported in part by Italian PRIN 2005 research project *DiCSI* (Knowledge Dynamics in Information Society).

We would like to thank one of the anonymous reviewers for the sensible caveat about semantics based on minima of functions.

References

1. Berners-Lee, T., Hendler, J., Lassila, O.: The semantic web. Scientific American, 28–37 (May 2001)
2. Berners-Lee, T.: What the semantic web can represent. W3C design issues, World Wide Web Consortium (September 1998)
3. Hayes, P.: RDF Semantics. Recommendation, W3C (February 10, 2004)
4. Bechhofeer, S., van Harmelen, F., Hendler, J., Horrocks, I., McGuinness, D., Patel-Schneider, P., Stein, L.A.: OWL web ontology language reference. Recommendation, W3C (February 10, 2004)
5. Horrocks, I., Patel-Schneider, P.F.: Reducing OWL entailment to description logic satisfiability. J. of Web Semantics 4(1) (2004)
6. Yen, J.: Generalizing term subsumption languages to fuzzy logic. In: Reiter, R., Myopoulos, J. (eds.) Proc. of the 12th Int. Joint Conf. on Artificial Intelligence (IJCAI 1991), pp. 472–477 (1991)
7. Straccia, U.: Reasoning within fuzzy description logics. J. of Artificial Intelligence Research 14, 137–166 (2001)
8. Straccia, U.: Towards a fuzzy description logic for the semantic web (preliminary report). In: Gómez-Pérez, A., Euzenat, J. (eds.) ESWC 2005. LNCS, vol. 3532, pp. 167–181. Springer, Heidelberg (2005)
9. Stoilos, G., Stamou, G., Tzouvaras, V., Pan, J., Horrocks, I.: Fuzzy OWL: Uncertainty and the semantic web. In: Proc. of the Int. Workshop of OWL: Experiences and Directions, Galway, Ireland (2005)

10. Tresp, C.B., Molitor, R.: A description logic for vague knowledge. In: Proc. of the 13th European Conf. on Artificial Intelligence (ECAI'98), pp. 361–365. J. Wiley and Sons, Chichester (1998)

11. Sánchez, D., Tettamanzi, A.: Generalizing quantification in fuzzy description logics. Computational Intelligence, Theory and Applications, 397–411 (2005)

12. Stoilos, G., Stamou, G., Tzouvaras, V., Pan, J., Horrocks, I.: A fuzzy description logic for multimedia knowledge representation. In: Int. Workshop on Multimedia and the Semantic Web (2005)

13. Hölldobler, S., Khang, T.D., Störr, H.P.: A fuzzy description logic with hedges as concept modifiers. In: Proc. InTech/VJFuzzy 2002, pp. 25–34 (2002)

14. Stoilos, G., Stamou, G., Tzouvaras, V., Pan, J., Horrocks, I.: The fuzzy description logic f-$\mathcal{SHIN}$. In: ISWC Workshop on Uncertainty Reasoning for the Semantic Web, Galway, Ireland, November 7 2005. CEUR Workshop Proceedings, vol. 173, pp. 67–76 (2005)

15. Gao, M., Liu, C.: Extending OWL by fuzzy descripion logic. In: Proc. of the 17th IEEE Int. Conf. on Tools with Artificial Intelligence (ICTAI 2005), Hong Kong, China, pp. 562–567. IEEE Computer Society, Los Alamitos (2005)

16. Stoilos, G., Straccia, U., Stamou, G., Pan, J.Z.: General concept inclusions in fuzzy description logics. In: Proc. of the 17th European Conf. on Artificial Intelligence (ECAI 2006), pp. 457–461. IOS Press, Amsterdam (2006)

17. Stoilos, G., Stamou, G., Pan, J.: Handling imprecise knowledge with fuzzy description logic. In: Proc. of DL 2006 (2006)

18. Stoilos, G., Stamou, G.: Extending fuzzy description logics for the semantic web-based. In: Proc. of the 3rd Int. Workshop on OWL: Experience and Direction, OWLED 2007 (2007)

19. Mazzieri, M.: A fuzzy RDF semantics to represent trust metadata. In: 1st Workshop on Semantic Web Applications and Perspectives (SWAP 2004), Ancona, Italy, December 10, 2004, pp. 83–89 (2004)

20. Mazzieri, M., Dragoni, A.F.: A fuzzy semantics for semantic web languages. In: ISWC Workshop on Uncertainty Reasoning for the Semantic Web, Galway, Ireland, November 7, 2005. CEUR Workshop Proceedings, vol. 173, pp. 12–22 (2005)

21. Udrea, O., Recupero, D.R., Subrahmanian, V.: Annotated RDF. ACM Transactions on Computational Logic (submitted)

22. Klyne, G., Carroll, J.J.: Resource Description Framework (RDF): Concepts and abstract syntax. W3C recommendation, World Wide Web Consortium (February 2004)

23. Berners-Lee, T., Fielding, R., Masinter, L.: Uniform Resource Identifiers (URI): Generic syntax. RFC 2396, IETF (August 1998)

24. Goldfarb, W.D. (ed.): Logical Writings of Jacques Herbrand. Harvard University Press, Cambridge (1971)

25. Zadeh, L.: A fuzzy set theoretic interpretation of linguistic hedges. J. of Cybernetics 2(3), 4–34 (1972)

26. Dubois, D., Prade, H.: A class of fuzzy measures based on triangular norms. Int. J. of General Systems 8, 43–61 (1982)

27. Zimmermann, H.J.: Fuzzy Set Theory— and Its Applications, 3rd edn. Kluwer Academic Publishers, Dordrecht (1996)

28. Nguyen, H.T., Walker, E.A.: A First Course in Fuzzy Logic, 2nd edn. Chapman & Hall/CRC, Boca Raton (2000)

29. Grant, J., Beckett, D.: RDF test cases. W3C recommendation, World Wide Web Consortium (February 10, 2004)

30. Beckett, D.: RDF/XML syntax specification (revised). W3C recommendation, World Wide Web Consortium (February 2004)
31. Zadeh, L.A.: Fuzzy sets. Information and Control 8, 338–353 (1965)
32. Dubois, D., Prade, H.: Fuzzy Sets and Systems. Academic Press, New York (1980)
33. Zadeh, L.A.: Calculus of Fuzzy Restrictions. In: Fuzzy Sets and Their Applications to Cognitive and Decision Processes, pp. 1–39. Academic Press, New York (1975)
34. Burillo, P., Frago, N., Fuentes, R.: Inclusion grade and fuzzy implication operators. Fuzzy Sets and Systems 114(3), 417–429 (2000)
35. Broekstra, J., Kampman, A., van Harmelen, F.: Sesame: A generic architecture for storing and quering RDF and RDF Schema. In: Horrocks, I., Hendler, J. (eds.) ISWC 2002. LNCS, vol. 2342, pp. 54–68. Springer, Heidelberg (2002)
36. Broekstra, J., Kampman, A.: Inferencing and truth maintenance in RDF Schema: exploring a naive practical approach. In: Workshop on Practical and Scalable Semantic Systems (PSSS 2003). 2nd Int. Semantic Web Conf. (ISWC 2003), Sanibel Island, Florida, USA, October 20–24 (2003)

Reasoning with the Fuzzy Description Logic f-$\mathcal{SHIN}$: Theory, Practice and Applications

Giorgos Stoilos[1], Giorgos Stamou[1], Jeff Z. Pan[2],
Nick Simou[1], and Vassilis Tzouvaras[1]

[1] Department of Electrical and Computer Engineering,
National Technical University of Athens
[2] Department of Computing Science,
The University of Aberdeen

Abstract. The last couple of years it is widely acknowledged that uncertainty and fuzzy extensions to ontology languages, like Description Logics (DLs) and OWL, could play a significant role in the improvement of many Semantic Web (SW) applications. Many of the tasks of SW like trust, matching, merging, ranking usually involve confidence or truth degrees that one requires to represent and reason about. Fuzzy DLs are able to represent vague concepts such as a "Tall" person, a "Hot" place, a "MiddleAged" person, a "near" destination and many more. In the current paper we present a fuzzy extension to the DL $\mathcal{SHIN}$. First, we present the semantics while latter a detailed reasoning algorithm that decides most of the key inference tasks of fuzzy-$\mathcal{SHIN}$. Finally, we briefly present the fuzzy reasoning system FiRE, which implements the proposed algorithm and two use case scenarios where we have applied fuzzy DLs through FiRE.

1 Introduction

The last decade a significant amount of research has been focused in the development of the Semantic Web [1]. Semantic Web actually consists of an extension of the current Web where information, that lies in databases, web pages, etc., would be semantically accessible, enabling complex tasks to be performed in an (semi)automatic way. For example, Semantic Web agents would be able to accomplish tasks like a holiday organization, a doctor appointment, the retrieval of images depicting specific events etc. in a semantic and (semi)automatic way. In order to accomplish this goal it is widely recognized that information on the web should be structured in a machine understandable way, by using knowledge representation languages and forming *ontologies* [1]. For those reasons W3C[1] has standardized a number of ontology (knowledge representation) languages for the Web. One of the most important and expressive ones is OWL [2]. The logical underpinnings of OWL consist of very expressive Description Logics [3] and more precisely, the OWL DL species of OWL is equivalent to $\mathcal{SHOIN}(\mathbf{D^+})$, while OWL Lite is equivalent to $\mathcal{SHIF}(\mathbf{D^+})$.

---

[1] http://www.w3.org/

P.C.G. da Costa et al. (Eds.): URSW 2005-2007, LNAI 5327, pp. 262–281, 2008.

Although, Description Logics are relatively expressive, they are based on two-valued (Boolean) logics, which consider everything either true or false, thus they are unable to represent truth degrees, which are important in representing vague (fuzzy) knowledge. For example, they are unable to correctly represent concepts like a "tall" man, a "fast" car, a "blue" sky and many more. Moreover, many Semantic Web applications, like knowledge based information retrieval and ontology matching [4], also involve degrees of equivalence or similarity, which are important to represent and reason about. For those reasons fuzzy Description Logics [5, 6] and fuzzy OWL [7] have been proposed as languages capable of representing and reasoning with vague knowledge in the Semantic Web. With fuzzy DLs one usually is able to provide the same schema information as classical (crisp) DLs. For example, one can still define the concept of a MiddleAged person as someone that is either in his/her Forties or Fifties with the following axiom:

$$\text{MiddleAged} \equiv \text{Forties} \sqcup \text{Fifties}$$

On the other hand, one is able to state that John is in his fifties to a degree at least 0.6 (since he is 46 years old) by writing $(john : \text{Fifties}) \geq 0.6$, while he is also tall to a degree at least 0.8 (since he is 190cm), writing $(john : \text{Tall}) \geq 0.8$. Apart from representing fuzzy knowledge one should be able to also reason about, and for example infer that John is middle aged to a degree at least 0.8.

In the current paper we report on some recent results obtained about reasoning in very expressive fuzzy Description Logics and more precisely about reasoning with the fuzzy DL f_{KD}-$\mathcal{SHIN}$ [6,8,9,10]. First, we present a tableaux reasoning algorithm for f_{KD}-$\mathcal{SHIN}$. Then, we report on an implementation of the algorithm which gave rise to the FiRE fuzzy DL system and consists of an extension of the tool presented in [8]. FiRE provides a graphical user interface that can be used to load and reason with fuzzy DL ontologies. Furthermore, FiRE is able to store a fuzzy knowledge into a triple store and query about it using very expressive fuzzy conjunctive queries [11]. The rest of the paper is organized as follows. Section 2 presents the syntax and semantics of the fuzzy extension of $\mathcal{SHIN}$. Then, in Section 3 we provide all the technical details for a reasoning algorithm that decides most of the inference problems of fuzzy-$\mathcal{SHIN}$. Subsequently, Section 4 provides a brief presentation of the FiRE system. Then, Section 5 presents two Use Case scenarios where we have applied FiRE and we discuss its potentials and future directions. Finally, Section 6 provides a discussion about state-of-the-art work in fuzzy DLs, while it also presents a list of important open problems related to the area of fuzzy Description Logics.

2 Syntax and Semantics of f-$\mathcal{SHIN}$

In this section we introduce the DL f-$\mathcal{SHIN}$. As usual we have an alphabet of distinct concept names (**C**), role names (**R**) and individual names (**I**). f-$\mathcal{SHIN}$-roles and f-$\mathcal{SHIN}$-concepts are defined as follows:

Definition 1. *Let $RN \in \mathbf{R}$ be a role name, R an f-$\mathcal{SHIN}$-role. f-$\mathcal{SHIN}$-roles are defined by the abstract syntax: $R ::= RN \mid R^-$. The inverse relation of roles is symmetric, and to avoid considering roles such as R^{--}, we define a function Inv, which returns the inverse of a role, more precisely $\mathsf{Inv}(R) := RN^-$ if $R = RN$ and $\mathsf{Inv}(R) := RN$ if $R = RN^-$.*

The set of f-$\mathcal{SHIN}$-concepts is the smallest set such that:

1. *every concept name $CN \in \mathbf{C}$ is an f-$\mathcal{SHIN}$-concept,*

2. *if C and D are f-$\mathcal{SHIN}$-concepts and R is an f-$\mathcal{SHIN}$-role, then $(C \sqcup D)$, $(C \sqcap D)$, $(\neg C)$, $(\forall R.C)$ and $(\exists R.C)$ are also f-$\mathcal{SHIN}$-concepts,*

3. *if R is a simple[2] f-$\mathcal{SHIN}$-role and $p \in \mathbb{N}$, then $(\geq pR)$ and $(\leq pR)$ are also f-$\mathcal{SHIN}$-concepts.*

Although the definition of $\mathcal{SHIN}$-concepts and roles is the same with the one of fuzzy-$\mathcal{SHIN}$-concepts and roles the semantics of f-$\mathcal{SHIN}$ are significantly extended. This is because semantically we have to provide a fuzzy meaning/interpretation to the building blocks of our language, like concepts, roles and constructors. For that reason the semantics of fuzzy DLs are defined with the help of *fuzzy interpretations* [5]. A fuzzy interpretation is a pair $\mathcal{I} = (\Delta^{\mathcal{I}}, \cdot^{\mathcal{I}})$ where the domain $\Delta^{\mathcal{I}}$ is a non-empty set of objects and $\cdot^{\mathcal{I}}$ is a *fuzzy interpretation function*, which maps:

1. an individual name $a \in \mathbf{I}$ to an element $a^{\mathcal{I}} \in \Delta^{\mathcal{I}}$,

2. a concept name $\mathsf{A} \in \mathbf{C}$ to a membership function $\mathsf{A}^{\mathcal{I}} : \Delta^{\mathcal{I}} \to [0,1]$,

3. a role name $RN \in \mathbf{R}$ to a membership function $R^{\mathcal{I}} : \Delta^{\mathcal{I}} \times \Delta^{\mathcal{I}} \to [0,1]$.

Intuitively, an object (pair of objects) can now belong to a fuzzy concept (role) to any degree between 0 and 1. For example, $\mathsf{HotPlace}^{\mathcal{I}}(\mathsf{Athens}^{\mathcal{I}}) = 0.7$, means that $\mathsf{Athens}^{\mathcal{I}}$ is a hot place to a degree equal to 0.7. Additionally, a fuzzy interpretation function can be extended in order to provide semantics to any complex f-$\mathcal{SHIN}$-concept and role by using the operators of fuzzy set theory. More precisely, in the current setting we use the Lukasiewicz negation ($c(a) = 1-a$), Gödel conjunction ($\min(a,b)$) and disjunction ($\max(a,b)$) and the Kleene-Dienes fuzzy implication ($\max(1-a,b)$). Then, since $C \sqcup D$ represents a disjunction (union) between concepts C and D we can use max and provide the semantic function for disjunction: $(C \sqcup D)^{\mathcal{I}}(a) = u(C^{\mathcal{I}}(a), D^{\mathcal{I}}(a))$. The complete set of semantics for f-$\mathcal{SHIN}$-concepts and roles is depicted in Table 1. We remark that due to the operators we use we call our language f_{KD}-$\mathcal{SHIN}$.

An f_{KD}-$\mathcal{SHIN}$ *TBox* $\mathcal{T}$ is a finite set of *terminological axioms*. Let C and D be two f_{KD}-$\mathcal{SHIN}$-concepts. Axioms of the form $C \sqsubseteq D$ are called *fuzzy concept inclusion axioms* or *fuzzy concept subsumptions* or simply subsumptions, while axioms of the form $C \equiv D$ are called *fuzzy concept equivalence axioms*. A fuzzy interpretation $\mathcal{I}$ satisfies an axiom $C \sqsubseteq D$ if $\forall a \in \Delta^{\mathcal{I}}$, $C^{\mathcal{I}}(a) \leq D^{\mathcal{I}}(a)$ while it satisfies an axiom $C \equiv D$ if $C^{\mathcal{I}}(a) = D^{\mathcal{I}}(a)$. Finally, a fuzzy interpretation $\mathcal{I}$

---

[2] A role is called *simple* if it is neither transitive nor has any transitive sub-roles.

Table 1. Semantics of f_{KD}-$\mathcal{SHIN}$-concepts and f_{KD}-$\mathcal{SHIN}$-roles

Constructor	Syntax	Semantics
top	$\top$	$\top^{\mathcal{I}}(a) = 1$
bottom	$\bot$	$\bot^{\mathcal{I}}(a) = 0$
general negation	$\neg C$	$(\neg C)^{\mathcal{I}}(a) = 1 - C^{\mathcal{I}}(a)$
conjunction	$C \sqcap D$	$(C \sqcap D)^{\mathcal{I}}(a) = \min(C^{\mathcal{I}}(a), D^{\mathcal{I}}(a))$
disjunction	$C \sqcup D$	$(C \sqcup D)^{\mathcal{I}}(a) = \max(C^{\mathcal{I}}(a), D^{\mathcal{I}}(a))$
exists restriction	$\exists R.C$	$(\exists R.C)^{\mathcal{I}}(a) = \sup_{b \in \Delta^{\mathcal{I}}} \{\min(R^{\mathcal{I}}(a, b), C^{\mathcal{I}}(b))\}$
value restriction	$\forall R.C$	$(\forall R.C)^{\mathcal{I}}(a) = \inf_{b \in \Delta^{\mathcal{I}}} \{\max(1 - R^{\mathcal{I}}(a, b), C^{\mathcal{I}}(b))\}$
at-most restriction	$\leq pR$	$(\leq pR)^{\mathcal{I}}(a) = \inf_{b_1, \ldots, b_{p+1} \in \Delta^{\mathcal{I}}} \max_{i=1}^{p+1} \{1 - R^{\mathcal{I}}(a, b_i)\}$
at-least restriction	$\geq pR$	$(\geq pR)^{\mathcal{I}}(a) = \sup_{b_1, \ldots, b_p \in \Delta^{\mathcal{I}}} \min_{i=1}^{p} \{R^{\mathcal{I}}(a, b_i)\}$
inverse roles	R^-	$(R^-)^{\mathcal{I}}(b, a) = R^{\mathcal{I}}(a, b)$

satisfies an f_{KD}-$\mathcal{SHIN}$ TBox $\mathcal{T}$ if it satisfies every axiom in $\mathcal{T}$. Then we say that $\mathcal{I}$ is a *model* of $\mathcal{T}$.

An f_{KD}-$\mathcal{SHIN}$ RBox $\mathcal{R}$ is a finite set of fuzzy role axioms. Axioms of the form $\mathsf{Trans}(R)$ are called *fuzzy transitive role axioms*, while axioms of the form $R \sqsubseteq S$ are called *fuzzy role inclusion axioms*. A fuzzy interpretation $\mathcal{I}$ satisfies an axiom $\mathsf{Trans}(R)$ if $\forall a, c \in \Delta^{\mathcal{I}}, R^{\mathcal{I}}(a, c) \geq \sup_{b \in \Delta^{\mathcal{I}}} \{\min(R^{\mathcal{I}}(a, b), R^{\mathcal{I}}(b, c))\}$ while it satisfies $R \sqsubseteq S$ if $\forall \langle a, b \rangle \in \Delta^{\mathcal{I}} \times \Delta^{\mathcal{I}}, R^{\mathcal{I}}(a, b) \leq S^{\mathcal{I}}(a, b)$. Finally, $\mathcal{I}$ satisfies an f_{KD}-$\mathcal{SHIN}$ RBox if it satisfies every axiom in $\mathcal{R}$. In that case we say that $\mathcal{I}$ is a model of $\mathcal{R}$. A set of fuzzy role inclusion axioms defines a *role hierarchy* $\mathcal{R}_h$. Additionally, we note that the semantics of role inclusion axioms imply that if $R \sqsubseteq S$, then also $\mathsf{Inv}(R) \sqsubseteq \mathsf{Inv}(S)$, like in the classical case.

An f_{KD}-$\mathcal{SHIN}$ ABox $\mathcal{A}$ is a finite set of *fuzzy assertions* [5] of the form $(a : C) \bowtie n$ or $((a, b) : R) \bowtie n$, where $\bowtie$ stands for $\geq, >, \leq$ and $<$, and $n \in [0, 1]$ or of the form $a \neq b$. Intuitively, a fuzzy assertion of the form $(a : C) \geq n$ means that the membership degree of the individual a to the concept C is at least equal to n. We call assertions defined using inequalities $\geq, >$ *positive*, while those using $\leq, <$ *negative*. Formally, given a fuzzy interpretation $\mathcal{I}$,

$$\mathcal{I} \text{ satisfies } (a : C) \geq n \text{ if } C^{\mathcal{I}}(a^{\mathcal{I}}) \geq n,$$
$$\mathcal{I} \text{ satisfies } (a : C) \leq n \text{ if } C^{\mathcal{I}}(a^{\mathcal{I}}) \leq n,$$
$$\mathcal{I} \text{ satisfies } ((a, b) : R) \geq n \text{ if } R^{\mathcal{I}}(a^{\mathcal{I}}, b^{\mathcal{I}}) \geq n,$$
$$\mathcal{I} \text{ satisfies } ((a, b) : R) \leq n \text{ if } R^{\mathcal{I}}(a^{\mathcal{I}}, b^{\mathcal{I}}) \leq n,$$
$$\mathcal{I} \text{ satisfies } a \neq b \text{ if } a^{\mathcal{I}} \neq b^{\mathcal{I}}.$$

The satisfiability of fuzzy assertions with $>, <$ is defined analogously. Observe that, we can also simulate assertions of the form $(a : C) = n$ by considering two assertions of the form $(a : C) \geq n$ and $(a : C) \leq n$. A fuzzy interpretation $\mathcal{I}$ satisfies an f_{KD}-$\mathcal{SHIN}$ ABox $\mathcal{A}$ iff it satisfies all fuzzy assertions in $\mathcal{A}$; in this case, we say that $\mathcal{I}$ is a model of $\mathcal{A}$.

Without loss of generality, we assume that no negative assertions exist. Negative assertions of the form $(a : C) \leq n$ and $(a : C) < n$ can be transformed into their *positive inequality normal form* (PINF), by applying a fuzzy complement in both sides getting, $(a : \neg C) \geq 1 - n$ and $(a : \neg C) > 1 - n$ (similarly with role assertions), respectively. Furthermore, we assume that a fuzzy ABox has been *normalized* [12], i.e. fuzzy assertions of the form $(a : C) > n$ are replaced by assertions of the form $(a : C) \geq n + \epsilon$, where ϵ is a small number converging to 0. Please note that in a normalized fuzzy KB with only positive inequalities degrees range over $[+\epsilon, 1 + \epsilon]$. Also note that a fuzzy ABox is consistent iff the normalized one is [13]. For a fuzzy ABox we define the set of *relative degrees* as

$$N^{\mathcal{A}} = \{0, 0.5, 1\} \cup \{1 - n, n \mid (a : C \geq n) \in \mathcal{A} \text{ or } ((a, b) : R) \geq n \in \mathcal{A}\}$$

An f_{KD}-$\mathcal{SHIN}$ *knowledge base* (KB) is defined as $\Sigma = \langle \mathcal{T}, \mathcal{R}, \mathcal{A} \rangle$. An interpretation $\mathcal{I}$ satisfies an f_{KD}-$\mathcal{SHIN}$ knowledge base Σ if it satisfies every axiom in $\mathcal{T}$, $\mathcal{R}$ and $\mathcal{A}$. In that case $\mathcal{I}$ is called a model of Σ.

Now we define the inference services of f_{KD}-$\mathcal{SHIN}$.

- **KB Satisfiability:** An f_{KD}-$\mathcal{SHIN}$ knowledge base $\Sigma = \langle \mathcal{T}, \mathcal{R}, \mathcal{A} \rangle$ is *satisfiable* (unsatisfiable) iff there exists (does not exist) a fuzzy interpretation $\mathcal{I}$ which satisfies all axioms in Σ.

- **Concepts n-satisfiabilty:** An f_{KD}-$\mathcal{SHIN}$-concept C is *n-satisfiable* w.r.t. Σ iff there exists a model $\mathcal{I}$ of Σ in which there exists some $a \in \Delta^{\mathcal{I}}$ such that $C^{\mathcal{I}}(a) = n$, and $n \in (0, 1]$.

- **Concept Subsumption:** A fuzzy concept C is subsumed by D w.r.t. Σ iff in every model $\mathcal{I}$ of Σ we have that $\forall d \in \Delta^{\mathcal{I}}, C^{\mathcal{I}}(d) \leq D^{\mathcal{I}}(d)$.

- **ABox Consistency:** An f_{KD}-$\mathcal{SHIN}$ $\mathcal{A}$ is *consistent* (*inconsistent*) w.r.t. a TBox $\mathcal{T}$ and an RBox $\mathcal{R}$ if there exists (does not exist) a model $\mathcal{I}$ of $\mathcal{T}$ and $\mathcal{R}$ which satisfies every assertion in $\mathcal{A}$.

- **Entailment:** Given a concept or role axiom or a fuzzy assertion, Ψ, we say that Σ *entails* Ψ, writing $\Sigma \models \Psi$ iff every model $\mathcal{I}$ of Σ satisfies Ψ.

- **Greater Lower Bound (glb):** The *greatest lower bound* of an assertion Φ w.r.t. Σ is defined as,

$$glb(\Sigma, \Phi) = \sup\{n \mid \Sigma \models \Phi \geq n\}, \text{ where } \sup \emptyset = 0.$$

As we note glb, actually consists of a set of entailment tests.

The problems of concept n-satisfiability, subsumption and entailment w.r.t. a knowledge base Σ can be reduced to the problem of knowledge base satisfiability Σ [5, 6]. Here, the reductions are slightly modified due to PINF and normalization. More precisely, a concept C is n-satisfiable w.r.t. $\mathcal{T}$ and $\mathcal{R}$ iff $\{(a : C) \geq n\}$ is consistent w.r.t. $\mathcal{T}$ and $\mathcal{R}$. Moreover, for $\Sigma = \langle \mathcal{T}, \mathcal{R}, \mathcal{A} \rangle$, and a PINF assertion $\phi \geq n$, where ϕ is a classical $\mathcal{SHIN}$ assertion, $\Sigma \models \phi \geq n$ iff $\Sigma = \langle \mathcal{T}, \mathcal{R}, \mathcal{A} \cup \{\neg \phi \geq 1 - n + \epsilon\}\rangle$ is unsatisfiable. Furthermore, $\Sigma \models C \sqsubseteq D$ iff $\langle \mathcal{T}, \mathcal{R}, \mathcal{A} \cup \{(a : C) \geq n, (a : \neg D) \geq 1 - n + \epsilon\}\rangle$ is unsatisfiable, for both $n \in \{n_1, n_2\}$, $n_1 \in (0, 0.5]$ and $n_2 \in (0.5, 1]$.

3   Reasoning with f_{KD}-$\mathcal{SHIN}$

In the previous section we show that all inference problems of fuzzy DLs, can be reduced to the problem of knowledge base satisfiability. Consequently, we have to construct an algorithm that decides such a reasoning problem. Our method will be based on tableaux algorithms.

Without loss of generality, we assume all concepts C occurring in assertions to be in their *negation normal form* (NNF) [3], denoted by $\sim C$; i.e., negations occur in front of concept names only. An f_{KD}-$\mathcal{SHIN}$-concept can be transformed into an equivalent one in NNF by pushing negations inwards making use of the De Morgan laws and the dualities between $\exists$ and $\forall$, and between concepts $\geq$ and $\leq$.

Definition 2. *For every concept D we inductively define the set of* sub-concepts *of $(sub(D))$ as,*

$$
\begin{aligned}
sub(A) &= \{A\} \text{ for every atomic concept } A \in \mathbf{C}, \\
sub(C \sqcap D) &= \{C \sqcap D\} \cup \{sub(C)\} \cup \{sub(D)\}, \\
sub(C \sqcup D) &= \{C \sqcup D\} \cup \{sub(C)\} \cup \{sub(D)\}, \\
sub(\exists R.C) &= \{\exists R.C\} \cup \{sub(C)\}, \\
sub(\forall R.C) &= \{\forall R.C\} \cup \{sub(C)\}, \\
sub(\geq nR) &= \{\geq nR\} \\
sub(\leq nR) &= \{\leq nR\}
\end{aligned}
$$

Definition 3. *For a fuzzy concept D and an RBox $\mathcal{R}$ we define $cl(D, \mathcal{R})$ as the smallest set of f_{KD}-$\mathcal{SI}$-concept which satisfies the following:*

- $D \in cl(D, \mathcal{R})$,
- $cl(D, \mathcal{R})$ *is closed under sub-concepts of D and $\sim D$, and*
- *if $\forall R.C \in cl(D, \mathcal{R})$ and $\mathsf{Trans}(P)$ with $P \sqsubseteq^* R$, then $\forall P.C \in cl(D, \mathcal{R})$*

Finally we define $cl(\mathcal{A}, \mathcal{R}) = \bigcup\limits_{(a:D) \geq n \in \mathcal{A}} cl(D, \mathcal{R})$.

When $\mathcal{R}$ is clear from the context we simply write $cl(\mathcal{A})$.

Definition 4. *If $\Sigma = \langle \mathcal{T}, \mathcal{R}, \mathcal{A} \rangle$ is an f_{KD}-SHIN knowledge base, $\mathbf{R}_\mathcal{A}$ is the set of roles occurring in Σ together with their inverses, $\mathbf{I}_\mathcal{A}$ is the set of individuals in $\mathcal{A}$, a fuzzy tableau T for Σ is defined to be a quadruple $(\mathbf{S}, \mathcal{L}, \mathcal{E}, \mathcal{V})$ such that: $\mathbf{S}$ is a set of elements, $\mathcal{L} : \mathbf{S} \times cl(\mathcal{A}) \to [0, 1]$ maps each element and concept to the membership degree of that element to the concept, $\mathcal{E} : \mathbf{R}_\mathcal{A} \times \mathbf{S} \times \mathbf{S} \to [0, 1]$ maps each role of $\mathbf{R}_\mathcal{A}$ and pair of elements to the membership degree of the pair to the role, and $\mathcal{V} : \mathbf{I}_\mathcal{A} \to \mathbf{S}$ maps individuals occurring in $\mathcal{A}$ to elements of $\mathbf{S}$. For all $s, t \in \mathbf{S}$, $C, E \in cl(\mathcal{A})$, $n \in [0, 1]$ and $R \in \mathbf{R}_\mathcal{A}$, T satisfies:*

1. *$\mathcal{L}(s, \perp) = 0$ and $\mathcal{L}(s, \top) = 1$ for all $s \in \mathbf{S}$,*

2. If $\mathcal{L}(s, \neg A) = n$, then $\mathcal{L}(s, A) = 1 - n$,

3. If $\mathcal{E}(\neg R, \langle s, t \rangle) = n$, then $\mathcal{E}(R, \langle s, t \rangle) = 1 - n$,

4. If $\mathcal{L}(s, C \sqcap E) \geq n$, then $\mathcal{L}(s, C) \geq n$ and $\mathcal{L}(s, E) \geq n$,

5. If $\mathcal{L}(s, C \sqcup E) \geq n$, then $\mathcal{L}(s, C) \geq n$ or $\mathcal{L}(s, E) \geq n$,

6. If $\mathcal{L}(s, \forall R.C) \geq n$, then either $\mathcal{E}(\neg R, \langle s, t \rangle) \geq n$ or $\mathcal{L}(t, C) \geq n$,

7. If $\mathcal{L}(s, \exists R.C) \geq n$, then there exists $t \in \mathbf{S}$ such that $\mathcal{E}(R, \langle s, t \rangle) \geq n$ and $\mathcal{L}(t, C) \geq n$,

8. If $\mathcal{L}(s, \forall R.C) \geq n$, then either $\mathcal{E}(\neg P, \langle s, t \rangle) \geq n$, for $P \sqsubseteq R$ with $\mathsf{Trans}(P)$ or $\mathcal{L}(t, \forall P.C) \geq n$,

9. $\mathcal{E}(R, \langle s, t \rangle) \geq n$ iff $\mathcal{E}(\mathsf{Inv}(R), \langle t, s \rangle) \geq n$,

10. If $\mathcal{E}(R, \langle s, t \rangle) \geq n$ and $R \sqsubseteq S$, then $\mathcal{E}(S, \langle s, t \rangle) \geq n$,

11. If $\mathcal{L}(s, \geq pR) \geq n$, then $\sharp R^T(s, \geq, n) \geq p$,

12. If $\mathcal{L}(s, \leq pR) \geq n$, then $\sharp R^T(s, \geq, 1 - n + \epsilon) \leq p$,

13. If $C \sqsubseteq D \in \mathcal{T}$, then either $\mathcal{L}(s, C) \geq 1 - n + \epsilon$ or $\mathcal{L}(s, D) \geq n$, for all $s \in \mathbf{S}$ and $n \in N^\mathcal{A}$,

14. If $(a : C) \geq n \in \mathcal{A}$, then $\mathcal{L}(\mathcal{V}(a), C) \geq n$,

15. If $((a, b) : R) \geq n \in \mathcal{A}$, then $\mathcal{E}(R, \langle \mathcal{V}(a), \mathcal{V}(b) \rangle) \geq n$,

16. If $a \not\approx b \in \mathcal{A}$, then $\mathcal{V}(a) \neq \mathcal{V}(b)$.

where $\sharp$ denotes the cardinality of a set, $R^T(s, \geq, n) = \{t \in \mathbf{S} \mid \mathcal{E}(R, \langle s, t \rangle) \geq n\}$ returns the set of elements $t \in \mathbf{S}$ that participate in R with some element s with a degree, greater or equal or greater than a given degree n.

Lemma 1. An f_{KD}-$\mathcal{SHIN}$ knowledge base Σ is satisfiable iff there exists a fuzzy tableau for Σ.

For a detailed proof of the above lemma as well as the intuition behind the properties of Definition 3 the reader is referred to [6] and [12].

The above lemma establishes a connection between the satisfiability of a knowledge base (existence of a model) and the existence of a fuzzy tableaux for Σ. Thus, it suggests that in order to decide the key inference problems of f_{KD}-$\mathcal{SHIN}$ we have to develop an algorithm that given an f_{KD}-$\mathcal{SHIN}$ KB Σ it constructs a fuzzy tableau for Σ.

3.1   The Tableaux Algorithm

In order to decide knowledge base satisfiability a procedure that constructs a fuzzy tableau for an f_{KD}-$\mathcal{SHIN}$ knowledge base has to be determined. In the current section we will provide the technical details for such an algorithm.

Definition 5. A completion-forest $\mathcal{F}$ for an f_{KD}-$\mathcal{SHIN}$ knowledge base is a collection of trees whose distinguished roots are arbitrarily connected by edges. Each node x is labelled with a set $\mathcal{L}(x) = \{\langle C, \geq, n \rangle\}$, where $C \in cl(\mathcal{A})$ and

$n \in [+\epsilon, 1 + \epsilon]$. Each edge $\langle x, y \rangle$ is labelled with a set $\mathcal{L}(\langle x, y \rangle) = \{\langle R, \geq, n \rangle\}$, where $S := R \mid \neg R$, and $R \in \mathbf{R}_\mathcal{A}$ is a (possibly inverse) role occurring in $\mathcal{A}$.

If nodes x and y are connected by an edge $\langle x, y \rangle$ with $\langle P, \geq, n \rangle \in \mathcal{L}(\langle x, y \rangle)$, and $P \stackrel{*}{\sqsubseteq} R$, then y is called an $R_{\geq n}$-successor of x and x is called an $R_{\geq n}$-predecessor of y. If y is an $R_{\geq n}$-successor or an $\mathsf{Inv}(R)_{\geq n}$-predecessor of x, then y is called an $R_{\geq n}$-neighbour of x. Let y be an $R_{>n}$-neighbour of x. Then, the edge $\langle x, y \rangle$ is conjugated with triples $\langle \neg R, \geq, m \rangle$ if $n + m \geq 1$. Similarly, we can extend it to the case of $R_{\geq n}$-neighbours. As usual, ancestor is the transitive closure of predecessor.

For two roles P, R, a node x in $\mathcal{F}$, an inequality $\geq$ and a membership degree $n \in [0, 1]$ we define: $R^{\mathcal{F}}_C(x, \geq, n) = \{y \mid y$ is an $R_{\geq n'}$-neighbour of x, and $\langle x, y \rangle$ is conjugated with $\langle \neg R, \geq, n \rangle\}$.

A node x is blocked iff it is not a root node and it is either directly or indirectly blocked. A node x is directly blocked iff none of its ancestors is blocked, and it has ancestors x', y and y' such that:

1. y is not a root node,
2. x is a successor of x' and y a successor of y',
3. $\mathcal{L}(x) = \mathcal{L}(y)$ and $\mathcal{L}(x') = \mathcal{L}(y')$ and,
4. $\mathcal{L}(\langle x', x \rangle) = \mathcal{L}(\langle y', y \rangle)$.

In this case we say that y blocks x. A node y is indirectly blocked iff one of its ancestors is blocked, or it is a successor of a node x and $\mathcal{L}(\langle x, y \rangle) = \emptyset$.

For a node x, $\mathcal{L}(x)$ is said to contain a clash if it contains one of the following:

− two conjugated pairs of triples,
− one of $\langle \bot, \geq, n \rangle$, with $n > 0$ or $\langle C, \geq, 1 + \epsilon \rangle$, or
− some triple $\langle \leq pR, \geq, n \rangle$ and x has $p + 1$ $R_{\geq n_i}$-neighbours $y_0, \ldots, y_p$, $\langle x, y_i \rangle$ is conjugated with $\langle \neg R, \geq, n \rangle$ and $y_i \neq y_j$, $n_i, n \in [0, 1]$, for all $0 \leq i < j \leq p$

Moreover, for an edge $\langle x, y \rangle$, $\mathcal{L}(\langle x, y \rangle)$ is said to contain a clash if (i) it contains two conjugated triples, or (ii) it contains the triple $\langle R, \geq, 1 + \epsilon \rangle$, or (iii) $\mathcal{L}(\langle x, y \rangle) \cup \{\langle \mathsf{Inv}(R), \geq, n \rangle \mid \langle R, \geq, n \rangle \in \mathcal{L}(\langle y, x \rangle)\}$, where x, y are root nodes, contains two conjugated triples.

For an $f_{KD}\text{-}\mathcal{SHIN}$ knowledge base, the algorithm initialises a forest $\mathcal{F}$ to contain

i. a root node x_{a_i}, for each individual $a_i \in \mathbf{I}_\mathcal{A}$ occurring in the ABox $\mathcal{A}$, labelled with $\mathcal{L}(x_{a_i})$ such that: $\mathcal{L}(x_{a_i}) = \{\langle C, \geq, n \rangle \mid (a_i : C) \geq n \in \mathcal{A}\}$,

ii. an edge $\langle x_{a_i}, x_{a_j} \rangle$, for each assertion $((a_i, a_j) : R) \geq n \in \mathcal{A}$, labelled with $\mathcal{L}(\langle x_{a_i}, x_{a_j} \rangle)$ such that: $\mathcal{L}(\langle x_{a_i}, x_{a_j} \rangle) = \{\langle R, \geq, n \rangle \mid \langle R, \geq, n \rangle \in \mathcal{A}\}$,

iii. the relation $\neq$ as $x_{a_i} \neq x_{a_j}$ if $a_i \neq a_j \in \mathcal{A}$ and the relation $\doteq$ to be empty.

Finally, the algorithm expands $\mathcal{R}$ by adding role inclusion axioms $\mathsf{Inv}(P) \sqsubseteq \mathsf{Inv}(R)$, for all $P \sqsubseteq R \in \mathcal{R}$ and by adding $\mathsf{Trans}(\mathsf{Inv}(R))$ for all $\mathsf{Trans}(R) \in \mathcal{R}$.

Table 2. Expansion rules for $f_{KD}\text{-}\mathcal{SHIN}$

Rule	Description
⊓	if 1. $\langle C_1 \sqcap C_2, \geq, n\rangle \in \mathcal{L}(x)$, x is not indirectly blocked, and 2. $\{\langle C_1, \geq, n\rangle, \langle C_2, \geq, n\rangle\} \not\subseteq \mathcal{L}(x)$ then $\mathcal{L}(x) \to \mathcal{L}(x) \cup \{\langle C_1, \geq, n\rangle, \langle C_2, \geq, n\rangle\}$
⊔	if 1. $\langle C_1 \sqcup C_2, \geq, n\rangle \in \mathcal{L}(x)$, x is not indirectly blocked, and 2. $\{\langle C_1, \geq, n\rangle, \langle C_2, \geq, n\rangle\} \cap \mathcal{L}(x) = \emptyset$ then $\mathcal{L}(x) \to \mathcal{L}(x) \cup \{C\}$ for some $C \in \{\langle C_1, \geq, n\rangle, \langle C_2, \geq, n\rangle\}$
∃	if 1. $\langle \exists R.C, \geq, n\rangle \in \mathcal{L}(x)$, x is not blocked, and 2. x has some $R_{\geq n}$-neighbour y with $\langle C, \geq, n\rangle \in \mathcal{L}(y)$ then create a new node y with $\mathcal{L}(\langle x, y\rangle) = \{\langle R, \geq, n\rangle\}$, $\mathcal{L}(y) = \{\langle C, \geq, n\rangle\}$
∀	if 1. $\langle \forall R.C, \geq, n\rangle \in \mathcal{L}(x)$, x is not indirectly blocked, 2. x has an $R_{\geq' n'}$-neighbour y with $\langle C, \geq, n\rangle \notin \mathcal{L}(y)$ and 3. $\langle x, y\rangle$ conjugates with $\langle \neg R, \geq, n\rangle$ then $\mathcal{L}(y) \to \mathcal{L}(y) \cup \{\langle C, \geq, n\rangle\}$
∀₊	if 1. $\langle \forall S.C, \geq, n\rangle \in \mathcal{L}(x)$, x is not indirectly blocked, 2. there exists some role R, with $\mathsf{Trans}(R)$ and $R \mathrel{\underline{\underline{\sqsubseteq}}} S$, 3. x has an $R_{\geq' n'}$-neighbour y with $\langle \forall R.C, \geq, n\rangle \notin \mathcal{L}(y)$, and 4. $\langle x, y\rangle$ conjugates with $\langle \neg R, \geq, n\rangle$ then $\mathcal{L}(y) \to \mathcal{L}(y) \cup \{\langle \forall R.C, \geq, n\rangle\}$
≥	if 1. $\langle \geq pR, \geq, n\rangle \in \mathcal{L}(x)$, x is not blocked, 2. there are no p $R_{\geq, n}$-neighbours $y_1, \ldots, y_p$ of x with $y_i \neq y_j$ for $1 \leq i < j \leq p$ then create p new nodes $y_1, \ldots, y_p$, with $\mathcal{L}(\langle x, y_i\rangle) = \{\langle R, \geq, n\rangle\}$ and $y_i \neq y_j$ for $1 \leq i < j \leq p$
≤	if 1. $\langle \leq pR, \geq, n\rangle \in \mathcal{L}(x)$, x is not indirectly blocked, 2. $\sharp R_C^{\mathcal{F}}(x, \geq, n) > p$, there are two of them y, z, with no $y \neq z$ and 3. y is neither a root node nor an ancestor of z then 1. $\mathcal{L}(z) \to \mathcal{L}(z) \cup \mathcal{L}(y)$ and 2. if z is an ancestor of x then $\mathcal{L}(\langle z, x\rangle) \longrightarrow \mathcal{L}(\langle z, x\rangle) \cup \mathsf{Inv}(\mathcal{L}(\langle x, y\rangle))$ else $\mathcal{L}(\langle x, z\rangle) \longrightarrow \mathcal{L}(\langle x, z\rangle) \cup \mathcal{L}(\langle x, y\rangle)$ 3. $\mathcal{L}(\langle x, y\rangle) \longrightarrow \emptyset$ and set $u \neq z$ for all u with $u \neq y$
≤ᵣ	if 1. $\langle \leq pR, \geq, n\rangle \in \mathcal{L}(x)$, 2. $\sharp R_C^{\mathcal{F}}(x, \geq, n) > p$, there are two of them y, z, both root nodes, with no $y \neq z$ and then 1. $\mathcal{L}(z) \to \mathcal{L}(z) \cup \mathcal{L}(y)$ and 2. For all edges $\langle y, w\rangle$: i. if the edge $\langle z, w\rangle$ does not exist, create it with $\mathcal{L}(\langle z, w\rangle) = \emptyset$ ii. $\mathcal{L}(\langle z, w\rangle) \longrightarrow \mathcal{L}(\langle z, w\rangle) \cup \mathcal{L}(\langle y, w\rangle)$ 3. For all edges $\langle w, y\rangle$: i. if the edge $\langle w, z\rangle$ does not exist, create it with $\mathcal{L}(\langle w, z\rangle) = \emptyset$ ii. $\mathcal{L}(\langle w, z\rangle) \longrightarrow \mathcal{L}(\langle w, z\rangle) \cup \mathcal{L}(\langle w, y\rangle)$ 4. Set $\mathcal{L}(y) = \emptyset$ and remove all edges to/from y 5. Set $u \neq z$ for all u with $u \neq y$ and set $y \doteq z$
⊑	if 1. $C \sqsubseteq D \in \mathcal{T}$, x is not indirectly blocked, and 2. $\{\langle \neg C, \geq, 1 - n + \epsilon\rangle, \langle D, \geq, n\rangle\} \cap \mathcal{L}(x) = \emptyset$ for $n \in N^{\mathcal{A}}$ then $\mathcal{L}(x) \to \mathcal{L}(x) \cup \{E\}$ for some $E \in \{\langle \neg C, \leq, 1 - n + \epsilon\rangle, \langle D, \geq, n\rangle\}$

$\mathcal{F}$ is then expanded by repeatedly applying the completion rules from Table 2. The completion-forest is complete when, for some node x, $\mathcal{L}(x)$ contains a clash, or none of the completion rules is applicable. The algorithm stops when a clash occurs; it answers 'Σ is satisfiable' iff the completion rules can be applied in such a way that they yield a complete and clash-free completion-forest, and 'Σ is unsatisfiable' otherwise.

Lemma 2. Let Σ be an f_{KD}-$\mathcal{SHIN}$ knowledge base. Then

1. when started for Σ the tableaux algorithm terminates
2. Σ has a fuzzy tableau if and only if the expansion rules can be applied to Σ such that they yield a complete and clash-free completion forest.

Finally, we conclude this section with an illustrative example that shows how the tableaux algorithm works.

Example 1. Consider the knowledge base $\Sigma = \langle \mathcal{T}, \mathcal{R}, \mathcal{A} \rangle$ where:

$$\mathcal{T} = \{\text{Arm} \sqsubseteq \exists\text{isPartOf.Body}, \text{Body} \sqsubseteq \exists\text{isPartOf.Human}\}$$
$$\mathcal{R} = \{\text{Trans(isPartOf)}\}$$
$$\mathcal{A} = \{((o_1, o_2) : \text{isPartOf}) \geq 0.8, ((o_2, o_3) : \text{isPartOf}) \geq 0.9,$$
$$(o_2 : \text{Body}) \geq 0.85, (o_1 : \text{Arm}) \geq 0.75\}$$

Now we want to use our reasoning algorithm to see if

$$\Sigma \models (o_3 : \exists\text{Inv(isPartOf).Body} \sqcap \exists\text{Inv(isPartOf).Arm}) < 0.75.$$

First we transform this negative assertion into its equivalent PINF form and then into its NNF form having finally the assertion $(o_3 : \forall\text{Inv(isPartOf).}\neg\text{Body} \sqcup \forall\text{Inv(isPartOf).}\neg\text{Arm}) > 0.25$. Subsequently, entailment checking is reduced to consistency of $\mathcal{A}' = \mathcal{A} \cup \{(o_3 : \forall\text{Inv(isPartOf).}\neg\text{Body} \sqcup \forall\text{Inv(isPartOf).}\neg\text{Arm}) > 0.25\}$, w.r.t. $\mathcal{R}$ and $\mathcal{T}$. According to Definition 5 the algorithm initializes a completion-forest to contain the following triples:

(1) $\langle\text{isPartOf}, \geq, 0.8\rangle \in \mathcal{L}(\langle x_{o_1}, x_{o_2}\rangle)$
(2) $\langle\text{isPartOf}, \geq, 0.9\rangle \in \mathcal{L}(\langle x_{o_2}, x_{o_3}\rangle)$
(3) $\langle\text{Body}, \geq, 0.85\rangle \in \mathcal{L}(x_{o_2})$
(4) $\langle\text{Arm}, \geq, 0.75\rangle \in \mathcal{L}(x_{o_1})$
(5) $\langle\forall\text{isPartOf}^-.\neg\text{Body} \sqcup \forall\text{isPartOf}^-.\neg\text{Arm}, >, 0.25\rangle \in \mathcal{L}(x_{o_3})$

Furthermore, the algorithm expands $\mathcal{R}$ by adding the axiom $\text{Trans(isPartOf}^-)$. Subsequently, by applying expansion rules from Table 2 we have the following steps:

(6) $\langle\forall\text{isPartOf}^-.\neg\text{Body}, >, 0.25\rangle \in \mathcal{L}(x_{o_3}) | \langle\forall\text{isPartOf}^-.\neg\text{Arm}, >, 0.25\rangle \in \mathcal{L}(x_{o_3}) \sqcup$

Hence at this point we have two possible completion forests. For the first one we have:

(6_1) $\langle \forall \mathsf{isPartOf}^-.\neg \mathsf{Body}, >, 0.25 \rangle \in \mathcal{L}(x_{o_3})$

(7_1) $\langle \neg \mathsf{Body}, >, 0.25 \rangle \in \mathcal{L}(x_{o_2})$ $\qquad \forall : (6_1), (2)$

(8_1) clash (7_1) and (3)

while for the second possible completion-forest we have:

(6_2) $\langle \forall \mathsf{isPartOf}^-.\neg \mathsf{Arm}, >, 0.25 \rangle \in \mathcal{L}(x_{o_3})$

(7_2) $\langle \neg \mathsf{Arm}, >, 0.25 \rangle \in \mathcal{L}(x_{o_2})$ $\qquad \forall : (6_2), (2)$

(8_2) $\langle \forall \mathsf{isPartOf}^-.\mathsf{Arm}, <, 0.75 \rangle \in \mathcal{L}(x_{o_2})$ $\quad \forall_+ : (6_2), (2)$

(9_2) $\langle \neg \mathsf{Arm}, >, 0.25 \rangle \in \mathcal{L}(x_{o_1})$ $\qquad \forall : (8_2), (1)$

(10_2) clash (9_2) and (4)

Thus, since all possible expansions result to a clash, $\mathcal{A}'$ is inconsistent and the knowledge base entails the fuzzy assertion.

4   FiRE: A Prototype f_{KD}-$\mathcal{SHIN}$ Reasoning System

FiRE is a JAVA implementation of a fuzzy DL reasoning engine for vague knowledge. Currently it implements the tableaux reasoning algorithm for f_{KD}-$\mathcal{SHIN}$ we presented in the previous section. Apart from the f_{KD}-$\mathcal{SHIN}$ reasoner, FiRE is also able to serialize a fuzzy KB into RDF triples and store it in the Sesame RDF triple store [14]. Then it is able to query Sesame using very expressive fuzzy conjunctive query languages [11]. In this section the graphical user interface, the syntax and the inference services of FiRE are briefly introduced.

4.1   FiRE Interface

FiRE can be found at http://www.image.ece.ntua.gr/~nsimou/FiRE together with installation instructions and examples. Figure 1 depicts the main GUI of FiRE. Its user interface consists of the *editor panel*, the *inference services panel* and the *output panel*. The user can create or edit an existing fuzzy knowledge base using the editor panel. The inference services panel allows the user to make different kinds of queries to the knowledge base (entailment, subsumption and glb) and also to query a Sesame repository using fuzzy conjunctive queries [11]. Finally, the output panel consists of four different tabs, each one displaying information depending on the user operation, like a trace of the tableaux expansion, possible syntax errors of the KB, classification of the KB (computing the subsumption hierarchy), and more.

4.2   FiRE Syntax

The current version of FiRE is using the Knowledge Representation System Specification (KRSS) proposal[3]. Since as we show in the previous sections we

---

[3] http://dl.kr.org/krss-spec.ps

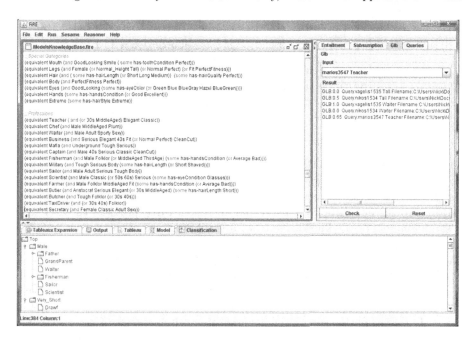

Fig. 1. The FiRE user interface: the editor panel (upper left), the inference services panel (upper right) and the output panel (bottom)

impose no syntactic changes to concept and role axioms, a user is capable of specifying concept and roles axioms using the standard KRSS syntax. So for example, one can define the concept MiddleAged with the following axiom:

```
(complete MiddleAged (or Forties Fifties))
```

using the keywords **complete** for specifying equivalence ($\equiv$) and **or** for specifying disjunction ($\sqcup$). Similarly we can specify subsumption axioms using the keyword **implies** or role axioms using the keywords **transitive**, **parent** and **inverse** for transitive role axioms, role inclusion axioms or specifying the inverse of a role, respectively.

On the other hand individual axioms (assertion) of KRSS need to be extended in order to capture confidence degrees. More precisely, fuzzy concept and role assertions are specified by using the following patterns:

```
(instance ind Concept ineqType n)
(related ind1 ind2 Role ineqType n)
```

where ineqType is one of ">=", ">", "<=", "<", "=", and $n \in (0,1]$ is a degree. Thus, in the first syntax we use the keyword **instance** to declare a fuzzy assertion between an individual and a concept with some inequality type and degree n; similarly with role assertions and keyword **related**.

Example 2. The syntax of the assertions *alice* : Female, (*paul* : (Tall ⊓ Thin) ≥ 0.8) and ((*frank*, *paul*) : has − friend) ≥ 0.7 are shown below in FiRE syntax.

```
(instance alice Female)
(instance paul (and Tall Thin) > 0.8)
(related frank paul has-friend >= 0.7)
```

4.3   Inference Services

FiRE offers all the fuzzy DL inference services we introduced in Section 2 plus a *global glb* service and answering *conjunctive queries* over RDF repositories, described below. More precisely, it allows to check ABox consistency. If the ABox is consistent w.r.t. to a TBox and an RBox, FiRE provides the user with a sample model of the knowledge base in the Model tab of the output panel. If the ABox is not consistent then a "**not satisfiable**" message is reported in the tableaux tab.

Then, FiRE offers a number of specialized tabs in the inference services panel that implement many services. More precisely, it offers an *Entailment* inference tab that allows users to ask for the entailment of fuzzy assertions. The syntax for such queries is the same as the syntax of specifying concept assertions. For example, in order to check whether $\Sigma \models (a : C) \geq n$ the user should enter the statement **instance a C >= n** in the entailment tab. On the other hand subsumption queries are specified in the *Subsumption* inference tab. Their syntax is of the following form (**concept1**) (**concept2**) where concept1 and concept2 are $f_{KD}\text{-}\mathcal{SHIN}$-concepts.

Subsequently, FiRE offers for computing the glb of an individual to a concept w.r.t. a knowledge base Σ. Glb queries are evaluated by FiRE performing entailment queries for all the degrees contained in the ABox, using the binary search algorithm in order to reduce the entailment tests. The syntax of glb queries is of the form **individual (concept)** where concept can be an $f_{KD}\text{-}\mathcal{SHIN}$-concept. Besides glb queries, FiRE offers for computing the *global glb* of a knowledge base. More precisely, it computes the glb of all the individuals in the ABox with all the defined concepts of the TBox. Roughly speaking, this process materializes (almost) all the relevant implied knowledge that is entailed by the knowledge base, i.e. the one that involves the defined concepts.

Finally, besides the standard inference services of fuzzy DLs, FiRE also offers the *Queries* inference tab, which can be used in order to issue expressive fuzzy conjunctive queries over a Sesame repository. More precisely, the user can issue *conjunctive threshold queries* (CTQs) or *generalized fuzzy conjunctive queries* (GFCQs), like *fuzzy threshold queries*, *fuzzy aggregation queries* and *fuzzy weighted t-norm queries*, as these have been defined and implemented for fuzzy-DL-Lite in [11]. An example GFCQ is the following:

```
x <- Goodlooking(x):0.6 ^ has-hairLength(x,y):1 ^ Long(y):0.8
```

asking for all x that are good looking and have long hair. We see that in such queries the user is capable of also specifying weights in query atoms.

(a) (b)

Fig. 2. Input image (left) and its segmentation (right)

5   Two Usage Scenarios

In the current Section we will present two application scenarios where we have tested FiRE and its potentials.

5.1   Multimedia Analysis and Scene Interpretation

One of the main research problems in multimedia analysis is how one could extract and represent all the underlying information and knowledge that exist within an image or a video. For example, an image could depict an event, a landscape, people, etc. that need to be represented in order for end-users to be able to query about them. Manual annotation is obviously very difficult and expensive hence (semi)automatic ways are explored. First, we apply image analysis algorithms, which are based on color, texture and shape criteria to group pixels and create segments which (possibly) depict an object. Subsequently, we apply a recognition system which ideally would be able to assign a semantic label to each region. Generally, this task is very difficult since moving from low-level features to high-level semantic descriptions, like complex objects is far from trivial. For those reasons proposals for knowledge-based multimedia analysis have been proposed [15, 16]. Using DLs one can provide definitions of high-level concepts and events that might exist in an image or video in order to assist the process of recognition. For example, we could have the following DL axioms:

$$
\begin{aligned}
\text{Leaves} &\equiv \text{GreenColored} \\
\text{Tree} &\equiv \text{BrownColored} \sqcap \exists \text{isConnected.Leaves} \\
\text{MuddyRoad} &\equiv \text{BrownColored} \sqcap \text{CoarseTextured}
\end{aligned}
$$

Image analysis is generally a process that involves a huge amount of uncertain and vague knowledge, hence we would prefer to use extended frameworks like fuzzy DLs as the underlying logical framework. Consider for example Figure 2(a) which shows a sample input image, while Figure 2(b) shows its segmentation. We see that the algorithm has identified several regions in the image for which we

Table 3. Semantic labelling

Region	Extracted Concept	Degree	Inferred Concept	Degree
$region_1$	GreenColored	0.80	Leaves	0.80
$region_2$	LightGreenColored	0.78	Grass	0.78
$region_3$	LightGreenColored	0.71	Grass	0.71
$region_4$	BrownColored	0.69	MuddyRoad	0.69
	CoarseTextured	0.80		
$region_5$	CoarseTextured	0.30	ClayRoad	0.30
	LightBrownColored	0.85		
$region_6$	BrownColored	0.67	MuddyRoad	0.67
	CoarseTextured	0.80		
$region_7$	LightGrayColored	0.72	TarRoad	0.70
	SmoothTextured	0.70		

can extract their MPEG-7 visual descriptors.[4] These are numerical values which provide information about the texture, shape and color of a region. Obviously, these values are very low-level and provide no semantic information. Nevertheless, one could use them in order to *move* from low-level descriptions to more high-level ones. For example, if reg_1's green component in the RGB color model was equal to 243, we can be based on a mapping (fuzzy partition) function [17] and deduce that reg_1 is GreenColored to a degree at least 0.8. On the other hand another region with a green component of 200 could be GreenColored to a degree 0.77. Similarly, we can extract additional fuzzy assertions using other MPEG-7 descriptors, like texture or shape. For example, in the leftmost part of Table 3 we see some fuzzy assertions extracted for a specific region and a concept, using MPEG-7 descriptors. Subsequently, we can use FiRE's global glb service in order to extract all the implied knowledge for the specific image [18]. The inferred assertions are depicted in the rightmost part of Table 3. We see that fuzzy DL reasoning can be used to provide more sophisticated labelling, but please note that these are still some *very* preliminary results and the current example is by no means complete.

5.2   Knowledge Based Information Retrieval and Recommendation

FiRE has been applied in an industrial strength Use Case scenario from a Greek National project. In this Use Case scenario we consider a production company, which has a knowledge base that consists of videos and images about persons (which usually are actors or models). This company wants to publish its content on the (Semantic) Web so as other advertisement or production companies can use this knowledge base to look for persons to be employed for advertisements (casting). Each entry in the knowledge base contains a photo or a video, and some specific information like body and face characteristics, age or profession-like characteristic. A user can query the knowledge base providing

---

[4] http://www.chiariglione.org/mpeg/standards/mpeg-7/mpeg-7.htm

information like the name, the height, the type of the hair, the body, age range, and more.

Usually casting people want to query such a knowledge base using some high level concepts like "Thirties", "MiddleAged", "Teen", "Kid", "Slim", "Tall", "StudentLooking", "TeacherLooking" and more, which can be used in commercials of respective context. Obviously, most of these concepts are vague (fuzzy) as for example the concepts of middle aged or tall persons cannot be precisely defined. In order to tackle the above Use Case scenario we have followed the next steps [18]:

1. *Database (DB) fuzzification:* First, we fuzzify fields of the database, in order to provide symbolic information from the existing numerical one. For example, the "age" field provides very low level information which can be used in order to define (fuzzy) concepts, like "Teen", "Twenties", "Thirties", "Old" etc. These concepts are defined as functions (fuzzy sets) that map the age value of a person a to the membership degree of a to them. Thus, we can crate fuzzy assertions. For example, the DB has that $john180$ is 34 years old, thus the function of "Thirties" tells us that $john180$: Thirties ≥ 0.6.

2. *Ontology construction:* Using the above concepts, together with additional ones of our domain, we can construct an ontology for human actors (models) focusing on appearance, that is important for casting tasks. For example, we can define the concept of student looking, tall child and scientist as:

$$
\begin{aligned}
\text{StudentLooking} &\sqsubseteq \text{Kid} \sqcup \text{Teen} \\
\text{TallChild} &\sqsubseteq \text{Child} \sqcap (\text{Short} \sqcup \text{Normal\_Height}) \\
\text{Scientist} &\sqsubseteq \text{Male} \sqcap \text{Classic} \sqcap (\text{50s} \sqcup \text{60s}) \sqcap \\
&\qquad \text{Serious} \sqcap \exists\text{has-eyeCondition.Glasses}
\end{aligned}
$$

using already defined concepts. Please note that if we hadn't created the fuzzy concepts Kid, Teen, Child, Short and Normal_Height in step 1, which initially did not exist in the database, we would not be able to define the above concepts. Similarly, we can define more concepts, like GrandParent, FishermanLooking and more.

3. *Extracting implied knowledge:* The ontology together with the fuzzy assertions that are produced by step 1, due to fuzzification, as well as the crisp assertions that exist in the database (e.g. $john180$ is a Male, Latin, etc.) is loaded into FiRE. Then we compute the global glb of the knowledge base in order to extract implied knowledge. Subsequently, knowledge is serialized and stored into Sesame.

4. *Querying the KB:* Finally, end-users can issue very expressive fuzzy conjunctive queries over Sesame through the FiRE platform in order to retrieve actors. For example, for a TV commercial for hair dyes one might want to retrieve all female models, that are in their twenties, have long, good quality hair and nice eyes, or for an MP3 player commercial one might want a student looking model.

6   Discussion and Open Problems

It has been widely approved that fuzzy DLs could play an important role in the Semantic Web by serving as a mathematical framework for knowledge representation and reasoning in applications that face vague knowledge, like image analysis and understanding [19], ontology searching [11], semantic portals [20] multimedia retrieval [21] and negotiation [22]. But still the full potential of fuzzy DLs has not been exhaustively explored, since they could be used in a wealth of tasks and applications in order to enhance automation and handle degrees of confidence, membership and truth that emerge by matching, retrieval, recommendation, negotiation or recognition systems.

After the first ideas about extending classical two-valued Description Logics with fuzzy Set Theory, by Yen in [23], Tresp and Molitor [24] and Straccia [5], there has been an increasing research effort on fuzzy Description Logics. The last couple of years research is focused on providing reasoning support for very expressive fuzzy DLs, in order to support reasoning in a full fuzzy extension of the OWL web ontology language. Towards this direction, recently Stoilos et. al. [6] presented a reasoning algorithm for the fuzzy DLs $f_{KD}\text{-}\mathcal{SI}$ and $f_{KD}\text{-}\mathcal{SHIN}$, while also in another work Stoilos et. al. [12] presented an algorithm for reasoning with General Concept Inclusion axioms, which was an open problem in fuzzy DLs. Interestingly, these results gave rise to the FiRE fuzzy DL systems, presented in section 4 (also a preliminary version was reported in [8]). Furthermore, the study of reasoning algorithms for fuzzy DLs that use other norm operation is also beginning to flourish, although still most results are focused on rather basic DLs like $\mathcal{ALC}$. More precisely, Straccia [25] presented an algorithm for $f_L\text{-}\mathcal{ALC}(\mathcal{D})$, and recently Bobillo and Straccia [26] a reasoning algorithm for $f_P\text{-}\mathcal{ALCf}(\mathcal{D})$ ($\mathcal{ALC}$ with functional role axioms). Also these algorithms are supported by the *fuzzyDL* system [27].

On the other hand, a recent trend in DL research is mainly focused in studying efficient and scalable (tractable) Description Logics, compared to the NEXPTIME-complete OWL DL. Following this trend Straccia proposed a fuzzy extension of DL-Lite [28]. DL-Lite [29] is an interesting lightweight ontology language, since it can answer conjunctive queries in a very efficient way, by using existing database technologies. Later Pan et al. [11] proposed some very expressive extensions to the conjunctive queries of f-DL-Lite. The algorithms for these queries were implemented in the system ONTOSEARCH2[5] and evaluation showed that these can still be answered in a very efficient way. Other interesting tractable DLs are those of the $\mathcal{EL}$ family, like $\mathcal{EL}+$ [30], which provide efficient algorithms for classifying big terminologies. Recently, Stoilos et. al. [31] presented an algorithm for $f_G\text{-}\mathcal{EL}+$ which classifies terminologies that also use fuzzy subsumption [25]. An overview of the field of fuzzy Description Logics can also be found in [32].

As we see from the above, regarding the theoretical side, fuzzy DLs have been studied relatively enough and their logical and mathematical properties are beginning to get quite understood. Another important side is the development

---

[5] http://dipper.csd.abdn.ac.uk/OntoSearch/

of tools and systems that would provide a flexible and efficient way to build and manage fuzzy knowledge. Although this aspect has not been explored much yet, there are again some first works towards this direction. We have reported about one such work in the current paper, and more precisely the FiRE system, which consists of (i) a beta fuzzy DL reasoner for f$_{KD}$-$\mathcal{SHIN}$, (ii) a GUI for editing and creating fuzzy KBs using the KRSS format and (iii) a module that provides persistent storage of large amounts of fuzzy knowledge bases in the RDF triple store Sesame and implements very expressive fuzzy conjunctive queries [11] over it, by extending Sesame's SeRQL query.

Still there is plenty of way to go until we can provide adequate support for fuzzy knowledge engineering and management. First, no support for parsing RDF/XML files that contain fuzzy assertions (as these have been described in [7]) exists. Moreover, there is currently no evidence about the scalability of the existing expressive fuzzy DL reasoning systems. In other words optimization techniques need to be investigated; some preliminary investigations have been carried out in [33] but still no evaluation or fuzzy DL system has been reported. Most important of all, besides the very basic manual support provided by current systems, there are currently no available graphical tools for assisting end users to (semi) automatically create fuzzy knowledge bases from raw numerical data. All these issues are very important in order for fuzzy DL technologies to be more widely adoptable in the Semantic Web.

Acknowledgements

The work of Giorgos Stoilos, Giorgos Stamou, Vassilis Tzouvaras and Nick Simou was partially supported by EU projects X-Media (FP6-26978) and BOEMIE (FP6-027538). We would also like to thank Thanos Athanasiadis for providing the image and segmentation figures.

References

1. Berners-Lee, T., Hendler, J., Lassila, O.: The semantic web. Scientific American (2001)
2. Horrocks, I., Patel-Schneider, P.F., van Harmelen, F.: From and RDF to OWL: The making of a web ontology language. Journal of Web Semantics 1 (2003)
3. Baader, F., McGuinness, D., Nardi, D., Patel-Schneider, P.: The Description Logic Handbook: Theory, implementation and applications. Cambridge Uni. Press, Cambridge (2002)
4. Euzenat, J., Shvaiko, P.: Ontology Matching. Springer, Heidelberg (2007)
5. Straccia, U.: Reasoning within fuzzy description logics. Journal of Artificial Intelligence Research 14, 137–166 (2001)
6. Stoilos, G., Stamou, G., Tzouvaras, V., Pan, J.Z., Horrocks, I.: Reasoning with very expressive fuzzy description logics. Journal of Artificial Intelligence Research 30, 273–320 (2007)
7. Stoilos, G., Stamou, G., Tzouvaras, V., Pan, J., Horrocks, I.: Fuzzy OWL: Uncertainty and the semantic web. In: Proc. of the International Workshop on OWL: Experiences and Directions (2005)

8. Stoilos, G., Simou, N., Stamou, G., Kollias, S.: Uncertainty and the semantic web. IEEE Intelligent Systems 21, 84–87 (2006)
9. Stoilos, G., Stamou, G., Tzouvaras, V., Pan, J., Horrocks, I.: A fuzzy description logic for multimedia knowledge representation. In: Proc. of the International Workshop on Multimedia and the Semantic Web (2005)
10. Stoilos, G., Stamou, G., Tzouvaras, V., Pan, J., Horrocks, I.: The fuzzy description logic . In: Proc. of the International Workshop on Uncertainty Reasoning for the Semantic Web, pp. 67–76 (2005)
11. Pan, J., Stamou, G., Stoilos, G., Thomas, E.: Scalable querying services over fuzzy ontologies. In: Proceedings of the International World Wide Web Conference (WWW 2008), Beijing (2008)
12. Stoilos, G., Straccia, U., Stamou, G., Pan, J.: General concept inclusions in fuzzy description logics. In: Proceedings of the 17th European Conference on Artificial Intelligence (ECAI 2006), pp. 457–461 (2006)
13. Li, Y., Xu, B., Lu, J., Kang, D.: Discrete tableau algorithms for $\mathcal{FSHI}$. In: Proceedings of the International Workshop on Description Logics (DL 2006), Lake District, UK (2006)
14. Broekstra, J., Kampman, A., van Harmelen, F.: Sesame: A generic architecture for storing and querying rdf and rdf schema. In: Horrocks, I., Hendler, J. (eds.) ISWC 2002. LNCS, vol. 2342, pp. 54–68. Springer, Heidelberg (2002)
15. Sciascio, E.D., Donini, F.: Description logics for image recognition: a preliminary proposal. In: International Workshop on Description Logics, DL 1999 (1999)
16. Neumann, B., Möller, R.: On scene interpretation with description logics. In: Christensen, H.I., Nagel, H.-H. (eds.) Cognitive Vision Systems. LNCS, vol. 3948, pp. 247–278. Springer, Heidelberg (2006)
17. Klir, G.J., Yuan, B.: Fuzzy Sets and Fuzzy Logic: Theory and Applications. Prentice-Hall, Englewood Cliffs (1995)
18. Simou, N., Stoilos, G., Pardalis, K., Tzouvaras, V., Stamou, G., Kollias, S.: Storing and querying fuzzy knowledge in the semantic web, Technical Report (2008)
19. Simou, N., Athanasiadis, T., Tzouvaras, V., Kollias, S.: Multimedia reasoning with . In: 2nd International Workshop on Semantic Media Adaptation and Personalization, London, December 17-18 (2007)
20. Holi, M., Hyvonen, E.: Fuzzy view-based semantic search. In: Asian Semantic Web Conference (2006)
21. Meghini, C., Sebastiani, F., Straccia, U.: A model of multimedia information retrieval. Journal of the ACM 48, 909–970 (2001)
22. Ragone, A., Straccia, U., Di Noia, T., Di Sciascio, E., Donini, F.: Vague knowledge bases for matchmaking in p2p e-marketplaces. In: Franconi, E., Kifer, M., May, W. (eds.) ESWC 2007. LNCS, vol. 4519, pp. 414–428. Springer, Heidelberg (2007)
23. Yen, J.: Generalising term subsumption languages to fuzzy logic. In: Proc of the 12th Int. Joint Conf on Artificial Intelligence (IJCAI 1991), pp. 472–477 (1991)
24. Tresp, C., Molitor, R.: A description logic for vague knowledge. In: Proc of the 13th European Conf. on Artificial Intelligence, ECAI 1998 (1998)
25. Straccia, U.: Description logics with fuzzy concrete domains. In: 21st Conf. on Uncertainty in Artificial Intelligence (UAI 2005), Edinburgh (2005)
26. Bobillo, F., Straccia, U.: A fuzzy description logic with product t-norm. In: Proceedings of the IEEE International Conference on Fuzzy Systems (Fuzz-IEEE 2007), London (2007)
27. Straccia, U.: FuzzyDl: An expressive fuzzy description logic reasoner. In: Proceedings of the International Conference on Fuzzy Systems, Fuzz-IEEE 2008 (2008)

28. Straccia, U.: Answering vague queries in fuzzy DL-Lite. In: Proceedings of the 11th International Conference on Information Processing and Management of Uncertainty in Knowledge-Based Systems (IPMU 2006), pp. 2238–2245 (2006)
29. Calvanese, D., Giacomo, G.D., Lembo, D., Lenzerini, M., Rosati, R.: Dl-lite: Tractable description logics for ontologies. In: Proceedings of the 20th National Conference on Artificial Intelligence, AAAI 2005 (2005)
30. Baader, F., Lutz, C., Suntisrivaraporn, B.: Is tractable reasoning in extensions of the description logic $\mathcal{EL}$ useful in practice? Journal of Logic, Language and Information, Special Issue on Method for Modality (to appear, 2008)
31. Stoilos, G., Stamou, G., Pan, J.: Efficient classification of fuzzy subsumption with fuzzy-$\mathcal{EL}+$. In: Proceedings of the 21st International Workshop on Description Logics (DL 2008), Dresden (2008)
32. Lukasiewicz, T., Straccia, U.: Managing uncertainty and vagueness in description logics for the semantic web. Journal of Web Semantics (2008)
33. Haarslev, V., Pai, H.I., Shiri, N.: Optimizing tableau reasoning in ALC extended with uncertainty. In: Proceedings of the 20th International Workshop on Description Logics (DL 2007), pp. 307–314 (2007)

Towards Machine Learning on the Semantic Web

Volker Tresp[1], Markus Bundschus[2], Achim Rettinger[3], and Yi Huang[1]

[1] Siemens AG, Corporate Technology, Information and Communications
Learning Systems, Munich, Germany
[2] Ludwig-Maximilian University Munich, Germany
[3] Technical University of Munich, Germany

Abstract. In this paper we explore some of the opportunities and challenges for machine learning on the Semantic Web. The Semantic Web provides standardized formats for the representation of both data and ontological background knowledge. Semantic Web standards are used to describe meta data but also have great potential as a general data format for data communication and data integration. Within a broad range of possible applications machine learning will play an increasingly important role: Machine learning solutions have been developed to support the management of ontologies, for the semi-automatic annotation of unstructured data, and to integrate semantic information into web mining. Machine learning will increasingly be employed to analyze distributed data sources described in Semantic Web formats and to support approximate Semantic Web reasoning and querying. In this paper we discuss existing and future applications of machine learning on the Semantic Web with a strong focus on learning algorithms that are suitable for the relational character of the Semantic Web's data structure. We discuss some of the particular aspects of learning that we expect will be of relevance for the Semantic Web such as scalability, missing and contradicting data, and the potential to integrate ontological background knowledge. In addition we review some of the work on the learning of ontologies and on the population of ontologies, mostly in the context of textual data.

1   Introduction

The world wide web (WWW) represents an ever increasing source of information. Until now the WWW is mostly accessible to humans via search engines and browsers whereas computers only have a very rudimentary understanding of web content. The vision behind the Semantic Web (SW) is that computers should also be able to understand and exploit information offered on the web [1]. In the near future, a web representation might contain human-readable parts and sections made available in SW-formats to be accessible for automated processing. The SW is based on two concepts. First, a formal ontology provides domain specific background information that is shared by several parties: It provides a common vocabulary for a given domain and describes object classes, predicate classes and their interdependencies, as well as additional background information formalized in logical statements. Second, web information is annotated by

P.C.G. da Costa et al. (Eds.): URSW 2005-2007, LNAI 5327, pp. 282–314, 2008.

statements readable and interpretable by machines via the common ontological background knowledge.

One of the prime SW applications will be context/user sensitive information retrieval where the result will still be in textual or multimedia format, to be interpreted by a human. But this information will be much more specific to the user's needs, since data can be integrated from multiple sites and smart information filters can be applied. Thus a search engine becomes more of an agent who knows the user, who has a deep understanding of the information request, who knows what to find where on the web and who presents the requested information in an appropriate user-friendly form. An immediate benefit from semantic annotation will be that annotated web pages might obtain a higher search rank since the match between query and page content can be evaluated with high confidence. In a second group of applications, the items to be searched for are not human readable texts or multimedia data but are machine readable information about an item or a web service. Semantic web services are of great interest both for academia and industry [2,3]. Service requests and service offerings can be formulated precisely based on SW standards and can be understood as precisely by semantic search engines and web applications. In the third family of applications the SW becomes the *web of data*. SW technologies will form the infrastructure for a standardized representation of information and for information exchange. Biomedicine is a forerunner here with almost 1000 databases publicly available today. If the data were published under a common SW ontological format, all this information would be accessible for querying and for analysis. As the WWW brought the knowledge of the world to our finger tips, the SW will bring the data of the world to our applications. Finally, in a fourth family of applications, SW technologies are being used in advanced expert systems to model complex industrial domains [4].

Reasoning plays an important role on the SW: Based on ontological background knowledge and the set of asserted statements, logical reasoning can derive new statements. But logical reasoning has its limitations. First, logical reasoning does not easily scale up to the size of the web as required by many applications; projects like the EU FP 7 Large-Scale Integrating Project LarKC are under way to address this issue [5]. Second, uncertain information is not suitable for logical reasoning. The representation of uncertain information on the SW and reasoning with uncertainty on the SW have only recently been addressed [6]. Third, logical reasoning is completely based on axiomatic prior knowledge and does not exploit regularities in the data that have not been formulated as ontological background knowledge. In contrast, and as it has been demonstrated in many application areas, successful solutions can often be achieved by induction, i.e., by learning from data. The analysis of the potential of machine learning for the SW is the topic of this contribution.

The most immediate application of machine learning is SW mining, enhancing traditional web mining applications. Web content mining, web structure mining, web usage mining and the learning of ranking functions for retrieval will all benefit from the additional information available on the SW [7]. In another group of applications, machine learning serves the SW by supporting ontology construction and management, ontology evaluation, ontology refinement, ontology evolution, as well as the mapping, merging and alignment of ontologies [8,9,10,11,12]. Mostly these tasks are addressed on the basis of unstructured or semi-structured textual data. After all, most current web pages contain textual information; but other types of input data will become increasingly important, as well [13]. Alternatively, researchers are concerned with learning of data already in SW formats. As already mentioned, the current trend is that an increasing amount of information is made available in SW formats and machine learning and data mining will be the basis for the analysis of the combined data sources. A particular aspect here is the learning of logical constraints that can then be formulated in the language of the employed ontology [14,15,16,17]. One can also contemplate that future ontologies should be extended to be able to represent learned information that cannot easily be formulated with current standards, e.g., represent the input-output mapping represented in probabilistic classifiers. The trained statistical models can then be used to estimate the probability that statements are true, which are neither explicitly asserted in the database nor can be proven to be true (or false) based on logical reasoning. Since the conclusions drawn from machine learning are typically probabilistic, this uncertainty needs to be represented [6,18,5]. Consequently, querying can include learned statements, e.g., : *Find all female persons that live in the southeastern US, are older than 21 years, own a house and are likely to own a sailboat* where the last information, i.e., the likelihood of owning a sailboat, was learned from data. Finally, in applications where the raw data is unstructured, machine learning can support the population of ontologies, i.e., the mapping of unstructured data to SW statements. Most work here concerns the population from textual data although the annotation of semi-structured data and multimedia data. e.g. images and video, is of great relevance as well. A goal here is to describe multimedia content semantically for fast content-based reasoning and retrieval.

In this paper we analyze algorithms from machine learning that are suitable for SW applications. First and foremost, SW data describe relationships between objects. Consequently, suitable learning approaches should be able to handel the relational character of the data. By far the majority of machine learning deals with non-relational feature-based representations (also referred to as propositional representation or attribute-value representation). Only recently statistical relational learning (SRL) is finding increasing interest in the ML community [19]. In Section 3 we present a novel discussion on feature-based learning in the SW and in Section 4 we relate this discussion to learning algorithms from inductive logic programming (ILP). In Section 5 we discuss matrix decomposition approaches and in Section 6 we present relational graphical models that are based on a joint probabilistic model of a relational domain. We discuss the machine learning approaches with respect to their applicability in a SW context, i.e., their scalability to the expected large size of the SW, their ability to integrate ontological background knowledge, their ability to handle the varying quality and reliability of data[1] and finally, their ability to deal with missing and contradictory data. In Section 7 we add a discussion on ontology learning and ontology population based on textual data. Ontology learning and ontology population are the most developed aspects of machine learning on the SW. In Section 8 we report first experiments based on the FOAF data set and in Section 9 we present conclusions. We will start the remaining part of the paper with an introduction into the SW as proposed by the W3C.

2   Components of the SW Languages

The World Wide Web Consortium (W3C) [21] is the main international standards organization for the WWW and develops recommendations for the SW. We will discuss here the main SW standards, i.e., RDF, RDFS and OWL [22,23]. RDF is useful for making statements about instances, RDFS defines schema and subclass hierarchies, and OWL can be used to formulate additional background knowledge. Very elegantly, the statements in RDF, RDFS and OWL can all be represented as one combined directed graph (Figure 1). A common semantics is based on the fact that some of the language components of RDFS and OWL have predefined domain-independent interpretations.

2.1   RDF: A Data Model for the SW

The recommended data model for the SW is the resource description framework (RDF). It has been developed to represent information about resources on the WWW (e.g., meta data/annotations), but might as well be used to describe other structured data, e.g., data from legacy systems. A resource stands for an object that can be uniquely identified via a uniform resource identifier, URI, which is sometimes referred to as a bar code for objects on the SW. The basic statement is a triple of the form *(subject, property, property value)* or, equivalently, *(subject, predicate, object)*. For example *(Eric, type, Person), (Eric, fullName, Eric Miller)* indicates that Eric is of the concept (or class) *Person* and that Eric's full name is Eric Miller. A triple can graphically be described as a directed arc, labeled by the property (predicate) and pointing from the subject node to the property value node. The subject of a statement is always a URI, the property value is either also a URI or a literal (e.g., String, Boolean, Float). In the first case, one denotes the property as object property and a statement as an object-to-object statement. In the latter case one speaks of a datatype property and of an object-to-literal statement. A complete database (triple store) can then be displayed as a directed graph, a semantic net (Figure 1). One might think of a triple as a tuple of a binary relation *property(subject, property values)*. A triple can only encode a binary relation involving the subject and the property value. Higher order relations are encoded using blank nodes. Consider the originally ternary relation *transaction(User, Item, Rating)*. The blank node might be

---

[1] Trust learning is an emerging field [20].

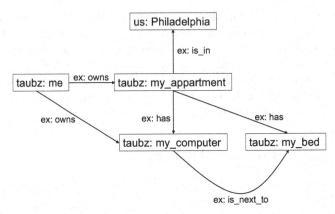

Fig. 1. An RDF-graph fragment. Redrawn from [24].

TransactionId with triples (binary relations): *(TransactionId, userRole, User)*, *(TransactionId, transactionObject, Item)* and *(TransactionId, evaluation, Rating)*. A blank node is treated as a regular resource with an identifier, only that it might be invisible from outside the file. Blank nodes are also helpful for defining containers such as bags (unordered container), sequences (ordered container) and collections (lists).

Each resource is associated with one or several concepts (i.e., classes) via the type-property. A concept can be interpreted as a property value in a type-of-statement. Conversely, one can think of a concept as representing all instances belonging to that concept. Concepts are defined in the RDF Vocabulary Description Language, also called RDF-Schema or RDFS. Both RDF and RDFS form a joint RDF/RDFS graph. In addition to defining all concepts, the RDFS also contains certain properties that have a predefined meaning, implementing specific constraints and entailment rules. First, there is the subclass property. If an instance is of type *Concept1* and *Concept1* is a subclass of *Concept2*, then the instance can be inferred to be also of type *Concept2*. Subclass relations are essential for generalization in reasoning and learning. Each property has a representation (node) in RDFS as well. A property can be a subproperty of another property. For example, the property *brotherOf* might be a subproperty of *relatedTo*. Thus if *A* is a brother of *B* one can infer that *A* is *relatedTo B*.

A property can have a domain respectively range constraint: *(marry, domain, Person)* and *(marry, range, Person)* states that if two resources are married then they must belong to the concept *Person*. Interestingly, RDF/RDFS statements cannot lead to contradictions in RDF/RDFS, one reason being that negation is missing. The same remains true for some less expressive ontologies.

2.2 Ontologies

Ontologies build on RDF/RDFS and add expressiveness. W3C developed standards for the web ontology language OWL, which comes in three dialects or profiles: the most expressive is OWL Full, which is a true superset of RDFS. A full inference procedure for OWL Full is not implementable with simple rule engines [23]. Some applications requiring OWL Full might build an application-specific reasoner instead of using a general one. OWL DL (description language) is included in OWL Full and OWL Lite is included in OWL DL. Both OWL DL and OWL Lite are decidable but are not true supersets of RDFS.

In OWL one can state that classes are equivalent or disjoint and that properties respectively instances are identical or different. The behavior of properties can be classified as being symmetric, transitive, functional or inverse functional, ... (e.g., teaches is the inverse of isTaughtby). In RDFS concepts are simply named. OWL allows the user to construct classes by enumerating their content (explicitly stating its members), through forming intersections, unions and complements of classes. Also classes can be defined via property restrictions. For example, the constraints that (1) first-year courses must be taught by professors, (2) mathematics courses are taught by David Billington, (3) all academic staff members must at least teach one undergraduate course, can all be expressed in OWL using the constructs allValuesFrom ($\forall$), hasValue, and someValuesfrom ($\exists$). Furthermore, cardinality constraints can be formulated (e.g., a course must be taught by someone, a department must have at least ten and at most 30 members) (Examples from [22]). Very attractive is that both instances and ontologies can be joined by simply joining the graphs: in fact the only real thing is the graph [23].

In some data rich applications ontologies will have no relevance beyond the definition of classes and properties. Conversely, in some domains, such as bioinformatics, medical informatics and some industrial applications [4], sophisticated ontologies have already been developed [23].

2.3   Reasoning

An ontology formulates logical statements, which can be used for analyzing data consistency and for deriving new implicit statements concerning instances and concepts. Total materialization denotes the calculation of all implicit triples at loading time, which might be preferred if query response time is critical [25]. Note, that total materialization is only feasible in some restricted ontologies.

2.4   Rules

RuleML (Rule Markup Language) is a rule language formulated in XML and is based on datalog, a function-free fragment of Horn clausal logic. RuleML allows the formulation of if-then-type rules. Both RuleML and OWL DL are different subsets of first-order logic (FOL). SWRL (Semantic Web Rule Language) is a proposal for a Semantic Web rules-language, combining sublanguages of OWL (OWL DL and Lite) with those of the Rule Markup Language (Unary/Binary Datalog). Datalog clauses are important for modeling background knowledge in cases where DL might be inappropriate, for example in many industrial applications.

2.5   Querying

The recommended RDF-query language for the SW is SPARQL (SPARQL Protocol and RDF Query Language). The SPARQL syntax is similar to SQL. A search pattern is a directed graph with variable nodes (i.e., a graph pattern). The result is is either in the form of a list of variable bindings or in the form of an RDF-graph.

3   Feature-Based Statistical Learning on the SW

3.1   Feature-Based Statistical Learning

Based on a long tradition, statistical learning has developed a large number of powerful analytical tools and it is highly desirable to make these tools available for the SW. Figure 2 (top) shows the main steps that are performed in statistical learning, analyzing, as example, students in a university. First, a *statistical unit* is defined, which is the entity that is the source of the variables or features of interest [26,27,28]. The goal is to generalize from observations on a few units to a statistical assembly of units. Typically a statistical unit is an object of a given type, here a student. In general one is not interested in all statistical units but only in a particular subset, i.e., the *population*. The population might be defined in various ways, for example it might concern all students in a particular country or, alternatively, all female students at a particular university.

In a statistical analysis only a subset of the population is available for investigation, i.e. a *sample*. Statistical inference is dependent on the details of the sampling process; the sampling process essentially defines the random experiment and, as a stationary process, allows the generalization from the sample to the population. In a *simple random sample* each unit is selected independently. Naturally, sometimes more complex sampling schemes are used, such as stratified random sampling, cluster sampling, and systematic sampling.

The quantities of interest of the statistical investigation are the *features* (or variables) that are derived from the statistical units. In the example, features are a student's IQ and a student's age. In the next step the *data matrix* is formed where each row corresponds to a statistical unit and each column corresponds to a feature. Finally, an appropriate statistical model is employed for modeling the data, i.e., the analysis of the features and the relationships between the features, and the final result is analyzed by the user. Naturally, all of this is typically an iterative process, e.g., based on a first analysis new features might be added and the statistical model might be modified.

In a supervised statistical analysis one partitions the features in explanatory variables (a.k.a. independent variables, predictor variables, regressors, controlled variables, input variables) and dependent variables (a.k.a response variables, the regressands, the responding variables, the explained variables, or the outcome/output variables). Note that it is often a design choice if one either defines a population based on the state of a variable or if one uses that variable as an independent variable. Consider a binary variable male/female. One choice

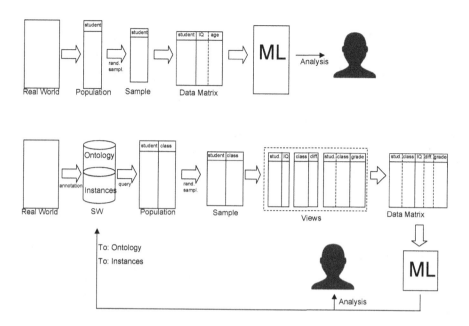

Fig. 2. Top: Standard machine learning. Bottom: Machine learning applied to the SW.

might be to partition the population into males and females and learn separate models for each population. Another option is to simply use gender as an independent variable and consider a joint population of males and females. The second choice is for example more appropriate if the sample is small. Hierarchical Bayesian modeling is a compromise in which statistical inference in different populations is coupled.

3.2 Feature-Based Statistical Learning on the SW

The main steps for statistical learning on the SW are displayed in Figure 2 (bottom). The first new aspect is that the statistical analysis is based on the world as it is represented on the SW and that all quantities of interest, i.e., statistical unit, population, sample and features, are defined in context of the SW.[2] As before, a *statistical unit* might be defined to be an object of a given type, e.g., a student. More generally a statistical unit might be composed of several objects that have a particular relationship to each other. In Figure 2, as an example, a statistical unit might be a composed entity consisting of a student and a class that the student attends, i.e., a registration.

A *population* might now be defined by a SW query that produces a table whose tuples (i.e., variable bindings) correspond to the objects that identify a

---

[2] Technically one needs to be aware that the generation of a sample with the help of a search engine or a crawler might introduce a bias, for example, if snowball sampling is employed.

statistical unit. In the example in Figure 2 we might define a query to generate a population table with objects student and class; a tuple then stands for the statement that a student registered in a particular class. Sampling, as before, selects a proper random subset of the population. A particular aspect of SW data is the dominance of relationships between objects. Thus, features that are calculated for a statistical unit might reflect this relationship structure.

Technically, one first generates a data matrix. The number of rows in the data matrix is identical to the number of tuples in the sample table, i.e., the number of statistical units in the sample. A statistical unit is a primary key for the table. The data matrix has a fixed number of columns corresponding to the number of features, which are derived for each unit. All matrix entries are initialized to be N/A (not available or missing) and will (partially) be replaced by feature values as described in the following two steps.

Next, database views[3] are generated that contain as attributes the objects in a statistical unit (respectively a subset of those objects) plus additional attributes. In Figure 2, the first view contains the student's ID and the student's IQ, the second view contains the class ID and the class difficulty and the third view contains the student ID, the class ID and the grade the student obtained in a class. Note that views can be generated from rather complex queries.

In the next step, relational features are calculated based on these views. In the simplest case each statistical unit is represented exactly in one tuple in each view and features are calculated based on the tuple attributes. The situation becomes more complex if a statistical unit is not represented in a view or if it is represented more than once. In the first case, i.e., a statistical unit is not represented in a view, one either enters zero or another default entry (e.g., the number of a person's children is zero) or one does not overwrite the corresponding N/A entry in the data matrix (e.g., when a student's IQ is unknown). In the second case, i.e., a statistical unit is represented in more than one tuple in a particular view —in the example if a student attended a class twice and got two grades— some form of aggregation can be applied (number-of, average, max, min, etc.). In domains like the SW, many-to-many relations often play a significant role and can lead to a large number of sparse features: The number of items a customer has acquired is typically still very small if compared to the total number of items. In the case that object IDs are used as features, the learning algorithm needs to be able to handle the potentially high-dimensional sparse data. Technically, it might be possible to execute the described steps, i.e., the generation of the sample, the views and the data matrix, in one SQL/SPARQL operation.

Finally, the statistical model can be applied beyond the sample to the population. It is important to note that we have a well-defined statistical problem as long as we restrict the analysis to the world in as much it is represented in the SW. Of course the SW can grow (and shrink) such that online learning and transfer learning might become applicable. To what degree the statistical model

---

[3] A view is a stored query accessible as virtual table composed of the result set of a query. Alternatively, one could also work with a temporary or persistent table.

can be generalized to the real world needs to be analyzed carefully since sometimes the SW data are generated by multiple parties for their own reasons and not for the purpose of a statistical analysis.

3.3 Search for the Best Features

So far it was assumed that the user would be able to define the features of interest. In particular in supervised learning one is often interested to automate the selection of the best input features. Popescul and Ungar [29] describe a relational learning approach based on a greedy search for optimal relational features derived from SQL queries (see also [30]). Features are dynamically generated by a refinement style search over SQL queries including aggregation, statistical operators, groupings, richer join conditions and argmax based queries. The features are used to predict the target relation using logistic regression. Additional features are generated by clustering, which leads to new "invented" attributes. The authors obtain good results on citation prediction and document classification. It is straightforward to implement a similar search procedure on SW data. Note that the automatic generation of candidate features is certainly attractive; on the other hand the computational burden is quite large; feature definition based on the experimenters insight and some pruning might be adequate in many applications.

3.4 Discussion

Statistical learning on the SW, as presented, is highly scalable since the determining factor is the number of statistical units in the sample, which basically is independent of the size of the SW. One needs to be aware that sampling with the help of a search engine or a crawler might introduce a bias. The queries, which need to be executed for the calculation of the features, can be executed efficiently with current technology [25]. Ontological background knowledge can be integrated in different ways. First, one might perform complete or partial materialization, which would derive statements from reasoning prior to training. Recall that total materialization is only feasible with less expressive ontologies. Second, since the ontology is part of the RDF-graph, features can be defined including ontological concepts of a statistical units, respecting the subclass restrictions. This has effectively been employed in [31]. It is conceivable that the trained statistical models could be added to an extended "probabilistic" ontology, indicated by the arrow at the bottom of Figure 2. In addition, the statistical models derive probabilistic statements about the truth values of triples. For example, if —based on a trained model— it can be derived that a person has a high IQ, this information could be added to the SW [6]. An option is a weighted RDF-triple, the weight reflecting the likelihood that the statement is true. Moreover, if it was found that particular features generated during learning are valuable, one could define corresponding statements and add those to the SW as "invented predicates". The same is true for the latent variables introduced in a cluster analysis or in a principle component analysis (PCA). We should emphasize again that statistical inference strictly speaking is only applicable within the experimental setting of a particular statistical unit, population and sampling approach. Thus if a statistical model allows the conclusion that *statement X is true with 90% probability*, this is only valid in a particular statistical context. Experiments have shown, for example, that predictive performance can depend to some degree on the object selected as statistical unit. An interesting aspect is that the results from a number of statistical models could be combined in a committee machine [32].

Feature generation is nontrivial and might exploit prior knowledge that is partially available in the domain ontology. For example it is relevant that a person only has exactly one age, exactly one mother, but zero or more children. In fact it would be desirable that the ontological information could be exploited in a way such that the statistical framework is automatically constructed requiring a minimum of additional domain background knowledge from the user. A problem with less expressive ontologies might be that one cannot express negation. Consider the example of gene-gene interactions where the literature primarily reports positive results, i.e., positive evidence for gene-gene interactions. Evidence that two genes do not interact would be important to report but might be difficult to represent in less expressive ontologies.

Maybe the most important issue in SW learning concerns missing or incomplete data. We can make a closed-world assumption and postulate that the world only exists in as much as it is represented in the SW: besides the statements that are known to be true or can be derived to be true, all statements are assumed false. Naturally, in many cases we are really interested to perform inference in the real world and it is more appropriate to assume that the truth values of some statements are unknown. Here we should distinguish, first, the case that statistical units are missing and, second, the case that due to missing information features cannot be calculated or features are biased. The first case is not a problem if statistical units are missing at random, e.g., if some of the students at a university are unknown. The situation is more complex if the fact that a statistical unit is missing is dependent on features of interest, e.g., if only smart students are in the data base. Then the missing data mechanism should be included in the statistical model. For the second case consider that the age of a person's father is an important feature that is not available: Either the age of a person's father might be unknown or a person father's ID might be unknown. Another example is that if the number of transactions is an important feature, the feature might be biased if not all transactions are recorded. If a closed-world assumption is not appropriate, one could deal with missing features using the appropriate procedures known from statistics [33]. Again, the missing data mechanism should be included in the statistical model. Also note that ontological information can be quite relevant for dealing with missing data. For example if we know that a person has brown eye color we know that all other statements about eye color must be false, since a person has only one eye color. Note that there are statistical models that can easily deal with missing data such as naive Bayes, many nearest neighbor methods, or kernel smoothers.

Naturally there are cases where simple missing data models are not appropriate, since missing data can render the independent sample assumption invalid. Consider objects of type *Person* and the properties *friendOf* and *income* and *age*. Furthermore assume that from the age of a person and from the income of a person's friends we can predict the income of a person with some certainty. If all features are available, then training an appropriate classifier is straightforward. If in training and testing the income of a person and of a person's friends are partially unknown, we have the situation that the income prediction for one person depends on the income prediction of the person's friends. The situation, where for the prediction of features of a statistical unit (here a person's income) the same features of linked statistical units are required, is typical for data defined on networks. In the analysis of social networks, this situation is referred to as a collective classification problem and a mechanism is added to propagate information using, e.g., Gibbs sampling, relaxation labeling, iterative classification or loopy belief propagation. Recent overviews are presented in [34,35]. One of the first papers demonstrating the benefits of collective classification in social networks is [36] and some important contributions are described in [37,38,39,40]. It is likely that collective classification will also concern SW applications. Interestingly, many social networks have been shown to exhibit homophily, which means that objects with similar attributes (e.g., persons with similar income) are linked (e.g., are friends). In networks exhibiting homophily, simple propagation models, for example based on Gaussian random field models employed in semi-supervised learning [41], give very competitive results. Collective classification is highly related to the relational graphical model approaches described in Section 6, in particular dependency networks [42,43]. Note, that in collective classification, features for different statistical units are not independent and a statistical analysis becomes more involved. Also recall, that we assumed previously that statistical units were selected randomly from the population. In contrast, in collective classification problems the statistical units (for both training and test) would typically be defined by the complete RDF-graph or a connected RDF-subgraph (compare Section 6).

4   Inductive Logic Programming

Inductive logic programming (ILP) encompasses a number of approaches that attempt to learn logical clauses[4] In the view of the discussion in the last section, ILP uses logical (binary) features derived from logical expressions, typically

---

[4] A (logical) literal is either an atomic sentence or a negated atomic sentence. A clause is a disjunction of literals: $l_1 \vee l_2 \ldots \vee l_n$. In a definite clause exactly one literal is positive. A definite clause can be written as an implication (if-then rule): $(\neg l_1 \wedge \neg l_2 \wedge \ldots \wedge \neg l_{n-1}) \Rightarrow l_n$ where l_n was assumed to be the positive literal. To the left of the implication sign is the rule body and l_n is the rule head. A Horn clause has at most one positive literal. A function-free definite clause is a datalog clause. A program clause can contain negative literals in the body.

conjunctions of (negated) atoms. Recent extensions on probabilistic ILP have also address uncertain domains.

4.1 ILP Overview

This section is on "strong" ILP, which covers the majority of ILP approaches and is concerned with the classification of statistical units and on predicate definition[5]. Strong ILP performs modeling in relational domains that is somewhat related to the approach discussed in the previous section. Let's consider FOIL (First Order Inductive Learner) as a typical representative [44]. The outcome of FOIL is a set of definite clauses (a particular if-then rule) with the same head (then-part).

Here is an example (modified from [45]). Let the statistical unit be a customer with ID *CID*. $VC = 1$, indicates that someone is a valuable customer, $GC = 1$ indicates that someone owns a golden credit card and $SB = 1$ indicates that someone would buy a sailboat. The first rule that FOIL might have learned is that a person is interested in buying a sailboat if this person owns a gold card. The second rule indicates that a person would buy a sailboat if this person is older than 30 and has at least once made a credit card purchase of more than 100 EURO:

$$sailBoat(CID, \ SB = 1) \leftarrow customer(CID,GC = 1) \tag{1}$$
$$sailBoat(CID, \ SB = 1) \leftarrow customer(CID, \ Age)$$
$$\wedge \ purchase(CID, \ PID, \ Value, \ PM)$$
$$\wedge \ PM = credit\text{-}card \wedge Value > 100 \wedge Age > 30.$$

In rule learning FOIL uses a covering paradigm. Thus the first rule is derived to correctly predict as many positive examples as possible (covering) with a minimum number of false positives. Subsequent rules then try to cover the remaining positive examples. The head of a rule (then-part) is a predicate and the body (the if-part) is a product of (negated) atoms containing constants and variables.[6] Naturally, there are many variants of FOIL. FOIL uses a top down search strategy for refining the rule bodies, PROGOL [46] a bottom up strategy and GOLEM [47] a combined strategy. Furthermore, FOIL uses a conjunction of atoms and negated atoms in the body, whereas other approaches use PROLOG constructs. The community typically discusses the different approaches in terms of language bias (which rules can the language express), search bias (which rules can be found) and validation bias (when does validation tell me to stop refining a rule). An advantage of ILP is that also non-grounded background knowledge can be taken into account (typically in form of a set of definite clauses that might be part of an ontology).

---

[5] A predicate definition is a set of program clauses with the same predicate symbol in their heads.

[6] FOIL learning is called learning from entailment in ILP terminology.

In view of the discussion in the last section, the statistical unit corresponds to a customer, and FOIL introduces a binary target feature (1) for the target predicate *sailBoat(CID, SB)*. The second feature (2) is one if the customer owns a golden credit card and zero otherwise. Then a view is generated with attribute CID. A CID is entered in that view each time the person has made a credit card purchase of more then 100 EURO, but only if that person is older than 30 years. The third feature (3) is binary and is equal to one if the CID is present in the view at least once and zero otherwise. FOIL then applies a very simple combination rule: if feature (2) or feature (3) is equal to one for a customer, then the target feature (1) is true.

4.2   Propositionalization, Upgrading and Lifting

ILP approaches like FOIL can be decomposed into the generation of binary features (based on the rule bodies) and a logical combination, which in case of FOIL is quite simple. As stated before, ILP approaches contain a complex search strategy for defining optimal rule bodies. If, in contrast, the generation of the rule bodies is performed as a preprocessing step, the process is referred to as propositionalization [48]. Instead of using the simple FOIL combination rule, other feature-based learners are often used. It has been proven that in some special cases, propositionalization is inefficient [49]. Still, propositionalization has produced excellent results. The binary features are often collected through simple joins of all possible attributes. An early approach to propositionalization is LINUS [50].

The inverse process to propositionalization is called *upgrading* (or lifting) [51] and turns a propositional feature-based learner into an ILP learner. The main differences to propositionalization is that the optimization of the features is guided by the improvement of the performance of the overall system. It turns out that many strong ILP systems can be interpreted as upgraded propositional learners: FOIL is an upgrade of the propositional rule-induction program CN2 and PROGOL can be viewed as upgrading the AQ approach to rule induction. Additional upgraded systems are Inductive Classification Logic (ICL [52]) that uses classification rules, TILDE [53] and S-CART that use classification trees, and RIBL [54] that uses nearest neighbor classifiers. nFOIL [55] combines FOIL with a naive Bayes (NB) classifier by changing the scoring function and by introducing probabilistic covering. nFoil was able to outperform FOIL and propositionalized NB on standard ILP problems. kFoil [56] is another variant that derives kernels from FOIL-based features.

4.3   Discussion

ILP algorithms can easily be applied to the SW if we identify atoms with basic statements. ILP fits well into the basically deterministic framework of the SW. In many ways, statistical SW learning as presented in Section 3 is related to ILP's propositionalization; the main difference is the principled statistical framework of the former. Thus most of the discussion on scalability in Section 3 carries over to ILP's propositionalization. When ILP's complex search strategy for defining optimal rule bodies is applied, training time increases but is still proportional to the number of samples. An interesting new aspect is that ILP produces definite clauses that can be integrated, maybe with some restrictions, into the Semantic Web Rule Language. ILP approaches that consider learning with description logic (and clauses) are described, for example, in [57,58,14,15,16,17]. An empirical study can be found in [59].

5   Learning with Relational Matrices

Another representation of a basic statement (RDF-triple) is a matrix entry. Consider the triple *(User, buys, Item)*. Recall that a standard relational representation would be the table *buys* with attributes *User* and *Item*. A relational adjacency matrix on the other hand has as many rows as there are users and as many columns as there are items and as many matrix entries as there are possibly true statements. A matrix entry is equal to one if the item was actually bought by a user and is equal to zero otherwise. Thus SW data can be represented as a set of matrices where the name of the matrix is the property of the relation under consideration. Matrix decomposition/reconstruction methods, e.g., the principle component analysis (PCA) and other more scalable approaches have been very successful in the prediction of unknown matrix entries [60]. Lippert *et al.* [61] have shown how several matrices can be decomposed/reconstructed jointly and have shown that this increases predictive performance if compared to single matrix decompositions. By filling in the unknown entries via matrix decomposition/reconstruction, the approach has an inherent way of dealing with data that is missing at random. Care must be taken if missing at random is not justified. In [61], one type of statement concerns gene-gene interactions where only positive statements are known. Reconstructed matrix entries can, as before, be entered into the SW, e.g., as weighted triples. Scalability of this approach has not been studied in depth but the decomposition scales approximately proportional to the number of known matrix entries. Note that the approach performs a prediction for all unknown statements in one global decomposition/reconstruction step. In contrast, the previous approaches would learn separate models for each statistical unit under consideration. Other approaches, which learn with the relational adjacency matrix, are described in [62] and [63].

6   Relational Graphical Models

The approaches described in Sections 3 and 4 aim at describing the statistical, respectively logical, dependencies between features derived from SW data. In contrast the matrix decomposition approach in the last section and the relational graphical models (RGMs) in this section predict the truth values of all basis statements (RDF-triples) in the SW. Unlike the matrix decomposition techniques in the last section, the RGMs are probabilistic models and statements are represented by random variables. RGMs can be thought of as upgraded versions of regular graphical models, e.g., Bayesian networks, Markov networks, dependency networks and latent variable models. RGMs have been developed in the context of frame-based logical representations, relational data models, plate models, entity-relationship models and first-order logic. Here, we attempt to relate the basic ideas of the different approaches to the SW framework.

6.1   Possible World Models on the SW

Consider all constants in the SW (i.e., all objects and literal values) and all statements that can possibly be true [7]. Now one introduces a binary random variable U for each possibly true statement (grounded atom), where $U = 1$ if the corresponding statement is true and $U = 0$ otherwise. In a graphical model, U would be identified with a node. These nodes should not be confused with the nodes in the RFD-graph, which represent URIs; rather U stands for a potential link in the RDF-graph. We can reduce the number of random variables if type constraints are available and if the truth value of some statements are assumed known in each world under consideration (e.g., if object-to-object statements are all assumed known, as in the basic PRM model in Subsection 6.2). If statements are mutually exclusive, e.g., the different blood types of a person, one might integrate several statements into one random variable using, e.g., multi-state multinomial variables or continuous variables (to encode, e.g., a person's height). An assignment of truth values to all random variables defines a possible world[8]. RGMs assign a probability distribution to each world in the form $P(U = u)$.[9] The approaches differ in how these probabilities are defined and mapped to random variables, and how they are learned.

6.2   Directed RGMs

The probability distribution in a directed RGM, i.e., relational Bayesian model, can be written as

$$P(U = u) = \prod_{U \in U} P(U|par(U)).$$

U is represented as a node in a Bayesian network and arcs are pointing from all parent nodes $par(U)$ to the node U. One now partitions all elements of U into node-classes. Each U belongs to exactly one node-class. The key property of all U in the same node-class is that their local distributions are identical, which means that $P(U|par(U))$ is the same for all nodes within a node-class and can be described by a truth-table or more complex representations such as decision trees. For example, all nodes representing the IQ-values of students

---

[7] We only consider a function-free case.
[8] RGM modeling would be termed learning from interpretation in ILP terminology.
[9] Our discussion includes the case that we are only interested in a conditional distribution of the form $P(U = u|V = v)$, as in conditional random fields [64].

in a university might form a node class, all nodes representing the difficulties of university courses might form a node class, and the nodes representing the grades of students in courses might form a node-class. Care must be taken, that no directed loops are introduced in the Bayesian network in modeling or structural learning.

Probabilistic Relational Models (PRMs): PRMs were one of the first published RGMs and found great interest in the statistical machine learning community [65,19]. PRMs combine a frame-based logical representation with probabilistic semantics based on directed graphical models. The nodes in a PRM model the probability distribution of object attributes whereas the relationships between objects are assumed known. Naturally, this assumption simplifies the model greatly. In context of the SW object attributes would primarily correspond to object-to-literal statements. In subsequent papers PRMs have been extended to also consider the case that relationships between objects (in context of the SW these would roughly be the object-to-object statements) are unknown, which is called *structural uncertainty* in the PRM framework [19]. The simpler case, where one of the objects in a statement is known, but the partner object is unknown, is referred to as *reference uncertainty*. In reference uncertainty the number of potentially true statements is assumed known, which means that only as many random nodes need to be introduced. The second form of structural uncertainty is referred to as *existence uncertainty*, where binary random variables are introduced representing the truth values of relationships between objects.

For some PRMs, regularities in the PRM structure can be exploited (encapsulation) and exact inference is possible. Large PRMs require approximate inference; commonly, loopy belief propagation is being used. Learning in PRMs is likelihood based or based on empirical Bayesian learning. Structural learning typically uses a greedy search strategy, where one needs to guarantee that the ground Bayesian network does not contain directed loops.

More Directed RGMs: A *Bayesian logic program* is defined as a set of Bayesian clauses [66]. A Bayesian clause specifies the conditional probability distribution of a random variable given its parents on a template level, i.e. in a node-class. A special feature is that, for a given random variable, *several* such conditional probability distributions might be given. As an example, $bt(X) \mid mc(X)$ and $bt(X) \mid pc(X)$ specify the probability distribution for blood type given the two different dispositions $mc(X)$ and $pc(X)$. The truth value for $bt(X) \mid mc(X), pc(X)$ can then be calculated based on various combination rules (e.g., noisy-or). In a Bayesian logic program, for each clause there is one conditional probability distribution and for each Bayesian predicate (i.e., node-class) there is one combination rule. *Relational Bayesian networks* [67] are related to Bayesian logic programs and use probability formulae for specifying conditional probabilities. *Relational dependency networks* [42] also belong to the family of directed RGMs and learn the dependency of a node given its Markov blanket using decision trees.

6.3   Undirected RGMs

The probability distribution of an undirected graphical model or Markov network can be written as

$$P(U = u) = \frac{1}{Z} \prod_k g_k(u_k)$$

where $g_k(.)$ is a potential function, u_k is the state of the k-th clique and Z is the partition function normalizing the distribution. One often prefers a more convenient log-linear representation of the form

$$P(U = u) = \frac{1}{Z} \exp \sum_k w_k f_k(u_k)$$

where the feature functions f_k can be any real-valued function and where $w_i \in \mathbb{R}$.

We will discuss two major approaches that use this representation: Markov logic networks and relational Markov models.

Markov Logic Networks (MLN): Let F_i be a formula of first-order and let $w_i \in \mathbb{R}$ be a weight attached to each formula. Then a MLN L is defined as a set of pairs (F_i, w_i) [68] [69]. One introduces a binary node for each possible grounding of each predicate appearing in L (i.e., in context of the SW we would introduce a node for each possible statement), given a set of constants $c_1, \ldots, c_{|C|}$. The state of the node is equal to 1 if the ground atom/statement is true, and 0 otherwise (for an N-ary predicate there are $|C|^N$ such nodes). A grounding of a formula is an assignment of constants to the variables in the formula (considering formulas that are universally quantified). If a formula contains N variables, then there are $|C|^N$ such assignments. The nodes in the Markov network $M_{L,C}$ are the grounded predicates. In addition the MLN contains one feature for each possible grounding of each formula F_i in L. The value of this feature is 1 if the ground formula is true, and 0 otherwise. w_i is the weight associated with F_i in L. A Markov network $M_{L,C}$ is a grounded Markov logic network of L with

$$P(U = u) = \frac{1}{Z} \exp \left(\sum_i w_i n_i(u) \right)$$

where $n_i(u)$ is the number of formula groundings that are true for F_i. MLN makes the unique names assumption, the domain closure assumption and the known function assumption, but all these assumptions can be relaxed.

A MLN puts weights on formulas: the larger the weight, the higher is the confidence that a formula is true. When all weights are equal and become infinite, one strictly enforces the formulas and all worlds that agree with the formulas have the same probability.

The simplest form of inference concerns the prediction of the truth value of a grounded predicate given the truth values of other grounded predicates (conjunction of predicates) for which the authors present an efficient algorithm. In the first phase, the minimal subset of the ground Markov network is returned that is required to calculate the conditional probability. It is essential that this subset is small since in the worst case, inference could involve alle nodes. In the second phase Gibbs sampling in this reduced network is used.

Learning consists of estimating the w_i. In learning, MLN makes a closed-world assumption and employs a pseudo-likelihood cost function, which is the product of the probabilities of each node given its Markov blanket. Optimization is performed using a limited memory BFGS algorithm.

Finally, there is the issue of structural learning, which, in this context, defines the employed first order formulae. Some formulae are typically defined by a domain expert *a priori*. Additional formulae can be learned by directly optimizing the pseudo-likelihood cost function or by using ILP algorithms. For the latter, the authors use CLAUDIAN [70], which can learn arbitrary first-order clauses (not just Horn clauses, as many other ILP approaches).

Relational Markov Networks (RMNs): RMNs generalize many concepts of PRMs to undirected RGMs [40]. RMNs use conjunctive database queries as clique templates. By default, RMNs define a feature function for each possible state of a clique, making them exponential in clique size. RMNs are mostly trained discriminately. In contrast to MLN, RMNs, as PRMs, do not make a closed-world assumption during learning.

6.4   Latent Class RGMs

The infinite hidden relational model (IHRM) [71] presented here is a directed RGM (i.e., a relational Bayesian model) with latent variables.[10] The IHRM is formed as follows. First, we partition all objects into classes $K_1, ... K_{|K|}$, using, for example, ontological class information. For each object in each class, we introduce a statement (*Object, hasHiddenState, H*). If *Object* belongs to class K_i, then $H \in \{1, \ldots, N_{K_i}\}$, i.e., the number of states of H is class dependent. As before, we introduce a random variable or node U for each grounded atom, respectively potentially true basic statement. Let Z_{Object} denote the random variables that involve *Object* and H. Z_{Object} is a latent variable or latent node since the true state of H is unknown. $Z_{Object} = j$ stand for the statement that (*Object, hasHiddenState, j*).

We now define a Bayesian network where the nodes Z_{Object} have no parents and the parents of the nodes for all other statement are the latent variables of the objects appearing in the statement. In other words, if U stands for the fact that (*Object$_1$, property, Object$_2$*) is true, then there are arcs from Z_{Object_1} and Z_{Object_2} to U. The object-classes of the objects in a statement together with the property define a node-class for U. If the property value is a literal, then the only parent of U is Z_{Object_1}.

In the IHRM we let the number of states in each latent node to be infinite and use the formalism of Dirichlet process mixture models. In inference, only a small number of the infinite states are occupied, leading to a clustering solution where

---

[10] Kemp et al. [72] presented an almost identical model independently.

the number of states in the latent variables N_{C_i} is automatically determined during inference.

Since the dependency structure in the ground Bayesian network is local, one might get the impression that only local information influences prediction. This is not true, since in the ground Bayesian network, common children U with evidence lead to interactions between the parent latent variables. Thus information can propagate in the network of latent variables. Training is based on various forms of Gibbs sampling (e.g., the Chinese restaurant process) or mean field approximations. Training only needs to consider random variables U corresponding to statements that received evidence, e.g., statements that are either known to be true or known not to be true; random variables that correspond to statements with an unknown truth value (i.e., without evidence) can completely be ignored.

The IHRM has a number of key advantages. First, no structural learning is required, since the directed arcs in the ground Bayesian network are directly given by the structure of the SW graph. Second, the IHRM model can be thought of as an infinite relational mixture model, realizing hierarchical Bayesian modeling. Third, the mixture model allows a cluster analysis providing insight into the relational domain.

The IHRM has been applied to recommender systems, for gene function prediction and to develop medical recommender systems. The IHRM was the first relational model applied to trust learning [20]. In [31] it was shown how ontological class information can be integrated into the IHRM.

6.5   Discussion

RGMs have been developed in the context of frame-based logical representations, relational data models, plate models, entity-relationship models and first-order logic but the main ideas can easily be adapted to the SW data model. One can distinguish two cases. In the first case, an RGM learns a joint probabilistic model over the complete SW or a segment of the SW. This might be the most elegant approach since there is only one (SW-) world and the dependencies between the variables are truthfully modeled, as discussed in Subsection 3.4. The draw back is that the computational requirements scale with the number of statements whose truth value is known or even the number of all potentially true statements. More appropriate for large-scale applications might be the second case where one applies the sampling approach as described in Section 3. As an example consider that the statistical unit is a student. A data point would then not correspond to a set of features but to a local subgraph that is anchored at the statistical unit, e.g., the student. As before sampling would make the training time essentially independent of SW-size. Ontological background knowledge can be integrated as discussed in Section 3. First, one can employ complete or partial materialization, which would derive statements from reasoning prior to training. Second, an ontological subgraph can be included in the subgraph of a statistical unit [31]. Also note that the MLN might be particularly suitable to exploit ontological background information: ontologies can formulate some of the first-order formulas that are the basis for the features in the MLN. PRMs have been extended to learn class hierarchies (PRM-CH), which can be a basis for ontology learning.

The RGM approaches typically make an open world assumption.[11] The corresponding random variables are assumed missing at random such that the approaches have an inherent mechanism to deal with missing data. If missing at random is not justified, then more complex missing data models need to be applied. As before, based on the estimated probabilities, weighted RDF-triples can be generated and added to the SW.

7   Unstructured Data and the SW

The realization of the SW heavily depends on (1) available ontologies and (2) the annotation of unstructured data with ontology-based meta data. Manual ontology development and manual annotation are two well known SW bottlenecks. Thus learning-based approaches for both tasks are finding increasing interest [2,9]. In this section, we will concentrate on two important tasks, namely ontology learning and semantic annotation (for a compilation of current work on ontology learning and population see, e.g., [73]). A particulary important source of information for these tasks is unstructured or semi-structured textual data. Note that there is a close relationship between textual data and SW data. Textual data describes, first, ontological concepts and relationships between concepts (e.g., a text might contain the sentence: *We all know that cats are mammals*) and, second, instances and relationships between instances (e.g., a document might inform us that: *Marry is married to Jack*). However, the input data for ontology learning and semantic annotation will not be limited to textual data; especially once the SW will be realized to a greater extent, other types of input data will become increasingly important. Learning ontologies from e.g., XML-DTDs, UML diagrams, database schemata or even raw RDF-graphs is also of great interest [74], but is out of scope here. The outline of this section is as follows: first, we consider the case, where a text corpus of interest is given and the task is to infer a prototype ontology. Second, given a text corpus and an ontology, we want to infer instances of the concepts and their relations.

7.1   Learning Ontologies from Text

Ontology learning, in general, consists of several subtasks. This includes the identification of terms, synonyms, polysems, concepts, concept hierarchies, properties, property hierarchies, domain and range constraints and class definitions. These tasks can be illustrated as the so-called ontology learning layer cake [74]. Different approaches differ mainly in the way a concept is defined and one distinguishes between formal ontologies, terminological ontologies and prototype-based ontologies [75]. In prototype-based ontologies, concepts are represented by collections of prototypical instances, which are arranged hierarchically in subclusters. An example would be the concept disease, which is defined by a set

---

[11] There are some exceptions, e.g., MLN make a closed-world assumption in training.

of diseases. Since prototype-based ontologies are defined by instances, they lack definitions and axiomatic grounding. In contrast, typical examples for terminological ontologies are WordNet and the Medical Subject Headings (MeSH[12]). Terminological ontologies are described by concept labels and both nouns and verbs are organized into hierarchies, defined by hypernym or subclass relationships. For example a disease is defined in WordNet as an impairment of health or a condition of abnormal functioning. Terminological ontologies typically also lack axiomatic grounding. A formal ontology such as OWL, in contrast, is seen as a conceptualization, whose categories are distinguished by axioms and definitions [76]. Most of the state-of-the-art approaches focus on learning prototype-based ontologies. Work on learning terminological or formal ontologies is still quite rare. Here, the big challenge is to deal with uncertain and often even contradicting extracted knowledge, introduced during the ontology learning process. This is addressed in [77], which presents a system that is able to transform a terminological ontology to a consistent formal OWL-DL ontology.

Prototype ontologies are often learned based on some type of hierarchical clustering techniques such as single-link, complete-link or average-link clustering. According to Harris' distributional hypothesis [78], semantic similarity between words can be assessed via the syntactic context, which they are sharing in a corpus. Thus most approaches base the semantic relatedness between words on some distributional similarity between the words. Usually, a vector-space model is used as input and the linguistic context of a term is described by, e.g., syntactic dependencies, which the term establishes in a corpus [79] The input vector for a term to be clustered can be, e.g., composed of syntactic expressions such as prepositional phrases following a verb or adjective modifiers. See [80] for an illustrative example for assessing the semantic similarity of terms. Hierarchical clustering, in its classical form, distinguishes between agglomerative (bottom-up) and divisive (top-down) clustering, whereas the agglomerative form is most commonly used due to its computational efficiency. Somewhat different from hierarchical clustering is the divisive bi-section-Kmeans algorithm, which yielded competitive results for document clustering [81] and has been applied to the task of learning concept hierarchies as well [82,83]. Another variant is the the Formal Concept Analysis (FCA) [84]. FCA is closely related to bi-clustering and tries to build a lattice of so-called formal concepts from a vector space model. FCA thereby makes use of order theory and analyzes the covariance between objects and their features. The reader is referred to [84] for more information.

Recently, [74] set up a benchmark to compare the above mentioned clustering techniques for learning concept hierarchies. While each of the methods had its own benefits, FCA performed better in terms of recall and precision. All the methods just mentioned, face the problem of not being able to appropriately label the resulting clusters, i.e., to determine the name of the concept. To overcome this limitation and to guide the clustering process, [85] either use hyponyms extracted from WordNet or use Hearst patterns [86] derived either from the corpus under investigation or from the WWW.

---

[12] http://www.nlm.nih.gov/mesh/

Another type of technique for learning prototype ontologies, comes from the topic modeling community, an active research area of machine learning [87,9]. Topic models are generative models based upon the idea that a document is made of a mixture of topics, where a topic is represented by a distribution over words. Powerful techniques such as Latent Semantic Analysis (LSA) [88], Probabilistic Latent Semantic Analysis (PLSA) [89] or Latent Dirichlet Allocation (LDA) [90] have been proposed for the automated extraction of useful information from large document collections. Applications include document annotation, query answering, document summarization, automatic topic extraction as well as trend analysis. Generative statistical models such as the ones mentioned, have been proven effective in addressing these problems. In general, the following advantages of topic models are highlighted in the context of document modeling: First, topics can be extracted in a complete unsupervised fashion, requiring no initial labeling of the topics. Second, the resulting representation of topics for a document collection is interpretable and last but not least, each document is usually expressed by a mixture of topics, thus capturing the topic combinations that arise in documents [89,90,91]. When applying topic modeling techniques in an ontology learning setting, a topic is referred to as concept. To satisfy the hierarchical structure of prototype ontologies, [87] extends the PLSA method to an hierarchical version, where super concepts are introduced. While yielding already impressive results with this kind of techniques, [87] concentrates on learning prototype ontologies, where no labeling of the concept is needed. Furthermore, the hierarchy of the ontology is assumed to be known a priori. Learning the hierarchical order in topic models is an area of growing interest. Here, [92] introduced hierarchical LDA, which models the setup of the tree-structure of the topics as a Chinese Restaurant Process (CRP). As a consequence, the hierarchy is not fixed a priori, instead it is a part of the learning process. To overcome the limitation of unlabeled topics or concepts, [93] tries to automatically infer an appropriate label for multinomial topic models. [9] discusses ontology learning based on topic models in context of the SW.

Ontology Merging, Alignment and Evolution: In many cases no dominant ontology will exist, which leads to the problem that several ontologies need to be merged and aligned. In [11] these tasks have been addressed with the support of machine learning. Another aspect is that an ontology is not a rigid and fixed construct — ontologies will evolve with time. Thus, the structure of an ontology will change and new concepts will be needed to be inserted into an existing ontology. This leads to another task, where machine learning can play a role in ontology engineering: ontology refinement and ontology evolution. This task is usually treated as classification task [76]. The reader is referred to [76,10] for more information.

7.2   Semantic Annotation

Besides ontological support, a second prerequisite to put the SW into practice, is the availability of machine-readable meta data. Producing human readable text from SW data is simple since an RDF triple can easily be formulated as a textual statement. However, even though the statement won't be powerfully eloquent, it will still serve its purpose. The inverse is much more difficult, i.e., the generation of triples from textual data. This process is called semantic annotation, knowledge markup or meta data generation [94]. Hereby, we are following the notion of semantic annotation as linguistic annotations (such as named entities, semantic classes, etc.) as well as user annotations like tags (see the ECIR 2008 workshop on 'Exploiting Semantic Annotations in Information Retrieval'[13]).

The Information Extraction (IE) community provides a number of approaches for these tasks. IE is traditionally defined as the process of filling the fields and records of a database from unstructured text and is seen as precursor to data mining [95]. Usually, the fields are filled with named entities (i.e., Named Entity Recognition (NER)), such as persons, locations or organizations. IE first populates a database from unstructured text and data mining then aims to find patterns. IE is, dependent on the task, made up of five subtasks: segmentation, classification, finding associations and last but not least normalization and deduplication [95]. Segmentation refers to the identification of text phrases, which describe entities of interest. Classification is the assignment to predefined types of entities, while finding associations is the identification of relations between the entities (i.e., relation extraction). Normalization and deduplication describe the task of merging different text descriptions with the same meaning (e.g., mapping entities to URIs).

NER is an active field of research and several evaluation conferences such as the Message Understanding Conference (MUC-6)[96], the Conference on Computational Natural Language Learning (CoNLL-2003) [97] and in the biomedical domain, the Critical Assessments of Information Extraction systems in Biology (BioCreAtIvE I+II[14]) [98] have attracted a lot of interest. While in MUC-6 the focus was NER for persons, locations, organizations in an English newswire domain, CoNLL-2003 focused on language-independent NER. BioCreAtIvE focused on the recognition of biomedical entities, in this case gene and protein mentions. The methods proposed for NER vary, in general, in their degree of reliance on dictionaries, and their different emphasis on statistical or rule-based approaches. Numerous machine learning techniques have been applied to NER tasks such as Support Vector Machines [99], Hidden Markov Models [100], Maximum Entropy Markov Models [101] and Conditional Random Fields [64].

An F-measure in the mid-90s can now be achieved for extracting persons, organizations and locations in the newswire domain [95]. For extracting gene and protein mentions, however, the F-measure lies currently in the mid- to high 80s (see the BioCreAtIvE II conference for details). So NER can provide high accuracy solutions for the SW, but typically only for a small number of classes, mostly because of a limited amount of labeled training data. However, when populating an existing ontology, there will often be the need to be able to extract hundreds of classes of entities. Thus, systems which are able to scale to a large

---

[13] http://www.yr-bcn.es/dokuwiki/doku.php?id=ecir08_entity_workshop_proposal
[14] http://biocreative.sourceforge.net/biocreative_2.html

number of classes on a large amount of unlabeled data are needed. Also flexible and domain-independent recognition of entities is an important and active field of research. State-of-the-art approaches try to extract hundreds of entity classes in an unsupervised fashion [102], but so far with a fairly low accuracy. Promising areas, which could help to overcome current limitations of supervised IE systems, are semi-supervised learning [103,104] as well as active learning [105].

The same entities can have different textual representation (e.g., 'Clark Kent', 'Kent Clark' and 'Mr. Clark' refer to the same person). Normalization is the process of standardizing the textual expressions. This task is usually also referred to as entity resolution, co-reference resolution or normalization and deduplication. The Stanford Entity Resolution Framework (SERF), e.g., has the goal to provide a framework for generic entity resolution [106]. Other techniques for entity resolution employ relational clustering [107] as well as probabilistic topic models [108].

Another important task is the identification of relations between instances of concepts (i.e., the association finding stage in the traditional IE workflow). Up to now, most of research on text information extraction has focused on tagging named entities. The Automatic Content Extraction (ACE) program provides annotation benchmark sets for the challenging task of relation extraction. At ACE, this task is called Relation Detection and Characterization (RDC). A representative system using an SVM with a rich set of features, reports results for Relation Detection (74.7% F-measure) and 68.0% F-measure for the RDC task [109]. Co-occurrence based relation extraction is a simple, effective and popular method [110], but usually suffers of a lower recall, since entities can co-occur for many other reasons. Other methods are kernel-based [111] or rule-based [112]. Recently, [113] propose a new method that treats relation extraction as sequential labeling task. They extend Conditional Random Fields (CRFs) towards the extraction of semantic relations. Hereby, they focus on the extraction of relations between genes and diseases (five types of relations) as well as between disease and treatment entities (eight types of relations). The work applies the authors' method to a biomedical textual database and provides the resulting network of genes and diseases in a machine-readable RDF graph. Thereby, gene and disease entities are normalized to Bio2RDF[15] URIs.

8   First Experiments in the Analysis of FOAF-Data

The purpose of the FOAF (Friend of a Friend) project [114] is to create a web of machine-readable pages describing people, the relationships between people and people's activities and interests, using W3C's RDF technology. The FOAF ontology is defined using RDFS/OWL and is formally specified in the FOAF Vocabulary Specification 0.91 [115]. In our study we employed the IHRM model as described in Section 6. The trained IHRM can, for instance, recommend new friendships, the affiliations of persons, and their interests and projects. Furthermore one might want to predict attributes of certain persons, like their gender

---

[15] http://bio2rdf.org/

or age. Finally, by interpreting the clustering results of the IHRM one can answer typical questions from social network analysis concerning the relationships between members of the FOAF social system.

In general FOAF data is either uploaded by each person individually or generated automatically from user profiles of community websites like Tribe.net, LiveJournal.com or my.opera.com. The resulting network of linked FOAF-files can be gathered using a FOAF harvester, a so called "scutter". Some scutter dumps are readily available for download, e.g., in one large rdf/xml-file or stored in a relational database.

Even though this use case only covers a very basic statistical inference problem on the SW, there still are major challenges to meet. First, there are characteristics of the FOAF-data that need special consideration: For instance, the actual data is extremely sparse. With more than 100000 users, there are far more potential links as actual links between persons.

Another typical characteristic of friendship data is that the topology of the *knows*-RDF-graph consists of a few barely connected star graphs, corresponding to a few active network users with a long list of friends as the "center" of the stars and the mass of users that don't specify their friends. Second, there are prevalent challenges of SW data in general that can also be observed in a FOAF analysis. For instance, there is a variety of additional untested and potentially conflicting ontologies specified by users. If this information is ignored by only considering data consistent with the FOAF ontology, most of the information specified by users is ignored. This also applies to the almost arbitrary use of literals by users. For instance the relation *interest* with range *Document* defined in the FOAF-schema is in reality mostly used with a literal instead. Consequently, this results in a loss of semantic information. To still make use of this information one would, e.g., need to use automated semantic annotation as described in Section 7. Another preprocessing step that needs to be considered in practice is the materialization of triples, which can be inferred deductively. For example there might be an instance of the relation *holdsAccount* with domain *Person* in the data, which is not given in the schema. However, from the ontology it can be inferred that *Person* is a *subClassOf Agent* which in turn has a property *holdsAccount*. As stated before, total materialization is only feasible in less expressive ontologies.

Considering these issues, it becomes clear that there are not only theoretical but also a large number of interesting practical challenges for learning on the SW.

9   Conclusions

Data in Semantic Web formats will bring many new opportunities and challenges to machine learning. Machine learning complements ontological background knowledge by exploiting regularities in the data while being robust against some of the inherent problems with Semantic Web data such as contradicting information and non-stationarity. A general issue with machine learning is that the problem of missing information needs to be carefully addressed in learning, in particular if either the selection of statistical units or the probability that a feature is missing depend on the features of interest, which is common in many-to-many relations.

We began with a section on feature-based statistical learning on the Semantic Web. This procedure is widely applicable, scales well with the size of the Semantic Web and provides a promising general purpose learning approach. The greatest challenge here is that most feature-based statistical learning approaches have no inherent way of dealing with missing data requiring additional missing data models. A common situation in social network data is that features in linked objects are mutually dependent and need to be modeled jointly. One can expect that this will also often occur in SW data and SW learning will benefit from ongoing research in social network modeling.

We then presented the main approaches in inductive logic programming. Inductive logic programming has the potential to learn deterministic constraints that can be integrated into the employed ontology. We presented a discussion on learning with relational matrices, which is quite attractive if multiple many-to-many relations are of interest, as in recommendation systems. We then studied relational graphical models. Although these approaches were originally defined in various frameworks, e.g., frame-based logical representation, relational data models, plate models, entity-relationship models and first-order logic, they can easily be modified to be applicable in context of the Semantic Web. Relational graphical models are capable of learning a global probabilistic Semantic Web model and inherently can deal with missing data. Scalability to the size of the Semantic Web might be a problem for RGMs and we discussed subgraph sampling as a possible solution. All approaches have means to include ontological background knowledge by complete or partial materialization. In addition, the ontological RDF-graph can be incorporated in learning and ontological features can be derived and exploited. Ontologically supported machine learning is an active area of research. It is conceivable that in future ontological standards, the developed statistical models could become in integral part of the ontology. Also, we have discussed that most presented approaches can be used to produce statements that are weighted by their probability value derived from machine learning, complementing statements that are derived form logical reasoning. An interesting opportunity is to include weighted triples in Semantic Web queries.

We reported about initial work on learning ontologies from textual data and on the semantic annotation of unstructured data. So far, this concerns the most advanced work in Semantic Web learning covering ontology construction and management, ontology evaluation, ontology refinement, ontology evolution, as well as the mapping, merging and alignment of ontologies. In addition there is growing work on Semantic Web mining extending the capabilities of standard web mining, although most of this work needs to wait for the Semantic Web to be realized on a large scale.

In summary, machine learning has the potential to realize a number of exciting applications on the Semantic Web and can complement axiomatic inference by exploiting regularities in the data.

Acknowledgements. We acknowledge funding by the German Federal Ministry of Economy and Technology (BMWi) under the THESEUS project and by the EU FP 7 Large-Scale Integrating Project LarKC. We thank Frank Harmelen, and two anonymous reviewers for their valuable comments.

References

1. Berners-Lee, T., Hendler, J., Lassila, O.: The semantic web. Scientific American (2001)
2. van Harmelen, F.: Semantische Techniken stehen kurz vor dem Durchbruch. C'T Magazine (2007)
3. Fensel, D., Hendler, J.A., Lieberman, H., Wahlster, W.: Spinning the Semantic Web: Bringing the World Wide Web to its Full Potential. MIT Press, Cambridge (2003)
4. Ontoprise: Neue Version des von Ontoprise entwickelten Ratgebersystems beschleunigt die Roboterwartung. Ontoprise Pressemitteilung (2007)
5. LarKC: The large Knowledge Collider. EU FP 7 Large-Scale Integrating Project (2008), http://www.larkc.eu/
6. Incubator Group: Uncertainty Reasoning for the World Wide Web. W3C (2005), http://www.w3.org/2005/Incubator/urw3/
7. Berendt, B., Hotho, A., Stumme, G.: Towards semantic web mining. In: Horrocks, I., Hendler, J. (eds.) ISWC 2002. LNCS, vol. 2342. Springer, Heidelberg (2002)
8. Grobelnik, M., Mladenic, D.: Automated knowledge discovery in advanced knowledge management. Library Hi Tech News incorporating Online and CD Notes 9(5) (2005)
9. Grobelnik, M., Mladenic, D.: Knowledge discovery for ontology construction. In: Davies, J., Studer, R., Warren, P. (eds.) Semantic Web Technologies. Wiley, Chichester (2006)
10. Bloehdorn, S., Haase, P., Sure, Y., Voelker, J.: Ontology evolution. In: Davies, J., Studer, R., Warren, P. (eds.) Semantic Web Technologies. Wiley, Chichester (2006)
11. Bruijn, J.d., Ehrig, M., Feier, C., Martin-Recuerda, F., Scharffe, F., Weiten, M.: Ontology mediation, merging, and aligning. In: Davies, J., Studer, R., Warren, P. (eds.) Semantic Web Technologies. Wiley, Chichester (2006)
12. Fortuna, B., Grobelnik, M., Mladenic, D.: Ontogen: Semi-automatic ontology editor. In: HCI (9) (2007)
13. Mladenic, D., Grobelnik, M., Foruna, B., Grcar, M.: Knowledge discovery for the semantic web (submitted, 2008)
14. Lisi, F.A.: Principles of inductive reasoning on the semantic web: A framework for learning in AL-Log. In: Fages, F., Soliman, S. (eds.) PPSWR 2005. LNCS, vol. 3703. Springer, Heidelberg (2005)
15. Lisi, F.A.: A methodology for building semantic web mining systems. In: The 16th International Symposium on Methodologies for Intelligent Systems (2006)
16. Lisi, F.A.: Practice of inductive reasoning on the semantic web: A system for semantic web mining. In: Alferes, J.J., Bailey, J., May, W., Schwertel, U. (eds.) PPSWR 2006. LNCS, vol. 4187. Springer, Heidelberg (2006)
17. Lisi, F.A.: The challenges of the semantic web to machine learning and data mining. In: Tutorial at ECML 2006 (2006)
18. Fukushige, Y.: Representing probabilistic relations in rdf. In: ISWC-URSW (2005)

19. Getoor, L., Friedman, N., Koller, D., Pferrer, A., Taskar, B.: Probabilistic relational models. In: Getoor, L., Taskar, B. (eds.) Introduction to Statistical Relational Learning. MIT Press, Cambridge (2007)
20. Rettinger, A., Nickles, M., Tresp, V.: A statistical relational model for trust learning. In: Proceeding of 7th International Conference on Autonomous Agents and Multiagent Systems, AAMAS 2008 (2008)
21. W3C: World Wide Web Consortium, http://www.w3.org/
22. Antoniou, G., van Harmelen, F.: A Semantic Web Primer. MIT Press, Cambridge (2004)
23. Herman, I.: Tutorial on the Semantic Web. W3C,
    http://www.w3.org/People/Ivan/CorePresentations/SWTutorial/Slides.pdf
24. Tauberer, J.: Resource Description Framework, http://rdfabout.com/
25. Kiryakov, A.: Measurable targets for scalable reasoning. Ontotext Technology White Paper (2007)
26. Fahrmeir, L., Künstler, R., Pigeot, I., Tutz, G., Caputo, A., Lang, S.: Arbeitsbuch Statistik, 4th edn. Springer, Heidelberg (2004)
27. Casella, G., Berger, R.L.: Statistical Inference. Duxbury Press (1990)
28. Trochim, W.: The Research Methods Knowledge Base, 2nd edn. Atomic Dog Publishing (2000)
29. Popescul, A., Ungar, L.H.: Feature generation and selection in multi-relational statistical learning. In: Getoor, L., Taskar, B. (eds.) Introduction to Statistical Relational Learning. MIT Press, Cambridge (2007)
30. Karalič, A., Bratko, I.: First order regression. Machine Learning 26(2-3) (1997)
31. Reckow, S., Tresp, V.: Integrating ontological prior knowledge into relational learning. Technical report, Siemens (2007)
32. Tresp, V.: Committee machines. In: Hu, Y.H., Hwang, J.N. (eds.) Handbook for Neural Network Signal Processing. CRC Press, Boca Raton (2001)
33. Little, R.J.A., Rubin, D.B.: Statistical Analysis with Missing Data, 2nd edn. Wiley, Chichester (2002)
34. Sen, P., Namata, G., Bilgic, M., Getoor, L., Gallagher, B., Eliassi-Rad, T.: Collective classification in network data. AI Magazine (Special Issue on AI and Networks), forthcoming (forthcoming, 2008)
35. Macskassy, S., Provost, F.: Classification in networked data: a toolkit and a univariate case study. Machine Learning (2007)
36. Chakrabarti, S., Dom, B., Indyk, P.: Enhanced hypertext categorization using hyperlinks. In: SIGMOD (1998)
37. Neville, J., Jensen, D.: Iterative classification in relational data. In: AAAI (2000)
38. Lu, Q., Getoor, L.: Link-based classification. In: ICML (2003)
39. Getoor, L., Friedman, N., Koller, D., Taskar, B.: Learning probabilistic models of link structure. journal of machine learning research. Journal of Machine Learning Research (2002)
40. Taskar, B., Abbeel, P., Koller, D.: Discriminative probabilistic models for relational data. In: Uncertainty in Artificial Intelligence, UAI (2002)
41. Zhu, X.: Semi-supervised learning literature survey. Technical report, Technical Report 1530, Department of Computer Sciences, University of Wisconsin (2005)
42. Neville, J., Jensen, D.: Dependency networks for relational data. In: ICDM 2004: Proceedings of the Fourth IEEE International Conference on Data Mining (2004)
43. Neville, J., Jensen, D.: Relational dependency networks. Journal of Machine Learning Research (2007)
44. Quinlan, J.R.: Learning logical definitions from relations. Machine Learning 5(3) (1990)

45. Džeroski, S.: Inductive logic programming in a nutshell. In: Getoor, L., Taskar, B. (eds.) Introduction to Statistical Relational Learning. MIT Press, Cambridge (2007)
46. Muggleton, S.: Inverse entailment and Progol. New Generation Computing, Special issue on Inductive Logic Programming 13(3-4) (1995)
47. Muggleton, S., Feng, C.: Efficient induction of logic programs. In: Proceedings of the 1st Conference on Algorithmic Learning Theory, Ohmsma, Tokyo, Japan (1990)
48. Kramer, S., Lavrac, N., Flach, P.: From propositional to relational data mining. In: Džeroski, S., Lavrac, L. (eds.) Relational Data Mining. Springer, Heidelberg (2001)
49. De Raedt, L.: Attribute-value learning versus inductive logic programming: The missing links (extended abstract). In: Page, D.L. (ed.) ILP 1998. LNCS, vol. 1446. Springer, Heidelberg (1998)
50. Lavrač, N., Džeroski, S., Grobelnik, M.: Learning nonrecursive definitions of relations with LINUS. In: Kodratoff, Y. (ed.) EWSL 1991. LNCS, vol. 482. Springer, Heidelberg (1991)
51. Van Laer, W., De Raedt, L.: How to upgrade propositional learners to first order logic: A case study. In: Machine Learning and Its Applications, Advanced Lectures (2001)
52. De Raedt, L., Van Laer, W.: Inductive constraint logic. In: Zeugmann, T., Shinohara, T., Jantke, K.P. (eds.) ALT 1995. LNCS, vol. 997. Springer, Heidelberg (1995)
53. Blockeel, H., De Raedt, L.: Top-down induction of first-order logical decision trees. Artificial Intelligence 101(1-2) (1998)
54. Emde, W., Wettschereck, D.: Relational instance based learning. In: Saitta, L. (ed.) Machine Learning - Proceedings 13th International Conference on Machine Learning (1996)
55. Landwehr, N., Kersting, K., De Raedt, L.: nFOIL: Integrating naïve bayes and FOIL. In: Veloso, M., Kambhampati, S. (eds.) Proceedings of the Twentieth National Conference on Artificial Intelligence (AAAI 2005) (2005)
56. Landwehr, N., Passerini, A., De Raedt, L., Frasconi: kFOIL: Learning simple relational kernels. In: National Conference on Artificial Intelligence (AAAI) (2006)
57. Cohen, W.W., Hirsh, H.: Learning the CLASSIC description logic: Theoretical and experimental results. In: Principles of Knowledge Representation and Reasoning: Proceedings of the Fourth International Conference (KR 1994) (1994)
58. Rouveirol, C., Ventos, V.: Towards learning in CARIN-ALN. In: International Workshop on Inductive Logic Programming (2000)
59. Edwards, P., Grimnes, G., Preece, A.: An empirical investigation of learning from the semantic web. In: ECML/PKDD, Semantic Web Mining Workshop (2002)
60. Takacs, G., Pilaszy, I., Nemeth, B., Tikk, D.: On the gravity recommendation system. In: Proceedings of KDD Cup and Workshop 2007 (2007)
61. Lippert, C., Huang, Y., Weber, S.H., Tresp, V., Schubert, M., Kriegel, H.P.: Relation prediction in multi-relational domains using matrix factorization. Technical report, Siemens (2008)
62. Yu, K., Chu, W., Yu, S., Tresp, V., Xu, Z.: Stochastic relational models for discriminative link prediction. In: Advances in Neural Information Processing Systems 19 (2006)
63. Yu, S., Yu, K., Tresp, V.: Soft clustering on graphs. In: Advances in Neural Information Processing Systems 18 (2005)

64. Lafferty, J., McCallum, A., Pereira, F.: Conditional random fields: Probabilistic models for segmenting and labeling sequence data. In: Proceedings of 18th International Conference on Machine Learning (2001)
65. Koller, D., Pfeffer, A.: Probabilistic frame-based systems. In: Proceedings of the National Conference on Artificial Intelligence (AAAI) (1998)
66. Kersting, K., De Raedt, L.: Bayesian logic programs. Technical report, Albert-Ludwigs University at Freiburg (2001)
67. Jaeger, M.: Relational bayesian networks. In: Proceedings of the 13th Conference on Uncertainty in Artificial Intelligence (UAI) (1997)
68. Richardson, M., Domingos, P.: Markov logic networks. Machine Learning 62(1-2) (2006)
69. Domingos, P., Richardson, M.: Markov logic: A unifying framework for statistical relational learning. In: Getoor, L., Taskar, B. (eds.) Introduction to Statistical Relational Learning. MIT Press, Cambridge (2007)
70. De Raedt, L., Dehaspe, L.: Clausal discovery. Machine Learning 26 (1997)
71. Xu, Z., Tresp, V., Yu, K., Kriegel, H.P.: Infinite hidden relational models. In: Uncertainty in Artificial Intelligence (UAI) (2006)
72. Kemp, C., Tenenbaum, J.B., Griffiths, T.L., Yamada, T., Ueda, N.: Learning systems of concepts with an infinite relational model. In: Poceedings of the National Conference on Artificial Intelligence (AAAI) (2006)
73. Buitelaar, P., Cimiano, P.: Ontology Learning and Population: Bridging the Gap between Text and Knowledge. IOS Press, Amsterdam (2008)
74. Cimiano, P.: Ontology Learning and Population from Text: Algorithms, Evaluation and Applications. Springer, Heidelberg (2006)
75. Sowa, J.F.: Ontology, metadata, and semiotics. In: International Conference on Computational Science (2000)
76. Biemann, C.: Ontology learning from text: A survey of methods. LDV Forum 20(2) (2005)
77. Völker, J., Haase, P., Hitzler, P.: Learning expressive ontologies. In: Buitelaar, P., Cimiano, P. (eds.) Ontology Learning and Population: Bridging the Gap between Text and Knowledge. IOS Press, Amsterdam (2008)
78. Harris, Z.S.: Mathematical Structures of Language. Wiley, Chichester (1968)
79. Hindle, D.: Noun classification from predicate-argument structures. In: Meeting of the Association for Computational Linguistics (1990)
80. Cimiano, P., Staab, S.: Learning concept hierarchies from text with a guided agglomerative clustering algorithm. In: Proceedings of the ICML 2005 Workshop on Learning and Extending Lexical Ontologies with Machine Learning Methods (2005)
81. Steinbach, M., Karypis, G., Kumar, V.: A comparison of document clustering techniques. In: KDD Workshop on Text Mining (2000)
82. Maedche, A., Staab, S.: Semi-automatic engineering of ontologies from text. In: Proceedings of the 12th International Conference on Software Engineering and Knowledge Engineering (2000)
83. Cimiano, P., Hotho, A., Staab, S.: Comparing conceptual, divise and agglomerative clustering for learning taxonomies from text. In: Proceedings of the 16th Eureopean Conference on Artificial Intelligence, ECAI 2004 (2004)
84. Ganter, B., Wille, R.: Formal Concept Analysis: Mathematical Foundations. Springer, Heidelberg (1997)

85. Cimiano, P., Staab, S.: Learning concept hierarchies from text with a guided agglomerative clustering algorithm. In: Biemann, C., Paas, G. (eds.) Proceedings of the ICML 2005 Workshop on Learning and Extending Lexical Ontologies with Machine Learning Methods (2005)
86. Hearst, M.A.: Automatic acquisition of hyponyms from large text corpora. In: Proceedings of the 14th conference on Computational linguistics (1992)
87. Paaß, G., Kindermann, J., Leopold, E.: Learning prototype ontologies by hierachical latent semantic analysis. In: Knowledge Discovery and Ontologies (2004)
88. Deerwester, S., Dumais, S.T., Furnas, G.W., Landauer, T.K., Harshman, R.: Indexing by latent semantic analysis. Journal of the American Society for Information Science 41 (1990)
89. Hofmann, T.: Probabilistic latent semantic analysis. In: Uncertainty in Artificial Intelligence (UAI) (1999)
90. Blei, D.M., Ng, A.Y., Jordan, M.I.: Latent dirichlet allocation. J. Mach. Learn. Res. 3 (2003)
91. Griffiths, T.L., Steyvers, M.: Finding scientific topics. Proc. Natl. Acad. Sci. USA (2004)
92. Blei, D.M., Griffiths, T.L., Jordan, M.I., Tenenbaum, J.B.: Hierarchical topic models and the nested chinese restaurant process. In: Advances in Neural Information Processing Systems (2003)
93. Mei, Q., Shen, X., Zhai, C.: Automatic labeling of multinomial topic models. In: Proceedings of the 13th ACM SIGKDD international conference on Knowledge discovery and data mining (2007)
94. Bontcheva, K., Cunningham, H., Kiryakov, A., Tablan, V.: Semantic annotation and human language technology. In: Davies, J., Studer, R., Warren, P. (eds.) Semantic Web Technologies. Wiley, Chichester (2006)
95. McCallum, A.: Information extraction: distilling structured data from unstructured text. Queue 3(9) (2005)
96. Grishman, R., Sundheim, B.: Design of the MUC-6 evaluation. In: MUC6 1995: Proceedings of the 6th conference on Message understanding (1995)
97. Erik, F., Sang, T., De Meulder, F.: Introduction to the CoNLL-2003 shared task: Language-independent named entity recognition. In: Daelemans, W., Osborne, M. (eds.) Proceedings of CoNLL 2003 (2003)
98. Yeh, A., Morgan, A., Colosimo, M., Hirschman, L.: Biocreative task 1a: gene mention finding evaluation. BMC Bioinformatics 6 (2005)
99. Mayfield, J., McNamee, P., Piatko, C.: Named entity recognition using hundreds of thousands of features. In: Proceedings of the seventh conference on natural language learning (2003)
100. Ray, S., Craven, M.: Representing sentence structure in hidden markov models for information extraction. In: Proceedings of the 17th International Joint Conference on Artificial Intelligence (2001)
101. Lin, Y.F., Tsai, T.H., Chou, W.C., Wu, K.P., Sung, T.Y., Hsu, W.L.: A maximum entropy approach to biomedical named entity recognition. In: Proceedings of 4th ACM SIGKDD Workshop on Data Mining in Bioinformatics (BioKDD) (2004)
102. Cimiano, P., Völker, J.: Towards large-scale, open-domain and ontology-based named entity classification. In: Angelova, G., Bontcheva, K., Mitkov, R., Nicolov, N. (eds.) Proceedings of the International Conference on Recent Advances in Natural Language Processing, RANLP (2005)
103. Chapelle, O., Schölkopf, B., Zien, A.: Semi-Supervised Learning. MIT Press, Cambridge (2006)

104. Ando, R.K., Zhang, T.: A high-performance semi-supervised learning method for text chunking. In: Proceedings of the 43rd Annual Meeting on Association for Computational Linguistics (2005)

105. Kristjansson, T.T., Culotta, A., Viola, P.A., McCallum, A.: Interactive information extraction with constrained conditional random fields. In: Nineteenth National Conference on Artificial Intelligence (AAAI) (2004)

106. Benjelloun, O., Garcia-Molina, H., Menestrina, D., Su, Q., Whang, S.E., Widom, J.: Swoosh: A generic approach to entity resolution. VLDB Journal (2008)

107. Bhattacharya, I., Getoor, L.: Collective entity resolution in relational data. ACM Transactions on Knowledge Discovery from Data 1(1) (2007)

108. Bhattacharya, I., Getoor, L.: A latent dirichlet model for unsupervised entity resolution. In: SIAM SDM 2006 (2006)

109. Zhou, G., Su, J., Zhang, J., Zhang, M.: Exploring various knowledge in relation extraction. In: Proceedings of the 43rd Annual Meeting on Association for Computational Linguistics (2005)

110. Ramani, A.K., Bunescu, R.C., Mooney, R.J., Marcotte, E.M.: Consolidating the set of known human protein-protein interactions in preparation for large-scale mapping of the human interactome. Genome Biol. 6(5) (2005)

111. Bunescu, R.C., Mooney, R.J.: Subsequence kernels for relation extraction. In: Advances in Neural Information Processing Systems (2005)

112. Ono, T., Hishigaki, H., Tanigami, A., Takagi, T.: Automated extraction of information on protein-protein interactions from the biological literature. Bioinformatics 17(2) (2001)

113. Bundschus, M., Dejori, M., Stetter, M., Tresp, V., Kriegel, H.: Extraction of semantic biomedical relations from text using conditional random fields. BMC Bioinformatics 9 (2008)

114. Brickley, D., Miller, L.: The Friend of a Friend (FOAF) project, http://www.foaf-project.org/

115. Brickley, D., Miller, L.: FOAF Vocabulary Specification, http://xmlns.com/foaf/spec/

Using Cognitive Entropy to Manage Uncertain Concepts in Formal Ontologies*

Joaquín Borrego-Díaz and Antonia M. Chávez-González

Departamento de Ciencias de la Computación e Inteligencia Artificial
E.T.S. Ingeniería Informática-Universidad de Sevilla
Avda. Reina Mercedes s.n. 41012-Sevilla
{jborrego,tchavez}@us.es

Abstract. A logical formalism to support the insertion of uncertain concepts in formal ontologies is presented. It is based on the search of extensions by means of two automated reasoning systems (ARS), and it is driven by what we call *cognitive entropy*.

1 Introduction

The challenge of data management with logical trust arose from the statement of the Semantic Web (SW). An important problem is the need for extending or revising ontologies. Such task is, from the point of view of companies, dangerous and expensive: since every change in ontology would affect the overall knowledge of the organization. It is also hard to be automated, because some criteria for revision cannot be fully formalized. Despite its importance, the tools designed to facilitate the syntactic extension or ontological mapping do not analyze, in general, their effect on the (automated) reasoning.

Our aim is to design tools for extending ontologies in a semi-automated way, that is one of the problems present in several methods for cleaning data in the SW, when it implies ontological revision (see e.g. [1] [3]). The method is based on the preservation by extensions of the notion of *ontology robustness*, see [8]. *lattice categoricity*, (described in sect. 3), is going to be applied in a special case: the change is induced by the user, who has detected the (cognitive) necessity of adding a *notion*. That is, a vague concept which comprises a set of elements with features roughly shaped by the existing concepts. In Ontological Engineering, careful consideration should be paid to the accurate classification of objects: the notion becomes a *concept* when its behavior is constrained by new axioms that relate it to the initial concepts. This scenario emphasizes the current need for an explanation of the reasoning behind cleaning programs. That is, a *formalized explanation* of the decisions made by systems. Note that such explanations are necessary for the desirable design of logical algorithms to be used by general-purpose cleaning agents [4]. It is evident that the task will need not only specific

* This work is partially supported by the project TIN2004-03884 of Spanish Ministry of Education and Science, cofinanced by FEDER founds.

P.C.G. da Costa et al. (Eds.): URSW 2005-2007, LNAI 5327, pp. 315–329, 2008.
© Springer-Verlag Berlin Heidelberg 2008

automated reasoning systems (ARSs) for SW, but also those for general purpose. The reason is that some tasks are not directly related to reasoning services for the SW [2] [17] [8]. Thus, we use ARSs for first order logic theories, in favor of one reaches major generality. Among the challenges the problem raises in a dynamic setting as the SW, there are three of them which are specially interesting from the point of view of automated reasoning. They seem to obstruct the design of a fully formalised methodology [4] from classical database field:

- We can not suppose the database to be stable (because new facts could be added in the future).
- Usually, the specification of an ontology is syntactically complex, so it is very likely that classical axiomatization of database theory becomes inconsistent, even if ontology itself is consistent.
- It is possible that the database does not contain facts about the whole relations of the language.

However, some limitations can be solved by weakening the requirements imposed in both database and ontological reasoning [8] [2].

The method proposed is based on the assistance of two ARS, McCune's OTTER and MACE4 (http://www-unix-mcs.anl.gov). The first one, OTTER, is an automated theorem prover (ATP) based on resolution and *support set* strategy. The program allows great autonomy: its auto2 mode suffices to find almost every automated proof that have been required. The second one, MACE4, is an automatic model finder sharing formula syntax with OTTER. It is based on Davis-Putnam-Loveland-Longemann's procedure to decide satisfiability. It has been useful for analyzing the models of the involved theories.

Finally, it would be good to add some information about MACE4. Despite it has not been formally verified to work correctly, once the result by MACE4 is determined, it is not difficult to certify that the models it gives are correct. It is necessary to use OTTER to prove that the list of models is exhaustive. Thus, MACE4 has been used as an automatic assistant to induce new results and investigate the effect of diverse axiomatizations, which must be certified later.

2   Logic-Based Ontological Extensions

Once the need for revision is accepted, the task can be seen, up to some extent -and specially when one designs her/his own logical theory-, from two points of view. The first one considers it like a task similar to belief revision, analyzing it by classic methods of AI. Nevertheless, the effort can be expensive, because it must study once again the impact of revision on the foundational features of the source ontology. The second one has a foundational character. The evolution of ontology should obey ground principles which are accepted on this matter. For example, preserving some sort of backward compatibility, if it is possible (extracted from [15]):

- *The ontology should be able to extend other ontologies with new terms and definitions.*

- *The revision of an ontology should not change the well-formedness of resources that commit themselves to an earlier version of the ontology.*

However, such principles are more adequate if the source ontology is *robust*, in the following sense [4]: *An ontology is robust if its core is clear, stable (except for extensions); if every model of its core exhibits similar properties w.r.t. the core language, and if it is capable of admitting (minor) changes made out of the core without committing core consistency.* By *core* we understand a portion of ontology that we consider as a sound theory with well known properties, and which is accepted as the best for the concepts involved. We can consider two kinds of extensions:

- *Extension by definition.* It produces conservative extensions. If definitions are not provided for the new elements, conservation can fail.
- *Ontological insertion:* Essentially new (nondefinable) concepts/relations are inserted. The task is to design good axioms to specify the new ones from core theory.

An interesting case occurs in the task of ATP-aided cleaning of logic databases. The *bottom-up change generation* in ontologies -due to the analysis of track interaction among the Knowledge Base, the ATP and the user- induces ontological revision. It can simulate new elements in ontology to be inserted (such as Skolem noise [2]). We analyze here a slightly different problem, which appears when the user is the person who decides to insert a new concept by collecting a set of data.

The *extension by definition* is the basis of *definitional methodologies* for building formal ontologies. It is based on the following principles [7]:

1. *Ontologies should be based upon a small number of primitive concepts.*
2. *These primitives should be given definite model theoretic semantics.*
3. *Axioms should only be given for the primitive concepts.*
4. *Categorical axiom sets should be sought.*
5. *The remaining vocabulary of the ontology (which may be very large), should be introduced purely by means of definitions.*

In this paper, the first three principles are assumed. The fourth one will be replaced by *lattice categoricity*. Categoricity is a strong requirement that can be hard to achieve and to preserve. Even when it is achieved, the resultant theory may be unmanageable (even undecidable) or unintuitive. This phenomenon might suggest that we restrict the analysis of completeness to coherent parts of the theory. However, it is not a *local* notion: since minor changes commit the categoricity and it is expensive to repeat the logical analysis.

With respect to the last principle, starting with a basic theory, it seems hard to define a new concept/relationship. It is better to consider it only as the starting point to build an ontology, thinking thus that we are in early steps of the process, where ontological insertions are necessary.

Finally (although it is not the topic of this paper), we would like to add that an ontological insertion should be supported by a good theory about its relationship with the original ontology. It should as well be supported by a nice way of expanding a representative class of models of the source theory to the new one. This class of models must contain the *intended* models (those that the ontology designer wants to represent). It can be required an interpretation of the new elements which should be formalised, and a re-interpretation of the older ones, which must be compatible with basic original principles.

3   Lattice Categorical Theories

In order to solve in practice the several logical problems that ontological insertion raises we will analyze the categoricity of the structure of the concepts of the ontology. We are going to take into account compatibility which has been previously mentioned, and we will try to obtain definitions of the concepts inserted in the new ontology. We will also analyze categoricity of structure of concepts of ontology. For the sake of clarity, we suppose that the set of concepts has a lattice structure. Actually, this is not a constraint: there are methods to extract ontologies from data which produce such structure (such as the Formal Concepts Analysis [14]) and, in general, the ontology is easy to be extended by definition, verifying lattice structure. Although we think about Description Logics [5] as ontological language (the logical basis for ontology languages as OWL, see http://www.w3.org/TR/owl-features/), the definitions are useful for full first order logic (FOL), so we give the definitions in FOL language.

On the one hand, a *lattice categorical* theory is the one that proves the lattice structure of its basic relationships. This notion is weaker than categoricity or completeness. On the other hand, lattice categoricity is a reasonable requirement: the theory must certify the basic relationships among the primitive concepts. In [8] we argued that completeness can be replaced by *lattice categoricity* to facilitate the design of feasible methods for extending ontologies. Let us summarize these ideas.

Given a fixed FOL language, let $\mathcal{C} = \{C_1, \ldots, C_n\}$ be a (finite) set of concept symbols, let T be a theory (in the general case, definable concepts in T can be considered). Given M a model of T, $M \models T$, we consider the structure $L(M, \mathcal{C})$, in the language $L_{\mathcal{C}} = \{\top, \bot, \leq\} \cup \{c_1, \ldots, c_n\}$, whose universe are the interpretations in M of the concepts (interpreting c_i as C_i^M), $\top$ is M, $\bot$ is $\emptyset$ and $\leq$ is the subset relation. We assume from now on that $L(M, \mathcal{C})$ is requested to have a lattice structure for every theory we consider. This requirement simplifies the examples.

The relationship between $L(M, \mathcal{C})$ and the model M itself is based in two facts. The first one, the lattice L can be characterized by a finite set of equations E_L, plus a set of formulas $\Theta_{\mathcal{C}}$ categorizing the lattice under completion, that is, $\Theta_{\mathcal{C}}$ includes the domain closure axiom, the unique names axioms and, additionally, the axioms of lattice theory. Thus, every model M of $E \cup \Theta_{\mathcal{C}}$ is finite. The second one, there exists a natural translation Π of these $L_{\mathcal{C}}$-equations into formulas in the FOL language so that if E is a set of equations characterizing $L(M, \mathcal{C})$ (so $L(M, \mathcal{C}) \models E$), then $M \models \Pi(E)$.

Definition 1. *Let E be a $L_\mathcal{C}$-theory. We say that E is a* **lattice skeleton** *(l.s.) for a theory T if E verifies that*

- *There is $M \models T$ such that $L(M, \mathcal{C}) \models E \cup \Theta_\mathcal{C}$, and*
- *$E \cup \Theta_\mathcal{C}$ has an unique model (modulo isomorphism).*

Every consistent theory has a lattice skeleton [8]. Roughly speaking, the existence of essentially different lattice skeletons makes difficult to reason with the ontology while the existence of only one would make it easy.

Definition 2. *T is called a* **lattice categorical (l.c.) theory** *if whatever pair of lattice skeletons for T are equivalent modulo $\Theta_\mathcal{C}$.*

Note that if T is l.c. and E is a l.s. of T, then $T \vdash \Pi(E)$. Note also that every consistent theory T has an extension T' which is lattice categorical: it suffices to consider a model $M \models T$, and then to find a set E of equations such that $\Theta_\mathcal{C} \cup E$ has $L(M, \mathcal{C})$ as only model. The theory $T \cup \Pi(E)$ (and any consistent extension of it) is l.c.

Finally, we can give a formalization of *robust ontological extension*, based in the categorical extension of the ontology:

Definition 3. *Given two pairs $(T_1, E_1), (T_2, E_2)$ we will say that (T_2, E_2) is a* **lattice categorical extension** *of (T_1, E_1) with respect to the sets of concepts $\mathcal{C}_1$ and $\mathcal{C}_2$ respectively, if $\mathcal{C}_1 \subseteq \mathcal{C}_2$ and $L(T_2, \mathcal{C}_2)$ is an E_1-conservative extension of $L(T_1, \mathcal{C}_1)$.*

For reasoning with the lattice of concepts it suffices to work with a lattice skeleton, so, to simplify, we suppose throughout that T is the self l.s.

3.1   Cognitive Support

Once formalized the notion of *lattice categorical extension*, we need to design several functions to advise how to select the best l.c. extension.

Assume that T is a theory, and L is the lattice defined by $\mathcal{C}$ in some $M \models T$. From the point of view of ontology designer, such a model M is the *intended* model that the ontology attempts to represent. Suppose that $\Delta = \{h_1, \ldots h_n\}$ is the set of facts on $\mathcal{C}$, and the user wants to classify some elements that occur in Δ by means of a new concept. We can suppose, to simplify the notation, that every fact explicit in T belongs to Δ. Let $U(\Delta)$ be the universe determined by Δ; that is, $\{a : \exists C \in \mathcal{C} \, [C(a) \in \Delta]\}$.

Given $C \in \mathcal{C}$ in Δ, we consider

$$|C|^\Delta := |\{a : C(a) \in \Delta\}| \text{ and } |C|_T^\Delta := |\{a \in U(\Delta) : T \cup \Delta \models C(a)\}|.$$

Definition 4. *The* **cognitive support** *of C with respect to Δ, T and L, is*

$$sup_{T,\Delta}^L(C) := \frac{|\{a \in U(\Delta) : \exists i[C_i \leq^L C \land T \cup \Delta \models C_i(a)]\}|}{|U(\Delta)|}$$

This support estimates the number of facts on the concept C entailed by T, normalized by the size of the universe $U(\Delta)$. Because of the computational complexity of logical reasoning, it can be hard in general to compute it: we need to seek, by logical entailment, the cone of concepts defined by C. However, this computation is trivial for lattice categorical theories:

Proposition 1. *If T is lattice categorical, then $sup^L_{T,\Delta}(C) = \dfrac{|C|^\Delta_T}{|U(\Delta)|}$*

The proposition holds because if $C_i \leq^L C$, then $T \models C_i \sqsubseteq C$. Thus, if $T \cup \Delta \models C_i(\boldsymbol{a})$, then $T \cup \Delta \models C(\boldsymbol{a})$.

From now on, we suppose that Δ is compounded by facts on atoms of the lattice of concepts (that is, about the most specific concepts). Note, also, that if T is l.c., then L is unique, and we will thus omit the superscript L in that case.

Corollary 1. *If $\mathcal{J} = \{C_1, \ldots, C_n\}$ is a Jointly Exhaustive and Pairwise Disjoint (JEPD) set of concepts in L, then $sup_{T,\Delta}(.)$ is a probability measure.*

Proof. It is easily seen that $\sum_{C \in \mathcal{J}} sup^L_{T,\Delta}(C) = 1$.

The **cognitive entropy** of $\mathcal{J}$ is $CH(\mathcal{J}) = -\displaystyle\sum_{C \in \mathcal{J}} sup_{T,\Delta}(C) \log sup_{T,\Delta}(C)$.

3.2   Entropy of Ontological Extensions

Suppose that the user decides that a set $\{a_1, \ldots, a_k\} \subseteq U(\Delta)$ induces a new concept D (provisionally, a notion). Such a notion might not be fully represented by those elements. Also, it is possible that some of them do not belong to the new concept, because of noise in the data. It might also be the case that the concept is constrained by a set Σ of axioms introduced by the user. Furthermore it is also possible that $T \cup \Sigma$ is not l. c., that is, this theory does not prove the intended lattice induced by $\mathcal{C} \cup \{D\}$. MACE4 provides the collection $\{L_1, \ldots, L_m\}$ of the lattices induced by the models of $T \cup \Sigma$. Let T_i be a lattice skeleton for L_i $(i = 1, \ldots, m)$.

Now, we focus our attention on a concrete *level* of the Ontology, where we intend to insert the new concept. The level will be a JEPD $\mathcal{J} = \{C_1, \ldots, C_k\}$ of the lattice L verifying that if the new concept D contains some of them,

$$\mathcal{J}^{L_i}_{\upharpoonright D} = \{C_i \in \mathcal{J} \; : \; C_i \leq^{L_i} D\} \neq \emptyset$$

then $\mathcal{J}_i = (\mathcal{J} \setminus \mathcal{J}^{L_i}_{\upharpoonright D}) \cup \{D\}$ is a JEPD in L_i. Since T_i is a l.c. extension of T, the support of D is easy to achieve:

Theorem 1. *In above conditions, $sup_{T_i,\Delta}(D) = \displaystyle\sum_{C \in \mathcal{J}^{L_i}_{\upharpoonright D}} sup^L_{T,\Delta}(C_i)$*

To estimate the conditional entropy of the new extension, we consider a natural definition of *conditional support*:

$$sup_{T_i,T,\Delta}(C'|C) := \frac{|\{\boldsymbol{a} \in U(\Delta) \; : \; T \cup \Delta \models C(\boldsymbol{a}) \wedge T_i \cup \Delta \models C'(\boldsymbol{a})\}|}{|C|^\Delta_T}$$

This support allows to estimate the amount of new information produced by the extension by standard methods; through the *conditional entropy* associated to the two probability measures. The **conditional cognitive entropy** is :

$$CH(\mathcal{J}\|\mathcal{J}_i) = -\sum_{\substack{C' \in \mathcal{J} \\ C \in \mathcal{J}_i}} sup_{T_i,\Delta}(C'|C) \log sup_{T_i,\Delta}(C'|C)$$

This sum can be simplified (assuming $0 \log 0 = 0$): if $C = C'$ or $C, C' \in \mathcal{J}$, then

$$sup_{T_i,T,\Delta}(C'|C) \log sup_{T_i,T,\Delta}(C'|C) = 0$$

and the following property holds:

Proposition 2. *In above conditions,* $sup_{T_i,T,\Delta}(C'|C) = \dfrac{|C'|_T^\Delta}{|C|_{T_i}^\Delta}$

This entropy is similar to Kullback-Leibler distance or relative entropy (see [16]), but using the entailment to classify the elements. It is known that it is minor than the initial entropy. In [13] similar entropies are used, but based on probabilistic assignation. Finally, in order to estimate what is the best extension for our purposes, it is necessary to compute the The **Shannon's diversity index** for each L_i. This index normalizes the amount of information produced by the extension, and is defined as

$$IH(\mathcal{J}_i) = \frac{CH(\mathcal{J}\|\mathcal{J}_i)}{\log |\mathcal{J}_i|}$$

The interpretation of the index is as follows: if we select L_i with minimum $IH(\mathcal{J}_i)$, the new information produced by the new concept is minor. This option is the *cautious* one: the reparation of the source ontology is *light* and we do not expect big changes in the representation of the intended model. If we select L_i with an upper $IH(J)$, the change of the information is more relevant; we select such an extension if we regard as robust the specification of the concept given by Σ together with the facts. In general, we have to chose the l.c. extension with minor index. Intuitively, in this way we do not change too much the information of the initial ontology.

4   An Example

We would like to show a short example in the field of Qualitative Spatial Reasoning (QSR). *Region Connection Calculus* (RCC) [12] is a well-known mereotopological approach to QSR, that we can consider to be a robust ontology. For RCC, the *spatial entities* are non-empty regular sets. The primary relation between them is *connection*, $C(x, y)$, with intended meaning: *"the topological closures of* x *and* y *intersect"*. The basic axioms of RCC are $A_1 := \forall x[C(x, x)]$ and $A_2 := \forall x, y[C(x, y) \rightarrow C(y, x)]$ jointly with a set of definitions on the main spatial relations (fig. 1), and other axioms not used here (see [12]).

$$DC(x,y) \leftrightarrow \neg C(x,y) \qquad\qquad (x \text{ is disconnected from } y)$$
$$P(x,y) \leftrightarrow \forall z[C(z,x) \rightarrow C(z,y)] \qquad\qquad (x \text{ is part of } y)$$
$$PP(x,y) \leftrightarrow P(x,y) \wedge \neg P(y,x) \qquad\qquad (x \text{ is proper part of } y)$$
$$EQ(x,y) \leftrightarrow P(x,y) \wedge P(y,x) \qquad\qquad (x \text{ is identical to } y)$$
$$O(x,y) \leftrightarrow \exists z[P(z,x) \wedge P(z,y)] \qquad\qquad (x \text{ overlaps } y)$$
$$DR(x,y) \leftrightarrow \neg O(x,y) \qquad\qquad (x \text{ is discrete from } y)$$
$$PO(x,y) \leftrightarrow O(x,y) \wedge \neg P(x,y) \wedge \neg P(y,x) \qquad\qquad (x \text{ partially overlaps } y)$$
$$EC(x,y) \leftrightarrow C(x,y) \wedge \neg O(x,y) \qquad\qquad (x \text{ is externally connected to } y)$$
$$TPP(x,y) \leftrightarrow PP(x,y) \wedge \exists z[EC(z,x) \wedge EC(z,y)] \quad (x \text{ is a tangential prop. part of } y)$$
$$NTPP(x,y) \leftrightarrow PP(x,y) \wedge \neg\exists z[EC(z,x) \wedge EC(z,y)] \ (x \text{ is a non-tang. prop. part of } y)$$

Fig. 1. Axioms of RCC

We have proved (by using MACE4 and OTTER) that the set of formulas E given in the figure 2 categorises under completion the lattice of the RCC-spatial relationships (given in fig. 3). The set of binary relations formed by the eight (JEPD) relations given in figure 3 is denoted by RCC8. If this set is thought to be a calculus, all possible unions of the basic relations are also used. Another interesting calculus is RCC5, based on $\{DR, PO, PP, PPi, EQ\}$.

$$\top \equiv C \sqcup DR \qquad\qquad PO \sqsubseteq \neg P \sqcap \neg Pi \sqcap \neg DR \qquad DR \equiv EC \sqcup DC$$
$$NTPP \sqsubseteq \neg TPP \sqcap \neg Pi \sqcap \neg DR \qquad C \equiv O \sqcup EC \qquad TPP \sqsubseteq \neg Pi \sqcap \neg DR$$
$$O \equiv PO \sqcup P \sqcup Pi \qquad\qquad EQ \sqsubseteq \neg PPi \sqcap \neg DR \qquad Pi \equiv EQ \sqcup PPi$$
$$TPPi \sqsubseteq \neg NTPPi \sqcap \neg DR \qquad P \equiv EQ \sqcup PP \qquad NTPPi \sqsubseteq \neg DR$$
$$PPi \equiv TPPi \sqcup NTPPi \qquad\qquad EC \sqsubseteq \neg DC \qquad PP \equiv TPP \sqcup NTPP$$

Fig. 2. A skeleton for RCC

Suppose that if we insert a new spatial uncertain relation D expressing "x and y have a isometric overlapping relation"; that is, D covers *partial overlapping PO* and *extensional equality EQ* relationships. That is, proper part is not possible between isometric objects. This is suggested by the study of spatial relationships among identical objects (e.g. the 2-D spatial configuration of a set of coins). Thus, we consider that the new relation D satisfies

$$RCC \cup \{\forall x \forall y (PO(x,y) \rightarrow D(x,y)), \forall x \forall y (EQ(x,y) \rightarrow D(x,y))\}$$

or, in terms of skeleton, $E \cup \{PO \sqsubseteq D, EQ \sqsubseteq D\}$. MACE4 produces seven l.c. extensions (classified according to their lattices in fig. 4). All these extensions can be mereotopologically interpreted [11]. Suppose that the set that motivates the extension is:

$$\Delta := \begin{cases} PO(m_1,m_2) \ EQ(m_2,m_3) \ EQ(m_3,m_4) \quad PO(m_1,m_3) \ DC(m_4,m_6) \\ DC(m_3,m_5) \ PO(m_5,m_1) \ NTPP(c_1,m_3) \ EC(c_2,m_1) \ TPP(c_2,c_5) \\ DC(c_1,c_2) \quad TPPi(c_5,c_2) \ NTPP(m_2,c_4) \ DC(m_1,c_3) \ TPP(c_1,c_3) \end{cases}$$

In this case, $|U(\Delta)| = 15$, and the basic JEPD is the set $\mathcal{J} = \{PO, PP, EQ, PPi, EC, DC\}$. In each L_i, $\mathcal{J}_i$ is a JEPD, so we can assign conditional entropy and

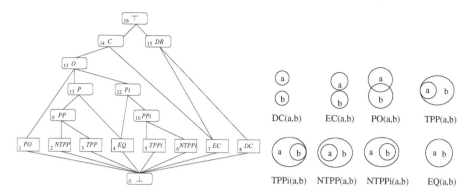

Fig. 3. The lattice of spatial relations of RCC (left) and he relations of RCC8 (right)

Shannon's diversity index to each extension.Thus, $T_2 \equiv E + \cup \{D \equiv PO \sqcup EQ\}$ is the selected l.c. extension because it has the minimum Shannon's index. On the other hand, the user's notion might be inconsistent. For instance, if the user's proposal for Σ' is $\{PO \sqsubseteq D, EQ \sqsubseteq D, P \sqsubseteq D, D \sqsubseteq O\}$, then there is not any l.c. extension, a fact that we have certified using MACE4 and OTTER.

5   Data-Driven Ontology Revision: Deficient Data Layout

In above sections, we have formalized the insertion of a concept that will remain well defined once the appropriate extension is selected. In that case, the computation of entropies is easier than the entropies defined in this section. Now, we aim to extend the ontology in a provisional way because the deficient classification of data induces the insertion of subconcepts for refining the classification of individuals which initially were misclassified[1]. In this case the new concepts will fall in the bottom level. Therefore, we aim to extend $\mathcal{J}_L$, the JEPD set of concepts which are the atoms of the lattice $L(T, \mathcal{C})$.

The following definition formalizes the notion of *insertion of a concept with certain degree of imprecision* as subconcept of a given concept C. It has to be determined whether there is a l.c. extension of the ontology with an (atomic) subconcept μC of C. Intuitively, the meaning of $\mu C(\mathbf{a})$ is "the concept $\mathbf{a}$ falls in the concept C, but we do not know any more specific information about $\mathbf{a}$". Formally,

Definition 5. *Let (T, E_0) be a l.c. core and $C \in \mathcal{C}$. We say that the ontology admits an undefinition at C $(T \leadsto_w C)$ if there is a l.c. extension of T, (T', E'), such that*

1. *T' is l.c. with respect to $\mathcal{C} \cup \{\mu C\}$, (where $\mu C \notin \mathcal{C}$).*
2. *$\{\mu C\}$ is an atom in the lattice $L' = L(T, \mathcal{C} \cup \{\mu C\})$.*
3. *There is not C' such that $\mu C <^{L'} C' <^{L'} C$.*

---

[1] This approach is inspired in the study presented at Eurocast 2007 [9].

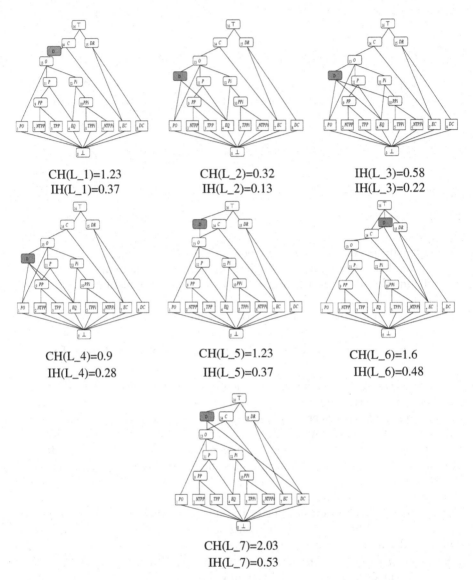

CH(L_1)=1.23
IH(L_1)=0.37

CH(L_2)=0.32
IH(L_2)=0.13

IH(L_3)=0.58
IH(L_3)=0.22

CH(L_4)=0.9
IH(L_4)=0.28

CH(L_5)=1.23
IH(L_5)=0.37

CH(L_6)=1.6
IH(L_6)=0.48

CH(L_7)=2.03
IH(L_7)=0.53

Fig. 4. The seven l.c. extensions by insertion. The grey box denotes the new relation.

Note that, in above conditions, $\mathcal{J}_L[\mu C] := \mathcal{J}_L \cup \{\mu C\}$ is a JEPD set for L' (see fig. 5, left). This requirement represents, in fact, that we have not any additional information about μC. For example, in figure 5 right, the relation $\mu C(a, b)$ means "the regions a and b are connected, but it is unknown if they overlap or they are externally connected".

The notation $T \models_\mu C(\boldsymbol{a})$ means $T \models C(\boldsymbol{a})$ and, for all $D <^L C$, $T \not\models D(\boldsymbol{a})$. In other words, $C(\boldsymbol{a})$ is the most specific knowledge on $\boldsymbol{a}$ entiled from T. It is easy to see that

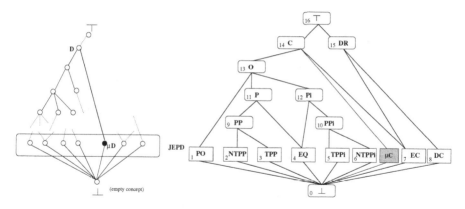

Fig. 5. The ontology (RCC, E) admits an undefinition in the concept C (*connection*) (right)

Proposition 3. *Whatever two extensions by undefinition at C of T have equivalent lattice skeletons modulo completion.*

Such a skeleton of the extension is denoted by $E[\mu C]$. We can also consider the iteration of this kind of extensions, namely $E[\mu C_1, \ldots, \mu C_k]$.

Corollary 2. $E[\{\mu C \; : \; C \in \mathcal{C} \wedge T \rightsquigarrow_w C\}]$ *is unique (modulo database completion axioms).*

5.1   Inserting Provisional Spatial Relationships in RCC

As we have already commented, the JEPD set named RCC8 is the representation of a precise classification for RCC. In order to build the adequate extension, we compute first the list of extensions obtained by inserting only one relation.

Theorem 2. *There are exactly eight extensions by undefinition of the lattice of RCC by insertion of a new relation D such that $RCC8 \cup \{D\}$ is a JEPD set.*

Such new relations can be mereotopologically interpreted [11]. A proof of this result appears at [11]. The lattices of extensions are detailed at [8]. For example, the lattice depicted in fig. 5 (right) has a skeleton $E[\mu C]$.

   The next step consists in deciding which is the best l.c. extension to classify data. Suppose that $\Delta = \{h_1, \ldots h_n\}$ is the set of facts. Assume that the user believes that the set of misclassified elements is $I = \{a_1, \ldots, a_k\} \subseteq U(\Delta)$ (according with user's ontology). In this case, the problem is not due to a new concept, because the user has not decided yet an insertion. Such elements are not falling on atomic concepts ($T \not\models C(a)$ for any $C \in \mathcal{J}_L$), because the user has not an specific definition of them, that is, he has got only unprecise information (as, instances of upper concepts).

   It is easy to provide an extension by undefinition with complete classification of data. For each $a_i \in I$, let $C^i \in \mathcal{C}$ such that $T \models_\mu C^i(a_i)$. Any extension by undefinition at the set $\{C^i \; : \; i = 1, \ldots k\}$ classifies every element of $U(\Delta)$ with a concept of the JEPD set $\mathcal{J}_{T'} := \mathcal{J} \cup \{\mu C^1, \ldots, \mu C^k\}$. Note also, that if we do not require C^i is the most specific one, the extension is not unique.

Definition 6. *Let T' be an extension by undefinition of T defined as in 5. The support of μC is defined as*

$$supp_{T',\Delta}(\mu C) = \frac{|\{a \in U(\Delta) : a \in I \wedge T \cup \Delta \models_\mu C(a)\}|}{|U(\Delta)|}$$

That is, the support of μC uses the number of elements for such that T proves they belong to C. In this way $supp_{T',\Delta}$ is also a probability measure on $\mathcal{J}_{T'}$. Note that this computation is equivalent to consider the support with respect to the theory $T' \cup \{\mu C(a) \ : \ T \cup \Delta \models_\mu C(a)\}$. To simplify, we consider throughout that T' is that theory.

Theorem 3. *The extension above defined exhibits the maximum cognitive entropy among every possible extension by undefinition classifying $U(\Delta)$.*

Sketch of proof: If T'' is other extension, then some a_i of I are classified with respect to a concept which is not the most specific one. Thus the result follows by the convexity of the function $p \log p$.

A l.c. extension by undefinition with maximum entropy gives little information on new concepts. This option is a *cautious* solution to the problem, because strong requirements for the new concepts are not been imposed.

Fig. 6. Map of United States

5.2   A Motivating Example

In order to understand the problem and its solution, let us suppose that a Geographical Information System (GIS) launches agents for finding, in the SW, information about several geographical objects in United States (see fig. 6). Suppose that the set Δ found by information agents is:

Overlap(*West, Mount Elbert*)
PartOf(*Mount Elbert, Colorado*)
PartOf(*Colorado, West*)
ProperPartOf(*Miami, Florida*)
Overlaps(*West, Colorado*)
Overlaps(*Basin of Platte River, Nebraska*)
Discrete(*Colorado, Basin of Missouri River*)
Overlaps(*East, Miami*)
PartialOverlaps(*Basin of Missouri River, West*)
ProperPart(*East, Colorado*)
PartOf(*Miami, Florida*)
ProperPartInverse(*Florida, Miami*)
TangentialProperPart(*Mount Elbert, Great Plains*)
Discrete(*West, Georgia*)
Part(*East, Georgia*)

Note that several facts do not provide the most specific spatial relation that it might be expressed with RCC ontology. That is the case of the fact Overlaps

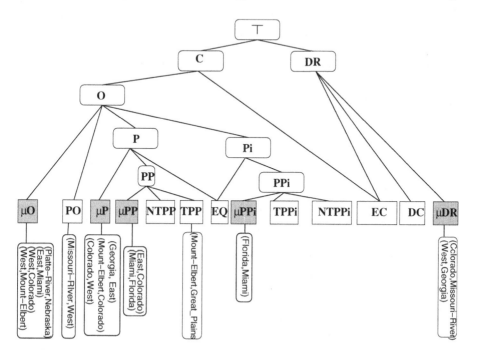

Fig. 7. Classification of data according to $E[\mu PP, \mu P, \mu PPi, \mu O, \mu DR]$

(*Basin of Platte River, Nebraska*). Both regions are overlapping, however there is no information about what level of overlapping relates these regions.

Since the GIS deals with concepts representing underspecified spatial relations such as `Overlaps`, `PartOf`, ..., it is hard to classify individual regions in an accurate way. They would be classify to work within a set of specific spatial-relations/concepts, a jointly exhaustive set of pairwise disjoint concepts to get the exhaustive intended classification.

The problem can be stated as follows: Given a set Δ of facts with respect to an ontology O, where most of specific information on some individuals can not entailed, to design a provisional robust extension of O to provisionally classify these concepts.

The extension of RCC for the running example will be a combination of some of the eight extensions. We are interested on finding an extension by undefinition of RCC that classifies the data and exhibits the highest entropy. According to data of the example, and th. 3, the selected extension has skeleton (see fig. 7): $E[\mu PP, \mu P, \mu PPi, \mu O, \mu DR]$. This l.c. extension reaches (by above theorem) maximum entropy, its value is 1.566. For example, $E[\mu P, \mu PPi, \mu O, \mu DR]$, shows entropy 1.326.

6   Closing Remarks

A formalization of data integration with unprecise information for the Semantic Web has been investigated. It presents a method to insert new concepts in an ontology with backward compatibility and preserving a weak form of completeness.

Although it is usual to study entropy for associating data to concepts in Ontology Learning, it is not usual to consider the *provability from ontology* like a factor, as we do. However, we think, that it will be a key issue in the SW. There are other approaches, but they deal with probabilistic objects. J. Calmet and A. Daemi also use entropy in order to revise or compare ontologies [10] [13]. This is based on the self taxonomy defined by the concepts but provability from specification is not regarded. Conditional entropy has already been considered in the similar task of Abductive Reasoning for learning qualitative relationships/concepts (usually in probabilistic terms, see e.g. [6]). The main difference between this approach and ours is that we work with probability mass distribution of *provable facts* from ontological specifications.

Finally, it should be noted that only some distributions of data will induce the user to decide an ontological insertion. Therefore, although once the distribution of data is determined, the method is fully formalized, the soundness of the extensions still depends on human decisions.

Future research lines are addressed, in the medium term, to implement the cognitive entropy into a representation of ontologies system as a tool to assist the extension of ontologies. In a long term, we will establish a theory, a formal theory in computational logic, to classify lattice categorical extensions.

References

1. Alonso-Jiménez, J.A., Borrego-Díaz, J., Chávez-González, A.M., Navarro-Marín, J.D.: A Methodology for the Computer–Aided Cleaning of Complex Knowledge Databases. In: 28th Conf. of IEEE Industrial Electronics Society IECON 2002, pp. 1806–1812 (2002)
2. Alonso-Jiménez, J., Borrego-Díaz, J., Chávez González, A.M., Gutiérrez-Naranjo, M.A., David Navarro-Marín, J.: Towards a Practical Argumentative Reasoning with Qualitative Spatial Databases. In: Chung, P.W.H., Hinde, C.J., Ali, M. (eds.) IEA/AIE 2003. LNCS (LNAI), vol. 2718, pp. 789–798. Springer, Heidelberg (2003)
3. Alonso-Jiménez, J.A., Borrego-Díaz, J., Chávez-González, A.M.: Ontology Cleaning by Mereotopological Reasoning. In: DEXA Workshop on Web Semantics WEBS 2004, pp. 132–137 (2004)
4. Alonso-Jiménez, J.A., Borrego-Díaz, J., Chávez-González, A.M., Martín-Mateos, F.J.: Foundational Challenges in Automated Data and Ontology Cleaning in the Semantic Web. IEEE Intelligent Systems 21(1), 42–52 (2006)
5. Baader, F., Calvanese, D., McGuinness, D., Nardi, D., Patel-Schneider, P. (eds.): The Description Logic Handbook. Theory, Implementation and Applications. Cambridge University Press, Cambridge (2003)
6. Bhatnagar, R., Kanal, L.N.: Structural and Probabilistic Knowledge for Abductive Reasoning. IEEE Trans. on Pattern Analysis and Machine Intelligence 15(3), 233–245 (1993)
7. Bennett, B.: The Role of Definitions in Construction and Analysis of Formal Ontologies. In: Doherty, P., McCarthy, J., Williams, M. (eds.) Logical Formalization of Commonsense Reasoning (2003 AAAI Spring Symp.), pp. 27–35. AAAI Press, Menlo Park (2003)
8. Borrego-Díaz, J., Chávez-González, A.M.: Extension of Ontologies Assisted by Automated Reasoning Systems. In: Moreno Díaz, R., Pichler, F., Quesada Arencibia, A. (eds.) EUROCAST 2005. LNCS, vol. 3643, pp. 247–253. Springer, Heidelberg (2005)
9. Borrego-Díaz, J., Chávez-González, A.M.: A Formal Foundation for Knowledge Integration of Deficient Information in the Semantic Web. In: Moreno Díaz, R., Pichler, F., Quesada Arencibia, A. (eds.) EUROCAST 2007. LNCS, vol. 4739, pp. 305–312. Springer, Heidelberg (2007)
10. Calmet, J., Daemi, A.: From entropy to ontology. In: 4th. Int. Symp. From Agent Theory to Agent Implementation AT2AI-4 (2004)
11. Chávez-González, A.M.: Automated Mereotopological Reasoning for Ontology Debugging, Ph.D. Thesis, University of Seville (2005)
12. Cohn, A.G., Bennett, B., Gooday, J.M., Gotts, N.M.: Representing and Reasoning with Qualitative Spatial Relations about Regions. In: Stock, O. (ed.) Spatial and Temporal Reasoning, ch. 4. Kluwer, Dordrecth (1997)
13. Daemi, A., Calmet, J.: From Ontologies to Trust through Entropy. In: Proc. of the Int. Conf. on Advances in Intelligent Systems - Theory and Applications (2004)
14. Ganter, B., Wille, R.: Formal Concept Analysis, Mathematical Foundations. Springer, Berlin (1999)
15. Heflin, J.: Towards the Semantic Web: Knowledge Representation in a Dynamic, Distributed Environment, Ph.D. Thesis, Univ. of Maryland, College Park (2001)
16. Kotz, S., Johnson, N.L. (eds.): Encyclopedia of Statistical Sciences, vol. 4, pp. 421–425. John Wiley and Sons, Chichester (1981)
17. Tsarkov, D., Riazanov, A., Bechhofer, S., Horrocks, I.: Using Vampire to Reason with OWL. In: McIlraith, S.A., Plexousakis, D., van Harmelen, F. (eds.) ISWC 2004. LNCS, vol. 3298, pp. 471–485. Springer, Heidelberg (2004)

Analogical Reasoning in Description Logics

Claudia d'Amato, Nicola Fanizzi, and Floriana Esposito

Dipartimento di Informatica, Università degli Studi di Bari,
Campus Universitario, Via Orabona 4, 70125 Bari, Italy
{claudia.damato,fanizzi,esposito}@di.uniba.it

Abstract. This work presents a method founded in instance-based learning for inductive (memory-based) reasoning on ABoxes. The method, which exploits a semantic dissimilarity measure between concepts and instances, can be employed both to answer class membership queries and to predict new assertions that may be not logically entailed by the knowledge base. These tasks may be the baseline for other inductive methods for ontology construction and evolution. In a preliminary experimentation, we show that the method is sound and it is actually able to induce new assertions that might be acquired in the knowledge base.

1 Introduction and Motivation

Most of the research on formal ontologies has been focused on methods based on deductive reasoning. However, important tasks, most of the times manually performed, such as ontology construction, ontology revision, ontology population and evolution could be likely supported by new generation of knowledge-based systems by exploiting inductive methods. In this paper, we investigate on the application of instance-based inductive learning methods to the standard ontological representation with the goal of making instance retrieval and ontology population tasks (semi-)automatic.

Currently, in order to support these tasks and to overcome the inherent complexity of classic logic-based inference, other forms of reasoning are being investigated, both deductive, such as *non-monotonic, paraconsistent* [23], *approximate* reasoning (see the discussion in [24]), *case-based reasoning* [15] and inductive-analogical forms such as inductive *generalization* [10] and *specialization* [18].

However, all these approaches require scalable and efficient reasoning. From this viewpoint, instance-based inductive methods [17] are particularly well suited. Indeed, they are known to be both very efficient and fault-tolerant compared to the classic logic-based methods. Indeed, differently from the deductive approach where a set of logic rules are applied to general axioms to infer conclusions of no greater generality than the premises, the inductive approach is applied to very specific facts/examples and produces conclusions that generalize/explain the premises and predicts new facts. Moreover, the presence of faults does not determine the failure of the reasoning process, since many other examples constitutes the basic from which the knowledge is generalized. Being this approach fault-tolerant, it can be suitably applied to shared knowledge bases coming from distributed sources and for this reason often characterized by the presence of noise, that is always a danger in any deductive and inductive process.

P.C.G. da Costa et al. (Eds.): URSW 2005-2007, LNAI 5327, pp. 330–347, 2008.

Instance-based algorithms have been mainly applied to attribute-value (also called feature-vector) representations to solve tasks such as classification, clustering, and case-based reasoning. Most of these algorithms are based on a notion of similarity or dissimilarity between the examples from which the learning task is performed. Upgrading these algorithms for working on multi-relational representations [17], specifically on the concept languages used in the Semantic Web (SW) and founded in Description Logics (DLs) [1] (see Sect. 2), requires novel similarity and/or dissimilarity measures[1] that are suitable for such First Order Logic fragments. However, as pointed out in [7], most of these measures focus on the similarity of atomic concepts within hierarchies or simple ontologies based on a few relations. Thus, the application of instance-based learning algorithms to SW representations requires to investigate the definition of similarity measures in more complex languages.

It has been observed that, adopting richer representations, the structural properties have less and less impact in assessing semantic similarity. Hence, the vision of similarity based only on the structural comparison of the representations (trees or graphs), such as in [9, 27], may fall short since it is not able to capture the concept semantics. In order to overcome this problem, we propose some dissimilarity measures for non trivial DL languages, based on the comparison of the concept semantics conveyed by the ABox assertions [11, 13]. These measures elicit the underlying semantics by querying the knowledge base for assessing the concept extensions, estimated through their *retrieval* [1] (as also hinted in [4]). Besides, the overall similarity is also influenced by the concepts that are related through role restrictions. Such measures can be applied to assess the dissimilarity between concepts and/or between individuals.

By combining an instance-based (analogical) approach with a similarity measure, this work intends to demonstrate the applicability of inductive reasoning to the ontological representation[2]. Specifically, an instance-based framework is proposed consisting in a classification procedure based on a *lazy learning* approach, namely a relational form of the k-*Nearest Neighbor* (k-NN) procedure [26]. Exploiting a dissimilarity measure, it can inductively derive (by analogy) both consistent consequences from the knowledge base (KB) and also new assertions which were not previously logically derivable.

The adaptation of the k-NN procedure to the context of DLs could not be straightforward. A theoretical problem is posed by the *Open World Assumption* (OWA) that is generally made in the target context, differently from data mining settings where the *Closed World Assumption* (CWA) is the standard. Besides, in the standard k-NN multi-class setting, different classes are generally assumed to be disjoint, which is not typical in a Semantic Web context where an individual can belong to more than one concept.

The semantic dissimilarity measures and the modified classification method have been implemented and some preliminary experimental results with real ontologies have been presented (Sect. 5). A reasoning procedure like this may be employed for answering class membership queries through analogical rather than deductive reasoning which, as discussed above, may be more robust with respect to noise and is likely to

---

[1] Since, given a dissimilarity measure it is always possible to obtain a similarity measure and vice-versa [6], in the sequel we will generally talk about similarity measures except when the distinction is necessary.

[2] This work is an extension of the paper [14].

suggest new knowledge (which was not logically derivable). Furthermore, it can be used for making semi-automatic the time-consuming ontology population task. Indeed, as we show in the experimentation, the newly induced assertions are quite accurate (commission errors, i.e. predicting a concept erroneously, were rarely observed). They can be utilized as recommendations to the knowledge engineer that has only to validate them rather than manually write them. In turn, the outcomes of the procedure may also trigger other related tasks such as induction (revision) of (faulty) knowledge bases.

The paper is organized as follows. In the next section, the representation language is briefly presented. Two concept dissimilarity measures are recalled (see Sect. 3) and exploited in a modified version of the k-NN classification procedure (see Sect. 4). The results of a preliminary experimental evaluation of the method are shown in Sect. 5 while possible developments of the work are examined in Sect. 6.

2   Representation and Logic Inference

Description Logics [1] are a family of logic languages with different expressive power, depending from the constructors that are allowed for building complex concept descriptions. We focus on $\mathcal{ALC}$ logic [28], since it comprises most of the constructors that are usually adopted for giving the ontological representation of a certain domain; moreover $\mathcal{ALC}$ logic is considered a good compromise between expressive power and computational complexity required by the inference operators.

In DLs, concept descriptions are defined in terms of a set N_C of *primitive concept names* and a set N_R of *primitive roles*. The semantics of the concept descriptions is defined by an *interpretation* $\mathcal{I} = (\Delta^{\mathcal{I}}, \cdot^{\mathcal{I}})$, where $\Delta^{\mathcal{I}}$ is a non-empty set representing the *domain* of the interpretation, and $\cdot^{\mathcal{I}}$ is the *interpretation function* that maps each $A \in N_C$ to a set $A^{\mathcal{I}} \subseteq \Delta^{\mathcal{I}}$ and each $R \in N_R$ to $R^{\mathcal{I}} \subseteq \Delta^{\mathcal{I}} \times \Delta^{\mathcal{I}}$. The *top* concept $\top$ is interpreted as the whole domain $\Delta^{\mathcal{I}}$, while the *bottom* concept $\bot$ corresponds to $\emptyset$. Complex descriptions can be built in $\mathcal{ALC}$ using the following constructors: *full negation*, denoted $\neg C$, given any concept description C, it amounts to $\Delta^{\mathcal{I}} \setminus C^{\mathcal{I}}$; *conjunction* of concepts, denoted with $C_1 \sqcap C_2$, yields an extension $C_1^{\mathcal{I}} \cap C_2^{\mathcal{I}}$; concept *disjunction*, denoted with $C_1 \sqcup C_2$, yields $C_1^{\mathcal{I}} \cup C_2^{\mathcal{I}}$; *existential role restriction*, denoted with $\exists R.C$, and interpreted as the set $\{x \in \Delta^{\mathcal{I}} \mid \exists y \in \Delta^{\mathcal{I}} : (x,y) \in R^{\mathcal{I}} \wedge y \in C^{\mathcal{I}}\}$; *value restriction*, denoted with $\forall R.C$, whose extension is $\{x \in \Delta^{\mathcal{I}} \mid \forall y \in \Delta^{\mathcal{I}} : (x,y) \in R^{\mathcal{I}} \rightarrow y \in C^{\mathcal{I}}\}$.

A *knowledge base* $\mathcal{K} = \langle \mathcal{T}, \mathcal{A} \rangle$ contains a *TBox* $\mathcal{T}$ and an *ABox* $\mathcal{A}$. $\mathcal{T}$ is a set of concept definitions[3] $C \equiv D$, meaning $C^{\mathcal{I}} = D^{\mathcal{I}}$, where C is atomic (the concept name) and D is an arbitrarily complex description defined as above. $\mathcal{A}$ contains assertions on the world state, e.g. $C(a)$ and $R(a,b)$, meaning that $a^{\mathcal{I}} \in C^{\mathcal{I}}$ and $(a^{\mathcal{I}}, b^{\mathcal{I}}) \in R^{\mathcal{I}}$.

In this context the most common inference is the *subsumption* between concepts:

Definition 2.1 (subsumption). *Given two concept descriptions C and D, C subsumes D, denoted by $D \sqsubseteq C$, iff for every interpretation $\mathcal{I}$ it holds that $D^{\mathcal{I}} \subseteq C^{\mathcal{I}}$. When $D \sqsubseteq C$ and $C \sqsubseteq D$ then they are* equivalent, *denoted with $C \equiv D$.*

---

[3] The cases of general axioms or cyclic definitions [1] will be not considered here.

Semantically equivalent (yet syntactically different) descriptions can be given for the same concept. Nevertheless, equivalent concepts can be reduced to a normal form by means of rewriting rules that preserve their equivalence [1]:

Definition 2.2 (normal form). *A concept description D is in $\mathcal{ALC}$ normal form iff $D = \bot$ or $D = \top$ or if $D = D_1 \sqcup \cdots \sqcup D_n$ ($\forall i = 1, \ldots, n$, $D_i \not\equiv \bot$) and*

$$D_i = \prod_{A \in \mathsf{prim}(D_i)} A \sqcap \prod_{R \in N_R} \left[\forall R.\mathsf{val}_R(D_i) \sqcap \prod_{E \in \mathsf{ex}_R(D_i)} \exists R.E \right]$$

- $\mathsf{prim}(D_i)$ *is the set of (negated) primitive concepts occurring at the top level of D_i;*
- $\mathsf{val}_R(D_i)$ *is the conjunction $C_1^i \sqcap \cdots \sqcap C_n^i$ in the value restriction of role R, if any (otherwise $\mathsf{val}_R(D_i) = \top$);*
- $\mathsf{ex}_R(D_i)$ *is the set of concepts in the existential restrictions on R.*

and $\forall R \in N_R \ \forall E \in \mathsf{ex}_R(D_i) \cup \{\mathsf{val}_R(D_i)\}$ E is in normal form.
$\mathcal{L}$ will denote the set of concepts in normal form ($\mathcal{L} = \mathcal{ALC}/_{\equiv}$).

It is worthwhile to note in the normal form that there is a conjunction of universal restriction on varying of the role R and a conjunction of existential restriction on varying of the role and on varying of the filler in the same role. This is because while it is possible to write $\forall R.C \sqcap \forall R.D = \forall R.(C \sqcap D)$ it is not possible to write the same for the existential restriction due to the open world assumption (see [8] for more details).

The *instance checking* is the reasoning procedure that decides if an individual is an instance of a concept or not [16, 1]. Another inference for reasoning with individuals requires finding the concepts to which an individual belongs to, especially the most specific one (w.r.t. the subsumption relationship):

Definition 2.3 (most specific concept). *Given an ABox $\mathcal{A}$ and an individual a, the most specific concept of a w.r.t. $\mathcal{A}$ is the concept C, denoted $MSC_{\mathcal{A}}(a)$, s.t. $\mathcal{A} \models C(a)$ and for any other concept D such that $\mathcal{A} \models D(a)$, it holds that $C \sqsubseteq D$.*

In a language endowed with existential (or numeric) restrictions, such as $\mathcal{ALC}$, the exact MSC may not be always expressed with a finite description [1], yet it may be approximated [10, 2].

The *Least Common Subsumer* is the inference operator that, given a set of concept descriptions, return the most specific concept (w.r.t. the subsumption relationship) that is able to subsumes all concept descriptions:

Definition 2.4 (Least Common Subsumer). *Let $\mathcal{L}$ be a description logic. A concept description E of $\mathcal{L}$ is the **least common subsumer (LCS)** of the concept descriptions $C_1, \cdots, C_n$ in $\mathcal{L}$ ($LCS(C_1, \cdots, C_n)$ for short) iff it satisfies:*

1. *$C_i \sqsubseteq E$ for all $i = 1, \cdots, n$ and*
2. *E is the least $\mathcal{L}$-concept description satisfying (1), i.e. if E' is an $\mathcal{L}$-concept description satisfying $C_i \sqsubseteq E'$ for all $i = 1, \cdots, n$, then $E \sqsubseteq E'$.*

Depending on the DL language, the LCS needs not always exist. If it exists, it is unique up to equivalence. In $\mathcal{ALC}$ logic, the LCS always exists [3, 1] and (as for every DL allowing for concept disjunction) it is given by the disjunction of the concepts.

3   Dissimilarity Measures in Description Logics

We recall two definition of dissimilarity measures for $\mathcal{ALC}$ descriptions expressed in normal form [11, 13]. These measures are based on both the structure and the semantics of the concept descriptions.

3.1   Overlap Dissimilarity Measure

The first measure is derived from a measure of the overlap between concepts that is obtained by exploiting the notion of *canonical interpretation*, namely the interpretation function that adopts the set of individuals in the ABox as domain and the identity as interpretation function [1]. The dissimilarity measure is formally defined as follows:

Definition 3.1 (overlap function). *Let $\mathcal{I}$ be the canonical interpretation of the ABox $\mathcal{A}$. The overlap function $f : \mathcal{L} \times \mathcal{L} \mapsto R^+$ defined for $\mathcal{ALC}$ normal form concept descriptions $C, D \in \mathcal{L}$, with $C = \bigsqcup_{i=1}^{n} C_i$ and $D = \bigsqcup_{j=1}^{m} D_j$ is formalized as:*
disjunctive level:

$$f(C, D) := f_{\sqcup}(C, D) = \begin{cases} \infty & C \equiv D \\ 0 & C \sqcap D \equiv \bot \\ \max_{i,j} f_{\sqcap}(C_i, D_j) & \text{otherwise} \end{cases}$$

conjunctive level:

$$f_{\sqcap}(C_i, D_j) := f_P(\text{prim}(C_i), \text{prim}(D_j)) + \lambda(f_{\forall}(C_i, D_j) + f_{\exists}(C_i, D_j))$$

with $\lambda \in [0, 1]$.
primitive concepts:

$$f_P(P_1, P_2) := \frac{|R(P_1) \cup R(P_2)|}{|(R(P_1) \cup R(P_2)) \setminus (R(P_1) \cap R(P_2))|}$$

where $R(P) = \bigcap_{A \in P} A^{\mathcal{I}}$ and $f_P(P_1, P_2) = \infty$ when $R(P_1) = R(P_2)$.
value restrictions:

$$f_{\forall}(C_i, D_j) := \sum_{R \in N_R} f_{\sqcup}(\text{val}_R(C_i), \text{val}_R(D_j))$$

existential restrictions:

$$f_{\exists}(C_i, D_j) := \sum_{R \in N_R} \sum_{k=1}^{N} \max_{p=1,\dots,M} f_{\sqcup}(C_i^k, D_j^p)$$

where $C_i^k \in \text{ex}_R(C_i)$ and $D_j^p \in \text{ex}_R(D_j)$ and we suppose w.l.o.g. that $N = |\text{ex}_R(C_i)| \geq |\text{ex}_R(D_j)| = M$, otherwise the indices N and M as well as C_i and D_j are to be exchanged in the formula above.

The function f represents a measure of the overlap between two descriptions expressed in $\mathcal{ALC}$ normal form. It measures the similarity of the input concepts on the ground of the similarity between their extensions (approximated with the retrieval) and on the ground of the similarity of the concepts reached by the role restrictions. The function is recursively defined, beginning from the top level of the descriptions (a disjunctive level) up to the bottom level represented by (conjunctions of) primitive concepts.

Overlap at the disjunctive level is treated as the maximum overlap between the disjunctive forms of the input concepts. At conjunctive levels, instead of simply considering the similarity like in Tversky's measure [29] (as we do for prim's), that could turn out to be null in case of disjoint retrieval sets for the input concepts, we distinguish the overlaps between the prims and those between the concepts in the scope of the various role restrictions[4]; the contribution of the overlap of concepts reached through role restrictions can be penalized by tweaking the parameter λ. The measure for primitive concepts resembles Tversky's measure and it represents a semantic baseline since it depends on the semantics of the knowledge base, as conveyed by the ABox assertions. This is in line with to the ideas in [4, 7], where semantics is elicited as a probability distribution over the domain of the interpretation.

As for the role restrictions overlap, for the universal restrictions we simply add the overlap measures varying the role, while for the existential restrictions the measure is trickier: borrowing the idea of the existential mappings we consider, per role, all possible matches between the concepts in the scope of existential restrictions, then we consider the maximal sum of overlaps resulting from all the matches.

Now, it is possible to derive a dissimilarity measure based on f as follows:

Definition 3.2 (overlap dissimilarity measure). *The* overlap dissimilarity measure *is a function* $d : \mathcal{L} \times \mathcal{L} \mapsto [0,1]$ *such that, given two concept descriptions* $C, D \in \mathcal{L}$, $C = \bigsqcup_{i=1}^{n} C_i$ *and* $D = \bigsqcup_{j=1}^{m} D_j$:

$$d(C, D) := \begin{cases} 1 & \text{if } f(C, D) = 0 \\ 0 & \text{if } f(C, D) = \infty \\ \frac{1}{f(C,D)} & \text{otherwise} \end{cases}$$

Function d simply measures the level of dissimilarity between two concepts as the inverse of the overlap function f. Particularly, if $f(C, D) = 0$, i.e. there is no overlap between the considered concepts, then d must indicate that the two concepts are totally different, indeed $d(C, D) = 1$, the maximum value of its range. If $f(C, D) = \infty$ this means that the two concepts are totally overlapped and consequently $d(C, D) = 0$ that means that the two concept are indistinguishable, indeed d assumes the minimum value of its range. If the considered concepts have a partial overlap then their dissimilarity is inversely proportional to their overlap, since in this case $f(C, D) > 1$ and consequently $0 < d(C, D) < 1$.

An example is reported to clarify the usage of the dissimilarity measure:

---

[4] We tried also other solutions, such as considering minima or products of the three overlap measures, yet experimentally this did not yield a better performance whereas the computation time was increased. For practical reasons, the maximal measure ∞ has been replaced with large numbers depending on the cardinality of the set of individuals in the ABox: $|\mathsf{Ind}(\mathcal{A})|$.

Example 3.1. Let be C and D two concepts in $\mathcal{ALC}$ normal form and defined as follows:

$C \equiv A_2 \sqcap \exists R.B_1 \sqcap \forall T.(\forall Q.(A_4 \sqcap B_5)) \sqcup A_1$

$D \equiv A_1 \sqcap B_2 \sqcap \exists R.A_3 \sqcap \exists R.B_2 \sqcap \forall S.B_3 \sqcap \forall T.(B_6 \sqcap B_4) \sqcup B_2$

where A_i and B_j are all primitive concepts.

Now we calculate the amount of dissimilarity among the two concepts C and D; first, it is necessary to compute $f(C, D)$. Known that neither C nor D are semantically equivalent nor are inconsistent, the value of f is estimated as in the third case of its definition. Let us denote with $C_1 := A_2 \sqcap \exists R.B_1 \sqcap \forall T.(\forall Q.(A_4 \sqcap B_5))$ and with $D_1 := A_1 \sqcap B_2 \sqcap \exists R.A_3 \sqcap \exists R.B_2 \sqcap \forall S.B_3 \sqcap \forall T.(B_6 \sqcap B_4)$. The computation of f is as follows

$$f(C, D) = \max\{\, f_\sqcap(C_1, D_1), f_\sqcap(C_1, B_2), \\ f_\sqcap(A_1, D_1), f_\sqcap(A_1, B_2) \,\}$$

For brevity, we consider the computation of $f_\sqcap(C_1, D_1)$. $f_\sqcap$ is computed as the sum of f_P, $f_\forall$, $f_\exists$ i.e., respectively, f applied to primitive concepts, f applied to concepts in the universal restrictions, f applied to concepts in the existential restrictions.

Suppose that $(A_2)^\mathcal{I} \neq (A_1 \sqcap B_2)^\mathcal{I}$. Then:

$$f_P(C_1, D_1) = f_P(\mathrm{prim}(C_1), \mathrm{prim}(D_1))$$
$$= f_P(A_2, A_1 \sqcap B_2)$$
$$= \frac{|I|}{|I \setminus ((A_2)^\mathcal{I} \cap (A_1 \sqcap B_2)^\mathcal{I})|}$$

where $I := (A_2)^\mathcal{I} \cup (A_1 \sqcap B_2)^\mathcal{I}$

In order to calculate $f_\forall$ it is necessary to note that there are two roles at the same level: T and S; so the summation over the different roles is made by two terms. Besides, the role S is only in D_1 and not in C_1, consequently $\mathrm{val}_R(C_1) = \top$. Thus in this case we have:

$$f_\forall(C_1, D_1) = \sum_{R \in N_R} f_\sqcup(\mathrm{val}_R(C_1), \mathrm{val}_R(D_1)) =$$
$$= f_\sqcup(\mathrm{val}_T(C_1), \mathrm{val}_T(D_1)) + f_\sqcup(\mathrm{val}_S(C_1), \mathrm{val}_S(D_1)) =$$
$$= f_\sqcup(\forall Q.(A_4 \sqcap B_5), B_6 \sqcap B_4) + f_\sqcup(\top, B_3)$$

The computation of $f_\sqcup(\forall Q.(A_4 \sqcap B_5), B_6 \sqcap B_4)$ and $f_\sqcup(\top, B_3)$ is the same reported above.

Now, by the definition of $f_\exists$, it is necessary to note that here is only a single one role R, so the first summation collapses in a single element. Then N and M, that are the number of conjunctive descriptions with existential restrictions w.r.t. the same role (S), are respectively $N = 2$ and $M = 1$, so we would have to find the max in a singleton, that can be simplified. So in this case we have:

$$f_\exists(C_1, D_1) = \sum_{k=1}^{2} f_\sqcup(\mathrm{ex}_R(C_1), \mathrm{ex}_R(D_1^k)) = f_\sqcup(B_1, A_3) + f_\sqcup(B_1, B_2)$$

Also in this case, the computation of $f_\sqcup$ is the same reported above.

In order to determine the dissimilarity value of C and D it is necessary to apply the same operations to the others elements in order to define the max value. After that the dissimilarity value is immediately defined as the inverse of the obtained max

3.2   A Dissimilarity Measure Based on Information Content

As discussed in [27], a measure of concept (dis)similarity can be derived from the notion of *Information Content* (IC) that, in turn, depends on the probability of an individual to belong to a certain concept. Now, differently from other works, which assume that a probability distribution for the concepts in an ontology is known, we derive it from the knowledge base, from the distribution that can be estimated therein [4, 7].

In order to approximate this probability for a certain concept C, we recur to its extension w.r.t. the considered ABox in a fixed interpretation. Namely, we chose the *canonical interpretation* $\mathcal{I}_A$, which is the one adopting the set of individuals mentioned in the ABox as its domain and the identity as its interpretation function [1]. Now, given a concept C its probability is estimated by:

$$pr(C) = |C^{\mathcal{I}_A}|/|\Delta^{\mathcal{I}_A}|$$

Finally, we can compute the information content of a concept, employing this probability:

$$IC(C) = -\log pr(C)$$

A measure of the concept dissimilarity is now formally defined [13]:

Definition 3.3 (IC dissimilarity). *Let A be an ABox with canonical interpretation $\mathcal{I}$. The* information content dissimilarity measure *is a function $g : \mathcal{L} \times \mathcal{L} \mapsto R^+$ defined recursively for any two normal form concept descriptions $C, D \in \mathcal{L}$, with $C = \bigsqcup_{i=1}^{n} C_i$ and $D = \bigsqcup_{j=1}^{m} D_j$*

disjunctive level:

$$g(C, D) := g_{\sqcup}(C, D) = \begin{cases} 0 & \text{if } C \equiv D \\ \infty & \text{if } C \sqcap D = \bot \\ \min_{\substack{1 \le i \le n \\ 1 \le j \le m}} g_{\sqcap}(C_i, D_j) & \text{otherwise} \end{cases}$$

conjunctive level:

$$g_{\sqcap}(C_i, D_j) := g_P(\text{prim}(C_i), \text{prim}(D_j)) + \lambda(g_{\forall}(C_i, D_j) + g_{\exists}(C_i, D_j))$$

with $\lambda \in [0, 1]$.
primitive concepts:

$$g_P(P_i, P_j) := \begin{cases} \infty & \text{if } P_i \sqcap P_j \equiv \bot \\ \dfrac{IC(P_i \sqcap P_j)+1}{IC(\text{LCS}(P_i, P_j))+1} & \text{otherwise} \end{cases}$$

value restrictions:

$$g_\forall(C_i, D_j) := \sum_{R \in N_R} g_\sqcup(\text{val}_R(C_i), \text{val}_R(D_j))$$

existential restrictions:

$$g_\exists(C_i, D_j) := \sum_{R \in N_R} \sum_{k=1}^{N} \min_{p=1,\dots,M} g_\sqcup(C_i^k, D_j^p)$$

where $C_i^k \in \text{ex}_R(C_i)$ and $D_j^p \in \text{ex}_R(D_j)$ and we suppose w.l.o.g. that $N = |\text{ex}_R(C_i)| \geq |\text{ex}_R(D_j)| = M$, otherwise the indices N and M are to be exchanged in the formula above.

The function g represents a measure of the dissimilarity between two descriptions expressed in $\mathcal{ALC}$ normal form. It is defined recursively beginning from the top level of the descriptions (a disjunctive level) up to the bottom level represented by (conjunctions of) primitive concepts.

Now g has values in $[0, \infty]$. It may be useful to derive a normalized dissimilarity measure as shown in the following.

Definition 3.4 (normalized IC dissimilarity). *Let $\mathcal{A}$ be an ABox with canonical interpretation $\mathcal{I}$. The* normalized information content dissimilarity measure *is the function $d : \mathcal{L} \times \mathcal{L} \mapsto [0, 1]$, such that given the concept descriptions in $\mathcal{ALC}$ normal form $C = \bigsqcup_{i=1}^{n} C_i$ and $D = \bigsqcup_{j=1}^{m} D_j$, let*

$$d(C, D) := \begin{cases} 0 & \text{if } g(C, D) = 0 \\ 1 & \text{if } g(C, D) = \infty \\ 1 - \frac{1}{g(C,D)} & \text{otherwise} \end{cases}$$

3.3 Measuring the Dissimilarity between Individuals

The notion of *Most Specific Concept* is commonly exploited for lifting individuals to the concept level.

Definition 3.5 (most specific concept). *Given an ABox $\mathcal{A}$ and an individual a, the* most specific concept *of a w.r.t. $\mathcal{A}$ is the concept C, denoted $MSC_{\mathcal{A}}(a)$, such that $\mathcal{A} \models C(a)$ and for any other concept D such that $\mathcal{A} \models D(a)$, it holds that $C \sqsubseteq D$.*

In case of cyclic ABoxes expressed in a DL with existential restrictions the MSC may not be expressed by a finite description [1], yet it may be often approximated.

On performing experiments related to another similarity measure exclusively based on concept extensions [12], we noticed that, recurring to the MSC for lifting individuals to the concept level, just falls short: indeed the MSCs may be too specific and unable to include other (similar) individuals in their extensions. By comparing descriptions reduced to the normal form we have given a more structural definition of dissimilarity. However, since MSCs are computed from the same ABox assertions, reflecting the current knowledge state, this guarantees that structurally similar representations will be obtained for semantically similar concepts.

Let us recall that, given the ABox, it is possible to calculate the most specific concept of an individual a w.r.t. the ABox, $MSC(a)$ or at least its approximation $MSC^k(a)$ up to a certain description depth k. In some cases these are equivalent concepts but in general we have that $MSC^k(a) \sqsupseteq MSC(a)$.

Given two individuals a and b in the ABox, we consider $MSC^k(a)$ and $MSC^k(b)$ (supposed in normal form). Now, in order to assess the dissimilarity between the individuals, d measure can be applied to these descriptions:

$$d(a, b) := d(MSC^k(a), MSC^k(b))$$

This may turn out to be handy in several tasks, namely both in inductive reasoning (construction, repairing of knowledge bases) and in information retrieval.

3.4   Discussion

We proved in [11, 13] that these measures are really dissimilarity measures according to the formal definition [5], considering that their input (what is actually compared) is equivalence classes in $\mathcal{L}$, i.e. $\mathcal{ALC}$ concept descriptions with the same normal form.

As previously mentioned, we have also attempted slightly different measure definitions (e.g. considering minima or products at the conjunctive level that sound more intuitive) which experimentally did not prove more effective than the simple (additive) definition given above.

The computational complexity of the measures depends on the complexity of the required reasoning services for computing the concepts retrieval. Namely both subsumption and instance-checking are P-space for the $\mathcal{ALC}$ logic[16], yet such inference can be computed once and preliminarily, before the measures are computed for the method we will present in the following.

Obviously when computing the dissimilarity measures for cases involving individuals, the extra cost of computing MSCs (or their approximations) has to be added.

Nevertheless, in practical applications, these computations may be efficiently carried out exploiting the statistics that are maintained by the DBMSs query optimizers. Besides, the counts that are necessary for computing the concept extensions could be estimated by means of the probability distribution over the domain.

4   A Nearest Neighbor Classification Procedure in Description Logics

We briefly review the basics of the k-Nearest Neighbor method (k-NN) and propose how to exploit the classification procedure for inductive reasoning. In this lazy approach to learning, a notion of distance (or dissimilarity) measure for the instance space is employed to classify a new instance.

Let x_q be the instance that requires a classification. Using a dissimilarity measure, the set of k nearest pre-classified instances is selected. The objective is to learn a discrete-valued target function $h : IS \mapsto V$ from a space of instances IS to a set of values

$V = \{v_1, \ldots, v_s\}$. In its simplest setting, the k-NN algorithm approximates h for new instances x_q on the ground of the value that h assumes in the neighborhood of x_q, i.e. the k closest instances to x_q in terms of a dissimilarity measure. Precisely, it is assigned according to the value which is *voted* by the majority of instances in the neighborhood. The the classification function $\hat{h}$ can be formally defined as:

$$\hat{h}(x_q) := \underset{v \in V}{\operatorname{argmax}} \sum_{i=1}^{k} \delta(v, h(x_i))$$

where δ is a function that returns 1 in case of matching arguments and 0 otherwise. Note that the function $\hat{h}$ is only extensionally defined, that is the k-NN method does not return an intensional classification model (e.g. a function or a concept definition), it merely gives an answer for new query instances to be classified.

This simple formulation does not take into account the similarity among instances, except when selecting the instances to be included in the neighborhood. Therefore a modified setting is generally adopted, weighting the vote on the grounds of the similarity of the training instances w.r.t. the test example:

$$\hat{h}(x_q) := \underset{v \in V}{\operatorname{argmax}} \sum_{i=1}^{k} w_i \delta(v, h(x_i)) \tag{1}$$

where, usually, $w_i = 1/d(x_i, x_q)$ or $w_i = 1/d(x_i, x_q)^2$.

This method is generally employed to classify vectors of features in some n-dimensional instance space (e.g. often $IS = \mathbf{R}^n$). For the best of our knowledge, there are no existing works able to apply the NN approach to DL representations. In the following, we adapt the k-NN algorithm to such rich representation language.

Preliminarily, it should be observed that a strong assumption of the classical feature vector representation setting is that it can be employed to assign a value (e.g. a class) to a query instance among a set of values which can be regarded as a set of pairwise disjoint concepts/classes. This is an assumption that cannot be always valid. In this case, indeed, an individual could be an instance of more than one concept.

Let us consider a new value set $V = \{C_1, \ldots, C_s\}$ of concepts C_j that may be assigned to a query instance x_q. If they were to be considered as disjoint (like in the standard machine learning setting), the decision procedure would adopt the hypothesis function defined in Eq. (1), with the query instance assigned the *single* concept voted by the weighted majority of instances in the neighborhood.

In the general case considered in this paper, when the disjointness of the classes cannot be assumed (unless explicitly stated in the TBox), one can adopt another answering procedure, decomposing the multi-class problem into smaller binary classification problems (one per concept). Therefore, a simple binary value set ($V = \{-1, +1\}$) is to be employed. Then, for each single concept (say C_j), a hypothesis $\hat{h}_j$ is computed on the fly during the classification phase:

$$\hat{h}_j(x_q) := \underset{v \in V}{\operatorname{argmax}} \sum_{i=1}^{k} \frac{\delta(v, h_j(x_i))}{d(x_q, x_i)^2} \quad \forall j \in \{1, \ldots, s\} \tag{2}$$

where each function h_j, simply indicates the occurrence $(+1)$ or absence (-1) of the corresponding assertion in the ABox for the k training instances x_i: $C_j(x_i) \in \mathcal{A}$. Alternately[5], h_j may return $+1$ when $C_j(x_i)$ is logically entailed by the knowledge base $\mathcal{K}$, and -1 otherwise.

The problem with non-explicitly disjoint concepts is also related to the *Closed World Assumption* usually made in the context of Information Retrieval and Machine Learning. That is the reason for adapting the standard setting to cope both with the case of non-disjoint concepts and with the OWA which is commonly made in the Semantic Web context. To deal with the OWA, the absence of information on whether a certain instance x belongs to the extension of concept C_j should not be interpreted negatively; rather, it should count as neutral information. Thus, one can still adopt the decision procedure in Eq. (2), however another value set has to be adopted for the h_j's, namely $V = \{-1, 0, +1\}$, where the three values denote, respectively, positive occurrence, absence and negative occurrence (positive for the concept negation). Formally:

$$h_j(x) = \begin{cases} +1 & C_j(x) \in \mathcal{A} \\ 0 & C_j(x) \notin \mathcal{A} \wedge \neg C_j(x) \notin \mathcal{A} \\ -1 & \neg C_j(x) \in \mathcal{A} \end{cases}$$

Again, a more complex procedure may be devised by simply substituting the notion of occurrence (absence) of assertions in (from) the ABox with one based on logic entailment (denoted with $\vdash$) from the knowledge base, i.e. $\mathcal{K} \vdash C_j(x)$, $\mathcal{K} \nvdash C_j(x)$ nor $\mathcal{K} \nvdash \neg C_j(x)$ and $\mathcal{K} \vdash \neg C_j(x)$, respectively. Although this may help reaching the precision of deductive reasoning, it is also much more computationally expensive, since the simple lookup in the ABox must be replaced with instance checking.

From a computational viewpoint this procedure could be implemented to provide an answer even more efficiently than with a standard deductive reasoner. Indeed, once the retrieval of the primitive concepts is computed, the dissimilarity measures can be easily computed by means of a dynamic programming algorithm. Besides, the various measures could be maintained in an ad hoc data structure which would allow for an efficient retrieval of the nearest neighbors, such as the kD-trees or ball trees [30]

5   Experiments

5.1   Experimental Setting

In order to assess the validity of the presented method with the dissimilarity measures proposed in Sect. 3, we have applied it to the instance classification problem, with four different ontologies represented in OWL: FSM, SURFACE-WATER-MODEL from the Protégé library[6], the FINANCIAL ontology[7] employed as a testbed for the PELLET reasoner and a small FAMILY ontology written in our lab. Although they are represented

---

[5] For the sake of simplicity and efficiency, this case will not be considered in the following.
[6] See the webpage:
   http://protege.stanford.edu/plugins/owl/owl-library
[7] See the webpage: http://www.cs.put.poznan.pl/alawrynowicz/financial.owl

in languages that are different from $\mathcal{ALC}$, we simply discarded these details when constructing the MSC approximations in order to be able to apply the presented measures.

FAMILY is an $\mathcal{ALCF}$ ontology describing *kinship* relationships. It is made up of 14 concepts (both primitive and defined and some of them declared to be disjoint), 5 object properties, 39 distinct individual names. Most of the individuals are asserted to be instances of more than one concept, and are involved in more than one role assertions. This ontology has been written to have a small yet more complex case w.r.t. the following ones. Indeed, while the other ontologies are more regular, i.e. only some concepts are employed in the assertions (the others are defined only intensionally), in the FAMILY ontology every concept has at least one instance asserted. The same happens for the assertions on roles; particularly, there are some cases where role assertions constitute a chain from an individual to another, by means of other intermediate assertions.

The FSM ontology describes the domain of *finite state machines* using the $\mathcal{SOF}(D)$ language. It is made up of 20 (both primitive and defined) concepts (some of them are explicitly declared to be disjoint), 10 object properties, 7 datatype properties, 37 distinct individual names. About half of the individuals are asserted as instances of a single concept and are not involved in any role assertion (object property).

SURFACE-WATER-MODEL is an $\mathcal{ALCOF}(D)$ ontology describing the domain of the surface water and the water quality models. It is based on the *Surface-water Models Information Clearinghouse* (SMIC) of the USGS. Namely, it is an ontology of numerical models for surface water flow and water quality simulation. The application domain of these models comprises lakes, oceans, estuaries etc.. These models are classified based on their availability, application domain, dimensions, partial differential equation solver, and characteristics types. It is made up of 19 concepts (both primitive and defined) without any specification about disjointness, 9 object properties, 115 distinct individual names; each of them is an instance of a single class and only some of them are involved in object properties.

FINANCIAL is an $\mathcal{ALCIF}$ ontology that describes the domain of eBanking. It is made up of 60 (both primitive and defined) concepts (some of them are declared to be disjoint), 17 object properties, and no datatype property. It contains 17941 distinct individual names. From the original ABox, we randomly extracted assertions for 652 individuals.

The classification method was applied to all the individuals in each ontology; namely, the individuals were checked to assess if they were instances of the concepts in the ontology through the analogical method. The performance was evaluated comparing its responses to those returned by a standard reasoner[8] as a baseline. Specifically, for each individual in the ontology, the MSC is computed and enlisted in the set of training (or test) examples. Each example is classified applying the adapted k-NN method presented in the previous section. As a value of k we chose $\sqrt{|\mathrm{Ind}(\mathcal{A})|}$, as advised in the instance-based learning literature [22].

The experiment has been repeated twice adopting the leave-one-out cross validation procedure with both the dissimilarity measures defined in Sect. 3. For each concept in the ontology, we measured the following parameters for the evaluation:

---

[8] We employed PELLET: `http://pellet.owldl.com`

Table 1. Average results of the experiments with the method employing the measure based on overlap

	Match Rate	Commission Rate	Omission Rate	Induction Rate
FAMILY	.654±.174	.000±.000	.231±.173	.115±.107
FSM	.974±.044	.026±.044	.000±.000	.000±.000
S.-W.-M.	.820±.241	.000±.000	.064±.111	.116±.246
FINANCIAL	.807±.091	.024±.076	.000±.001	.169±.076

- *match rate*: number of cases of individuals that got exactly the same classification by both classifiers and the standard reasoner w.r.t. the overall number of individuals;
- *omission error rate*: amount of unlabeled individuals (our method could not determine whether it was an instance or not) while they were classified as instances of the considered concepts;
- *commission error rate*: amount of individuals (analogically) labeled as instances of a concept, while they (logically) belong to the negation of that concept or vice-versa
- *induction rate*: amount of individuals that were found to belong to a concept or its negation, while this information is not logically derivable from the knowledge base

We report the average rates obtained over all the concepts in each ontology and also their standard deviation.

5.2 Experiments Employing the Overlap Measure

By looking at Tab. 1 reporting the experimental outcomes with the dissimilarity measure based on the overlap (see Def. 3.2), preliminarily it is important to note that, for every ontology, the commission error was quite low. This means that the classifier did not make critical mistakes i.e. cases when an individual is deemed as an instance of a concept while it really is an instance of another disjoint concept.

In particular, by looking at the outcomes related to the FAMILY ontology, it can be observed that the match rate is the lowest while the highest omission error rate was reported. This may be due to two facts: 1) very few individuals were available w.r.t. the number of concepts[9]; 2) a sparse data situation is dominant in this ontology, namely instances are irregularly *spread* over the concepts, that is, some individuals are instances of most of the concepts in the ontology while other individuals are instances of a very few concepts. Hence, the computed MSC approximations also resulted very different one from another, thus reducing the possibility of significantly matching similar MSCs. Nevertheless, as mentioned above, it is important to note that the algorithm did not make any commission error and it is able to infer new knowledge (11%).

As regards the FSM ontology, we have observed the maximum match rate with respect to the classification given by the logic reasoner. Moreover, differently from the other ontologies, both the omission error rate and the induction rate were null. A very limited percentage of incorrect classification cases was observed. These outcomes were

---

[9] Instance-based methods make an intensive use of the information about the individuals and improve their performance with the increase of the number of instances considered.

Table 2. Average results of the experiments with the method employing the measure based on information content

	Match Rate	Commission Rate	Omission Rate	Induction Rate
FAMILY	.608±.230	.000±.000	.330±.216	.062±.217
FSM	.899±.178	.096±.179	.000±.000	.005±.024
S.-W.-M.	.820±.241	.000±.000	.064±.111	.116±.246
FINANCIAL	.807±.091	.024±.076	.000±.001	.169±.046

probably due to the fact that individuals in this ontology are quite regularly divided by the assertions on concepts and roles, so the computed MSCs are all very similar to each other and consequently, the amount of information they convey is very low. A choice of a lower number k of neighbors could probably help committing those residual errors.

For the same reasons, also for the SURFACE-WATER-MODEL ontology a quite high rate of matching classifications was reported (yet less then the previous ontology); moreover, some cases of omission error (6%) were observed. The induction rate was about 12%. Hence, for this ontology our classifier almost always assigned individuals to the correct concepts and, in some cases, it could also induce new assertions. Since this rate represents assertions that were not logically deducible from the ontology and yet they were inferred inductively by the analogical classifier, these figures would be a positive outcome (provided this knowledge were deemed as correct by an expert). Particularly, in this case the increase of the induction rate has been due to the presence of assertions of mutual disjointness for some of the concepts.

Results are no different also for the case of the experiments with FINANCIAL ontology that largely exceeds the others in terms of number of concepts and individuals. The observed match rate is again above the 80% and the rest of the cases are comprised in the induction rate (17%), leaving a limited margin to residual errors. Actually, performing a 10-fold cross validation we obtained almost the same results.

Looking at Tab. 1, it is possible to assert that, generally, the classifications results improve with the increasing of the available individuals in the considered ontologies. This corroborates a fact about the NN learners, that is their reaching better and better performance in the limit, as long as new training instances become available. Hence, the quite high registered match rate is very significant, since it has been obtained by the use of rather small ontologies that makes harder the classification task.

However, since inductive conclusions are less certain than the deductive ones, the best usage of the presented classifier is, in our opinion, as additional layer on top of a standard deductive reasoner in order to enriching the results of the instance retrieval and query answering tasks.

5.3 Experiments with the Information Content Measure

The average results obtained by adopting the procedure with the measure based on information content (see Def. 3.4) are reported in Table 2.

By analyzing this table it is possible to note that no sensible variation was observed in the classifications performed using the first dissimilarity measure. This is opposite to the theory (not experimentally proved) presented in [7] for which (dis-)similarity measures based on the notion of Information Content should ensure a higher accuracy in determining the (dis-)similarity values between objects w.r.t. other existing approaches such as the structural comparison and the feature matching.

Particularly, with both measures the method correctly classified all the individuals, without commission errors. The reason is that, in most of the cases, the individuals of these ontologies are instances of one concept only and they are involved in a few roles (object properties). Some of the figures are slightly lower than those observed in the other experiment: this is confirmed by a higher variability.

Surprisingly, the results on the larger ontologies (S.-W.-M. and FINANCIAL) perfectly match those obtained with the other measure which is probably due to the fact that we used a leave-one-out cross-validation mode which yielded a high value for the number k of training instances for the neighborhood and it is well known that the NN procedure becomes more and more precise as more instances can be considered. The price to be paid was a longer computation time.

6   Conclusions and Future Work

In this work we have coupled with a method for inductive inference on ABoxes with two concept dissimilarity measures. The method classifies individuals w.r.t. a query concept in analogy with the instantiation of the majority of neighbor training individuals. As experimentally shown, it is able to induce assertions that are not logically derivable; as such, it can be naturally exploited for predicting/suggesting missing information about individuals, thus enabling a sort of inductive instance retrieval and query answering. An increase in the classification accuracy was observed when instances are numerous and homogeneously spread w.r.t. the concepts in the KB. To further improve the accuracy of the method, the employed dissimilarity measures could be refined by introducing a tweaking weighting factor that decreases the impact of the dissimilarity between nested sub-concepts in the descriptions on the determination of the overall value.

The presented method could be extended from several point of views. One of this is to make possible its application to DL languages more expressive than $\mathcal{ALC}$. In order to do this, similarity or dissimilarity measures able to cope with higher expressive power languages need to be defined. Lately, we have defined a similarity measure for $\mathcal{ALN}$ [20] logic. Moving from this result and the measures presented above, a natural extension may concern the definition of (dis-)similarity measures for $\mathcal{ALCN}$ language.

Another extension may concern the classifications results. Currently the classification algorithm is able to assess if an individual is instance of a certain concept, if it is instance of the negated concept or if there are no information for classifying the individual. A useful extension could be to return the degree of confidence of the classification results. This additional information may be helpful for the knowledge engineer that has to validate the induced assertions. Such an extension could be realized by the use of a different (yet still computationally tractable) answering procedures grounded on statistical inference (non-parametric tests based on ranked distances) methods.

Furthermore, the rationale of the presented method can be exploited for realizing an inductive classifier grounded on kernel methods that are efficient learning algorithms which are able to work in high dimensional feature space. This algorithms are based on the use of kernel functions that are another means to express a notion of similarity is some unknown feature space. On the ground of this idea, we are working at the definition of kernel functions on DLs representations [19, 21], thus allowing the exploitation of kernel methods efficiency (e.g. the support vector machines) in a multi-relational setting.

References

[1] Baader, F., Calvanese, D., McGuinness, D., Nardi, D., Patel-Schneider, P. (eds.): The Description Logic Handbook. Cambridge University Press, Cambridge (2003)

[2] Baader, F., Küsters, R.: Non-standard inferences in description logics: The story so far. In: Gabbay, D., Goncharov, S.S., Zakharyaschev, M. (eds.) Mathematical Problems from Applied Logic. New Logics for the XXIst Century. International Mathematical Series, vol. 4. Kluwer/Plenum Publishers (2005)

[3] Baader, F., Küsters, R., Molitor, R.: Computing least common subsumers in description logics with existential restrictions. In: Dean, T. (ed.) Proceedings of the 16th International Joint Conference on Artificial Intelligence, pp. 96–101. Morgan Kaufmann, San Francisco (1999)

[4] Bacchus, F.: Lp, a logic for representing and reasoning with statistical knowledge. Computational Intelligence 6, 209–231 (1990)

[5] Bock, H.: Analysis of Symbolic Data: Exploratory Methods for Extracting Statistical Information from Complex Data. Springer, Heidelberg (1999)

[6] Bock, H., Diday, E.: Analysis of symbolic data: exploratory methods for extracting statistical information from complex data. Springer, Heidelberg (2000)

[7] Borgida, A., Walsh, T., Hirsh, H.: Towards measuring similarity in description logics. In: Working Notes of the International Description Logics Workshop, CEUR Workshop Proceedings, Edinburgh, UK (2005)

[8] Brandt, S., Küsters, R., Turhan, A.-Y.: Approximation and difference in description logics. In: Fensel, D., Giunchiglia, F., McGuinness, D., Williams, M.-A. (eds.) Proceedings of the International Conference on Knowledge Representation, pp. 203–214. Morgan Kaufmann, San Francisco (2002)

[9] Bright, M.W., Hurson, A.R., Pakzad, S.H.: Automated resolution of semantic heterogeneity in multidatabases. ACM Transaction on Database Systems 19(2), 212–253 (1994)

[10] Cohen, W., Hirsh, H.: Learning the CLASSIC description logic. In: Torasso, P., Doyle, J., Sandewall, E. (eds.) Proceedings of the 4th International Conference on the Principles of Knowledge Representation and Reasoning, pp. 121–133. Morgan Kaufmann, San Francisco (1994)

[11] d'Amato, C., Fanizzi, N., Esposito, F.: A dissimilarity measure for the $\mathcal{ALC}$ description logic. In: Bouquet, P., Tummarello, G. (eds.) Semantic Web Applications and Perspectives, 2nd Italian Semantic Web Workshop SWAP 2005, Trento, Italy, vol. 166, CEUR (2005)

[12] d'Amato, C., Fanizzi, N., Esposito, F.: A semantic similarity measure for expressive description logics. In: Pettorossi, A. (ed.) Proceedings of Convegno Italiano di Logica Computazionale, CILC 2005, Rome, Italy (2005)

[13] d'Amato, C., Fanizzi, N., Esposito, F.: A dissimilarity measure for $\mathcal{ALC}$ concept descriptions. In: Haddad, H. (ed.) Proceedings of the 21st Annual ACM Symposium of Applied Computing, SAC 2006, Dijon, France, pp. 1695–1699. ACM, New York (2006)

[14] d'Amato, C., Fanizzi, N., Esposito, F.: Reasoning by analogy in description logics through instance-based learning. In: Tummarello, G., Bouquet, P., Signore, O. (eds.) Proceedings of Semantic Web Applications and Perspectives, 3rd Italian Semantic Web Workshop, SWAP 2006, Pisa, Italy, vol. 201, CEUR (2006)

[15] d'Aquin, M., Lieber, J., Napoli, A.: Decentralized case-based reasoning for the Semantic Web. In: Gil, Y., Motta, E., Benjamins, V.R., Musen, M.A. (eds.) ISWC 2005. LNCS, vol. 3729, pp. 142–155. Springer, Heidelberg (2005)

[16] Donini, F.M., Lenzerini, M., Nardi, D., Schaerf, A.: Deduction in concept languages: From subsumption to instance checking. Journal of Logic and Computation 4(4), 423–452 (1994)

[17] Emde, W., Wettschereck, D.: Relational instance-based learning. In: Saitta, L. (ed.) Proceedings of the Thirteenth International Conference, ICML 1996, pp. 122–130. Morgan Kaufmann, San Francisco (1996)

[18] Esposito, F., Fanizzi, N., Iannone, L., Palmisano, I., Semeraro, G.: Knowledge-intensive induction of terminologies from metadata. In: McIlraith, S.A., Plexousakis, D., van Harmelen, F. (eds.) ISWC 2004. LNCS, vol. 3298, pp. 441–455. Springer, Heidelberg (2004)

[19] Fanizzi, N., d'Amato, C.: A declarative kernel for $\mathcal{ALC}$ concept descriptions. In: Esposito, F., Raś, Z.W., Malerba, D., Semeraro, G. (eds.) ISMIS 2006. LNCS, vol. 4203, pp. 322–331. Springer, Heidelberg (2006)

[20] Fanizzi, N., d'Amato, C.: A similarity measure for the $\mathcal{ALN}$ description logic. In: Proceedings of Convegno Italiano di Logica Computazionale, CILC 2006, Bari, Italy (2006), http://cilc2006.di.uniba.it/download/camera/15_Fanizzi_CILC06.pdf

[21] Fanizzi, N., d'Amato, C.: Inductive concept retrieval and query answering with semantic knowledge bases through kernel methods. In: Apolloni, B., Howlett, R.J., Jain, L. (eds.) KES 2007, Part I. LNCS, vol. 4692, pp. 148–155. Springer, Heidelberg (2007)

[22] Gora, G., Wojna, A.: Riona: A classifier combining rule induction and k-nn method with automated selection of optimal neighbourhood. In: Elomaa, T., Mannila, H., Toivonen, H. (eds.) ECML 2002. LNCS, vol. 2430, pp. 111–123. Springer, Heidelberg (2002)

[23] Haase, P., van Harmelen, F., Huang, Z., Stuckenschmidt, H., Sure, Y.: A framework for handling inconsistency in changing ontologies. In: Gil, Y., Motta, E., Benjamins, V.R., Musen, M.A. (eds.) ISWC 2005. LNCS, vol. 3729, pp. 353–367. Springer, Heidelberg (2005)

[24] Hitzler, P., Vrandečić, D.: Resolution-based approximate reasoning for OWL DL. In: Gil, Y., Motta, E., Benjamins, V.R., Musen, M.A. (eds.) ISWC 2005. LNCS, vol. 3729, pp. 383–397. Springer, Heidelberg (2005)

[25] Mantay, T.: Commonality-based ABox retrieval. Technical Report FBI-HH-M-291/2000, Dept. of Computer Science, University of Hamburg, Germany (2000)

[26] Mitchell, T.: Machine Learning. McGraw-Hill, New York (1997)

[27] Resnik, P.: Semantic similarity in a taxonomy: An information-based measure and its application to problems of ambiguity in natural language. Journal of Artificial Intelligence Research 11, 95–130 (1999)

[28] Schmidt-Schauß, M., Smolka, G.: Attributive concept descriptions with complements. Artificial Intelligence 48(1), 1–26 (1991)

[29] Tversky, A.: Features of similarity. Psycological Review 84(4), 327–352 (1997)

[30] Witten, I.H., Frank, E.: Data Mining, 2nd edn. Morgan Kaufmann, San Francisco (2005)

Approximate Measures of Semantic Dissimilarity under Uncertainty

Nicola Fanizzi, Claudia d'Amato, and Floriana Esposito

Dipartimento di Informatica
Università degli studi di Bari
Campus Universitario
Via Orabona 4, 70125 Bari, Italy
{fanizzi,claudia.damato,esposito}@di.uniba.it

Abstract. We propose semantic distance measures based on the criterion of approximate discernibility and on evidence combination. In the presence of incomplete knowledge, the distance functions measure the degree of belief in the discernibility of two individuals by combining estimates of basic probability masses related to a set of discriminating features. We also suggest ways to extend this distance for comparing individuals to concepts and concepts to other concepts. Integrated within a k-Nearest Neighbor algorithm, the measures have been experimentally tested on a task of inductive concept retrieval demonstrating the effectiveness of their application.

1 Introduction

In the context of reasoning in the Semantic Web, a growing interest is being committed to alternative inductive procedures extending the scope of the methods that can be applied to concept representations. Most of these methods are based on a notion of similarity; some examples are: *case-based reasoning* [6], *retrieval* [4], *conceptual clustering* [10] and *ontology matching* [9]. However this notion is not easily captured in a definition, especially in presence of uncertainty into the data. Moreover, as pointed out in the seminal paper [3] concerning similarity in Description Logics (DL), most of the existing measures focus on the similarity between atomic concepts within simple hierarchies. Alternative approaches for computing concept similarity are based on related notions of *feature* similarity or *information content*.

Very few works [4] focused on assessing similarity between individuals asserted in DL knowledge bases. Moreover, they are still partly based on structural criteria which also determine the main weakness of the measures proposed, that is, they are hardly scalable to complex languages since the constructors of a specific DL language (specifically $\mathcal{ALC}$ logic) are considered.

The ability to assess individual (dis-)similarities can be very important for several tasks such as ontology matching and ontology learning, but also for applying new inductive methods (besides of the usual deductive one) to the

P.C.G. da Costa et al. (Eds.): URSW 2005-2007, LNAI 5327, pp. 348–365, 2008.

ontological representation. Indeed, these methods can be useful for managing the inherent uncertainty that characterizes the distributed knowledge of the Semantic Web context.

In the perspective of crafting inductive methods for the aforementioned tasks, the need for a definition of a semantic similarity measure for *individuals* arises. We devised a new family of dissimilarity measures for semantically annotated resources, which can overcome the aforementioned limitations of the structural approach [11]. Our measures are mainly based on Minkowski's measures for Euclidean spaces [21] induced by means of a method developed in the context of relational *machine learning* [17]. We extend the idea with a notion of *discernibility* borrowed from *rough sets* theory [16] that aims at the formal definition of vague sets (concepts) by means of their approximations. Hence, in this paper, we propose dissimilarity measures based on semantic discernibility and on evidence combination [19, 8, 18]. Namely, the measures are based on the degree of discernibility of the input individuals with respect to a set of features (in the following *committee of features*), which are represented by concept descriptions expressed in the concept language of choice. For the best of our knowledge, these are the only dissimilarity measures that do not rely on a particular DL language, consequently they can be directly applied to expressive ontology languages.

We also propose a way to extend the presented dissimilarity measures to the case of assessing concept similarity by means of the notion of *medoid* [14] that is the most centrally located individual in a concept extension computed by the use of a given metric.

Experimentally, it may be shown that the measures that use large committee of features (e.g. including all primitive and defined concepts) can be sufficiently accurate (i.e. properly discriminating) when employed for classification tasks even though the committee of features employed were not the optimal one or if the concepts therein were partially redundant. Nevertheless, this has led us to investigate on a method to optimize the committee of features that serve as dimensions for the computation of the measure. To this purpose, the employment of genetic programming and randomized search procedures was considered. Finally, we opted an optimization procedure based on *simulated annealing* [7], that is a randomized approach which can overcome the problem of the local minima, i.e. finding a good solution with respect to the fitness function that is not globally optimal.

The remainder of the paper is organized as follows. In Sect. 2, we recall the basics of approximate semantic distance measures for individuals in a DL knowledge base. Hence, in Sect. 3, we extend the measures with a more principled treatment of uncertainty based on evidence combination. Integrated within a k-Nearest Neighbor algorithm (see Sect. 4), the measures have been experimentally tested on a task of inductive concept retrieval, demonstrating the effectiveness of their application. In Sect. 5, conclusions discuss the applicability of these measures in further works.

2   Semantic Distance Measures

Since our method is not intended for a particular representation, in the following we assume that resources, concepts and their relationships may be defined in terms of a generic representation that may be mapped to some DL language with the standard model-theoretic semantics (see the handbook [1] for a thorough reference).

In this context, a *knowledge base* $\mathcal{K} = \langle \mathcal{T}, \mathcal{A} \rangle$ contains a *TBox* $\mathcal{T}$ and an *ABox* $\mathcal{A}$. $\mathcal{T}$ is a set of concept definitions. $\mathcal{A}$ contains assertions concerning the world state. The set of the individuals (resources) occurring in $\mathcal{A}$ will be denoted with $\mathsf{Ind}(\mathcal{A})$. Each individual can be assumed to be identified by its own URI. The *unique names assumption* on such individuals can be also made.

As regards the inference services, our procedure requires performing *instance-checking*[1] and the related service of *retrieval*[2], which will be used for the approximations.

2.1   A Simple Semantic Metric for Individuals

Aiming at distance-based tasks such as clustering or similarity search and following ideas borrowed from metric learning in clausal spaces [17], we have developed a new family of dissimilarity measures with a definition that totally depends on semantic aspects of the individuals in the knowledge base [11].

Indeed, for our purposes, we needed functions to measure the (dis)similarity of individuals. However individuals do not have a syntactic (or algebraic) structure that can be compared. In order to exploit the semantics of individuals of an ontological knowledge base, we propose a family of dissimilarity measures that is based on the intuition that, on a semantic level, similar individuals should behave similarly with respect to the same concepts. On the ground of this intuition, we compare the individual semantics along a number of dimensions represented by a set of concept descriptions (henceforth also referred to as *committee of features*). Particularly, the rationale of the measures is to compare individuals on the grounds of their behavior w.r.t. a given set of concept descriptions, say $\mathsf{F} = \{F_1, F_2, \ldots, F_k\}$, which stands as a group of discriminating *features* expressed in the language taken into account.

We begin with defining the behavior of an individual with respect to a certain concept in terms of projecting it in this dimension:

Definition 2.1 (projection function). *Given a concept* $F_i \in \mathsf{F}$*, the related* projection function

$$\pi_i : \mathsf{Ind}(\mathcal{A}) \mapsto \{0, \frac{1}{2}, 1\}$$

---

[1] *Instance checking* is the inference procedure which assesses if an individual is instance of a considered concept or not.

[2] *Instance retrieval* is the inference procedure that, given a concept, returns all the individuals that are instances of the considered concept.

is defined:
$\forall a \in \mathsf{Ind}(\mathcal{A})$

$$\pi_i(a) := \begin{cases} 1 & \mathcal{K} \models F_i(a) \\ 0 & \mathcal{K} \models \neg F_i(a) \\ 1/2 & otherwise \end{cases}$$

The case of $\pi_i(a) = 1/2$ corresponds to the case when a reasoner cannot give the truth value for a certain membership query. This is due to the *Open World Assumption* normally made in this context. Hence, as in the classic probabilistic models, uncertainty is coped with by considering a uniform distribution over the possible cases.

Now the discernibility function related to the committee concepts that are used for comparing individuals through the projection functions is defined:

Definition 2.2 (discernibility function). *Given a feature concept* $F_i \in \mathsf{F}$, *the related* discernibility function

$$\delta_i : \mathsf{Ind}(\mathcal{A}) \times \mathsf{Ind}(\mathcal{A}) \mapsto [0,1]$$

is defined as follows:
$\forall (a,b) \in \mathsf{Ind}(\mathcal{A}) \times \mathsf{Ind}(\mathcal{A})$

$$\delta_i(a,b) = |\pi_i(a) - \pi_i(b)|$$

Finally, a whole family of distance functions for individuals inspired to Minkowski's metrics L_p [21] can be defined as follows [11]:

Definition 2.3 (dissimilarity measures). *Let* $\mathcal{K} = \langle \mathcal{T}, \mathcal{A} \rangle$ *be a knowledge base. Given a set of concept descriptions* $\mathsf{F} = \{F_1, F_2, \ldots, F_k\}$, *a family of dissimilarity measures* $\{d_p^{\mathsf{F}}\}_{p \in \mathbb{N}}$, *contains functions*

$$d_p^{\mathsf{F}} : \mathsf{Ind}(\mathcal{A}) \times \mathsf{Ind}(\mathcal{A}) \mapsto [0,1]$$

defined
$\forall (a,b) \in \mathsf{Ind}(\mathcal{A}) \times \mathsf{Ind}(\mathcal{A})$:

$$d_p^{\mathsf{F}}(a,b) := \frac{L_p(\pi_i(a), \pi_i(b))}{|\mathsf{F}|} = \frac{1}{k} \sqrt[p]{\sum_{i=1}^{k} \delta_i(a,b)^p}$$

Note that k depends on F and the effect of the factor $1/k$ is just to normalize the norms with respect to the number of features that are involved. Since the defined function depends from the choice of F, the definition above formalize a family of dissimilarity functions rather than a unique one. Moreover, due to the dependency of the measure from F, the comparison of the dissimilarity values computed across different feature committees may not be meaningful. For instance, larger committees are likely to decrease the dissimilarity values because of the normalizing factor; furthermore these values are affected also by the degree of redundancy of the features employed.

2.2 Example

Let us consider a knowledge base in a DL language made up of a TBox:
$T = \{$ Female $\equiv \neg$Male,

      Parent $\equiv \exists$hasChild.Being $\sqcap$ $\forall$hasChild.Being,

      Father $\equiv$ Male $\sqcap$ Parent,

      FatherWithoutSons $\equiv$ Father $\sqcap$ $\forall$hasChild.Female

    $\}$

and of an ABox:
$A = \{$ Being(ZEUS), Being(APOLLO), Being(HERCULES), Being(HERA),

      Male(ZEUS), Male(APOLLO), Male(HERCULES),

      Parent(ZEUS), Parent(APOLLO), $\neg$Father(HERA),

      God(ZEUS), God(APOLLO), God(HERA), $\neg$God(HERCULES),

      hasChild(ZEUS, APOLLO), hasChild(HERA, APOLLO),

      hasChild(ZEUS, HERCULES),

    $\}$

Suppose $F = \{F_1, F_2, F_3, F_4\} = \{$Male, God, Parent, FatherWithoutSons$\}$. Let us compute the distances (with $p = 1$):

$d_1^F(\text{ZEUS}, \text{HERA}) = (|1 - 0| + |1 - 1| + |1 - 1| + |0 - 0|)/4 = 1/4$

$d_1^F(\text{HERA}, \text{APOLLO}) = (|0 - 1| + |1 - 1| + |1 - 1| + |0 - 1/2|)/4 = 3/8$

$d_1^F(\text{APOLLO}, \text{HERCULES}) = (|1 - 1| + |1 - 0| + |1 - 1/2| + |1/2 - 1/2|)/4 = 3/8$

$d_1^F(\text{HERCULES}, \text{ZEUS}) = (|1 - 1| + |0 - 1| + |1/2 - 1| + |1/2 - 0|)/4 = 1/2$

$d_1^F(\text{HERA}, \text{HERCULES}) = (|0 - 1| + |1 - 0| + |1 - 1/2| + |0 - 1/2|)/4 = 3/4$

$d_1^F(\text{APOLLO}, \text{ZEUS}) = (|1 - 1| + |1 - 1| + |1 - 1| + |1/2 - 0|)/4 = 1/8$

2.3 Discussion

In the following we prove that these dissimilarity functions have the standard properties for semi-distances [11]:

Proposition 2.1 (semi-distance). *Let d_p be the dissimilarity function defined above, let F a fixed feature set and $p > 0$. d_p is a semi-distance, that is for all $a, b, c \in$ Ind(A), d_p satisfies the following formal properties:*

1. $d_p^F(a, b) \geq 0$ *and* $d_p^F(a, b) = 0$ *if* $a = b$
2. $d_p^F(a, b) = d_p^F(b, a)$
3. $d_p^F(a, c) \leq d_p^F(a, b) + d_p^F(b, c)$

Proof. 1. and 2. are trivial. As for 3., note that

$$(d_p(a, c))^p = \frac{1}{m^p} \sum_{i=1}^m | \pi_i(a) - \pi_i(c) |^p =$$

$$= \frac{1}{m^p} \sum_{i=1}^m | \pi_i(a) - \pi_i(b) + \pi_i(b) - \pi_i(c) |^p$$

$$\leq \frac{1}{m^p} \sum_{i=1}^m | \pi_i(a) - \pi_i(b) |^p + \frac{1}{m^p} \sum_{i=1}^m | \pi_i(b) - \pi_i(c) |^p$$

$$\leq (d_p(a, b))^p + (d_p(b, c))^p \leq (d_p(a, b) + d_p(b, c))^p$$

then the property follows for the monotonicity of the power function.

This measure is not a distance since it does not hold that $a = b$ if $d_p^F(a,b) = 0$. This is the case of *indiscernible* individuals with respect to the given committee of features F. However, if the *unique names assumption* were made then one may define a supplementary dimension for the committee (a sort of meta-feature F_0) based on equality, such that:
$\forall (a,b) \in \text{Ind}(\mathcal{A}) \times \text{Ind}(\mathcal{A})$

$$\delta_0(a,b) := \begin{cases} 0 & a = b \\ 1 & a \neq b \end{cases}$$

and then

$$d_p^F(a,b) := \frac{1}{k+1} \sqrt[p]{\sum_{i=0}^{k} \delta_i(a,b)^p}$$

The resulting measures are distance measures.

Compared to other proposed dissimilarity measures [3, 4], the presented functions do not depend on the constructors of a specific language, rather they require only (retrieval or) instance-checking for computing the projections through class-membership queries to the knowledge base.

The complexity of measuring the dissimilarity of two individuals depends on the complexity of such inferences (see [1], Ch. 3). Note also that the projections that determine the measure can be computed (or derived from statistics maintained on the knowledge base) before the actual distance application, thus determining a speed-up in the computation of the measure. This is very important for algorithms that massively use this distance, such as all instance-based methods.

So far we made the assumption that F may represent a sufficient number of (possibly redundant) features that are able to discriminate really different individuals. The choice of the concepts to be included – *feature selection* – may be crucial. Therefore, we have devised specific optimization algorithms founded in *randomized search* which are able to find optimal choices of discriminating concept committees [11, 10].

The fitness function to be optimized is based on the *discernibility factor* [16] of the committee. Given the whole set of individuals $\text{Ind}(\mathcal{A})$ (or just a hold-out sample to be used to induce an optimal measure) $HS \subseteq \text{Ind}(\mathcal{A})$ the fitness function to be maximized is:

$$\text{DISCERNIBILITY}(\mathsf{F}, HS) := \sum_{(a,b) \in HS^2} \sum_{i=1}^{k} \delta_i(a,b)$$

However, the results obtained so far with knowledge bases drawn from ontology libraries [10, 12] show that (a selection) of the (primitive and defined) concepts is often sufficient to induce satisfactory dissimilarity measures.

3   Dissimilarity Measures Based on Uncertainty

The measure defined in the previous section deals with uncertainty in a uniform way: in particular, the degree of discernibility of two individuals is null when they have the same behavior with respect to the same feature, even in the presence of total uncertainty of class-membership for both. When uncertainty regards only one projection, then they are considered partially (possibly) similar.

We would like to make this uncertainty more explicit[3]. One way to deal with uncertainty would be considering intervals rather than numbers in [0,1] as a measure of dissimilarity. This is similar to the case of imprecise probabilities [20].

In order to extend the measure, we propose an epistemic definition based on rules for combining evidence [8, 18]. The ultimate aim is to assess the distance between two individuals as a combination of the evidence that they differ based on some selected features (as in the previous section).

The distance measure that is to be defined is again based on the degree of belief of discernibility of individuals with respect to such features. To this purpose the probability masses of the basic events (class-membership) have to be assessed. However, in this case we will not treat uncertainty in the classic probabilistic way (uniform probability). Rather, we intend to take into account uncertainty in the computation.

The new dissimilarity measure will be derived as a combination of the degree of belief in the discernibility of the individuals with respect to each single feature. Before introducing the combination rule (that will have the measure as a specialization), the basic probability assignments have to be considered, especially for the cases when instance-checking is not able to provide a certain answer.

As in previous works [4], we may estimate the concept extensions recurring to their retrieval [1], i.e. the individuals of the ABox that can be proved to belong to a concept. Thus, in case of uncertainty, the basic probabilities masses for each feature concept can be approximated[4] in the following way:
$\forall i \in \{1, \ldots, k\}$

$$m_i(\mathcal{K} \models F_i(a)) \approx |\mathsf{retrieval}(F_i, \mathcal{K})| / |\mathsf{Ind}(\mathcal{A})|$$
$$m_i(\mathcal{K} \models \neg F_i(a)) \approx |\mathsf{retrieval}(\neg F_i, \mathcal{K})| / |\mathsf{Ind}(\mathcal{A})|$$
$$m_i(\mathcal{K} \models F_i(a) \vee \mathcal{K} \models \neg F_i(a)) \approx 1 - m_i(\mathcal{K} \models F_i(a)) - m_i(\mathcal{K} \models \neg F_i(a))$$

where the $\mathsf{retrieval}(\cdot, \cdot)$ operator returns the individuals which can be proven to be members of the argument concept in the context of the current knowledge base [1]. The rationale is that the larger the (estimated) extension the more likely is for individuals to belong to the concept. This rationale can be easily understood thinking about a simple ontology describing the kinship relationships. Supposing that, in this ontology, the concepts Human, Male, Female, Man and

---

[3] We are referring to a notion of *epistemic* (rather than *aleatory*) probability [18], which seems more suitable for our purposes. See Shafer's introductory chapter in [19] on this distinction.

[4] In case of a certain answer received from the reasoner, the probability mass amounts to 0 or 1.

Woman are introduced, it is highly probable that an individual in the ontology (or a new individual) is instance of the concept Human. Hence, the approximated probability masses become more precise as more information (namely more instances) is acquired.

As in the previous section, we define a discernibility function related to a fixed concept which measures the amount of evidence that two input individuals may be separated by that concept:

Definition 3.1 (discernibility function). *Given a feature concept* $F_i \in \mathsf{F}$, *the related* discernibility function

$$\delta_i : \mathsf{Ind}(\mathcal{A}) \times \mathsf{Ind}(\mathcal{A}) \mapsto [0, 1]$$

is defined as follows:
$\forall (a, b) \in \mathsf{Ind}(\mathcal{A}) \times \mathsf{Ind}(\mathcal{A})$

$$\delta_i(a, b) := \begin{cases} m_i(\mathcal{K} \models \neg F_i(b)) & \textit{if } \mathcal{K} \models F_i(a) \\ m_i(\mathcal{K} \models F_i(b)) & \textit{else if } \mathcal{K} \models \neg F_i(a) \\ \delta_i(b, a) & \textit{else if } \mathcal{K} \models F_i(b) \vee \mathcal{K} \models \neg F_i(b) \\ 2 \cdot m_i(\mathcal{K} \models F_i(a)) \cdot m_i(\mathcal{K} \models \neg F_i(b)) & \textit{otherwise} \end{cases}$$

The extreme values $\{0, 1\}$ are returned when the answers from the instance-checking service are certain for both individuals. If the first individual is an instance of the i-th feature (respectively, its complement) then the discernibility depends on the belief of class-membership to the complement concept of the other individual. Otherwise, if there is uncertainty for the former individual but not for the latter, the function changes its perspective, swapping the roles of the two individuals. Finally, in case there were uncertainty for both individuals, the discernibility is computed as the chance that they may belong one to the feature concept and one to its complement,

The combined degree of belief in the case of discernible individuals, assessed using the *mixing combination rule* [15, 18], can give a measure of the semantic distance between them.

Definition 3.2 (weighted average measure). *Given an ABox $\mathcal{A}$, a dissimilarity measure for the individuals in $\mathcal{A}$*

$$d^{\mathsf{F}}_{avg} : \mathsf{Ind}(\mathcal{A}) \times \mathsf{Ind}(\mathcal{A}) \mapsto [0, 1]$$

is defined as follows:
$\forall (a, b) \in \mathsf{Ind}(\mathcal{A}) \times \mathsf{Ind}(\mathcal{A})$

$$d^{\mathsf{F}}_{avg}(a, b) := \sum_{i=1}^{k} w_i \delta_i(a, b)$$

The choices for the weights are various. The most straightforward one is, of course, considering uniform weights: $w_i = 1/k$. Another one is

$$w_i = \frac{u_i}{u}$$

where

$$u_i = \frac{1}{|\mathsf{Ind}(A) \setminus \mathsf{retrieval}(F_k, \mathcal{K})|} \text{ and } u = \sum_{i=1}^{k} u_i$$

It is easy to see that this can be considered as a generalization of the measure defined in the previous section (for $p = 1$).

3.1   Discussion

It can be proved that function has the standard properties for semi-distances:

Proposition 3.1 (semi-distance). *For a fixed choice of weights $\{w_i\}_{i=1}^{k}$, function d_{avg}^{F} is a semi-distance.*

This proposition can be easily proved, by analogy with the proof of Prop. 2.1.

The underlying idea for the measure is to combine the belief of the dissimilarity of the two input individuals provided by several sources, that are related to the feature concepts. In the original framework for evidence composition the various sources are supposed to be independent, which is generally unlikely to hold. Yet, from a practical viewpoint, overlooking this property for the sake of simplicity may still lead to effective methods, as the Naïve Bayes approach in Machine Learning demonstrates.

It could also be criticized that the subsumption hierarchy has not been explicitly involved. However, this may be actually yielded as a side-effect of the possible partial redundancy of the various concepts, which has an impact on their extensions and thus on the related projection function. A tradeoff is to be made between the number of features employed and the computational effort required for computing the related projection functions, since with the increasing of the feature in the *committee*, a higher number of instance checking has to computed.

The discriminating power of each feature concept can be weighted in terms of information and entropy measures. The weights should reflect the impact of the single feature concept with respect to the overall dissimilarity. As mentioned, this can be determined by the quantity of information conveyed by a feature, which can be measured as its entropy. Namely, the degree of information yielded by each of these features can be estimated as follows:

$$H_i(a, b) = - \sum_{A \subseteq \Theta} m_i(A) \log(m_i(A))$$

where 2^{Θ}, w.r.t. the *frame of discernment*[5] [19, 18] $\Theta = \{D, \overline{D}\}$. then, the sum

$$\sum_{(a,b) \in HS} H_i(a, b)$$

---

[5] Here D stands for the case of discernible individuals w.r.t. F_i, $\overline{D}$ for the opposite case, and some probability mass may be assigned also to the uncertain case represented by $\{D, \overline{D}\}$.

provides a measure of the utility of the discernibility function related to each feature which can be used in randomized optimization algorithms.

Alternatively, the extension of a feature F_i with respect to the whole domain of objects may be quantified as $P_F = |F_i^{\mathcal{I}}|/|\Delta^{\mathcal{I}}|$ (w.r.t. the canonical interpretation $\mathcal{I}$). This can be roughly approximated with: $P_{F_i} = |\text{retrieval}(F_i)|/|\text{Ind}(\mathcal{A})|$. Hence, considering also the probability $P_{\neg F_i}$ related to its negation and that related to the unclassified individuals (w.r.t. F_i), denoted P_U, we may give an entropic measure for the feature:

$$H(F_i) = -\left(P_{F_i} \log(P_{F_i}) + P_{\neg F_i} \log(P_{\neg F_i}) + P_U \log(P_U)\right).$$

These measures may be normalized for providing a good set of weights for the distance measures.

3.2   Extensions

Following the rationale of the average link criterion used in agglomerative clustering [14], the measures can be extended to the case of concepts, by recurring to the notion of medoids.

The *medoid* of a group of individuals is the individual that has the highest similarity with respect to the others. Formally, given a group $G = \{a_1, a_2, \ldots, a_n\}$, the medoid is defined:

$$\text{medoid}(G) = \operatorname*{argmin}_{a \in G} \sum_{j=1}^{n} d(a, a_j)$$

Now, given two concepts C_1, C_2, we can consider the two corresponding groups of individuals obtained by retrieval $R_i = \{a \in \text{Ind}(\mathcal{A}) \mid \mathcal{K} \models C_i(a)\}$, and their respectively medoids $m_i = \text{medoid}(R_i)$ for $i = 1, 2$ with respect to a given measure d_p^{F} (for some $p > 0$ and committee F). Then the function for concepts can be defined as follows:

$$d_p^{\mathsf{F}}(C_1, C_2) := d_p^{\mathsf{F}}(m_1, m_2)$$

Similarly, if the distance of an individual a to a concept C has to be assessed, one could consider the nearest (respectively farthest) individual in the concept extension or its medoid. Let $m = \text{medoid}(\text{retrieval}(C))$ with respect to a given measure d_p^{F}. Then the measure for this case can be defined as follows:

$$d_p^{\mathsf{F}}(a, C) := d_p^{\mathsf{F}}(a, m)$$

Of course these approximate measures become more and more precise as the knowledge base is populated with an increasing number of individuals.

4   Experimentation

A k-Nearest Neighbor classification procedure exploiting the metrics proposed in the previous sections has been tested at solving a number of retrieval problems [5].

In its simplest setting, a k-NN algorithm approximates a hypothesis function h_Q for classifying a query individual x_0 with respect to a query concept Q on the grounds of the value that h_Q is known to assume for the training instances in its neighborhood $NN(x_0)$, i.e. the k closest instances to x_0 in terms of a similarity measure. Precisely, the value is decided by means of a weighted majority voting procedure: it is simply the most *voted* value by the instances in $NN(x_0)$ weighted by the similarity of the neighbor individual.

The answer of the estimated hypothesis function for the query individual is:

$$\hat{h}_Q(x_0) := \operatorname*{argmax}_{v \in V} \sum_{i=1}^{k} w_i \delta(v, h_Q(x_i)) \tag{1}$$

where δ returns 1 in case of matching arguments and 0 otherwise, and, given a dissimilarity measure d, the weights are determined by $w_i = 1/d(x_i, x_0)$.

The task can be cast as follows: given a query concept Q, determine the membership of an instance x_0 through the voting procedure (Eq. 1) where $V = \{-1, 0, +1\}$ and the hypothesis function values for the training instances are determined by the entailment of the corresponding assertion from the knowledge base, as follows:

$$h_Q(x) = \begin{cases} +1 & \mathcal{K} \models Q(x) \\ -1 & \mathcal{K} \models \neg Q(x) \\ 0 & otherwise \end{cases}$$

It should be noted that the inductive inference made by the procedure above is not guaranteed to be deductively valid. Indeed, inductive inference naturally yields a certain degree of uncertainty. In order to measure the likelihood of the decision made by the procedure (individual x_0 belongs to the query concept denoted by value v maximizing the argmax argument in Eq. 1), given the nearest training individuals in $NN(x_0, k) = \{x_1, \ldots, x_k\}$, the quantity that determined the decision should be normalized by dividing it by the sum of such arguments over the (three) possible values:

$$l(class(x_0) = v | NN(x_0, k)) = \frac{\sum_{i=1}^{k} w_i \cdot \delta(v, h_Q(x_i))}{\sum_{v' \in V} \sum_{i=1}^{k} w_i \cdot \delta(v', h_Q(x_i))} \tag{2}$$

Hence the likelihood of the assertion $Q(x_0)$ corresponds to $l(class(x_0) = 1 | NN(x_0, k))$. By adding the likelihood $l(class(x_0) = 0 | NN(x_0, k))$ we may obtain a suggested upper bound for the likelihood of the membership assertion.

4.1   Experimental Setting

A number of ontologies for different domains represented in OWL was selected, namely: SURFACE-WATER-MODEL (SWM), NEWTESTAMENTNAMES (NTN) from the Protégé library[6], the Semantic Web Service Discovery dataset[7]

---

[6] http://protege.stanford.edu/plugins/owl/owl-library
[7] https://www.uni-koblenz.de/FB4/Institutes/IFI/AGStaab/Projects/xmedia/dl-tree.htm

Table 1. Facts concerning the ontologies employed in the experiments

Ontology	DL language	#concepts	#object prop.	#data prop.	#individuals
SWM	$\mathcal{ALCOF}(D)$	19	9	1	115
BioPAX	$\mathcal{ALCHF}(D)$	28	19	30	323
LUBM	$\mathcal{ALR^+HI}(D)$	43	7	25	555
NTN	$\mathcal{SHIF}(D)$	47	27	8	676
SWSD	$\mathcal{ALCH}$	258	25	0	732
Financial	$\mathcal{ALCIF}$	60	17	0	1000

(SWSD), the University0.0 ontology generated by the Lehigh University Benchmark[8] (LUBM), the BioPax glycolysis ontology[9] (BioPax) and the Financial ontology[10]. Tab. 1 summarizes details concerning these ontologies; specifically the number of concepts, object properties, data type properties and individuals.

For each ontology, 20 queries were randomly generated by composition (conjunction and/or disjunction) of (2 through 8) primitive and defined concepts in each knowledge base. Query concepts were constructed so that each one offered both positive and negative instances among the ABox individuals. The performance of the inductive method was evaluated by comparing its responses to those returned by a standard reasoner[11] as a baseline.

Experimentally, it was observed that large training sets make the distance measures (and consequently the NN procedure) very accurate. In order to make the problems more difficult, we selected limited training sets (TS) that amount to only 4% of the individuals occurring in each ontology. Then the parameter k was set to $\log(|TS|)$ depending on the number of individuals in the training set. Again, we found experimentally that much smaller values could be chosen, resulting in the same classification.

The simpler distances (d_1^F) were employed from the *original* family (uniform weights) and *entropic* family (weighted on the feature entropy), using all the concepts in the knowledge base for determining the set F with no further optimization.

4.2   Results

Standard measures. Initially the standard measures precision, recall, F_1-measure were employed to evaluate the system performance, especially when selecting the positive instances (individuals that should belong to the query concept). The outcomes are reported in Tab. 2. For each knowledge base, we report the average values obtained over the 20 random queries as well as their standard deviation and minimum-maximum ranges of values.

---

[8] http://swat.cse.lehigh.edu/projects/lubm
[9] http://www.biopax.org/Downloads/Level1v1.4/biopax-example-ecocyc-glycolysis.owl
[10] http://www.cs.put.poznan.pl/alawrynowicz/financial.owl
[11] We employed Pellet v. 1.5.1. See http://pellet.owldl.com

Table 2. Experimental results in terms of standard measures: averages ± standard deviations and [min,max] intervals

	ORIGINAL MEASURE		
	precision	recall	F-measure
SWM	89.1 ± 27.3 [16.3;100.0]	84.4 ± 30.6 [11.1;100.0]	78.7 ± 30.6 [20.0;100.0]
BIOPAX	99.2 ± 1.9 [93.8;100.0]	97.3 ± 11.3 [50.0;100.0]	97,8 ± 7.4 [66.7;100.0]
LUBM	100.0 ± 0.0 [100.0;100.0]	71.7 ± 38.4 [9.1;100.0]	76.2 ± 34.4 [16.7;100.0]
NTN	98.8 ± 3.0 [86.9;100.0]	62.6 ± 42.8 [4.3;100.0]	66.9 ± 37.7 [8.2;100.0]
SWSD	74.7 ± 37.2 [8.0;100.0]	43.4 ± 35.5 [2.2;100.0]	54.9 ± 34.7 [4.3;100.0]
FINANCIAL	99.6 ± 1.3 [94.3;100.0]	94.8 ± 15.3 [50.0;100.0]	97.1 ± 10.2 [66.7;100.0]
	ENTROPIC MEASURE		
	precision	recall	F-measure
SWM	99.0 ± 4.3 [80.6;100.0]	75.8 ± 36.7 [11.1;100.0]	79.5 ± 30.8 [20.0;100.0]
BIOPAX	99.9 ± 0.4 [98.2;100.0]	97.3 ± 11.3 [50.0;100.0]	98,2 ± 7.4 [66.7;100.0]
LUBM	100.0 ± 0.0 [100.0;100.0]	81.6 ± 32.8 [11.1;100.0]	85.0 ± 28.4 [20.0;100.0]
NTN	97.0 ± 5.8 [76.4;100.0]	40.1 ± 41.3 [4.3;100.0]	45.1 ± 35.4 [8.2;97.2]
SWSD	94.1 ± 18.0 [40.0;100.0]	38.4 ± 37.9 [2.4;100.0]	46.5 ± 35.0 [4.5;100.0]
FINANCIAL	99.8 ± 0.3 [98.7;100.0]	95.0 ± 15.4 [50.0;100.0]	96.6 ± 10.2 [66.7;100.0]

As an overall consideration we may observe that generally the outcomes obtained adopting the extended measure improve on those with the other one and appear also more stable (with some exceptions). Besides, it is possible to note that precision and recall values are generally quite good for all ontologies except for SWSD, where especially recall is significantly lower. This ontology turned out to be more difficult (also in terms of precision) for two reasons: 1) a very limited number of individuals per concept was available; 2) the number of different concepts is larger with respect to the other knowledge bases. For the other ontologies, values are much higher, as testified also by the F-measure values. The results in terms of precision are also more stable than those for recall, as proved by the limited variance observed, whereas single queries happened to turn out quite difficult as regards the correctness of the answer.

The reason for precision being generally higher is probably due to the *Open World Assumption* (OWA). Indeed, in many cases it was observed that the NN procedure deemed some individuals as relevant for the query issued while the DL reasoner was not able to assess this relevance, namely was not able to give any reply. This different behavior of the two approaches was computed as a mistake by the metric mentioned above, while it may likely turn out to be a correct inference of the NN procedure when judged by a human agent.

Because of these problems in the evaluation with the standard indices, especially due to the cases on unknown answers from the reference system (the reasoner) we thought to make this case more explicit by measuring both the rate of inductively classified individuals and the nature of the mistakes.

Alternative measures. Due to the OWA, cases were observed when, it could not be (deductively) ascertained whether a resource was relevant or not for a given query. Hence, we introduced the following indices for a further evaluation:

- *match rate*: number of individuals that got exactly the same classification ($v \in V$) by both the inductive and the deductive classifier with respect to the overall number of individuals (v vs. v);
- *omission error rate*: number of individuals for which inductive method could not determine whether they were relevant to the query or not while they were actually relevant according to the reasoner (0 vs. ± 1);
- *commission error rate*: number of individuals (analogically) found to be relevant to the query concept, while they (logically) belong to its negation or vice-versa (+1 vs. −1 or −1 vs. +1);
- *induction rate*: number of individuals found to be relevant to the query concept or to its negation, while either case is not logically derivable from the knowledge base (± 1 vs. 0);

Tab. 3 reports the outcomes in terms of these indices. Preliminarily, it is important to note that, in each experiment, the commission error was quite low or absent. This means that the inductive search procedure is quite accurate, namely it did not make critical mistakes attributing an individual to a concept that is disjoint with the right one. Also the omission error rate was generally quite low, yet more frequent than the previous type of error.

The usage of all concepts for the set F of d_1^F made the measure quite accurate, which is the reason why the procedure resulted quite conservative as regards inducing new assertions. In many cases, it matched rather faithfully the reasoner decisions. From the retrieval point of view, the cases of induction are interesting because they suggest new assertions which cannot be logically derived by using a deductive reasoner yet they might be used to complete a knowledge base [2, 5], e.g. after being validated by an ontology engineer. For each candidate new assertion, Eq. 2 may be employed to assess the likelihood and hence decide on its inclusion (see next section).

If we compare these outcomes with those reported in other works on instance retrieval and inductive classification [4], where the highest average match rate observed was around 80%, we find a significant increase of the performance due to the accuracy of the new measure. Also the elapsed time (not reported here) was much less because of the different dissimilarity measure: once the values for

Table 3. Results with alternative indices: averages ± standard deviations and [min,max] intervals

ORIGINAL MEASURE				
	match	commission	omission	induction
SWM	93.3 ± 10.3 [68.7;100.0]	0.0 ± 0.0 [0.0;0.0]	2.5 ± 4.4 [0.0;16.5]	4.2 ± 10.5 [0.0;31.3]
BioPax	99.9 ± 0.2 [99.4;100.0]	0.2 ± 0.2 [0.0;0.06]	0.0 ± 0.0 [0.0;0.0]	0.0 ± 0.0 [0.0;0.0]
LUBM	99.2 ± 0.8 [98.0;100.0]	0.0 ± 0.0 [0.0;0.0]	0.8 ± 0.8 [0.0;0.2]	0.0 ± 0.0 [0.0;0.0]
NTN	98.6 ± 1.5 [93.9;100.0]	0.0 ± 0.1 [0.0;0.4]	0.8 ± 1.1 [0.0;3.7]	0.6 ± 1.4 [0.0;6.1]
SWSD	97.5 ± 3.7 [84.6;100.0]	0.0 ± 0.0 [0.0;0.0]	1.8 ± 2.6 [0.0;9.7]	0.8 ± 1.5 [0.0;5.7]
FINANCIAL	99.5 ± 0.8 [97.3;100.0]	0.3 ± 0.7 [0.0;2.4]	0.0 ± 0.0 [0.0;0.0]	0.2 ± 0.2 [0.0;0.6]

ENTROPIC MEASURE				
	match	commission	omission	induction
SWM	97.5 ± 3.2 [89.6;100.0]	0.0 ± 0.0 [0.0;0.0]	2.2 ± 3.1 [0.0;10.4]	0.3 ± 1.2 [0.0;5.2]
BioPax	99.9 ± 0.2 [99.4;100.0]	0.1 ± 0.2 [0.0;0.06]	0.0 ± 0.0 [0.0;0.0]	0.0 ± 0.0 [0.0;0.0]
LUBM	99.5 ± 0.7 [98.2;100.0]	0.0 ± 0.0 [0.0;0.0]	0.5 ± 0.7 [0.0;1.8]	0.0 ± 0.0 [0.0;0.0]
NTN	97.5 ± 1.9 [91.3;99.3]	0.6 ± 0.7 [0.0;1.6]	1.3 ± 1.4 [0.0;4.9]	0.6 ± 1.7 [0.0;7.1]
SWSD	98.0 ± 3.0 [88.3;100.0]	0.0 ± 0.0 [0.0;0.0]	1.9 ± 2.9 [0.0;11.3]	0.1 ± 0.2 [0.0;0.5]
FINANCIAL	99.7 ± 0.2 [99.4;100.0]	0.0 ± 0.0 [0.0;0.1]	0.0 ± 0.0 [0.0;0.0]	0.2 ± 0.2 [0.0;0.6]

the projection functions are pre-computed, the efficiency of the classification, which depends on the computation of the dissimilarity, was also improved.

As mentioned, we also found that a lower value for the parameter k can be considered still ensuring quite high correct classification results. This yielded also that the likelihood of the inferences made (see Eq. 2) turned out quite high.

5   Concluding Remarks

We have proposed the definition of dissimilarity measures over spaces of individuals in a knowledge base. The measures are not language-dependent, differently from other previous proposals [4], yet they are parameterized on a committee of concepts. Optimal committees can be found via randomized search methods [11]. Besides, we have extended the measures to cope with cases of uncertainty by means of a simple evidence combination method.

One of the advantages of the measures is that their computation can be very efficient in cases when statistics (on class-membership) are maintained by the KBMS [13]. As previously mentioned, the subsumption relationships among concepts in the committee is not explicitly exploited in the measure for making the relative distances more accurate. The extension to the case of concept distance may also be improved. Hence, scalability should be guaranteed as far as a good committee has been found and does not change also because of the locality properties observed for instances in several domains (e.g. social or biological networks).

A refinement of the committee may become necessary only when a degradation of the discernibility factor is detected due to the availability of somewhat new individuals. This may involve further tasks such as *novelty* or *concept drift* detection.

5.1   Applications

The measures have been integrated in an instance-based learning system implementing a nearest-neighbor learning algorithm: an experimentation on performing semantic-based retrieval proved the effectiveness of the new measures, compared to the outcomes obtained adopting other measures [4]. It is worthwhile to mention that results where not particularly affected by feature selection: often using the very concepts defined in the knowledge base provides good committees which are able to discern among the different individuals [12].

We are also exploiting the implementation of these measures for performing conceptual clustering [14], where (a hierarchy of) clusters is created by grouping instances on the grounds of their similarity, possibly triggering the induction of new emerging concepts [10].

5.2   Extensions

The measure may have a wide range of application of distance-based methods to knowledge bases. For example, logic approaches to ontology matching [9] may be backed up by the usage of our measures, especially when concepts to be matched across different terminologies are known to share a common set of individuals. Ontology matching could be a phase in a larger process aimed at data integration. Moreover metrics could also support a process of *(semi-)automated classification* of new data also as a first step towards ontology evolution.

Another problem that could be tackled by means of dissimilarity measures could be the *ranking* of the answers provided by a *matchmaking* algorithm based on the similarity between the concept representing the query and the retrieved individuals.

References

[1] Baader, F., Calvanese, D., McGuinness, D., Nardi, D., Patel-Schneider, P. (eds.): The Description Logic Handbook. Cambridge University Press, Cambridge (2003)

[2] Baader, F., Ganter, B., Sertkaya, B., Sattler, U.: Completing description logic knowledge bases using formal concept analysis. In: Veloso, M. (ed.) Proceedings of the 20th International Joint Conference on Artificial Intelligence, Hyderabad, India, pp. 230–235 (2007)

[3] Borgida, A., Walsh, T.J., Hirsh, H.: Towards measuring similarity in description logics. In: Horrocks, I., Sattler, U., Wolter, F. (eds.) Working Notes of the International Description Logics Workshop, Edinburgh, UK. CEUR Workshop Proceedings, vol. 147 (2005)

[4] d'Amato, C., Fanizzi, N., Esposito, F.: Reasoning by analogy in description logics through instance-based learning. In: Tummarello, G., Bouquet, P., Signore, O. (eds.) Proceedings of Semantic Web Applications and Perspectives, 3rd Italian Semantic Web Workshop, SWAP2006, Pisa, Italy. CEUR Workshop Proceedings, vol. 201 (2006)

[5] d'Amato, C., Fanizzi, N., Esposito, F.: Query answering and ontology population: an inductive approach. In: Bechhofer, S., Hauswirth, M., Hoffmann, J., Koubarakis, M. (eds.) ESWC 2008. LNCS, vol. 5021. Springer, Heidelberg (forthcoming, 2008)

[6] d'Aquin, M., Lieber, J., Napoli, A.: Decentralized case-based reasoning for the Semantic Web. In: Gil, Y., Motta, E., Benjamins, V.R., Musen, M.A. (eds.) ISWC 2005. LNCS, vol. 3729, pp. 142–155. Springer, Heidelberg (2005)

[7] Dowsland, K.A., Soubeiga, E., Burke, E.K.: A simulated annealing based hyper-heuristic for determining shipper sizes for storage and transportation. European Journal of Operational Research 179(3), 759–774 (2007)

[8] Dubois, D., Prade, H.: On the combination of evidence in various mathematical frameworks. In: Flamm, J., Luisi, T. (eds.) Reliability Data Collection and Analysis, pp. 213–241 (1992)

[9] Euzenat, J., Shvaiko, P.: Ontology matching. Springer, Heidelberg (2007)

[10] Fanizzi, N., d'Amato, C., Esposito, F.: Evolutionary conceptual clustering of semantically annotated resources. In: Proceedings of the 1st International Conference on Semantic Computing, IEEE-ICSC2007, Irvine, CA, pp. 783–790. IEEE Computer Society Press, Los Alamitos (2007)

[11] Fanizzi, N., d'Amato, C., Esposito, F.: Induction of optimal semi-distances for individuals based on feature sets. In: Working Notes of the International Description Logics Workshop, DL 2007, Bressanone, Italy. CEUR Workshop Proceedings, vol. 250 (2007)

[12] Fanizzi, N., d'Amato, C., Esposito, F.: Instance-based query answering with semantic knowledge bases. In: Basili, R., Pazienza, M.T. (eds.) AI*IA 2007. LNCS, vol. 4733, pp. 254–265. Springer, Heidelberg (2007)

[13] Horrocks, I.R., Li, L., Turi, D., Bechhofer, S.K.: The Instance Store: DL reasoning with large numbers of individuals. In: Haarslev, V., Möller, R. (eds.) Proceedings of the 2004 Description Logic Workshop, DL 2004. CEUR Workshop Proceedings, vol. 104, pp. 31–40. CEUR (2004)

[14] Jain, A.K., Murty, M.N., Flynn, P.J.: Data clustering: A review. ACM Computing Surveys 31(3), 264–323 (1999)

[15] Murphy, C.K.: Combining belief functions when evidence conflicts. Decision Support Systems 29(1-9) (2000)

[16] Pawlak, Z.: Rough Sets: Theoretical Aspects of Reasoning About Data. Kluwer Academic Publishers, Dordrecht (1991)

[17] Sebag, M.: Distance induction in first order logic. In: Džeroski, S., Lavrač, N. (eds.) ILP 1997. LNCS (LNAI), vol. 1297, pp. 264–272. Springer, Heidelberg (1997)

[18] Sentz, K., Ferson, S.: Combination of evidence in Dempster-Shafer theory. Technical Report SAND2002-0835, SANDIATech (April 2002)

[19] Shafer, G.: A Mathematical Theory of Evidence. Princeton University Press, Princeton (1976)

[20] Walley, P.: Statistical reasoning with imprecise probabilities. Chapman and Hall, London (1991)

[21] Zezula, P., Amato, G., Dohnal, V., Batko, M.: Similarity Search – The Metric Space Approach. In: Advances in Database Systems. Springer, Heidelberg (2007)

Ontology Learning and Reasoning — Dealing with Uncertainty and Inconsistency

Peter Haase and Johanna Völker

Institute AIFB, University of Karlsruhe, Germany
{haase,voelker}@aifb.uni-karlsruhe.de

Abstract. Ontology learning aims at generating domain ontologies from various kinds of resources by applying natural language processing and machine learning techniques. It is inherent to the ontology learning process that the acquired ontologies represent uncertain and possibly contradicting knowledge. From a logical perspective, the learned ontologies are potentially inconsistent knowledge bases, that as such do not allow for meaningful reasoning. In this paper, we present an approach to generating consistent OWL ontologies from automatically generated or enriched ontology models, which takes into account the uncertainty of the acquired knowledge. We illustrate and evaluate the application of our approach with two experiments in the scenarios of consistent evolution of learned ontologies and enrichment of ontologies with disjointness axioms.

1 Introduction

Ontology Learning aims at generating domain ontologies from a given collection of resources by applying natural language processing and machine learning techniques. Due to an increasing demand for efficient support in knowledge acquisition, a number of tools for automatic or semi-automatic ontology learning have been developed during the last decade. Common to all of them is the need for handling the uncertainty that is inherent to any kind of knowledge acquisition process. Moreover, ontology-based applications have to face the challenge of reasoning with large amounts of imperfect information resulting from automatic ontology acquisition systems.

According to [25], such imperfection can be due to *imprecision*, *inconsistency* or *uncertainty*. Imprecision and inconsistency are properties of the information itself. Either more than one world (in the case of ambiguous, vague or approximate information) or no world (if contradictory conclusions can be derived from the information) is compatible with the given information. Uncertainty means that an agent, i.e. a computer or a human, has only partial knowledge about the truth value of a given piece of information. One can distinguish between objective and subjective uncertainty. Whereas *objective* uncertainty relates to randomness referring to the propensity or disposition of something to be true, *subjective* uncertainty depends on an agent's opinion about the truth value of information. In particular, the agent can consider information as unreliable or irrelevant.

In ontology learning, (subjective) uncertainty is the most prominent form of imperfection. This is due to the fact that the results of the different algorithms have to be considered as unreliable or irrelevant due to imprecision and errors introduced during

P.C.G. da Costa et al. (Eds.): URSW 2005-2007, LNAI 5327, pp. 366–384, 2008.

the ontology generation process. There exist different approaches to the representation of uncertainty: Uncertainty can, for instance, be represented as part of the learned ontologies, e.g., using probabilistic extensions to the target knowledge representation formalism [7,9], or at a meta-level as application-specific information associated with the learned structures [34].

In our approach to ontology learning we target the Web Ontology Language OWL, which is today the standard for representing ontologies on the web. With its grounding in description logics, reasoning with OWL ontologies is very well understood and efficient reasoning algorithms have been devised. Because of the uncertain and thus potentially contradicting information, the learned ontologies typically are highly inconsistent and thus do not allow for meaningful reasoning. Even approaches to the acquisition of very simple axioms such as subsumption between atomic classes can provoke inconsistencies if they are used for the enrichment of an existing ontology.

In order to deal with this problem, we suggest the following approach: We apply ontology learning algorithms to generate ontologies, which also contain information about uncertainty in the form of annotations to ontology elements and axioms. These annotations capture the confidence about the correctness or relevance of the acquired knowledge. In doing so, we apply the techniques of *consistent ontology evolution* to ensure that the learned ontologies are kept consistent as the ontology learning procedures generate changes to the ontology over time. The process of consistent ontology evolution makes use of the confidence annotations to produce an ontology that is (1) consistent and (2) "most likely to be correct", or certain. We illustrate and evaluate the application of our approach by means of two application scenarios.

Scenario 1: Consistent Evolution of Learned Ontologies We apply our approach to incrementally learn an ontology from scratch based a corpus of text documents in a digital library. Intelligent search over document corpora in digital libraries is an application scenario that shows the immediate benefit of the ability to reason over ontologies automatically learned from text. While search in digital libraries nowadays is restricted to structured queries against the bibliographic metadata (author, title, etc.) and to unstructured keyword-based queries over the full text documents, complex queries that involve reasoning over the knowledge present in the documents are not possible. Ontology learning enables obtaining the required formal representations of the knowledge available in the corpus to be able to support such advanced types of search. This application scenario is the subject of a case study within the digital library of BT (British Telecom) as part of the SEKT[1] project. One of the key elements of the case study is to automatically learn ontologies to enhance search and finally be able to support queries of the kind "Find knowledge management applications that support Peer-to-Peer knowledge sharing." To validate the work presented in this paper, we performed experiments with data from the BT Digital Library.

Scenario 2: Enriching Ontologies with Disjointness Axioms In the second scenario, we target the enrichment of an existing ontology with disjointness axioms to make the ontology more precise. As shown in [28], semantically rich primitives such as disjointness of concepts can be used for effective semantic clarification in ontologies and thus enable

---

[1] http://www.sekt-project.com

drawing more meaningful conclusions. We apply the enrichment process to the EKAW ontology – an ontology covering the domain of conference organization – with disjointness axioms automatically generated by an ontology learning tool. We show how our approach of handling uncertainty and inconsistency in the learning process indeed leads to an improvement of quality with respect to a gold standard.

Overview of the paper. The rest of the paper is organized as follows. In Section 2 we recapitulate the foundations of the OWL ontology language, query answering with OWL ontologies and the role of logical inconsistencies. Section 3 introduces a set of complementary approaches to ontology learning and enrichment, that we applied within the subsequently described experiments. It is followed by a detailed discussion of our methods for handling inconsistencies in learned ontologies (cf. Section 4). The experimental evaluation of these methods is presented in Section 5. In Section 6, we give an overview of related work before concluding with Section 7.

2   Reasoning with OWL

Traditionally, a number of different knowledge representation paradigms have competed to provide languages for representing ontologies, including most notably description logics and frame logics. With the advent of the Web Ontology Language OWL, developed by the Web Ontology Working Group and recommended by the World Wide Web Consortium (W3C), a standard for the representation of ontologies has been created. Adhering to this standard, we base our work on the OWL language (in particular OWL DL, as discussed below) and describe the developed formalisms in its terms.

2.1   OWL as a Description Logic

The OWL ontology language is based on description logics, a family of class-based knowledge representation formalisms. In description logics, the important notions of a domain are described by means of concept descriptions that are built from *concepts* (also referred to as *classes*), *roles* (also called *properties* or *relations*), denoting relationships between things, and *individuals* (or *instances*). It is now possible to state facts about the domain in the form of axioms. Terminological axioms (also referred to as $\mathcal{T}$-Box axioms) make statements about how concepts or roles are related to each other, assertional axioms (sometimes called *facts* or $\mathcal{A}$-Box axioms) make statements about the properties of individuals.

We here informally introduce the language constructs of $\mathcal{SHOIN}$, the description logic underlying OWL DL. For the correspondence between our notation and various OWL DL syntaxes, see [17]. In the description logic $\mathcal{SHOIN}$, we can build complex classes from atomic ones using the following constructors:

- $C \sqcap D$ (intersection), denoting the concept of individuals that belong to both C and D,
- $C \sqcup D$ (union), denoting the concept of individuals that belong to either C or D,
- $\neg C$ (complement), denoting the concept of individuals that do not belong to C,

- $\forall R.C$ (universal restriction), denoting the concept of individuals that are related via the role R only with individuals belonging to the concept C,
- $\exists R.C$ (existential restriction), denoting the concept of individuals that are related via the role R with some individual belonging to the concept C,
- $\geq n\,R$, $\leq n\,R$ (qualified number restriction), denoting the concept of individuals that are related with at least (at most) n individuals via the role R.
- $\{c_1, \ldots, c_n\}$ (enumeration), denoting the concept of individuals explicitly enumerated.

Based on these class descriptions, axioms of the following types can be formed:

- concept inclusion axioms $C \sqsubseteq D$, stating that the concept C is a subconcept of the concept D,
- transitivity axioms $\mathsf{Trans}(R)$, stating that the role R is transitive,
- role inclusion axioms $R \sqsubseteq S$ stating that the role R is a subrole of the role S,
- concept assertions $C(a)$ stating that the individual a is in the extension of the concept C,
- role assertions $R(a, b)$ stating that the individuals a, b are in the extension of the role R,
- individual (in)equalities $a \approx b$, and $a \not\approx b$, respectively, stating that a and b denote the same (different) individuals.

With the constructs above, we can make complex statements, e.g., expressing that two concepts are disjoint using the axiom $A \sqsubseteq \neg B$. This axioms literally states that A is a subconcept of the complement of B, which intuitively means that there must not be any overlap in the extensions of A and B.

In the design of description logics, emphasis is put on retaining decidability of key reasoning problems and the provision of sound and complete reasoning algorithms. As the name suggests, description logics are logics, i.e. they have well-defined semantics. Typically, the semantics of a description logic is specified via *model theoretic semantics*, which explicates the relationship between the language syntax and the models of a domain using *interpretations*.

An interpretation consists of a domain of interpretation (essentially, a set) and an interpretation function which maps from individuals, concepts and roles to elements, subsets and binary relations on the domain of interpretation, respectively. A description logic knowledge base consists of a set of axioms which act as constraints on the interpretations. The meaning of a knowledge base derives from features and relationships that are common in all possible interpretations. An interpretation is said *to satisfy a knowledge base*, if it satisfies each axiom in the knowledge base. Such an interpretation is called a *model* of the knowledge base. If the relationship specified by some axiom (which may not be part of the knowledge base) holds in all models of a knowledge base, the axiom is said to be *entailed* by the knowledge base. Checking consistency and entailment are two standard reasoning tasks for description logics. Other standard reasoning tasks include computing the concept hierarchy and answering conjunctive queries.

2.2 Approaches to Dealing with Inconsistencies

Generally, inconsistencies refer to logical contradictions in an ontology. In the description logics community, different notions of inconsistency have been defined and used. In the classical definition, i.e. using first-order logic interpretations, a knowledge base is said to be *inconsistent*, if it has no models (satisfying interpretations). Incoherence of a knowledge base is defined as the existence of unsatisfiable concepts, i.e. concepts that have an empty extension in all models of the knowledge base. These unsatisfiable concepts are also referred to as *incoherent concepts*.

Logical contradictions hamper the effective use of an ontology. For example, the standard entailment as defined above is explosive, i.e. an inconsistent ontology has *all* axioms as consequences. Formally, if an ontology O is inconsistent, then for all axioms α we have $O \models \alpha$. In other words, query answers for inconsistent ontologies are completely meaningless, since for all queries the query answer will be *true*. To deal with the issue of potential inconsistencies in ontologies, we can choose from a number of alternative approaches [13]:

Consistent Ontology Evolution is the process of managing ontology changes by preserving the consistency of the ontology with respect to a given notion of consistency. The consistency of an ontology is defined in terms of consistency conditions, or invariants that must be satisfied by the ontology. The approach of consistent ontology evolution imposes certain requirements with respect to its applicability. For example, it requires that the ontology is consistent in the first place and that changes to the ontology can be controlled. In certain application scenarios, these requirements may not hold, and consequently, other means for dealing with inconsistencies in changing ontologies may be required.

Repairing Inconsistencies involves a process of diagnosis and repair: first the cause (or a set of potential causes) of the inconsistency needs to be determined, which can subsequently be repaired. Unlike the approach of consistent ontology evolution, repairing inconsistencies does not require to start with a consistent ontology and is thus adequate if the ontology is already inconsistent in the first place.

Reasoning with Inconsistent Ontologies does not try to avoid or repair the inconsistency (as in the previous two approaches), but simply tries to "live with it" by trying to return meaningful answers to queries, even though the ontology is inconsistent. In some cases, consistency cannot be guaranteed at all and inconsistencies cannot be repaired, but still one wants to derive meaningful answers when reasoning.

While in principle all the above alternatives may be applicable to deal, in this article we illustrate how ontology learning can be combined with the approach of *consistent ontology evolution* to guarantee that the learned ontologies are kept consistent as the ontology learning procedures generate changes to the ontology over time.

3 Ontology Learning

The automatic acquisition of ontological elements from lexical or logical resources, most commonly referred to as ontology learning, seems a promising way to facilitate the tedious task of ontology engineering. In the following, we describe two complementary ontology learning approaches. Section 3.1 gives an overview of Text2Onto, a framework developed to support the generation of ontologies from scratch. A specialized approach to the automatic enrichment of ontologies by disjointness axioms is presented in Section 3.2. Both approaches generate sets of axioms associated with uncertainty values and causing a fairly large number of inconsistencies or incoherences. An approach to handling and resolving these kinds of logical contradictions will be presented in the following Section 4.

3.1 Learning Ontologies from Text

We believe that linguistic evidence with respect to an ontology can be appropriately measured by ontology learning techniques which try to capture the ontological commitment in human language. Since ontology learning algorithms such as implemented in Text2Onto [5] consider the relation of individual ontology elements with the data the ontology has been engineered from, they allow to assess how well the ontology reflects the underlying corpus of data. This is especially relevant for an application scenario as introduced in Section 1, which involves question answering in the context of a digital library. In the following, we give a brief overview of the algorithms provided by Text2Onto for acquiring various kinds of ontology elements. Each of these algorithms aims to deliver evidence such as term occurrence statistics or lexical relationships present in WordNet from which we compute confidence or relevance values. For further details with regards to the architecture and the internal ontology representation model of Text2Onto, please refer to our original paper [5].

Algorithms. We now describe for each kind of ontology element supported by Text2Onto the algorithms that we used for our experiments. In particular, we give an intuition of how Text2Onto calculates the confidence and relevance ratings for the learned ontology elements. An overview of the primitives that can be acquired by Text2Onto is given by Table 1.

Concepts and Instances. The extraction of concepts and instances is based on a statistical approach to the identification of domain-relevant noun phrases. Different term weighting measures are used to compute the relevance of each term with respect to the corpus: Relative Term Frequency (RTF), TFIDF, Entropy and the C-value/NC-value method [20]. A distinction between terms denoting concepts and those referring to instances is made based on the syntactical categories of the head nouns – common or proper nouns, respectively.

Subconcept-of Relations. In order to learn subconcept-of relations, we have implemented a variety of different algorithms exploiting the hypernym structure of WordNet [11], matching Hearst patterns [16] in the corpus as well as in the WWW and applying linguistic heuristics mentioned in [35]. The respective confidence values are computed, e.g., based on the frequency of matched patterns, and finally combined through combination strategies as described in [4].

Instance-of Relations. In order to assign instances or named entities appearing in the corpus to a concept in the ontology, Text2Onto relies on a similarity-based approach extracting context vectors for instances and concepts from the text collection. Each instance is then assigned to those concepts whose vectors are most similar to its own one [6], with the respective similarities being interpreted as a confidence values. In addition, we also implemented a pattern-matching algorithm similar to the one used for discovering part-of relations.

General Relations. To learn general relations, Text2Onto employs a shallow parsing strategy to extract subcategorization frames (e.g. hit(subj,obj,pp(with))), transitive verb and prepositional complement) enriched with information about the frequency of the terms appearing as arguments [23]. These subcategorization frames are mapped to relations such as hit(*person,thing*) and hit_with(*person,object*). The confidence is estimated on the basis of the frequency of the subcategorization frame as well as of the frequency with which a certain term appears at the argument position.

Disjointness. For the extraction of disjointness axioms Text2Onto uses a simple heuristic based on lexico-syntactic patterns. Given an enumeration of noun phrases $NP_1, NP_2, ...(and|or)NP_n$ we conclude that the concepts $C_1, C_2, ...C_k$ denoted by these noun phrases are pairwise disjoint, where the confidence for the disjointness of two concepts is obtained from the number of evidences found for their disjointness in relation to the total number of evidences for the disjointness of these concepts with other concepts.

Text2Onto's approach to the acquisition of disjointness axioms (see Section 3.1) is completely unsupervised, hence computationally cheap, but lacks the precision of a more specialized method for learning disjointness. In the following Section 3.2, we therefore introduce a supervised, classification-based approach particularly developed to support the enrichment of ontologies (e.g. automatically acquired by Text2Onto) with disjointness axioms.

3.2   Learning Disjointness Axioms

Our approach to enriching learned or manually engineered ontologies by disjointness axioms relies on a machine learning classifier that determines disjointness of any two

Table 1. Ontology elements learned by Text2Onto

Learned Element	Explanation	OWL
concept	A concept C. Example: *man, person*	C
instance	An instance a. Example: *John, Mary*	a
subconcept-of	Concept inheritance. Example: subconcept-of(*man,person*)	$C_1 \sqsubseteq C_2$
instance-of	Concept instantiation. Example: instance-of(*John,person*).	$C(a)$
relation	A relation R between C_1 and C_2. Example: love(*person,person*)	$C_1 \sqsubseteq \forall R.C_2$
relation	A relation R between C_1 and C_2. Example: drive(*person,car*)	$\top \sqsubseteq \forall R.C_2$ $\top \sqsubseteq \forall R^{-1}.C_1$
disjointness	Disjointness of concepts C_1 and C_2. Example: disjointness(*man,woman*)	$C_1 \sqsubseteq \neg C_2$

classes. The classifier is trained based on a "Gold Standard" of manually created disjointness axioms, i.e. pairs of classes each of which is associated with a label – "disjoint" or "not disjoint" – and a vector of feature values. As in our earlier experiments [36], we used a variety of lexical and logical features, which we believe to provide a solid basis for learning disjointness. These features are used to build an overall classification model on whose basis the classifier can predict disjointness for previously unseen pairs of classes. The classification probabilities generated, e.g., by a Naive Bayes approach are thereby interpreted as confidence values indicating the reliability of the predicted labels.

In the following, we briefly describe the set of methods, which we used for extracting the values of the classification features. All of the methods introduced in the following are implemented within the open-source tool LeDA[2] (*Learning Disjointness Axioms*).

Logical Features

Taxonomic Overlap. In description logics, two classes are disjoint *iff* their "taxonomic overlap", i.e. the set of common individuals *must* be empty. Note that the individuals of a class do not necessarily have to be explicitly named in the ontology. Hence, the taxonomic overlap of two classes is considered not empty as long as there *could* be common individuals within the domain that is modeled by the ontology. Following these considerations, we developed several methods to assess the actual or possible overlap of two extensions. Both of the following formulas are based on the Jaccard similarity coefficient [19]. C denotes the set of all the atomic classes in the ontology, while I refers to the set of named individuals.

$$f_{overlap_i}(c_1, c_2) = \frac{|\{i \in I | c_1(i) \wedge c_2(i)\}|}{|\{i \in I | c_1(i) \vee c_2(i)\}|} \qquad f_{overlap_c}(c_1, c_2) = \frac{|\{c \in C | c \sqsubseteq c_1 \sqcap c_2\}|}{|\{c \in C | c \sqsubseteq c_1 \sqcup c_2\}|}$$

These two features are complemented by f_{sub}, that represents a particular case of taxonomic overlap, while at the same time capturing negative information such as class complements or already existing disjointness contained in the ontology. The value of f_{sub} for any pair of classes c_1 and c_2 is 1 for $c_1 \sqsubseteq c_2 \vee c_2 \sqsubseteq c_1$, 0 for $c_1 \sqsubseteq \neg c_2$ and *undefined* otherwise.

Semantic Distance. The semantic distance between two classes c_1 and c_2 is the minimum length of a path consisting of subsumption relationships between atomic classes that connects c_1 and c_2 (as defined in [36]).

Object Properties. This feature encodes the semantic relatedness of two classes, c_1 and c_2, based on the number of object properties they share. More precisely, we divided the number of properties p with $p(c_1, c_2)$ or $p(c_2, c_1)$ by the number of all properties whose domain subsumes c_1 whereas their range subsumes c_2 or vice-versa. This measure can be seen as a variant of the Jaccard similarity coefficient with object properties considered as undirected edges.

---

[2] http://ontoware.org/projects/leda/

Lexical Features

Label Similarity. The semantic similarity of two classes is in many cases reflected by their labels – especially, in case their labels share a common prefix or postfix. This is because the right-most constituent of an English noun phrase can be assumed to be the lexical head, that determines the syntactic category and usually indicates the semantic type of the noun phrase. A common prefix, on the other hand, often represents a nominal or attribute adjunct which describes some semantic characteristics of the noun phrase referent. In order to compute the lexical similarity of the two class labels, we therefore used three different similarity measures: Levenshtein, QGrams and Jaro-Winkler.

WordNet Similarity. In order to compute the lexical similarity of two classes (their labels, to be precise), we applied two variants of a WordNet-based similarity measure by Patwardhan and Pedersen [27].[3] This similarity measure computes the cosine similarity between vector-based representations of the glosses, that are associated with the two synsets.[4] We omitted any sort of word sense disambiguation at this point, assuming that every class label refers to the most frequently used synset it is contained in.

Features based on Learned Ontology. As an additional source of background knowledge about the classes in our input ontology we used an automatically acquired corpus of Wikipedia articles. By querying Wikipedia for each class label we obtained an initial set of articles some of which were disambiguation pages. We followed all content links and applied a simple word sense disambiguation method in order to obtain the most relevant article for each class: For each class label we considered the article to be most relevant, which had, relative to its length, the highest "terminological overlap" with all of the labels used in the ontology. The resulting corpus of Wikipedia articles was fed into Text2Onto (cf. Section 3.1) to generate an additional background ontology for each of the original ontologies in our data set, consisting of classes, individuals, subsumption and class membership axioms.

Based on this newly acquired background knowledge, we defined the following features: *subsumption, taxonomic overlap of subclasses* and *individuals* – all of these are defined as their counterparts described above – as well as document-based *lexical context similiarity*, which we computed by comparing the Wikipedia article associated with the two classes. This type of similarity is in line with Harris' distributional hypothesis [15] claiming that two words are semantically similar to the extent to which they share syntactic contexts.

3.3   Uncertainty in Ontology Learning

Considering the outcome of an ontology learning process, no matter if it aims to generate lightweight or more expressive ontologies, subjective uncertainty clearly is the most prominent form of imperfection. This uncertainty may arise, e.g., from imprecision of input data and background knowledge, or the inaccuracy of ontology learning

---

[3] http://www.d.umn.edu/~tpederse/similarity.html

[4] In WordNet, a synset is a set of (almost) synonymous words, roughly corresponding to a class or concept in an ontology. A gloss is a textual description of a synset's meaning, that most often also contains usage examples.

methods, which leads to incorrect interpretations of the input data. In any case, an appropriate treatment of both imprecision and uncertainty can have a significant impact on the overall quality and usefulness of a learned ontology as explained in the following.

- Handling uncertainties in the first hand and representing them at the level of intermediate ontology learning results prevents the propagation of errors introduced by crisp modeling decisions.
- Moreover, uncertainty values may help to integrate the acquired knowledge, in particular if multiple methods are used to accomplish identical or overlapping ontology learning tasks.
- Uncertainty values facilitate the inspection and revision of learned ontologies by allowing the user to focus on those parts of the ontology which are considered the least reliable.
- For some applications, e.g., in the domains of medicine or e-government, information about the reliability of acquired knowledge is crucial for reasons of health or security.
- An explicit representation of uncertainty can support the process of handling logical inconsistencies in learned ontologies (cf. Section 4).

There are two principal approaches to modeling uncertainty: as part of the learned ontologies, e.g., using probabilistic extensions to the target knowledge representation formalism, or at a meta-level as application-specific information associated with the learned structures. In any case, certainty (or confidence) values are a kind of self-assessment of ontology learning methods and as such not always correlated with a human rating. However, they are important means to support both, global and local evaluation of the ontology – no matter, if this evaluation is performed automatically or by a human expert.

In our work, we follow the second approach to representing uncertainty: In order to capture provence and certainty information about ontology elements in the learning process, we use our framework described in [33] for representing metalevel information. This framework is based on the observation that domain and metalevel information have distinct universes of discourse. We store the metalevel statements in the domain ontology using *axiom annotations*—ontology statements that are akin to comments in a programming language and that do not affect the semantics of the domain information. We give semantics to this information by translating the metalevel statements from the domain ontology into a *metaview*—an ontology that explicitly talks about the facts in the domain ontology.

In the context of ontology learning, we use the annotations to model the certainty of the system about the correctness of a particular ontology element. In particular, we use a special annotation property *confidence* to indicate how confident the system is about the correctness of an ontology element. The confidences are calculated based on different kinds of evidences provided by the ontology learning algorithms that indicate the correctness and the relevance of ontology elements for the domain in question. They can be considered as a corpus-based support for ontology elements.

Confidence annotations can change if new evidence is detected by the ontology learning algorithms, which increases, for example, the certainty of the system about the correctness or relevance of a particular modeling primitive. Inconsistencies in the ontology which might result from these changes must be resolved as described in Section 4.

4   Handling Inconsistencies in Learned Ontologies

One of the major problems of learning ontologies is the potential introduction of inconsistencies. These inconsistencies are a consequence of the fact that in any ontology learning process the acquired ontologies represent *imperfect* information. Ignoring the fact that learned ontologies contain uncertain and thus potentially contradicting information would result in highly inconsistent ontologies, which do not allow meaningful reasoning. In the following we show how inconsistencies can be dealt with in the process of ontology learning. In particular, we show how the concept of consistent ontology evolution can be applied in the context of ontology learning. The goal of this consistent ontology evolution process is to obtain an ontology that is (1) consistent (to allow meaningful reasoning), and (2) captures the most certain information while disregarding the potentially erroneous information.

Logical consistency addresses the question whether the ontology is "semantically correct", i.e. does not contain contradicting information. We say *logical consistency* is satisfied for an ontology O if O is satisfiable, i.e. if O has a model. Please note that because of the monotonicity of the considered logic, an ontology can only become logically inconsistent by adding axioms: If a set of axioms is satisfiable, it will still be satisfiable when any axiom is deleted. Therefore, we only need to check the consistency for ontology change operations that add axioms to the ontology. Effectively, if $O \cup \{\alpha\}$ is inconsistent, in order to keep the resulting ontology consistent some of the axioms in the ontology O have to be removed or altered.

As we have already discussed in Section 2.2, the most adequate approach to dealing with inconsistencies in ontology learning is by realizing a consistent evolution of the ontology. The goal of consistent ontology evolution is the resolution of a given ontology change in a systematic manner by ensuring the consistency of the whole ontology. It is realized in two steps:

1. *Inconsistency Localization*: This step is responsible for checking the consistency of an ontology with the respect to the ontology consistency definition. Its goal is to find "parts" of the ontology that do not meet consistency conditions;
2. *Change Generation*: This step is responsible for ensuring the consistency of the ontology by generating additional changes to resolve detected inconsistencies.

The first step essentially is a diagnosis process. There are different approaches how to perform the diagnosis step [12]. A typical way to diagnose an inconsistent ontology is to identify a minimal inconsistent subontology, i.e. a minimal set of contradicting axioms. Formally, we call an ontology O' a *minimal inconsistent subontology* of O, if $O' \subseteq O$ and O' is inconsistent and for all O'' with $O'' \subset O'$, O'' is consistent. Intuitively, this definition states that the removal of any axiom from O' will result in a consistent ontology. A simple way of finding a minimal inconsistent subontology is as follows: We start with one candidate ontology containing initially only the axiom that was added to the ontology as part of the change operation. As long as we have not found an inconsistent subontology, we create new candidate ontologies by adding axioms (one at a time) that are in some way connected with the axioms in the candidate ontology. One simple, but useful notion of connectedness is structural connectedness: We say that axioms are structurally connected if they refer to shared ontology entities. Once the minimal inconsistent ontology is found, it is by definition sufficient to remove any of the axioms to resolve the inconsistency.

While the removal of any of the axioms from a minimal inconsistent subontology will resolve the inconsistency, the important question of course is deciding *which* axiom to remove. This problem of only removing dispensable axioms requires some semantic selection functions capturing the relevance of particular axioms. These semantic selection functions can for example exploit information about the confidence in the axioms that allows us to remove "less correct" axioms. In the resolution of the changes we may decide to remove the axioms that have the lowest confidence, i.e. those axioms that are most likely incorrect. We are thus able to incrementally evolve an ontology that is (1) consistent and (2) captures the information with the highest confidence. For details of such a process and evaluation results, we refer the reader to [14].

Based on the discussions above, we can now outline an algorithm (cf. Algorithm 1) to ensure the consistent evolution of a learned ontology.

Algorithm 1. Algorithm for consistent ontology learning

Require: A consistent ontology O
Require: A set of ontology changes OC
 1: **for all** $\alpha \in OC, \text{confidence}(\alpha) \geq t$ **do**
 2:     $O := O \cup \{\alpha\}$
 3:     **while** O is inconsistent **do**
 4:         $O' := \text{minimal\_inconsistent\_subontology}(O, \alpha)$
 5:         $\alpha^- := \alpha$
 6:         **for all** $\alpha' \in O'$ **do**
 7:             **if** $\text{confidence}(\alpha') \leq \text{confidence}(\alpha)$ **then**
 8:                 $\alpha^- := \alpha'$
 9:             **end if**
10:         **end for**
11:         $O := O \setminus \{\alpha^-\}$
12:     **end while**
13: **end for**

Starting with some consistent ontology O, we incrementally add all axioms generated by the ontology learning process – contained in the set of ontology changes OC – whose confidence is equal to or greater than a given threshold t. If adding the axioms leads to an inconsistent ontology, we localize the inconsistency by identifying a minimal inconsistent subontology. Within this minimal inconsistent subontology we then identify the axiom that is most uncertain, i.e. has the lowest confidence value. This axiom will be removed from the ontology, thus resolving the inconsistency. It may be possible that one added axiom introduced multiple inconsistencies. For this case, the above inconsistency resolution has to be applied iteratively.

Besides the general notion of confidence used above, we may rely on various other information to obtain a ranking of the axioms for the resolution of inconsistencies. This information can, for example, be contextual information about *provenance* that may indicate the reliability of the results depending on the trustworthiness and quality of the learning resources. Other alternatives are based on measuring and ranking how much a particular axioms contributes to logical contradictions in the ontology. The intuition here is to minimize the changes needed by first removing axioms that are causing the most inconsistencies, as for example proposed by [21].

5   Evaluation and Experimental Results

In what follows, we describe the experimental evaluation of our approach to consistent ontology evolution as introduced in the previous Section 4. We thereby focus on two different evaluation scenarios: In Section 5.1 we consider the common task of ontology learning from scratch. We specifically look at the relationship between particular degrees of uncertainty and inconsistency, as well as different kinds of logical inconsistencies. Section 5.2 describes the application of our approach to inconsistency handling in the context of ontology enrichment. We add automatically created disjointness axioms to an ontology and analyze the impact the of inconsistency diagnosis and repair on the accuracy of these axioms.

5.1   Consistent Evolution of Learned Ontologies

We applied the ontology learning framework Text2Onto (see Section 3.1) to a corpus of 1,700 abstracts of the BT Digital Library,[5] all of them taken from documents about knowledge management. The learned ontology consists of 938 concepts and 125 instances. For the concepts, 406 subconcept-of relations as well as 2,322 disjoint-concepts relations were identified, while 143 instance-of relations could be obtained for the extracted instances.[6] All these ontology entities and axioms were annotated with confidence values as described in Section 3.3.

We then started our experimental evaluation by studying the impact of an uncertainty threshold on the consistency of the ontology. The results listed by Table 2 show the connection between the level of uncertainty and inconsistency introduced: With an increasing threshold for the certainty, the number of encountered inconsistencies decreases.

A low threshold t results in more uncertain information being allowed in the target ontology. As a result, the chances for inconsistencies increase. How to choose the "right" threshold t for the transformation process will very much depend on the application scenario, as it essentially means finding a trade-off between the amount of information learned and the confidence in the correctness of the learned information.

In the following, we will discuss typical types of inconsistencies and present examples of such inconsistencies that were detected and resolved. The first type of inconsistency involves unsatisfiable concepts (often called incoherent concepts) in the

---

[6] Note that we allowed for multiple classifications of each instance.

Table 2. Influence of certainty threshold t on transformation process

Threshold t	# of Inconsistencies	# of Axioms in Result
0.1	40	1706
0.2	8	705
0.4	3	389
0.8	0	197

$\mathcal{T}$-Box of the ontology. This can for example happen if two concepts are identified to be disjoint, but at the same time these concepts are in a subconcept-relation (either explicitly asserted or inferred). Interestingly, this type of inconsistency often occurred for pairs of concepts whose correct relationship is hard to identify even for a domain expert, as the following example shows:

Example 1. The relationship between the concepts *Data*, *Information*, and *Knowledge* is a very subtle (often philosophical) one, for which one will encounter different definitions depending on the context. The (inconsistent) definitions learned from our data set stated that $Data$ is a subconcept of both $Information$ and $Knowledge$, while $Information$ and $Knowledge$ are disjoint concepts:

Axiom t	Confidence
$Data \sqsubseteq Information$	1.0
$Data \sqsubseteq Knowledge$	1.0
$Information \sqsubseteq \neg Knowledge$	0.7

The inconsistency was resolved by removing the disjointness axiom, as its confidence value was the lowest.

The second type of inconsistencies involves $\mathcal{A}$-Box assertions. Here, typically instances were asserted to be instances of two concepts that were identified to be disjoint. We again present an example:

Example 2. Here $KaViDo$ was identified to be both an instance of $Application$ and a $Tool$ (based on the abstract of [32]), however, $Application$ and $Tool$ were learned to be disjoint concepts:

Axiom t	Confidence
$Application(kavido)$	0.46
$Tool(kavido)$	0.46
$Tool \sqsubseteq \neg Application$	0.3

This inconsistency was again resolved by removing the disjointness axiom.

Other types of inconsistencies involving, for example, domain and range restrictions were not considered in our current experiments, thus being left for future work. Nevertheless, this evaluation showed that inconsistency is an important issue in ontology learning.

5.2   Enriching Ontologies with Disjointness Axioms

For the second part of our experimental evaluation, we enriched EKAW, one of the OntoFarm[7] ontologies, with disjointness axioms automatically created by LeDA (cf. Section 3.2. The EKAW ontology covers the domain of conference organization, containing 77 concepts like publications, researchers, conferences, etc. We enriched the original EKAW ontology (without disjointness axioms) with three complete sets of disjointness axioms each of them automatically generated by a machine learning classifier as described in Section 3.2. These classifiers (we used the Weka implementation of Naive Bayes in all of our experiments) had previously been trained on CMT, CRS and SIGKDD – three other ontologies in our gold standard and fully equipped with manually created disjointness.

Table 3. Results of debugging learned disjointness axioms

Test	Training	Axioms	Acc	Acc_{base}	Acc_{debug}	Unsatisfiable Concepts	Removed Axioms
EKAW	CMT	2,659	0.925	0.851	0.934	27	20
EKAW	CRS	2,712	0.892	0.851	0.933	130	89
EKAW	SIGKDD	2,418	0.904	0.851	0.941	43	20

When adding the learned disjointness axioms to the ontology, we applied the method of consistent ontology evolution (cf. Section 4). Table 3 shows the results for the enriched ontologies, including the number of unsatisfiable concepts identified and the number of axioms that had to be removed in order to maintain consistency. The table also shows that the accuracy of the resulting disjointness axioms (Acc_{debug}) in the ontology obtained by the consistent ontology evolution is significantly better than the accuracy of the entire set of disjointness axioms (Acc). It increased, e.g., from 90.4% to 94.1% in the case training was performed on SIGKDD. The majority baseline (Acc_{base}), however, is beaten by both the (Acc) and Acc_{debug}.

As in our previous experiments [36], we chose a majority baseline for accuracy (Acc_{base}), which is defined as the number of examples in the majority class (e.g. "not disjoint") divided by the overall number of examples. The majority baseline represents the performance level that would be achieved by a naïve classifier that labels all entities in the test set with the majority class, i.e. "disjoint" for all ontologies in our data set. This simple, yet efficient strategy is hard to beat, especially for data sets that are relatively unbalanced and biased towards one of the target classes.

Example. The variant of EKAW that was generated after previous training on CRS contains 130 unsatisfiable concepts, among them $Contributed\_Talk$. The following Table 4 shows the three diagnoses (minimal sets of conflicting axioms) we obtained for this concept as well as associated confidence values, i.e. probabilities generated by our NaiveBayes classifier. Please note that only the learned disjointness axioms are annotated with confidence values and are considered as candidates for removal in the repair process.

---

[7] http://nb.vse.cz/~svatek/ontofarm.html

Table 4. Example for unsatisfiable concept

Axiom t	Confidence
$Abstract \sqsubseteq \neg Camera\_Ready\_Paper$	0.9999992
$presentationOfPaper \equiv paperPresentedAs^{-1}$	–
$\top \sqsubseteq paperPresentedAs^{-1}.(Abstract \sqcup Paper)$	–
$Camera\_Ready\_Paper \sqsubseteq \neg Paper$	0.9996211
$Contributed\_Talk \sqsubseteq \exists presentationOfPaper.Camera\_Ready\_Paper$	–
$Camera\_Ready\_Paper \sqsubseteq Paper$	–
$Camera\_Ready\_Paper \sqsubseteq \neg Paper$	0.9996211
$Contributed\_Talk \sqsubseteq \exists presentationOfPaper.Camera\_Ready\_Paper$	–
$Abstract \sqsubseteq \neg Camera\_Ready\_Paper$	0.9999992
$\top \sqsubseteq \exists presentationOfPaper.(Abstract \sqcup Paper)$	–
$Camera\_Ready\_Paper \sqsubseteq \neg Paper$	0.9996211
$Contributed\_Talk \sqsubseteq \exists presentationOfPaper.Camera\_Ready\_Paper$	–

In order to resolve these incoherences, we removed the following axiom from the ontology: $Camera\_Ready\_Paper \sqsubseteq \neg Paper$. Note that all decisions upon the axioms to be removed were made based on their confidence values, our algorithm removed this axiom due to its comparatively lower probability of 0.9996211. This decision seems plausible as every camera ready paper should naturally be a paper.

6   Related Work

Since building an ontology for large amounts of data is a difficult and time consuming task a number of ontology learning frameworks such as TextToOnto [24], the ASIUM system [10], the Mo'k Workbench [2], OntoLearn [35] or OntoLT [3] have been developed in order to support the user in constructing ontologies from a given set of textual data. So far, very few of these frameworks, with the notable exception of TextToOnto's successor, Text2Onto (cf. Section 3.1), for example, explicitly address the problem of uncertainty. But the problem of dealing with uncertainty in taxonomy induction and the acquisition of lexical knowledge has been dealt with in various publications (see e.g. [30]). Moreover, several researchers have already tackled the issue of integrating and reasoning with probabilities in knowledge representation formalisms. Ding and Peng [8] for example present a probabilistic extension of the Ontology Language OWL which relies on Bayesian Networks for reasoning. Other researchers have integrated probabilities into first-order logic [1] or description logics [22]. Fuzzy extensions of OWL have been proposed, e.g., in [31].

The approach to dealing with inconsistencies presented in this work is based on the idea of using consistent ontology evolution in order to be able to perform meaningful reasoning. A very related approach is that of diagnosis and repair of inconsistencies. The problem of diagnosing inconsistent ontologies has received increased attention, which is a prerequisite for a successful repair. For example, Schlobach and colleagues [29] propose a diagnosis based on identifying MIPS (Minimal Incoherence Preserving Sub-$\mathcal{T}$-Boxes) and MUPS (Minimal Unsatisfiability Preserving Sub-$\mathcal{T}$-Boxes), which are used to generate explanations for unsatisfiable concepts. A similar approach is also taken by the Pellet reasoner within the SWOOP ontology editor [26,21], which offers a debugging mode in which explanations of inconsistencies are provided as a result of the reasoning process. After performing the diagnosis, the important question is how the inconsistency should be resolved, i.e. which axioms should be removed. For this problem [21] proposes a number of different ways to rank axioms. The confidence values used in our approach could in principle be used for such a ranking.

Another related approach is that of reasoning with inconsistent ontologies. A typical technique is the selection of a consistent subontology for a given query, which yields a consistent query answer (cf. [18]). Also here the question is how to select the *right* subontology. While current techniques often rely on syntactic selection functions, it would again be possible to rely on the confidence annotations to guide the selection function.

7   Conclusion and Future Work

Ontology learning is a promising technique for automated knowledge acquisition. However, uncertainty and inconsistency are issues that need to be dealt with in order to enable meaningful reasoning over the learned ontologies. While the challenge of dealing with inconsistencies is a general one for expressive ontologies, it is especially important in the context of ontology learning. This is essentially because all ontology learning approaches generate knowledge that is afflicted with various forms of imperfection. To address this problem, we described a way to represent uncertainty resulting from ontology learning methods and proposed an approach to handling inconsistency in learned ontologies. In our approach, ontology learning is combined with consistent ontology evolution, in order to ensure that the learned ontologies are kept consistent as the ontology learning procedures generate changes to the ontology over time. Our experiments with both learned and enriched ontologies show the feasibility and usefulness of the approach. The results confirm the intuition that the less certain the generated knowledge is, the more logical contradictions are produced and need to be resolved. At the same time, exploiting the uncertainty information in the evolution process considerably improves the quality of the resulting ontologies.

As future work, we intend to rely on various other forms of contextual information – such as information about trust and provenance – in order to obtain a ranking of the axioms for the resolution of inconsistencies. In addition, we plan to investigate ways to feed back information about logical inconsistencies into the ontology learning process, e.g., to improve an existing classification model. An approach as described in Section 3.2, for instance, could benefit from such information, in that multiple iterations of learning and debugging would improve the quality of the acquired disjointness axioms.

Acknowledgements. Research reported in this paper has been financed by the EU in the IST projects SEKT (IST-2003-506826) and NeOn (IST-2006-027595).

References

1. Bacchus, F.: Representing and Reasoning with Probabilistic Knowledge. MIT Press, Cambridge (1990)
2. Bisson, G., Nedellec, C., Canamero, L.: Designing clustering methods for ontology building - The Mo'K workbench. In: Proc. of the ECAI Ontology Learning WS (2000)

3. Buitelaar, P., Olejnik, D., Sintek, M.: OntoLT: A protégé plug-in for ontology extraction from text. In: Proceedings of the International Semantic Web Conference (ISWC) (2003)

4. Cimiano, P., Pivk, A., Schmidt-Thieme, L., Staab, S.: Learning taxonomic relations from heterogeneous sources of evidence. In: Ontology Learning from Text: Methods, Applications and Evaluation. IOS Press, Amsterdam (2005)

5. Cimiano, P., Völker, J.: Text2onto - a framework for ontology learning and data-driven change discovery. In: Montoyo, A., Muñoz, R., Métais, E. (eds.) NLDB 2005. LNCS, vol. 3513, pp. 227–238. Springer, Heidelberg (2005)

6. Cimiano, P., Völker, J.: Towards large-scale, open-domain and ontology-based named entity classification. In: Proceedings of the International Conference on Recent Advances in Natural Language Processing (RANLP 2005) (September 2005)

7. da Costa, P.C.G., Laskey, K.B.: PR-OWL: A framework for probabilistic ontologies. In: Proceedings of the International Conference on Formal Ontology in Information Systems (2006)

8. Ding, Z., Peng, Y.: A probabilistic extension to ontology language OWL. In: Proceedings of the 37th Hawaii International Conference on System Sciences (2004)

9. Ding, Z., Peng, Y., Pan, R.: BayesOWL: Uncertainty Modeling in Semantic Web Ontologies. Studies in Fuzziness and Soft Computing, p. 27. Springer, Heidelberg (2005)

10. Faure, D., Nedellec, C.: A corpus-based conceptual clustering method for verb frames and ontology. In: Proceedings of the LREC Workshop on Adapting lexical and corpus resources to sublanguages and applications (1998)

11. Fellbaum, C.: WordNet, an electronic lexical database. MIT Press, Cambridge (1998)

12. Haase, P., Qi, G.: An analysis of approaches to resolving inconsistencies in dl-based ontologies. In: Proceedings of International Workshop on Ontology Dynamics (IWOD 2007) (June 2007)

13. Haase, P., van Harmelen, F., Huang, Z., Stuckenschmidt, H., Sure, Y.: A framework for handling inconsistency in changing ontologies. In: Gil, Y., Motta, E., Benjamins, V.R., Musen, M.A. (eds.) ISWC 2005. LNCS, vol. 3729, pp. 353–367. Springer, Heidelberg (2005)

14. Haase, P., Völker, J.: Ontology learning and reasoning – dealing with uncertainty and inconsistency. In: da Costa, P.C.G., Laskey, K.B., Laskey, K.J., Pool, M. (eds.) Proceedings of the Workshop on Uncertainty Reasoning for the Semantic Web (URSW), pp. 45–55 (November 2005)

15. Harris, Z.: Distributional structure. In: Katz, J.J. (ed.) The Philosophy of Linguistics, New York, pp. 26–47. Oxford University Press, Oxford (1985)

16. Hearst, M.A.: Automatic acquisition of hyponyms from large text corpora. In: Proceedings of the 14th International Conference on Computational Linguistics, pp. 539–545 (1992)

17. Horrocks, I., Patel-Schneider, P.F.: Reducing OWL Entailment to Description Logic Satisfiability. Journal of Web Semantics 1(4) (2004)

18. Huang, Z., van Harmelen, F., ten Teije, A.: Reasoning with inconsistent ontologies. In: Proceedings of IJCAI 2005 (August 2005)

19. Jaccard, P.: The distribution of flora in the alpine zone 11, 37–50 (1912)

20. Tsuji, J., Frantzi, K., Ananiadou, S.: The c-value/nc-value method of automatic recognition for multi -word terms. In: Proceedings of the ECDL, pp. 585–604 (1998)

21. Kalyanpur, A., Parsia, B., Sirin, E., Grau, B.C.: Repairing unsatisfiable concepts in owl ontologies. In: Sure, Y., Domingue, J. (eds.) ESWC 2006. LNCS, vol. 4011, pp. 170–184. Springer, Heidelberg (2006)

22. Koller, D., Levy, A., Pfeffer, A.: P-classic: A tractable probabilistic description logic. In: Proceedings of AAAI 1997, pp. 390–397 (1997)

23. Maedche, A., Staab, S.: Discovering conceptual relations from text. In: Horn, W. (ed.) Proceedings of the 14th ECAI 2000 (2000)

24. Maedche, A., Staab, S.: Ontology learning. In: Staab, S., Studer, R. (eds.) Handbook on Ontologies, pp. 173–189. Springer, Heidelberg (2004)

25. Motro, A., Smets, P.: Uncertainty Management In Information Systems. Springer, Heidelberg (1997)
26. Parsia, B., Sirin, E., Kalyanpur, A.: Debugging OWL ontologies. In: Proceedings of the 14th international conference on World Wide Web, WWW 2005, Chiba, Japan, May 10-14, 2005, pp. 633–640 (2005)
27. Patwardhan, S., Banerjee, S., Pedersen, T.: Using measures of semantic relatedness for word sense disambiguation. In: Proceedings of the Fourth International Conference on Intelligent Text Processing and Computational Linguistics, pp. 241–257 (February 2003)
28. Schlobach, S.: Debugging and semantic clarification by pinpointing. In: Gómez-Pérez, A., Euzenat, J. (eds.) ESWC 2005. LNCS, vol. 3532, pp. 226–240. Springer, Heidelberg (2005)
29. Schlobach, S.: Diagnosing terminologies. In: Veloso, M.M., Kambhampati, S. (eds.) AAAI, pp. 670–675. AAAI Press / The MIT Press (2005)
30. Snow, R., Jurafsky, D., Ng, A.Y.: Semantic taxonomy induction from heterogenous evidence. In: Proceedings of the 21st International Conference on Computational Linguistics and the 44th annual meeting of the ACL, Morristown, NJ, USA, pp. 801–808. Association for Computational Linguistics (2006)
31. Straccia, U.: Towards a fuzzy description logic for the semantic web (preliminary report). In: Proceedings of the Second European Semantic Web Conference, 2005, pp. 167–181 (2005)
32. Tamine, O., Dillmann, R.: Kavido: a web-based system for collaborative research and development processes. Computers in Industry 52(1), 29–45 (2003)
33. Tran, D.T., Haase, P., Motik, B., Grau, B.C., Horrocks, I.: Metalevel information in ontology-based applications. In: Proceedings of the 23th AAAI Conference on Artificial Intelligence (AAAI 2008), Chicago, USA (July 2008)
34. Thanh Tran, D., Haase, P., Motik, B., Cuenca Grau, B., Horrocks, I.: Metalevel information in ontology-based applications. In: Proceedings of the 23th AAAI Conference on Artificial Intelligence (AAAI 2008), Chicago, USA (July 2008)
35. Velardi, P., Navigli, R., Cuchiarelli, A., Neri, F.: Evaluation of ontolearn, a methodology for automatic population of domain ontologies. In: Ontology Learning from Text: Methods, Applications and Evaluation. IOS Press, Amsterdam (2005)
36. Völker, J., Vrandecic, D., Sure, Y., Hotho, A.: Learning disjointness. In: Franconi, E., Kifer, M., May, W. (eds.) ESWC 2007. LNCS, vol. 4519, pp. 175–189. Springer, Heidelberg (2007)

Uncertainty Reasoning for Ontologies with General TBoxes in Description Logic

Volker Haarslev, Hsueh-Ieng Pai, and Nematollaah Shiri

Concordia University
Dept. of Computer Science & Software Engineering
Montreal, Quebec, Canada
{haarslev,hsueh_pa,shiri}@cse.concordia.ca

Abstract. We present a reasoning procedure for ontologies with uncertainty described in Description Logic (DL) which include General TBoxes, i.e., include cycles and General Concept Inclusions (GCIs). For this, we consider the description language $\mathcal{ALC}_U$, in which uncertainty parameters are associated with ABoxes and TBoxes, and which allows General TBoxes. Using this language as a basis, we then present a tableau algorithm which encodes the semantics of the input knowledge base as a set of assertions and linear and/or nonlinear arithmetic constraints on certainty variables. By tuning the uncertainty parameters in the knowledge base, different notions of uncertainty can be modeled and reasoned with, within the same framework. Our reasoning procedure is deterministic, and hence avoids possible empirical intractability in standard DL with General TBoxes. We further illustrate the need for blocking when reasoning with General TBoxes in the context of $\mathcal{ALC}_U$.

1 Introduction

Over the last few years, a number of ontology languages have been developed to help make Web resources more machine-interpretable by giving Web resources a well-defined meaning. Among these languages, the OWL Web Ontology Language [26] is the most recent W3C Recommendation. One of its species, OWL DL, is named because of its correspondence with *Description Logics* (DLs) [1].

The family of DLs is mostly a subset of first-order logic (FOL) that is considered to be attractive because it keeps a good compromise between the expressive power and the computational tractability [1]. The standard DLs, such as the one that OWL DL is based on, focus on the classical logic, which is more suitable to describe concepts that are crisp and well-defined in nature. However, in real-world applications, uncertainty is everywhere. In the context of this paper, *uncertainty* refers to a form of deficiency or imperfection in the information for which the truth of such information is not established definitely [15]. The need to model and reason with uncertainty in DLs has been found in many different Semantic Web contexts, such as Semantic Web services, multimedia annotation, and bioinformatics. For example, in an online medical diagnosis system, one

P.C.G. da Costa et al. (Eds.): URSW 2005-2007, LNAI 5327, pp. 385–402, 2008.

might want to express that the certainty of an obese person having a parent who is obese lies between 0.7 and 1, and John is an obese person with a degree between 0.8 and 1. Such knowledge cannot be expressed nor be reasoned with the standard DLs.

Over the last decade, a number of frameworks have been proposed which extend the standard DLs with uncertainty [5,6,7,13,14,19,20,21,22,23,24,25]. Some of them deal with only vagueness while others deal with only probabilistic knowledge. Since different applications may require to model different notions of uncertainty, and there may be situations in which one needs to model different notions of uncertainty within the same application, we are interested in developing a framework such that different forms of uncertainty knowledge can be represented and reasoned with, in a generic way.

In this paper, we propose a reasoning procedure for uncertainty knowledge bases with General TBoxes by taking a generic approach. As in the standard description languages, a General TBox augments the expressive power of the language by allowing cycles and General Concept Inclusions (GCIs) in the TBox, which in turn allows statements such as domain and range restrictions to be expressed. The proposed reasoning procedure is based on the DL $\mathcal{ALC}_U$, which extends the standard DL $\mathcal{ALC}$ with uncertainty. Inspired by the approach of the parametric framework [16] which incorporates uncertainty in the standard logic programming and deductive databases, the interesting feature of our approach is that, by tuning the uncertainty parameters that are associated with the axioms and assertions in the $\mathcal{ALC}_U$ knowledge bases, different notions of uncertainty can be modeled and reasoned with, using a single reasoning procedure. In addition, since the proposed reasoning procedure is deterministic, it avoids possible empirical intractability in standard DL reasoning when dealing with General TBoxes.

This paper is an extension of our previous work as follows. In [8], we presented a generic framework for representing DLs with uncertainty. That work was further extended in [10] with a core reasoning procedure. A reasoning procedure for dealing with acyclic uncertainty knowledge bases was presented in [9]. In this paper, we further extend [9] by presenting a reasoning procedure that supports uncertainty knowledge bases with General TBoxes.

The rest of this paper is organized as follows. We next review the standard DL $\mathcal{ALC}$ and related work. Section 3 presents the DL $\mathcal{ALC}_U$ and the proposed tableau reasoning algorithm. A detailed example is provided in Section 4 to illustrate the need for blocking when reasoning with $\mathcal{ALC}_U$ knowledge bases which contain General TBoxes. Concluding remarks together with future works are discussed in Section 5.

2   Background and Related Work

In this section, we first review the standard DL language $\mathcal{ALC}$, which is the basis for $\mathcal{ALC}_U$, a generic DL language we proposed which unifies DL frameworks with uncertainty. We will also review related work in this context.

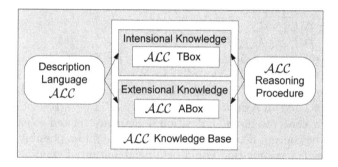

Fig. 1. The $\mathcal{ALC}$ Framework

2.1 Overview of the DL $\mathcal{ALC}$

Description logics form a family of knowledge representation languages that can be used to represent the knowledge of an application domain using concept *descriptions* and have *logic*-based semantics [1,3]. The DL fragment on which we focus in this paper is called $\mathcal{ALC}$, which corresponds to the propositional multi-modal logic $K_{(m)}$ [18]. As shown in Fig. 1, the $\mathcal{ALC}$ framework consists of three main components – the description language, the knowledge base, and the reasoning procedure.

1. $\mathcal{ALC}$ *Description Language:* Every description language has elementary descriptions which include atomic concepts (unary predicates) and atomic roles (binary predicates). Complex descriptions can then be built inductively from concept constructors. The description language $\mathcal{ALC}$ consists of a set of language constructors that are of practical interest. Specifically, let R be a role name, the syntax of a concept description (denoted C or D) in $\mathcal{ALC}$ is described as follows, where the name of each rule is given in parenthesis.

    $C, D \rightarrow A$ (Atomic Concept) |
    $\qquad \neg C$ (Concept Negation) |
    $\qquad C \sqcap D$ (Concept Conjunction) |
    $\qquad C \sqcup D$ (Concept Disjunction) |
    $\qquad \exists R.C$ (Role Exists Restriction) |
    $\qquad \forall R.C$ (Role Value Restriction)

    For example, let *Person* be an atomic concept and *hasParent* be a role. Then $\forall hasParent.Person$ is a concept description. We use $\top$ as a synonym for $A \sqcup \neg A$, and $\bot$ as a synonym for $A \sqcap \neg A$.

    The semantics of the description language is defined using the notion of interpretation. An interpretation $\mathcal{I}$ is a pair $\mathcal{I} = (\Delta^{\mathcal{I}}, \cdot^{\mathcal{I}})$, where $\Delta^{\mathcal{I}}$ is a non-empty domain of the interpretation, and $\cdot^{\mathcal{I}}$ is an interpretation function that maps each atomic concept A to a set $A^{\mathcal{I}} \subseteq \Delta^{\mathcal{I}}$, each atomic role R to a binary relation $R^{\mathcal{I}} \subseteq \Delta^{\mathcal{I}} \times \Delta^{\mathcal{I}}$, and each individual name a to an element $a \in \Delta^{\mathcal{I}}$. The interpretations of concept descriptions are shown below:

$$(\neg C)^{\mathcal{I}} = \Delta^{\mathcal{I}} \backslash C^{\mathcal{I}}$$
$$(C \sqcap D)^{\mathcal{I}} = C^{\mathcal{I}} \cap D^{\mathcal{I}}$$
$$(C \sqcup D)^{\mathcal{I}} = C^{\mathcal{I}} \cup D^{\mathcal{I}}$$
$$(\exists R.C)^{\mathcal{I}} = \{a \in \Delta^{\mathcal{I}} | \ \exists b \in \Delta^{\mathcal{I}} : (a,b) \in R^{\mathcal{I}} \wedge b \in C^{\mathcal{I}}\}$$
$$(\forall R.C)^{\mathcal{I}} = \{a \in \Delta^{\mathcal{I}} | \ \forall b \in \Delta^{\mathcal{I}} : (a,b) \in R^{\mathcal{I}} \rightarrow b \in C^{\mathcal{I}}\}$$

2. $\mathcal{ALC}$ *Knowledge Base:* The knowledge base is composed of a Terminological Box (TBox) and an Assertional Box (ABox). A TBox $\mathcal{T}$ is a set of statements about how concepts in an application domain are related to each other. Let C and D be concept descriptions. The TBox is a finite, possibly empty, set of terminological axioms that could be a combination of *concept inclusions* of the form $\langle C \sqsubseteq D \rangle$ (that is, C is subsumed by D) and *concept equations* of the form $\langle C \equiv D \rangle$ (that is, C is equivalent to D). A General Concept Inclusion (GCI) is a special kind of concept inclusion where the left hand side of the axiom is not restricted to be a concept name but can be an arbitrary concept description. An interpretation $\mathcal{I}$ satisfies $\langle C \sqsubseteq D \rangle$ if $C^{\mathcal{I}} \subseteq D^{\mathcal{I}}$, and it satisfies $\langle C \equiv D \rangle$ if $C^{\mathcal{I}} = D^{\mathcal{I}}$. An interpretation $\mathcal{I}$ satisfies a TBox $\mathcal{T}$ iff $\mathcal{I}$ satisfies every axiom in $\mathcal{T}$.

   An ABox is a set of statements that describe a specific state of affairs in an application domain, with respect to some individuals, in terms of concepts and roles. Let a and b be individuals, C be a concept, R be a role, and let ":" denote "is an instance of". An ABox includes of a set of assertions that could be a combination of concept assertions of the form $\langle a : C \rangle$ and role assertions of the form $\langle (a,b) : R \rangle$. An interpretation $\mathcal{I}$ satisfies $\langle a : C \rangle$ if $a^{\mathcal{I}} \in C^{\mathcal{I}}$, and it satisfies $\langle (a,b) : R \rangle$ if $(a^{\mathcal{I}}, b^{\mathcal{I}}) \in R^{\mathcal{I}}$. An interpretation $\mathcal{I}$ satisfies an ABox $\mathcal{A}$, iff it satisfies every assertion in $\mathcal{A}$ with respect to a TBox $\mathcal{T}$.

   An interpretation $\mathcal{I}$ *satisfies* (or is a *model* of) a knowledge base $\Sigma = \langle \mathcal{T}, \mathcal{A} \rangle$ (denoted $\mathcal{I} \models \Sigma$), iff it satisfies both components of Σ. The knowledge base Σ is *consistent* if there exists an interpretation $\mathcal{I}$ that satisfies Σ. We say that Σ is *inconsistent* otherwise.

3. $\mathcal{ALC}$ *Reasoning Procedure:* Most DL systems use tableau-based reasoning procedure (called tableau algorithm) to provide reasoning services [1]. The main reasoning services include (i) the consistency problem which checks if the ABox is consistent with respect to the TBox, (ii) the entailment problem which checks if an assertion is entailed by a knowledge base, (iii) the concept satisfiability problem which checks if a concept is satisfiable with respect to a TBox, and (iv) the subsumption problem which checks if a concept is subsumed by another concept with respect to a TBox. All these reasoning services can be reduced to the consistency problem [1]. The tableau algorithm can be used to check consistency of the knowledge base Σ. It tries to construct a model by iteratively applying a set of so-called completion rules in arbitrary order. Each completion rule application adds one or more additional inferred assertions to the ABox to make it explicit the knowledge that was previously present implicitly. The algorithm terminates when no further completion rule is applicable. If one could arrive a completion that contains no contradiction (also known as clash), then the knowledge base is consistent. Otherwise, the knowledge base is inconsistent.

2.2   Related Work

Incorporating uncertainty reasoning in DL frameworks has been the topic of numerous research for more than a decade [5,6,7,13,14,19,20,21,22,23,24,25]. Some research extended the tableau-based reasoning procedure used in standard DLs, some transformed the uncertainty knowledge bases into standard DL knowledge bases, while others employed completely different reasoning procedures such as the inference algorithm developed for Bayesian networks. A survey of these frameworks can be found in Chapter 6 of [1] and in [12].

Although General TBoxes were supported in [4,5,19,20,24], there are some major differences between these works and the one we present in this paper. Unlike our reasoning procedure which encode the semantics of the knowledge base as uncertainty constraints, the one proposed in [4] transforms the fuzzy knowledge bases into standard knowledge bases. On the other hand, the reasoning procedure presented in [19,20] deal with the certainty values directly within the tableau algorithm. Finally, although constraint-based reasoning procedures were also proposed in [5,24], the main difference between these two and ours is that, while our approach is to develop one reasoning procedure for dealing with uncertainty with different mathematical foundations, these works mainly considered one form. Specifically, [5] supports only product t-norm, and [24] supports only Lukasiewicz semantics.

3   A Reasoning Procedure for $\mathcal{ALC}_U$ Knowledge Bases with General TBoxes

In this section, we present a reasoning procedure for $\mathcal{ALC}_U$ knowledge bases with General TBoxes. For this, we first introduce the DL $\mathcal{ALC}_U$, which extends the standard DL $\mathcal{ALC}$ with uncertainty, by presenting the syntax and semantics of its description language and knowledge base. We then present the proposed $\mathcal{ALC}_U$ reasoning procedure. After that, we illustrate through an example the various extended components of the $\mathcal{ALC}_U$ framework.

3.1   The Description Language $\mathcal{ALC}_U$

The description language refers to the language used for building concepts. The syntax of the $\mathcal{ALC}_U$ description language is identical to that of the standard $\mathcal{ALC}$, while the corresponding semantics is extended with uncertainty.

In order to model different notions of uncertainty, we assume that the certainty values form a complete lattice $\mathcal{L} = \langle \mathcal{V}, \preceq \rangle$, where $\mathcal{V}$ is the certainty domain, and $\preceq$ is the partial order on $\mathcal{V}$. Also, $\prec, \succeq, \succ$, and $=$ are used with their obvious meanings. We use l to denote the least element in $\mathcal{V}$, t for the greatest element in $\mathcal{V}$, $\oplus$ for the join operator (the least upper bound) in $\mathcal{L}$, and $\otimes$ for the meet operator (the greatest lower bound). We also assume that there is only one underlying certainty lattice for the entire knowledge base.

The semantics of the description language is based on the notion of an interpretation. An interpretation $\mathcal{I}$ is defined as a pair $(\Delta^{\mathcal{I}}, \cdot^{\mathcal{I}})$, where $\Delta^{\mathcal{I}}$ is the domain and $\cdot^{\mathcal{I}}$ is an interpretation function that maps each

Table 1. Syntax and Semantics of the Description Language $\mathcal{ALC}_U$

Name	Syntax	Semantics ($a \in \Delta^{\mathcal{I}}$)
Top Concept	$\top$	$\top^{\mathcal{I}}(a) = t$
Bottom Concept	$\bot$	$\bot^{\mathcal{I}}(a) = l$
Concept Negation	$\neg C$	$(\neg C)^{\mathcal{I}}(a) =\sim C^{\mathcal{I}}(a)$
Concept Conjunction	$C \sqcap D$	$(C \sqcap D)^{\mathcal{I}}(a) = f_c(C^{\mathcal{I}}(a), D^{\mathcal{I}}(a))$
Concept Disjunction	$C \sqcup D$	$(C \sqcup D)^{\mathcal{I}}(a) = f_d(C^{\mathcal{I}}(a), D^{\mathcal{I}}(a))$
Role Exists Restriction	$\exists R.C$	$(\exists R.C)^{\mathcal{I}}(a) = \oplus_{b \in \Delta^{\mathcal{I}}}\{f_c(R^{\mathcal{I}}(a,b), C^{\mathcal{I}}(b))\}$
Role Value Restriction	$\forall R.C$	$(\forall R.C)^{\mathcal{I}}(a) = \otimes_{b \in \Delta^{\mathcal{I}}}\{f_d(\sim R^{\mathcal{I}}(a,b), C^{\mathcal{I}}(b))\}$

- atomic concept A into a certainty function CF_C, where $CF_C : \Delta^{\mathcal{I}} \to \mathcal{V}$
- atomic role R into a certainty function CF_R, where $CF_R : \Delta^{\mathcal{I}} \times \Delta^{\mathcal{I}} \to \mathcal{V}$
- individual name a to an element $a \in \Delta^{\mathcal{I}}$

where $\mathcal{V}$ is the certainty domain. For example, let *John* be an individual name and *Obese* be an atomic concept. Then, $Obese^{\mathcal{I}}(John^{\mathcal{I}})$ gives the certainty that *John* is an instance of the concept *Obese*. The syntax and semantics of the description language $\mathcal{ALC}_U$ are summarized in Table 1.

The interpretation of the top concept $\top$ is the greatest element in the certainty domain $\mathcal{V}$, that is, $\top^{\mathcal{I}}(a) = t$, for all $a \in \Delta^{\mathcal{I}}$. For instance, the interpretation of $\top$ is 1 (or true) in the standard logic with $\mathcal{V} = \{0, 1\}$, as well as in other logics with $\mathcal{V} = [0, 1]$. Similarly, the interpretation of the bottom concept $\bot$ is the least element in the certainty domain $\mathcal{V}$, that is, $\bot^{\mathcal{I}}(a) = l$, for all $a \in \Delta^{\mathcal{I}}$. For example, this corresponds to 0 (or false) in the standard logic with $\mathcal{V} = \{0, 1\}$, as well as in other logics with $\mathcal{V} = [0, 1]$.

The semantics of the concept negation $\neg C$ is defined as $(\neg C)^{\mathcal{I}}(a) =\sim C^{\mathcal{I}}(a)$, for all $a \in \Delta^{\mathcal{I}}$. The symbol $\sim$ denotes the negation function, where $\sim: \mathcal{V} \to \mathcal{V}$ must satisfy the following properties:

- Boundary Conditions: $\sim l = t$ and $\sim t = l$.
- Double Negation: $\sim(\sim \alpha) = \alpha$, for all $\alpha \in \mathcal{V}$.

For example, a common interpretation of $\neg C$ is $1 - C^{\mathcal{I}}(a)$.

In addition, f_c and f_d in Table 1 denote the conjunction and disjunction functions, respectively, both of which we refer as the *combination functions*. They are used to specify how one should interpret a given description language. A combination function f is a binary function from $\mathcal{V} \times \mathcal{V}$ to $\mathcal{V}$. This function combines a pair of certainty values into one. A combination function must satisfy some properties as listed in Table 2 [16].

A *conjunction function* f_c is a combination function that satisfies properties P_1, P_2, P_5, P_6, P_7, and P_8 as described in Table 2. The monotonicity property asserts that increasing the certainties of the arguments in f improves the certainty that f returns. The bounded value and boundary condition properties are included so that the interpretation of the certainty values makes sense. The commutativity property allows reordering of the arguments of f, say for optimization purposes. Finally, the associativity of f ensures that different evaluation orders of

Table 2. Combination Function Properties

ID	Property Name	Property Definition
P_1	Monotonicity	$f(\alpha_1, \alpha_2) \preceq f(\beta_1, \beta_2)$ if $\alpha_i \preceq \beta_i$, for $i = 1, 2$
P_2	Bounded Above	$f(\alpha_1, \alpha_2) \preceq \alpha_i$, for $i = 1, 2$
P_3	Bounded Below	$f(\alpha_1, \alpha_2) \succeq \alpha_i$, for $i = 1, 2$
P_4	Boundary Condition (Above)	$\forall \alpha \in \mathcal{V},\ f(\alpha, l) = \alpha$ and $f(\alpha, t) = t$
P_5	Boundary Condition (Below)	$\forall \alpha \in \mathcal{V},\ f(\alpha, t) = \alpha$ and $f(\alpha, l) = l$
P_6	Continuity	f is continuous w.r.t. each of its arguments
P_7	Commutativity	$\forall \alpha, \beta \in \mathcal{V},\ f(\alpha, \beta) = f(\beta, \alpha)$
P_8	Associativity	$\forall \alpha, \beta, \delta \in \mathcal{V},\ f(\alpha, f(\beta, \delta)) = f(f(\alpha, \beta), \delta)$

concept conjunctions will not yield different results. Some common conjunction functions are the well-known minimum function and the algebraic product.

A *disjunction function* f_d is a combination function that satisfies properties P_1, P_3, P_4, P_6, P_7, and P_8 as described in Table 2. These properties are enforced for similar reasons as in the conjunction case. Some common disjunction functions are the maximum and the probability independent functions.

In Table 1, the semantics of the Role Exists Restriction $\exists R.C$ is defined as $(\exists R.C)^{\mathcal{I}}(a) = \oplus_{b \in \Delta^{\mathcal{I}}} \{f_c(R^{\mathcal{I}}(a, b), C^{\mathcal{I}}(b))\}$, for all $a \in \Delta^{\mathcal{I}}$. The intuition here is that $\exists R.C$ is viewed as the open first order formula $\exists b.\ R(a, b) \wedge C(b)$, where $\exists$ is viewed as a disjunction over certainty values associated with $R(a, b) \wedge C(b)$. Specifically, the semantics of $R(a, b) \wedge C(b)$ is captured using the conjunction function $f_c(R^{\mathcal{I}}(a, b), C^{\mathcal{I}}(b))$, and $\exists b$ is captured using the join operator in the certainty lattice $\oplus_{b \in \Delta^{\mathcal{I}}}$.

Similarly, the semantics of the Role Value Restriction $\forall R.C$ is defined as $(\forall R.C)^{\mathcal{I}}(a) = \otimes_{b \in \Delta^{\mathcal{I}}} \{f_d(\sim R^{\mathcal{I}}(a, b), C^{\mathcal{I}}(b))\}$, for all $a \in \Delta^{\mathcal{I}}$. The intuition here is that $\forall R.C$ is viewed as the open first order formula $\forall b.\ R(a, b) \rightarrow C(b)$, where $R(a, b) \rightarrow C(b)$ is equivalent to $\neg R(a, b) \vee C(b)$, and $\forall$ is viewed as a conjunction over certainty values associated with the implication $R(a, b) \rightarrow C(b)$. To be more precise, the semantics of $R(a, b) \rightarrow C(b)$ is captured using the disjunction and the negation functions as $f_d(\sim R^{\mathcal{I}}(a, b), C^{\mathcal{I}}(b))$, and $\forall b$ is captured using the meet operator in the certainty lattice $\otimes_{b \in \Delta^{\mathcal{I}}}$.

We say a concept is in *negation normal form* (NNF) if the negation operator appears only in front of concept names. The following two inter-constructor properties allow the transformation of concept descriptions into NNFs.

- De Morgan's Rule: $\neg(C \sqcup D) \equiv \neg C \sqcap \neg D$ and $\neg(C \sqcap D) \equiv \neg C \sqcup \neg D$.
- Negating Quantifiers Rule: $\neg \exists R.C \equiv \forall R.\neg C$ and $\neg \forall R.C \equiv \exists R.\neg C$.

3.2 $\mathcal{ALC}_U$ Knowledge Base

The *knowledge base* Σ in the $\mathcal{ALC}_U$ framework is a pair $\langle \mathcal{T}, \mathcal{A} \rangle$, where $\mathcal{T}$ is a TBox and $\mathcal{A}$ is an ABox. An interpretation $\mathcal{I}$ *satisfies* (or is a *model* of) Σ (denoted $\mathcal{I} \models \Sigma$), if and only if it satisfies both $\mathcal{T}$ and $\mathcal{A}$. The knowledge base Σ is *consistent* if there exists an interpretation $\mathcal{I}$ that satisfies Σ, and is *inconsistent* otherwise.

$\mathcal{ALC}_U$ TBox. An $\mathcal{ALC}_U$ TBox $\mathcal{T}$ consists of a set of terminological axioms defining how concepts are related to each other. Each axiom is associated with a certainty value as well as a conjunction function and a disjunction function which are used to interpret the concept descriptions in the axiom. Specifically, an $\mathcal{ALC}_U$ TBox consists of axioms that could be a combination of GCIs of the form $\langle C \sqsubseteq D \mid \alpha, f_c, f_d \rangle$ and concept equations of the form $\langle C \equiv D \mid \alpha, f_c, f_d \rangle$, where C and D are concept descriptions, $\alpha \in \mathcal{V}$ is the certainty that the axiom holds, f_c is the conjunction function used as the semantics of concept conjunction and part of the role exists restriction, and f_d is the disjunction function used as the semantics of concept disjunction and part of the role value restriction. In case the choice of the combination function in the current axiom is immaterial, "$-$" is used as a place holder. The concept equation $\langle C \equiv D \mid \alpha, f_c, f_d \rangle$ is equivalent to the two concept inclusions $\langle C \sqsubseteq D \mid \alpha, f_c, f_d \rangle$ and $\langle D \sqsubseteq C \mid \alpha, f_c, f_d \rangle$.

For example, the axiom $\langle Rich \sqsubseteq ((\exists owns.ExpensiveCar \sqcup \exists owns.Airplane) \sqcap Golfer) \mid [0.8, 1], min, max \rangle$ states that the concept $Rich$ is subsumed by owning expensive car or owning an airplane, and being a golfer. The certainty of this axiom is at least 0.8, with all the concept conjunctions interpreted using min function, and all the concept disjunctions interpreted using max.

In order to transform an axiom of the form $\langle C \sqsubseteq D \mid \alpha, f_c, f_d \rangle$ into its normal form, $\langle \top \sqsubseteq \neg C \sqcup D \mid \alpha, f_c, f_d \rangle$, the semantics of the concept subsumption is restricted to be $f_d(\sim C^{\mathcal{I}}(a), D^{\mathcal{I}}(a))$, for all $a \in \Delta^{\mathcal{I}}$, where $\sim C^{\mathcal{I}}(a)$ captures the semantics of $\neg C$, and f_d captures the semantics of $\sqcup$ in $\neg C \sqcup D$. Hence, an interpretation $\mathcal{I}$ satisfies $\langle C \sqsubseteq D \mid \alpha, f_c, f_d \rangle$ if $f_d(\sim C^{\mathcal{I}}(a), D^{\mathcal{I}}(a)) = \alpha$, for all $a \in \Delta^{\mathcal{I}}$.

$\mathcal{ALC}_U$ ABox. An $\mathcal{ALC}_U$ ABox $\mathcal{A}$ consists of a set of assertions, each of which is associated with a certainty value and a pair of combination functions used to interpret the concept description(s) in the assertion. Specifically, these assertions could include concept assertions of the form $\langle a : C \mid \alpha, f_c, f_d \rangle$ and role assertions of the form $\langle (a, b) : R \mid \alpha, -, - \rangle$, where a and b are individuals, C is a concept, R is a role, $\alpha \in \mathcal{V}$, f_c is the conjunction function, f_d is the disjunction function, and $-$ denotes that the corresponding combination function is not applicable.

For instance, the assertion "Mary is either tall or thin, and smart with probability between 0.6 and 0.8" can be expressed as $\langle Mary : (Tall \sqcup Thin) \sqcap Smart \mid [0.6, 0.8], \times, ind \rangle$. Here, the concept conjunction is interpreted using the algebraic product ($\times$), and the disjunction function is interpreted using the probability independent function (ind). Note that, since reasoning with probability often requires extra information/knowledge about the events and facts in the world, we are investigating ways to model knowledge base with more general probability theory, such as positive/negative correlation, ignorance, and conditional independence.

In terms of the semantics of the assertions, an interpretation $\mathcal{I}$ satisfies $\langle a : C \mid \alpha, f_c, f_d \rangle$ if $C^{\mathcal{I}}(a^{\mathcal{I}}) = \alpha$, and $\mathcal{I}$ satisfies $\langle (a, b) : R \mid \alpha, -, - \rangle$ if $R^{\mathcal{I}}(a^{\mathcal{I}}, b^{\mathcal{I}}) = \alpha$.

There are two types of individuals that could be in an ABox - defined individuals and generated individuals, defined as follows. We also introduce the notion of predecessor and ancestor in Definition 2.

Definition 1. *(Defined/Generated Individual) Let* **I** *be the set of all individuals in an ABox. We call individuals whose names explicitly appear in the input ABox "defined individuals"* ($\mathbf{I_D}$)*, and those generated by the reasoning procedure "generated individuals"* ($\mathbf{I_G}$)*.*

Definition 2. *(Predecessor/Ancestor) An individual a is a "predecessor" of an individual b (or b is a R-successor of a) if the ABox $\mathcal{A}$ contains the assertion $\langle(a,b) : R \mid \alpha, -, -\rangle$. An individual a is an "ancestor" of b if it is either a predecessor of b or there exists a chain of assertions $\langle(a,b_1) : R_1 \mid \alpha_1, -, -\rangle$, $\langle(b_1,b_2) : R_2 \mid \alpha_2, -, -\rangle$,..., $\langle(b_k,b) : R_{k+1} \mid \alpha_{k+1}, -, -\rangle$ in $\mathcal{A}$.*

Note 1. Since each axiom/assertion in the $\mathcal{ALC}_U$ knowledge base is associated with a pair of combination functions, different notions of uncertainty (such as fuzzy and simple probability) can be modeled within the same knowledge base, if desired by the user.

3.3 $\mathcal{ALC}_U$ Reasoning Procedure

Let $\Sigma = \langle \mathcal{T}, \mathcal{A} \rangle$ be an $\mathcal{ALC}_U$ knowledge base. Fig. 2 gives an overview of the $\mathcal{ALC}_U$ tableau reasoning procedure. The rectangles represent data or knowledge bases, the arrows show the data flow, and the gray rounded boxes show where data processing is performed.

In what follows, we present the $\mathcal{ALC}_U$ tableau algorithm in detail. We first introduce the reasoning services offered, and then present the pre-processing phase and the completion rules.

$\mathcal{ALC}_U$ Reasoning Services. The $\mathcal{ALC}_U$ reasoning services include the consistency, the entailment, and the subsumption problems as described below.

Consistency Problem: To check if an $\mathcal{ALC}_U$ knowledge base $\Sigma = \langle \mathcal{T}, \mathcal{A} \rangle$ is consistent, we first apply the pre-processing steps (see Section 3.3) to obtain the initial extended ABox, $\mathcal{A}_0^{\mathcal{E}}$. In addition, the constraints set $\mathcal{C}_0$ is initialized to the empty set $\{\}$. We then apply the completion rules (see Section 3.3) to derive implicit knowledge from explicit ones. Through the application of each rule, we add any assertions that are derived to the extended ABox $\mathcal{A}_i^{\mathcal{E}}$. In addition, constraints which denote the semantics of the assertions are added to the constraints set $\mathcal{C}_j$, in the form of linear or nonlinear inequations. The completion rules are applied in arbitrary order as long possible, until either $\mathcal{A}_i^{\mathcal{E}}$ contains a clash or no further rule could be applied to $\mathcal{A}_i^{\mathcal{E}}$. If $\mathcal{A}_i^{\mathcal{E}}$ contains a clash, the knowledge base is inconsistent. Otherwise, the system of inequations in $\mathcal{C}_j$ is fed into the constraint solver to check its solvability. If the system of inequations is unsolvable, the knowledge base is inconsistent. Otherwise, the knowledge base is consistent.

Entailment Problem: Given an $\mathcal{ALC}_U$ knowledge base Σ, the entailment problem determines the degree to which an assertion X is true. Like in standard DLs, the entailment problem can be reduced to the consistency problem. That is, let

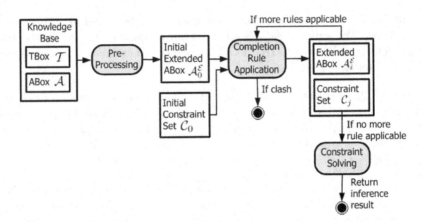

Fig. 2. Reasoning Procedure for $\mathcal{ALC}_U$

X be an assertion of the form $\langle a : C \mid x_{a:C}, f_c, f_d \rangle$. The degree that Σ entails X is the degree of $x_{a:C}$ such that $\Sigma \cup \{\langle a : \neg C,\, x_{a:\neg C}\rangle\langle f_c, f_d\rangle\}$ is consistent.

Subsumption Problem: Let $\Sigma = \langle \mathcal{T}, \mathcal{A}\rangle$ be an $\mathcal{ALC}_U$ knowledge base, and $\langle C \sqsubseteq D \mid x_{C \sqsubseteq D}, f_c, f_d\rangle$ be the subsumption relationship to be checked. The subsumption problem determines the degree to which C is subsumed by D with respect to the TBox $\mathcal{T}$. Like in standard DLs, this problem can be reduced to the consistency problem by finding the degree of $x_{a:\neg C \sqcup D}$ such that $\Sigma \cup \{\langle a : C \sqcap \neg D \mid x_{a:C \sqcap \neg D}, f_c, f_d\rangle\}$ is consistent, where a is a new, generated individual name.

As in standard DLs, the model being constructed by the $\mathcal{ALC}_U$ tableau algorithm can be thought of as a forest.

Definition 3. *(Forest, Node Label, Node Constraint, Edge Label) A "forest" is a collection of trees, with nodes corresponding to individuals, edges corresponding to relationships/roles between individuals, and root nodes corresponding to individuals present in the initial extended ABox. Each node is associated with a "node label", $\mathcal{L}(individual)$, to show the concept assertions associated with a particular individual, as well as a "node constraint", $\mathcal{C}(individual)$, for the corresponding constraints. Unlike in the standard DL where each element in the node label is a concept, each element in our node label is a quadruple, $\langle Concept, Certainty, f_c, f_d\rangle$. Finally, unlike in the standard DL where each edge is labeled with a role name, each edge in our case is associated with an "edge label", $\mathcal{L}(\langle individual_1, individual_2\rangle)$ which consists of a pair of elements $\langle Role, Certainty\rangle$. In case the certainty is a variable, "$-$" is used as a place holder.*

Pre-processing Phase. The $\mathcal{ALC}_U$ tableau algorithm starts by applying the following pre-processing steps, which maintains the equivalence of the result with the original knowledge base.

1. Replace each axiom of the form $\langle C \equiv D \mid \alpha, f_c, f_d\rangle$ with the following two axioms: $\langle C \sqsubseteq D \mid \alpha, f_c, f_d\rangle$ and $\langle D \sqsubseteq C \mid \alpha, f_c, f_d\rangle$.

2. Transform every axiom in the TBox into its normal form. That is, replace each axiom of the form $\langle C \sqsubseteq D \mid \alpha, f_c, f_d \rangle$ with $\langle \top \sqsubseteq \neg C \sqcup D \mid \alpha, f_c, f_d \rangle$.
3. Transform every concept (the TBox and the ABox) into its NNF. Let C and D be concepts, and R be a role. The NNF can be obtained by applying the following rules:
   - $\neg\neg(C) \equiv C$
   - $\neg(C \sqcup D) \equiv \neg C \sqcap \neg D$
   - $\neg(C \sqcap D) \equiv \neg C \sqcup \neg D$
   - $\neg \exists R.C \equiv \forall R.\neg C$
   - $\neg \forall R.C \equiv \exists R.\neg C$
4. Augment the ABox $\mathcal{A}$ with respect to the TBox $\mathcal{T}$. That is, for each individual a in $\mathcal{A}$ and each axiom $\langle \top \sqsubseteq \neg C \sqcup D \mid \alpha, f_c, f_d \rangle$ in $\mathcal{T}$, we add to $\mathcal{A}$, the assertion $\langle a : \neg C \sqcup D \mid \alpha, f_c, f_d \rangle$.

We call the resulting ABox after the pre-processing phase the *initial extended ABox*, denoted by $\mathcal{A}_0^{\mathcal{E}}$.

$\mathcal{ALC}_U$ Completion Rules. Let $\mathcal{T}$ be a TBox obtained after the pre-processing phase, $\mathcal{A}_0^{\mathcal{E}}$ be the initial extended ABox, and $\mathcal{C}_0$ be the initial constraints set. Also, let α and β be certainty values, and Γ be either a certainty value in the certainty domain or the variable x_X denoting the certainty of assertion X. The $\mathcal{ALC}_U$ completion rules are listed in Fig. 3. Since the application of the completion rules may lead to nontermination (see Section 4 for an example), we introduce the notion of blocking to handle this situation.

Definition 4. *(Blocking) Let $a, b \in \mathbf{I_G}$ be generated individuals in the extended ABox $\mathcal{A}_i^{\mathcal{E}}$, $\mathcal{A}_i^{\mathcal{E}}(a)$ and $\mathcal{A}_i^{\mathcal{E}}(b)$ be all the concept assertions for a and b in $\mathcal{A}_i^{\mathcal{E}}$. An individual b is blocked by some ancestor a (or a is the blocking individual for b) if $\mathcal{A}_i^{\mathcal{E}}(b) \subseteq \mathcal{A}_i^{\mathcal{E}}(a)$.*

The purpose of the Clash Triggers is to detect possible inconsistencies in the knowledge base. For example, suppose the certainty domain is $\mathcal{V} = \mathcal{C}[0, 1]$, i.e., the set of closed subintervals $[\alpha, \beta]$ in $[0, 1]$ where $\alpha \preceq \beta$. If a knowledge base contains both assertions $\langle John : Tall \mid [0, 0.2], -, - \rangle$ and $\langle John : Tall \mid [0.7, 1], -, - \rangle$, then the third clash trigger will detect this as an inconsistency.

The Concept Assertion and Role Assertion rules simply add the certainty value of each atomic concept/role assertion and its negation to the constraints set $\mathcal{C}_j$. For example, suppose we have the assertion $\langle John : Tall \mid [0.6, 1], -, - \rangle$ in the extended ABox. If the certainty domain is $\mathcal{V} = \mathcal{C}[0, 1]$ and if the negation function is $\sim(x) = t - x$, where t is the top certainty in the lattice, then we add the constraints $(x_{John:Tall} = [0.6, 1])$ and $(x_{John:\neg Tall} = [0, 0.4])$ to the set of constraints $\mathcal{C}_j$.

The intuition behind the Negation Rule is that, if we know an assertion has a certainty value Γ, then the certainty of its negation can be obtained by applying the negation function to Γ. For example, suppose the certainty domain is $\mathcal{V} = [0, 1]$, and the negation function is defined as $\sim(x) = 1 - x$. If we

Clash Triggers:
$\langle a : \bot \mid \alpha, -, - \rangle \in \mathcal{A}_i^{\mathcal{E}}$, with $\alpha \succ l$
$\langle a : \top \mid \alpha, -, - \rangle \in \mathcal{A}_i^{\mathcal{E}}$, with $\alpha \prec t$
$\{\langle a : A \mid \alpha, -, - \rangle, \langle a : A \mid \beta, -, - \rangle\} \subseteq \mathcal{A}_i^{\mathcal{E}}$,
  with $\otimes(\alpha, \beta) = l$
$\{\langle (a, b) : R \mid \alpha, -, - \rangle, \langle (a, b) : R \mid \beta, -, - \rangle\} \subseteq \mathcal{A}_i^{\mathcal{E}}$,
  with $\otimes(\alpha, \beta) = l$

Concept Assertion Rule:
if $\langle a : A \mid \Gamma, -, - \rangle \in \mathcal{A}_i^{\mathcal{E}}$
then if $(x_{a:A} = \Gamma) \notin C_j$ and Γ is not a variable
  then $C_{j+1} = C_j \cup \{(x_{a:A} = \Gamma)\}$
  if $(x_{a:\neg A} = \sim \Gamma) \notin C_j$
  then $C_{j+1} = C_j \cup \{(x_{a:\neg A} = \sim \Gamma)\}$

Role Assertion Rule:
if $\langle (a, b) : R \mid \Gamma, -, - \rangle \in \mathcal{A}_i^{\mathcal{E}}$
then if $(x_{(a,b):R} = \Gamma) \notin C_j$ and Γ is not a variable
  then $C_{j+1} = C_j \cup \{(x_{(a,b):R} = \Gamma)\}$
  if $(x_{\neg(a,b):R} = \sim \Gamma) \notin C_j$
  then $C_{j+1} = C_j \cup \{(x_{\neg(a,b):R} = \sim \Gamma)\}$

Negation Rule:
if $\langle a : \neg A \mid \Gamma, -, - \rangle \in \mathcal{A}_i^{\mathcal{E}}$
then if $\langle a : A \mid \sim \Gamma, -, - \rangle \notin \mathcal{A}_i^{\mathcal{E}}$
  then $\mathcal{A}_{i+1}^{\mathcal{E}} = \mathcal{A}_i^{\mathcal{E}} \cup \{\langle a : A \mid \sim \Gamma, -, - \rangle\}$

Conjunction Rule:
if $\langle a : C \sqcap D \mid \Gamma, f_c, f_d \rangle \in \mathcal{A}_i^{\mathcal{E}}$
then for each $\Psi \in \{C, D\}$
  if Ψ is atomic and $\langle a : \Psi \mid x_{a:\Psi}, -, - \rangle \notin \mathcal{A}_i^{\mathcal{E}}$
  then $\mathcal{A}_{i+1}^{\mathcal{E}} = \mathcal{A}_i^{\mathcal{E}} \cup \{\langle a : \Psi \mid x_{a:\Psi}, -, - \rangle\}$
  else if Ψ is not atomic and $\langle a : \Psi \mid x_{a:\Psi}, f_c, f_d \rangle \notin \mathcal{A}_i^{\mathcal{E}}$
    then $\mathcal{A}_{i+1}^{\mathcal{E}} = \mathcal{A}_i^{\mathcal{E}} \cup \{\langle a : \Psi \mid x_{a:\Psi}, f_c, f_d \rangle\}$
  if $(f_c(x_{a:C}, x_{a:D}) = \Gamma) \notin C_j$,
  then $C_{j+1} = C_j \cup \{(f_c(x_{a:C}, x_{a:D}) = \Gamma)\}$

Disjunction Rule:
if $\langle a : C \sqcup D \mid \Gamma, f_c, f_d \rangle \in \mathcal{A}_i^{\mathcal{E}}$
then for each $\Psi \in \{C, D\}$
  if Ψ is atomic and $\langle a : \Psi \mid x_{a:\Psi}, -, - \rangle \notin \mathcal{A}_i^{\mathcal{E}}$

then $\mathcal{A}_{i+1}^{\mathcal{E}} = \mathcal{A}_i^{\mathcal{E}} \cup \{\langle a : \Psi \mid x_{a:\Psi}, -, - \rangle\}$
else if Ψ is not atomic and $\langle a : \Psi \mid x_{a:\Psi}, f_c, f_d \rangle \notin \mathcal{A}_i^{\mathcal{E}}$
  then $\mathcal{A}_{i+1}^{\mathcal{E}} = \mathcal{A}_i^{\mathcal{E}} \cup \{\langle a : \Psi \mid x_{a:\Psi}, f_c, f_d \rangle\}$
if $(f_d(x_{a:C}, x_{a:D}) = \Gamma) \notin C_j$,
then $C_{j+1} = C_j \cup \{(f_d(x_{a:C}, x_{a:D}) = \Gamma)\}$

Role Exists Restriction Rule:
if $\langle a : \exists R.C \mid \Gamma, f_c, f_d \rangle \in \mathcal{A}_i^{\mathcal{E}}$ and a is not blocked
then if $\nexists$ individual b such that
  $(f_c(x_{(a,b):R}, x_{b:C}) = x_{a:\exists R.C}) \in C_j$
then let b be a new individual
  $\mathcal{A}_{i+1}^{\mathcal{E}} = \mathcal{A}_i^{\mathcal{E}} \cup \{\langle (a, b) : R \mid x_{(a,b):R}, -, - \rangle\}$
  if C is atomic
  then $\mathcal{A}_{i+1}^{\mathcal{E}} = \mathcal{A}_i^{\mathcal{E}} \cup \{\langle b : C \mid x_{b:C}, -, - \rangle\}$
  else $\mathcal{A}_{i+1}^{\mathcal{E}} = \mathcal{A}_i^{\mathcal{E}} \cup \{\langle b : C \mid x_{b:C}, f_c, f_d \rangle\}$
  $C_{j+1} = C_j \cup \{(f_c(x_{(a,b):R}, x_{b:C}) = x_{a:\exists R.C})\}$
  for each axiom $\langle \top \sqsubseteq \neg C \sqcup D \mid \alpha, f_c, f_d \rangle$ in $\mathcal{T}$
    $\mathcal{A}_{i+1}^{\mathcal{E}} = \mathcal{A}_i^{\mathcal{E}} \cup \{\langle b : \neg C \sqcup D \mid \alpha, f_c, f_d \rangle\}$
  if Γ is not the variable $x_{a:\exists R.C}$
  then if $(x_{a:\exists R.C} = \Gamma') \in C_j$
    then if $\Gamma \neq \Gamma'$ and Γ is not an element in Γ'
      then $C_{j+1} = C_j \setminus \{(x_{a:\exists R.C} = \Gamma')\}$
        $\cup \{(x_{a:\exists R.C} = \oplus(\Gamma, \Gamma'))\}$
    else $C_{j+1} = C_j \cup \{(x_{a:\exists R.C} = \Gamma)\}$

Role Value Restriction Rule:
if $\{\langle a : \forall R.C \mid \Gamma, f_c, f_d \rangle, \langle (a, b) : R \mid \Gamma', -, - \rangle\} \subseteq \mathcal{A}_i^{\mathcal{E}}$
then if C is atomic and $\langle b : C \mid x_{b:C}, -, - \rangle \notin \mathcal{A}_i^{\mathcal{E}}$
  then $\mathcal{A}_{i+1}^{\mathcal{E}} = \mathcal{A}_i^{\mathcal{E}} \cup \{\langle b : C \mid x_{b:C}, -, - \rangle\}$
  else if C is not atomic and $\langle b : C \mid x_{b:C}, f_c, f_d \rangle \notin \mathcal{A}_i^{\mathcal{E}}$
    then $\mathcal{A}_{i+1}^{\mathcal{E}} = \mathcal{A}_i^{\mathcal{E}} \cup \{\langle b : C \mid x_{b:C}, f_c, f_d \rangle\}$
  if $(f_d(x_{\neg(a,b):R}, x_{b:c}) = x_{a:\forall R.C}) \notin C_j$
  then $C_{j+1} = C_j \cup \{(f_d(x_{\neg(a,b):R}, x_{b:c}) = x_{a:\forall R.C})\}$
  if Γ is not the variable $x_{a:\forall R.C}$
  then if $(x_{a:\forall R.C} = \Gamma'') \in C_j$
    then if $\Gamma \neq \Gamma''$ and Γ is not an element in Γ''
      then $C_{j+1} = C_j \setminus \{(x_{a:\forall R.C} = \Gamma'')\}$
        $\cup \{(x_{a:\forall R.C} = \otimes(\Gamma, \Gamma''))\}$
    else $C_{j+1} = C_j \cup \{(x_{a:\forall R.C} = \Gamma)\}$

Fig. 3. Completion Rules for $\mathcal{ALC}_U$

have the assertion $\langle John : \neg Tall \mid 0.8, -, - \rangle$ in the ABox, we could also infer $\langle John : Tall \mid 0.2, -, - \rangle$, which is added to the extended ABox.

The Conjunction and Disjunction rules capture the semantics of concept conjunction (resp. disjunction) by applying the conjunction (resp. disjunction) function to the interpretations of $a : C$ and $a : D$. An interesting thing to note is that, in the standard DL, the Disjunction Rule is non-deterministic, since it can be applied in different ways to the same ABox. However, the $\mathcal{ALC}_U$ disjunction rule is deterministic. This is because the semantics of the concept disjunction is encoded in the disjunction function in the form of a constraint. For example, suppose the extended ABox includes the assertion $\langle Mary : Tall \sqcup Thin \mid 0.8, min, max \rangle$, then we infer that $\langle Mary : Tall \mid x_{Mary:Tall}, -, - \rangle$ and $\langle Mary : Thin \mid x_{Mary:Thin}, -, - \rangle$. Moreover, the constraint $max(x_{Mary:Tall}, x_{Mary:Thin}) = 0.8$ must be satisfied, which means that either $x_{Mary:Tall} = 0.8$, or $x_{Mary:Thin} = 0.8$, or $x_{Mary:Tall} = x_{Mary:Thin} = 0.8$.

Note 2. Like in standard DLs, our tableau algorithm treats each axiom in $\mathcal{T}$ as a meta constraint. That is, for each individual a in $\mathcal{A}$ and each axiom $\langle C \sqsubseteq D \mid \alpha, f_c, f_d \rangle$ in $\mathcal{T}$, we add $\langle a : \neg C \sqcup D \mid \alpha, f_c, f_d \rangle$ to $\mathcal{A}$. This results in a large number of assertions with concept disjunction be added to $\mathcal{A}$. In standard DLs, this would dramatically extend the search space and is the main cause for empirical intractability, since the disjunction rule in standard DLs is non-deterministic. Therefore, optimization technique like lazy unfolding was introduced in [2]. However, since the disjunction rule in $\mathcal{ALC}_U$ is deterministic, large number of assertions with concept disjunction does not cause problems in our context.

The Role Exists Restriction and Role Value Restriction rules have a similar structure, although they have different semantics. The intuition behind the Role Exists Restriction Rule is that, if we know that an individual a is in $\exists R.C$, there must exist at least one individual, say b, such that a is related to b through the relationship R, and b is in the concept C. If no such individual b exists in the extended ABox, then we create such a new individual, and assert that this individual satisfies all the axioms in the TBox. As usual, the semantics of the Role Exists Restriction is encoded as constraints. On the other hand, the intuition behind the Role Value Restriction Rule is that, if we know that an individual a is in $\forall R.C$, and if there is an individual b such that a is related to b through the relationship R, then b must be in the concept C. The semantics of the Role Value Restriction is also encoded as constraints. For example, suppose the assertion $\langle Tom : \exists hasDisease.Diabetes \mid [0.4, 0.6], min, max \rangle$ is in the extended ABox and the axiom $\langle \top \sqsubseteq \neg Obese \sqcup \exists hasDisease.Diabetes \mid [0.7, 1], \times, ind \rangle$ is in the TBox. Assume also that the ABox originally does not contain any individual b such that Tom is related to b through the role $hasDisease$, and b is in the concept $Diabetes$. Then, we could infer $\langle (Tom, d1) : hasDisease \mid x_{(Tom,d1):hasDisease}, -, - \rangle$ and $\langle d1 : Diabetes \mid x_{d1:Diabetes}, -, - \rangle$, where $d1$ is a new individual. In addition, since $d1$ must satisfy the axioms in the TBox, the assertion $\langle d1 : \neg Obese \sqcup \exists hasDisease.Diabetes \mid [0.7, 1], \times, ind \rangle$ is added to the extended ABox. Finally, the constraints $(min(x_{(Tom,d1):hasDisease}, x_{d1:Diabetes}) = x_{Tom:\exists hasDisease.Diabetes})$ as well as $(x_{Tom:\exists hasDisease.Diabetes} = [0.4, 0.6])$ must be satisfied.

The correctness of the $\mathcal{ALC}_U$ tableau algorithm can be established by showing that it is sound, complete, and it terminates, as shown in [17].

4   An Illustrative Example

To illustrate the $\mathcal{ALC}_U$ tableau algorithm and the need for blocking, let us consider a cyclic fuzzy knowledge base $\Sigma = \langle \mathcal{T}, \mathcal{A} \rangle$, where:

$\mathcal{T} = \{\langle ObesePerson \sqsubseteq \exists hasParent.ObesePerson \mid [0.7, 1], min, max \rangle\}$

$\mathcal{A} = \{\langle John : ObesePerson \mid [0.8, 1], -, - \rangle\}$

Note that the fuzzy knowledge bases can be expressed in $\mathcal{ALC}_U$ by setting the certainty lattice as $\mathcal{L} = \langle \mathcal{V}, \preceq \rangle$, where $\mathcal{V} = \mathcal{C}[0,1]$ is the set of closed subintervals $[\alpha, \beta]$ in $[0, 1]$ such that $\alpha \preceq \beta$. We also set the meet operator in the lattice as *inf* (infimum), the join operator as *sup* (supremum), and the negation function as $\sim(x) = t - x$, where $t = [1,1]$ is the greatest value in the certainty lattice. Finally, the conjunction function is set to min, and the disjunction function is set to max.

To find out if Σ is consistent, we first apply the pre-processing steps. For this, we transform the axiom into its normal form:

$$\mathcal{T} = \{\langle \top \sqsubseteq (\neg ObesePerson \sqcup \exists hasParent.ObesePerson \mid [0.7,1], min, max\rangle\}$$

We then augment the ABox with respect to the TBox. That is, for each individual a in the ABox (in this case, we have only $John$) and for each axiom of the form $\langle \top \sqsubseteq \neg C \sqcup D \mid \alpha, f_c, f_d\rangle$ in the TBox, we add an assertion $\langle a : \neg C \sqcup D \mid \alpha, f_c, f_d\rangle$ to the ABox. Hence, in this step, we add the following assertion to the ABox:

$$\langle John : (\neg ObesePerson \sqcup \exists hasParent.ObesePerson \mid [0.7,1], min, max\rangle\}$$

Now, we can initialize the extended ABox to be:

$$\mathcal{A}_0^{\mathcal{E}} = \{\langle John : ObesePerson \mid [0.8,1], -, -\rangle,$$
$$\langle John : (\neg ObesePerson \sqcup \exists hasParent.ObesePerson \mid [0.7,1], min, max\rangle\}$$

and the constraints set to be $\mathcal{C}_0 = \{\}$.

Once the pre-processing phase is over, we are ready to apply the completion rules. The first assertion is $\langle John : ObesePerson \mid [0.8,1], -, -\rangle$. Since $ObesePerson$ is an atomic concept, we apply the Concept Assertion Rule, which yields:

$$\mathcal{C}_1 = \mathcal{C}_0 \cup \{(x_{John:ObesePerson} = [0.8,1])\}$$
$$\mathcal{C}_2 = \mathcal{C}_1 \cup \{(x_{John:\neg ObesePerson} = t - x_{John:ObesePerson})\}, \text{ where } t \text{ is the greatest}$$
element in the lattice, $[1,1]$.

The other assertion in $\mathcal{A}_0^{\mathcal{E}}$ is $\langle John : (\neg ObesePerson \sqcup \exists hasParent.ObesePerson \mid [0.7,1], min, max\rangle\}$. Since this assertion includes a concept disjunction, the Disjunction Rule applies. This yields:

$$\mathcal{A}_1^{\mathcal{E}} = \mathcal{A}_0^{\mathcal{E}} \cup \{\langle John : \neg ObesePerson \mid x_{John:\neg ObesePerson}, -, -\rangle\}$$
$$\mathcal{A}_2^{\mathcal{E}} = \mathcal{A}_1^{\mathcal{E}} \cup \{\langle John : \exists hasParent.ObesePerson \mid x_{John:\exists hasParent.ObesePerson},$$
$$min, max\rangle\}$$
$$\mathcal{C}_3 = \mathcal{C}_2 \cup \{(max(x_{John:\neg ObesePerson}, x_{John:\exists hasParent.ObesePerson}) = [0.7,1])\}$$

The assertion $\langle John : \neg ObesePerson \mid x_{John:\neg ObesePerson}, -, -\rangle$ in $\mathcal{A}_1^{\mathcal{E}}$ triggers the Negation Rule, which yields:

$$\mathcal{A}_3^{\mathcal{E}} = \mathcal{A}_2^{\mathcal{E}} \cup \{\langle John : ObesePerson \mid x_{John:ObesePerson}, -, -\rangle\}$$

The application of the Concept Assertion Rule to the assertion $\langle John : Obese Person \mid x_{John:ObesePerson}, -, -\rangle$ in $\mathcal{A}_3^{\mathcal{E}}$ does not derive any new assertion nor constraint. Next, we apply the Role Exists Restriction Rule to the assertion in $\mathcal{A}_2^{\mathcal{E}}$, and obtain:

$$\mathcal{A}_4^{\mathcal{E}} = \mathcal{A}_3^{\mathcal{E}} \cup \{\langle (John, ind1) : hasParent \mid x_{(John,ind1):hasParent}, -, -\rangle\}$$

$$\mathcal{A}_5^{\mathcal{E}} = \mathcal{A}_4^{\mathcal{E}} \cup \{\langle ind1 : ObesePerson \mid x_{ind1:ObesePerson}, -, -\rangle\}$$

$$\mathcal{C}_4 = \mathcal{C}_3 \cup \{(min(x_{(John,ind1):hasParent}, x_{ind1:ObesePerson}) = x_{John:\exists hasParent.ObesePerson})\}$$

$$\mathcal{A}_6^{\mathcal{E}} = \mathcal{A}_5^{\mathcal{E}} \cup \{\langle ind1 : (\neg ObesePerson \sqcup \exists hasParent.ObesePerson \mid [0.7,1], min, max\rangle\}$$

The application of the Role Assertion Rule to the assertion in $\mathcal{A}_4^{\mathcal{E}}$ yields:

$$\mathcal{C}_5 = \mathcal{C}_4 \cup \{(x_{(John,ind1):\neg hasParent} = t - x_{(John,ind1):hasParent})\}$$

After applying the Concept Assertion Rule to the assertion $\langle ind1 : ObesePerson \mid x_{ind1:ObesePerson}, -, -\rangle\}$ in $\mathcal{A}_5^{\mathcal{E}}$, we obtain:

$$\mathcal{C}_6 = \mathcal{C}_5 \cup \{(x_{ind1:\neg ObesePerson} = t - x_{ind1:ObesePerson})\}$$

The assertion in $\mathcal{A}_6^{\mathcal{E}}$ triggers the Disjunction Rule, which yields:

$$\mathcal{A}_7^{\mathcal{E}} = \mathcal{A}_6^{\mathcal{E}} \cup \{\langle ind1 : \neg ObesePerson \mid x_{ind1:\neg ObesePerson}, -, -\rangle\}$$

$$\mathcal{A}_8^{\mathcal{E}} = \mathcal{A}_7^{\mathcal{E}} \cup \{\langle ind1 : \exists hasParent.ObesePerson \mid x_{ind1:\exists hasParent.ObesePerson}, min, max\rangle\}$$

$$\mathcal{C}_7 = \mathcal{C}_6 \cup \{(max(x_{ind1:\neg ObesePerson}, x_{ind1:\exists hasParent.ObesePerson}) = [0.7,1])\}$$

Next, the application of the Negation Rule to the assertion in $\mathcal{A}_7^{\mathcal{E}}$ yields:

$$\mathcal{A}_9^{\mathcal{E}} = \mathcal{A}_8^{\mathcal{E}} \cup \{\langle ind1 : ObesePerson \mid x_{ind1:ObesePerson}, -, -\rangle\}$$

We then apply the Concept Assertion Rule to the assertion in $\mathcal{A}_9^{\mathcal{E}}$, and obtain:

$$\mathcal{C}_8 = \mathcal{C}_7 \cup \{(x_{ind1:\neg ObesePerson} = t - x_{ind1:ObesePerson})\}$$

The application of the Role Exists Restriction Rule to the assertion in $\mathcal{A}_8^{\mathcal{E}}$ yields:

$$\mathcal{A}_{10}^{\mathcal{E}} = \mathcal{A}_9^{\mathcal{E}} \cup \{\langle (ind1, ind2) : hasParent \mid x_{(ind1,ind2):hasParent}, -, -\rangle\}$$

$$\mathcal{A}_{11}^{\mathcal{E}} = \mathcal{A}_{10}^{\mathcal{E}} \cup \{\langle ind2 : ObesePerson \mid x_{ind2:ObesePerson}, -, -\rangle\}$$

$$\mathcal{C}_9 = \mathcal{C}_8 \cup \{(min(x_{(ind1,ind2):hasParent}, x_{ind2:ObesePerson}) = x_{ind1:\exists hasParent.ObesePerson})\}$$

$$\mathcal{A}_{12}^{\mathcal{E}} = \mathcal{A}_{11}^{\mathcal{E}} \cup \{\langle ind2 : (\neg ObesePerson \sqcup \exists hasParent.ObesePerson \mid [0.7,1], min, max\rangle\}$$

Next, the Role Assertion Rule is applied to the assertion in $\mathcal{A}_{10}^{\mathcal{E}}$ yields:

$$\mathcal{C}_{10} = \mathcal{C}_9 \cup \{(x_{(ind1,ind2):\neg hasParent} = t - x_{(ind1,ind2):hasParent})\}$$

After applying the Concept Assertion Rule to the assertion in $\mathcal{A}_{11}^{\mathcal{E}}$, we obtain:

$$\mathcal{C}_{11} = \mathcal{C}_{10} \cup \{(x_{ind2:\neg ObesePerson} = t - x_{ind2:ObesePerson})\}$$

The assertion in $\mathcal{A}_{12}^{\mathcal{E}}$ triggers the Disjunction Rule, which yields:

$$\mathcal{A}_{13}^{\mathcal{E}} = \mathcal{A}_{12}^{\mathcal{E}} \cup \{\langle ind2 : \neg ObesePerson \mid x_{ind2:\neg ObesePerson}, -, -\rangle\}$$

$$\mathcal{A}_{14}^{\mathcal{E}} = \mathcal{A}_{13}^{\mathcal{E}} \cup \{\langle ind2 : \exists hasParent.ObesePerson \mid x_{ind2:\exists hasParent.ObesePerson}, min, max\rangle\}$$

$$\mathcal{C}_{12} = \mathcal{C}_{11} \cup \{(max(x_{ind2:\neg ObesePerson}, x_{ind2:\exists hasParent.ObesePerson}) = [0.7, 1])\}$$

Next, the application of the Negation Rule to the assertion in $\mathcal{A}_{13}^{\mathcal{E}}$ yields:

$$\mathcal{A}_{15}^{\mathcal{E}} = \mathcal{A}_{14}^{\mathcal{E}} \cup \{\langle ind2 : ObesePerson \mid x_{ind2:ObesePerson}, -, -\rangle\}$$

We then apply the Concept Assertion Rule to the assertion in $\mathcal{A}_{15}^{\mathcal{E}}$, and obtain:

$$\mathcal{C}_{13} = \mathcal{C}_{12} \cup \{(x_{ind2:\neg ObesePerson} = t - x_{ind2:ObesePerson})\}$$

Next, consider the assertion in $\mathcal{A}_{14}^{\mathcal{E}}$. Since $ind1$ is an ancestor of $ind2$ and $\mathcal{L}(ind2) \subseteq \mathcal{L}(ind1)$, individual $ind2$ is blocked. Therefore, we will not continue applying the Role Exists Restriction Rule to the assertion in $\mathcal{A}_{14}^{\mathcal{E}}$, and the completion rule application terminates at this point. Note that without blocking, the tableau algorithm would never terminate since new individual will be generated for each application of the Role Exists Restriction Rule.

Since there is no more rule applicable, the set of constraints in $\mathcal{C}_{13}$ is fed into the constraint solver to check its solvability. Since the constraints are solvable, the knowledge base is consistent.

5   Conclusion and Future Work

In this paper, we presented a tableau reasoning procedure for the DL $\mathcal{ALC}_U$ that is capable of handling General TBoxes. The proposed tableau algorithm derives a set of assertions as well as linear/nonlinear constraints that encode the semantics of the knowledge base. The advantage of this approach is that it makes the design of the $\mathcal{ALC}_U$ tableau algorithm generic and uniform for computing different semantics. That is, by simply tuning the uncertainty parameters that are associated with the axioms and assertions in the knowledge base, different notions of uncertainty can be modeled and reasoned with, using a single reasoning procedure. In addition, the proposed reasoning procedure can handle large number of concept disjunctions caused by the introduction of General TBoxes without blowing up the search space because our Disjunction Rule is deterministic. We also illustrated through a detailed example the need for blocking in order to ensure termination of the reasoning procedure when General TBoxes are present.

The optimization aspect of the $\mathcal{ALC}_U$ reasoning procedure is beyond the scope of this paper. However, a preliminary study in this regard can be found in [11]. As future work, we plan to extend $\mathcal{ALC}_U$ to support a more expressive portion of DL (e.g., $\mathcal{SHOIN}$, which the popular Web ontology language OWL DL [26] is based on) so that constructors such as number restrictions and transitive properties can be supported. It is also promising to extend the syntax of the description language to support other forms of uncertainty, such as the conditional probability. These additional expressivity would give us more power to handle real-world applications.

Acknowledgement. This work was supported in part by Natural Sciences and Engineering Research Council of Canada, and by ENCS Concordia University.

References

1. Baader, F., Calvanese, D., McGuinness, D.L., Nardi, D., Patel-Schneider, P.F. (eds.): The Description Logic Handbook: Theory, Implementation, and Applications. Cambridge University Press, Cambridge (2003)
2. Baader, F., Hollunder, B., Nebel, B., Profitlich, H.-J., Franconi, E.: An empirical analysis of optimization techniques for terminological representation systems or "making KRIS get a move on". In: Nebel, B., Swartout, W., Rich, C. (eds.) Principles of Knowledge Representation and Reasoning: Proceedings of the 3rd International Conference, San Mateo, pp. 270–281. Morgan Kaufmann, San Francisco (1992)
3. Baader, F., Horrocks, I., Sattler, U.: Description Logics. In: van Harmelen, F., Lifschitz, V., Porter, B. (eds.) Handbook of Knowledge Representation. Elsevier, Amsterdam (2007)
4. Bobillo, F., Delgado, M., Gomez-Romero, J.: A crisp representation for fuzzy $\mathcal{SHOIN}$ with fuzzy nominals and general concept inclusions. In: Proceedings of the 2nd Workshop on Uncertainty Reasoning for the Semantic Web (URSW), Athens, Georgia, USA (November 2006)
5. Bobillo, F., Straccia, U.: A fuzzy description logic with product t-norm. In: Proceedings of the IEEE International Conference on Fuzzy Systems (Fuzz IEEE 2007), pp. 652–657. IEEE Computer Society, Los Alamitos (2007)
6. Dürig, M., Studer, T.: Probabilistic ABox reasoning: Preliminary results. In: Proceedings of the International Workshop on Description Logics (DL 2005), pp. 104–111 (2005)
7. Giugno, R., Lukasiewicz, T.: P-$\mathcal{SHOQ}$(D): A probabilistic extension of $\mathcal{SHOQ}$(D) for probabilistic ontologies in the Semantic Web. In: Flesca, S., Greco, S., Leone, N., Ianni, G. (eds.) JELIA 2002. LNCS, vol. 2424, pp. 86–97. Springer, Heidelberg (2002)
8. Haarslev, V., Pai, H.I., Shiri, N.: A generic framework for description logics with uncertainty. In: Proceedings of the 2005 Workshop on Uncertainty Reasoning for the Semantic Web (URSW) at the 4th International Semantic Web Conference, Galway, Ireland, pp. 77–86 (November 2005)
9. Haarslev, V., Pai, H.I., Shiri, N.: Completion rules for uncertainty reasoning with the description logic $\mathcal{ALC}$. In: Proceedings of the Canadian Semantic Web Working Symposium - Semantic Web and Beyond: Computing for Human Experience, Quebec City, Canada, vol. 4, pp. 205–225. Springer, Heidelberg (2006)
10. Haarslev, V., Pai, H.I., Shiri, N.: Uncertainty reasoning in description logics: A generic approach. In: Proceedings of the 19th International FLAIRS Conference, Melbourne Beach, Florida, pp. 818–823. AAAI Press, Menlo Park (2006)
11. Haarslev, V., Pai, H.I., Shiri, N.: Optimizing tableau reasoning in $\mathcal{ALC}$ extended with uncertainty. In: Proceedings of the International Workshop on Description Logics (DL 2007), Brixen-Bressanone, Italy, pp. 307–314 (June 2007)
12. Haarslev, V., Pai, H.I., Shiri, N.: Semantic web uncertainty management. In: Encyclopedia of Information Science and Technology, 2nd edn., Information Science Reference (2008)
13. Jaeger, M.: Probabilistic reasoning in terminological logics. In: Proceedings of the 4th International Conference on Principles of Knowledge Representation and Reasoning (KR 1994), pp. 305–316 (1994)
14. Koller, D., Levy, A.Y., Pfeffer, A.: P-CLASSIC: A tractable probablistic description logic. In: Proceedings of the 14th National Conference on Artificial Intelligence, Providence, Rhode Island, pp. 390–397. AAAI Press, Menlo Park (1997)

15. Lakshmanan, L.V.S., Shiri, N.: Logic programming and deductive databases with uncertainty: A survey. In: Encyclopedia of Computer Science and Technology, vol. 45, pp. 153–176. Marcel Dekker, Inc., New York (2001)
16. Lakshmanan, L.V.S., Shiri, N.: A parametric approach to deductive databases with uncertainty. IEEE Transactions on Knowledge and Data Engineering 13(4), 554–570 (2001)
17. Pai, H.I. Uncertainty Management for Description Logic-Based Ontologies. PhD thesis, Concordia University (2008)
18. Schild, K.: A correspondence theory for terminological logics: preliminary report. In: Proceedings of IJCAI 1991, 12th International Joint Conference on Artificial Intelligence, Sidney, AU, pp. 466–471 (1991)
19. Stoilos, G., Stamou, G., Pan, J.Z., Tzouvaras, V., Horrocks, I.: Reasoning with very expressive fuzzy description logics. Journal of Artificial Intelligence Research 30, 273–320 (2007)
20. Stoilos, G., Straccia, U., Stamou, G., Pan, J.Z.: General concept inclusions in fuzzy description logics. In: Proceedings of the 17th European Conference on Artificial Intelligence (ECAI 2006) (2006)
21. Straccia, U.: Reasoning within fuzzy description logics. Journal of Artificial Intelligence Research 14, 137–166 (2001)
22. Straccia, U.: Uncertainty in description logics: a lattice-based approach. In: Proceedings of the 10th International Conference on Information Processing and Management of Uncertainty in Knowledge-Based Systems, pp. 251–258 (2004)
23. Straccia, U. Fuzzy description logic with concrete domains. Technical Report 2005-TR-03, Istituto di Elaborazione dell'Informazione (January 2005)
24. Straccia, U., Bobillo, F.: Mixed integer programming, general concept inclusions and fuzzy description logics. In: Proceedings of the 5th Conference of the European Society for Fuzzy Logic and Technology (EUSFLAT 2007), Ostrava, Czech Republic, vol. 2, pp. 213–220. University of Ostrava (2007)
25. Tresp, C., Molitor, R.: A description logic for vague knowledge. In: Proceedings of ECAI 1998, Brighton, UK, pp. 361–365. John Wiley and Sons, Chichester (1998)
26. W3C. OWL web ontology language overview (2004), http://www.w3.org/TR/owl-features/

Author Index

Lecture Notes in Artificial Intelligence (LNAI)